FOUNDATIONS OF
TOPOLOGY

The Jones and Bartlett Publishers Series in Mathematics

Geometry

Euclidean and Transformational Geometry: A Deductive Inquiry
Libeskind (978-0-7637-4366-6) © 2008

A Gateway to Modern Geometry: The Poincaré Half-Plane, Second Edition
Stahl (978-0-7637-5381-8) © 2008

Lebesgue Integration on Euclidean Space, Revised Edition
Jones (978-0-7637-1708-7) © 2001

Geometry with an Introduction to Cosmic Topology
Hitchman (978-0-7637-5457-0) © 2009

Understanding Modern Mathematics
Stahl (978-0-7637-3401-5) © 2007

Precalculus

Precalculus with Calculus Previews (Essentials Version), Fourth Edition
Zill/Dewar (978-0-7637-3779-5) © 2007

Precalculus with Calculus Previews (Expanded Volume), Fourth Edition
Zill/Dewar (978-0-7637-6631-3) © 2010

Precalculus: A Functional Approach to Graphing and Problem Solving, Sixth Edition
Smith (978-0-7637-5177-7) © 2010

Calculus

Calculus of a Single Variable: Early Transcendentals, Fourth Edition
Zill/Wright (978-0-7637-4965-1) © 2010

Multivariable Calculus, Fourth Edition
Zill/Wright (978-0-7637-4966-8) © 2010

Calculus: Early Transcendentals, Fourth Edition
Zill/Wright (978-0-7637-5995-7) © 2010

Calculus: The Language of Change
Cohen/Henle (978-0-7637-2947-9) © 2005

Applied Calculus for Scientists and Engineers
Blume (978-0-7637-2877-9) © 2005

Exploring Calculus with MATLAB: Topics and Applications
Smaldone (978-0-7637-7002-0) © 2010

Calculus: Labs for Mathematica
O'Connor (978-0-7637-3425-1) © 2005

Calculus: Labs for MATLAB
O'Connor (978-0-7637-3426-8) © 2005

Linear Algebra

Linear Algebra: Theory and Applications
Cheney/Kincaid (978-0-7637-5020-6) © 2009

Linear Algebra with Applications, Sixth Edition
Williams (978-0-7637-5753-3) © 2008

Advanced Engineering Mathematics

Advanced Engineering Mathematics, Third Edition
Zill/Cullen (978-0-7637-4591-2) © 2006

An Elementary Course in Partial Differential Equations, Second Edition
Amaranath (978-0-7637-6244-5) © 2009

Complex Analysis

A First Course in Complex Analysis with Applications, Second Edition
Zill/Shanahan (978-0-7637-5772-4) © 2009

Classical Complex Analysis
Hahn (978-0-8672-0494-0) © 1996

Complex Analysis for Mathematics and Engineering, Fifth Edition
Mathews/Howell (978-0-7637-3748-1) © 2006

Real Analysis

Closer and Closer: Introducing Real Analysis
Schumacher (978-0-7637-3593-7) © 2008

The Way of Analysis, Revised Edition
Strichartz (978-0-7637-1497-0) © 2000

Topology

Foundations of Topology, Second Edition
Patty (978-0-7637-4234-8) © 2009

Discrete Math and Logic

Essentials of Discrete Mathematics
Hunter (978-0-7637-4892-0) © 2009

Discrete Structures, Logic, and Computability, Third Edition
Hein (978-0-7637-7206-2) © 2010

Logic, Sets, and Recursion, Second Edition
Causey (978-0-7637-3784-9) © 2006

Numerical Methods

Numerical Mathematics
Grasselli/Pelinovsky (978-0-7637-3767-2) © 2008

Exploring Numerical Methods: An Introduction to Scientific Computing Using MATLAB
Linz (978-0-7637-1499-4) © 2003

Advanced Mathematics

Clinical Statistics: Introducing Clinical Trials, Survival Analysis, and Longitudinal Data Analysis
Korosteleva (978-0-7637-5850-9) © 2009

Harmonic Analysis: A Gentle Introduction
DeVito (978-0-7637-3893-8) © 2007

Beginning Number Theory, Second Edition
Robbins (978-0-7637-3768-9) © 2006

A Gateway to Higher Mathematics
Goodfriend (978-0-7637-2733-8) © 2006

For more information on any of the titles above please visit us online at http://www.jbpub.com/math. Qualified instructors, contact your Publisher's Representative at 1-800-832-0034 or info@jbpub.com to request review copies for course consideration.

SECOND EDITION

FOUNDATIONS OF TOPOLOGY

C. WAYNE PATTY
Virginia Polytechnic Institute
and State University
Blacksburg, Virginia

JONES AND BARTLETT PUBLISHERS
Sudbury, Massachusetts
BOSTON   TORONTO   LONDON   SINGAPORE

World Headquarters
Jones and Bartlett Publishers
40 Tall Pine Drive
Sudbury, MA 01776
978-443-5000
info@jbpub.com
www.jbpub.com

Jones and Bartlett Publishers
Canada
6339 Ormindale Way
Mississauga, Ontario L5V 1J2
Canada

Jones and Bartlett Publishers
International
Barb House, Barb Mews
London W6 7PA
United Kingdom

Jones and Bartlett's books and products are available through most bookstores and online booksellers. To contact Jones and Bartlett Publishers directly, call 800-832-0034, fax 978-443-8000, or visit our website www.jbpub.com.

Substantial discounts on bulk quantities of Jones and Bartlett's publications are available to corporations, professional associations, and other qualified organizations. For details and specific discount information, contact the special sales department at Jones and Bartlett via the above contact information or send an email to specialsales@jbpub.com.

Copyright © 2009 by Jones and Bartlett Publishers, LLC

All rights reserved. No part of the material protected by this copyright may be reproduced or utilized in any form, electronic or mechanical, including photocopying, recording, or by any information storage and retrieval system, without written permission from the copyright owner.

Production Credits
Acquisitions Editor: Timothy Anderson
Production Director: Amy Rose
Production Intern: Ashlee Hazeltine
Senior Marketing Manager: Andrea DeFronzo
V.P., Manufacturing and Inventory Control: Therese Connell
Composition: ATLIS Graphics
Cover Design: Kristin E. Ohlin
Cover Image: © Ekl/Dreamstime.com
Printing and Binding: Malloy, Inc.
Cover Printing: Malloy, Inc.

Library of Congress Cataloging-in-Publication Data

Patty, C. Wayne.
   Foundations of topology / C. Wayne Patty.—2nd ed.
     p. cm.
   Includes bibliographical references and index.
   ISBN-13: 978-0-7637-4234-8 (casebound)
   ISBN-10: 0-7637-4234-1 (casebound)
   1. Topology.   I. Title.
   QA611.P33 2008
   514—dc22

2008027310

6048
Printed in the United States of America
12 11 10 09 08   10 9 8 7 6 5 4 3 2 1

Contents

	Preface	ix
	Acknowledgments	xiii
	Interdependence Diagram	xv

Chapter 1 **Topological Spaces** — 1
- 1.1 Metric Spaces — 1
- 1.2 Topological Spaces: The Definition and Examples — 10
- 1.3 Basis for a Topology — 15
- 1.4 Closed Sets, Closures, and Interiors of Sets — 26
- 1.5 Metric Spaces Revisited — 37
- 1.6 Convergence — 43
- 1.7 Continuous Functions and Homeomorphisms — 53

Chapter 2 **New Spaces from Old Ones** — 63
- 2.1 Subspaces — 63
- 2.2 The Product Topology on $X \times Y$ — 71
- 2.3 The Product Topology — 78
- 2.4 The Weak Topology and the Product Topology — 86
- 2.5 The Uniform Metric — 91
- 2.6 Quotient Spaces — 96

Chapter 3 **Connectedness** — 109
- 3.1 Connected Spaces — 109
- 3.2 Pathwise and Local Connectedness — 119
- 3.3 Totally Disconnected Spaces — 126

Chapter 4 **Compactness** — 131
- 4.1 Compactness in Metric Spaces — 131
- 4.2 Compact Spaces — 140
- 4.3 Local Compactness and the Relation Between Various Forms of Compactness — 149
- 4.4 The Weak Topology on a Topological Space — 154
- 4.5 Equicontinuity — 159

Chapter 5 **The Separation and Countability Axioms** — 163
- 5.1 T_0-, T_1-, and T_2-Spaces — 163
- 5.2 Regular and Completely Regular Spaces — 168
- 5.3 Normal and Completely Normal Spaces — 177
- 5.4 The Countability Axioms — 182
- 5.5 Urysohn's Lemma and the Tietze Extension Theorem — 186
- 5.6 Embeddings — 191

viii Contents

Chapter 6	**Special Topics**	**197**
	6.1 Contraction Mappings in Metric Spaces	197
	6.2 Normed Linear Spaces	200
	6.3 The Fréchet Derivative	205
	6.4 Manifolds	212
	6.5 Fractals	224
	6.6 Compactifications	232
	6.7 The Alexander Subbase and the Tychonoff Theorems	240

Chapter 7	**Metrizability and Paracompactness**	**245**
	7.1 Urysohn's Metrization Theorem	245
	7.2 Paracompactness	249
	7.3 The Nagata–Smirnov Metrization Theorem	257

Chapter 8	**The Fundamental Group and Covering Spaces**	**265**
	8.1 Homotopy of Paths	265
	8.2 The Fundamental Group	275
	8.3 The Fundamental Group of the Circle	279
	8.4 Covering Spaces	283
	8.5 Applications and Additional Examples of Fundamental Groups	287
	8.6 Knots	293

Chapter 9	**Applications of Homotopy**	**301**
	9.1 Inessential Maps	301
	9.2 The Fundamental Theorem of Algebra	303
	9.3 Homotopic Maps	305
	9.4 The Jordan Curve Theorem	308

Appendices

	A	Logic and Proofs	315
	B	Sets	321
	C	Functions	327
	D	Indexing Sets and Cartesian Products	335
	E	Equivalence Relations and Order Relations	339
	F	Countable Sets	345
	G	Uncountable Sets	349
	H	Ordinal and Cardinal Numbers	353
	I	Algebra	359

Bibliography   **367**

Index   **369**

Preface

This text is designed for a one- or two-semester introduction to topology at the undergraduate and beginning graduate levels. While there is no specific prerequisite for the study of elementary topology, a certain amount of mathematical maturity is necessary since this text presents the basic principles of topology in a rigorous manner. A bit of analysis (rigorous calculus) is needed to understand the motivation for some of the concepts, and a knowledge of elementary group theory is necessary for the last two chapters. The background in set theory that a student needs to study this text is provided in the appendices, and thus is available for reference.

Topology developed in a natural way from geometry and analysis. It is not only a powerful tool in many branches of mathematics, but it also has a beauty of its own since it is a natural, geometric, and intuitively appealing branch of mathematics. Topology is an excellent subject for the student to develop the skill to write clear and precise proofs.

This text is written with the student in mind. The topics are well motivated, the thoroughness of proofs makes them easy to follow, and the notation is clear and direct. Historical comments are dispersed throughout the text, and there are many exercises of varying degrees of difficulty. Many of the applications to the real line are in the exercises. Some proofs and other details are omitted. The instructor may choose to supply some of these and assign the remaining ones for in-class presentation or as written assignments.

We have organized the book with as much flexibility as possible. This enables the instructor to follow his/her own purposes and preferences, and for this reason an interdependence diagram is provided on page xv.

The first seven chapters cover the usual topics of point-set or general topology. Chapter 1 is central to the text. It begins with the study of metric spaces. Open sets and continuous functions are introduced in this setting. The concept of an abstract topological space is then introduced, and the student quickly sees how a metric on a set generates a topology on the set. Thus the reader immediately encounters the concept of a topology on a set but quickly sees familiar examples, such as the real line and the plane, of topological spaces. Concepts such as basis, closed set, the closure of a set, limit point, and interior point are introduced. This is followed by a revisit to metric spaces, where the product metric space is defined and concepts such as boundedness and diameter are discussed. Then convergence and complete metric spaces are studied. First and second countable spaces, separable spaces, and Hausdorff spaces are also introduced in Chapter 1, and the chapter concludes with the study of continuous functions and homeomorphisms.

The sections on subspaces, product spaces, and quotient spaces in Chapter 2 are also core material. The student is introduced to the product of two topological spaces before encountering the more abstract and difficult concept of the product of an arbitrary collection of topological spaces. While we have labeled both sections on product spaces as core material, it is possible to tailor a one-semester course in such a way that the product of an arbitrary collection of topological spaces is omitted. As shown in the interdependence diagram, this material is used throughout the text. One can, however, simply omit the material in each section that refers to the product of an arbitrary collection of topological spaces. The material on quotient spaces emphasizes the geometric aspect of topology, and, while this section is highly recommended, it can also be omitted in a one-semester course. The section on the uniform metric is available for those who wish to emphasize metric spaces. As illustrated, however, in the interdependence diagram, most of the material in the text does not depend on either Section 4 or 5.

Connectedness (Chapter 3), while perhaps not as important as compactness, belongs in every introductory topology course. Certainly everyone should know about the Cantor set. As illustrated early in this chapter, the concept of connectedness allows one to give a very simple proof of the Intermediate-Value Theorem. This short chapter also introduces the related concepts of the fixed-point property, pathwise connectedness, local pathwise connectedness, local connectedness, and total disconnectedness.

The importance of compactness (Chapter 4) cannot be overemphasized. We begin by studying the concept in a metric setting. The Bolzano-Weierstrass property, countable compactness, and sequential compactness are introduced, and it is shown that in a metric space, these concepts are equivalent to compactness. Sections 2 and 3 provide a study of compactness in a general setting and introduce local compactness. With the exception of the definition of compactness, Section 1 is not a prerequisite for Section 2. As seen in the interdependence diagram these first three sections are prerequisites for much of the material in Chapters 5 and 7. The last two sections on the weak topology and equicontinuity are optional. For those who wish to emphasize metric spaces, Section 5 (equicontinuity) does not depend on Sections 3 or 4.

Chapter 5 presents a rather comprehensive study of the separation and countability axioms, culminating in Urysohn's Lemma, the Tietze Extension Theorem, and a characterization of completely regular spaces in terms of an embedding theorem (Sections 5 and 6). While the interdependence diagram shows that the first three sections are prerequisites for Section 5.5 and that the first two sections are prerequisites for Section 5.6, this text is written with sufficient flexibility to allow the instructor to omit a large amount of material in Sections 1 and 3 and still obtain the three results mentioned above. This can be accomplished by covering Theorems 5.1, 5.7, and 5.8 and Corollary 5.9 in Section 1 together with the definition of a normal space and Theorem 5.24 in Section 3. It will be necessary to cover most of the material in Section 2, but the rather complicated example of a regular space that is not completely regular can be omitted. Sections 4, 5, and 6 are mutually independent of each other.

Chapters 6 and 7 are non-core material. Chapter 6 provides a study of some independent special topics. The first three sections are available for those who wish to emphasize metric spaces. The first two sections, on contraction mappings and normed linear spaces, are independent and provide the necessary background for the study of the Frechet derivative in Section 3. Section 4 provides a brief introduction to manifolds and gives a brief list of references for those who wish to pursue the subject. Section 5 provides a brief introduction to fractals (a "hot" topic today), with special emphasis on the Hausdorff metric, and gives a small list of references for those who wish to continue the study of fractals. Section 6 discusses the possible embedding of a non-compact topological space into a compact space, and Section 7 provides a proof of the Tychonoff Theorem. The interdependence diagram indicates that Section 5.3 is a prerequisite for Section 7. This is because of the example of a normal space that is not completely normal. If you omit this example, Section 2.4 is sufficient.

Chapter 7 introduces paracompactness and gives conditions, stated in terms of the topology, which guarantee that there is a metric that generates the given topology. Many of the results in this chapter were discovered in the 1950s.

Chapters 8 and 9 are designed to provide an introduction and motivation for the study of algebraic topology. These two chapters emphasize geometric applications. For example, the fundamental group of certain familiar spaces is calculated, and the Brouwer fixed-point theorem and the Jordan curve theorem are proved. As shown in the interdependence diagram, one can go directly from Section 4.1 to Chapter 8. Thus, under the assumption that the students have had a brief introduction to groups, some of this material can be included in a one-semester course.

In order to fully emphasize the flexibility of this text, we present two extremes for the content of a one-semester course: one for those who wish to emphasize metric spaces and one for those who wish to introduce algebraic topology. There are, of course, many possibilities for students between these two extremes.

I: Chapter 1, Chapter 2 (with the exception of Section 4), Chapter 3, Sections 1, 2, and 5 of Chapter 4, and the first three sections of Chapter 6.

II: Chapter 1, Chapter 2 (with the exceptions of Sections 4 and 5), Chapter 3, Sections 4.1 and 4.2, Chapter 8, and Chapter 9.

List of Symbols
- $\mathbb{N}$ The set of all natural numbers
- $\mathbb{Z}$ The set of all integers
- $\mathbb{R}$ The set of all real numbers
- $\mathbb{Q}$ The set of all rational numbers

Acknowledgments

It is a pleasure to acknowledge the advice and assistance we have received from our colleague Robert A. McCoy, and we are grateful to our colleague Charles E. Aull for his assistance with dates. We have incorporated the ideas and insights of our reviewers from the following colleges and universities and thank them for sharing their experiences with us:

Smith College

University of Wisconsin–Milwaukee

Russell Sage College

Macalester College

University of Bridgeport

California Polytechnic Institute–San Luis Obispo

DePauw University

Ball State University

University of Missouri–Columbia

University of Puget Sound

Youngstown State University

A special thank you to Jones and Bartlett Publishers. We are most grateful to the editorial team for their professionalism, commitment, and hard work on this edition.

Interdependence Diagram

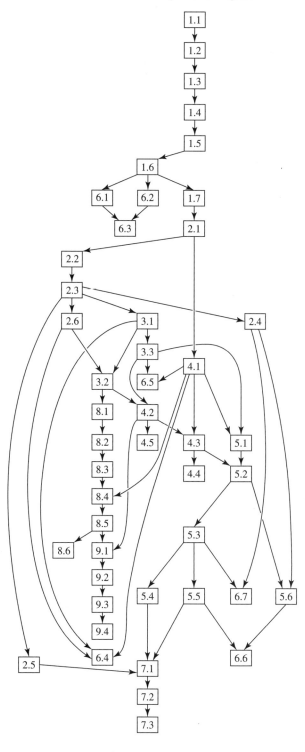

Chapter 1
Topological Spaces

Continuous functions are central to the study of analysis, and as functions are defined on more complicated spaces, a general notion of continuity becomes important. Historically, there were two major steps in defining continuity other than on the real line and the Euclidean plane.

Maurice Fréchet (1878–1973) was the first to extend topological considerations beyond Euclidean spaces. He introduced metric spaces in 1906 in a context that permitted one to consider abstract objects and not just real numbers or n-tuples of real numbers. Topology emerged as a coherent discipline in 1914 when Felix Hausdorff (1868–1942) published his classic treatise *Grundzüge der Mengenlehre*. Hausdorff defined a topological space in terms of neighborhoods of members of a set. These concepts were introduced immediately after Georg Cantor (1845–1918) had developed a general theory of sets in the 1880s. But even before Cantor, Bernard Riemann (1826–1866) had foreseen the study of abstract spaces.

1.1 Metric Spaces

We begin the study of topology with a brief motivational introduction to metric spaces. The ideas of metric and metric space are abstractions of the concept of distance in Euclidean space. These abstractions are fundamental and useful in all branches of mathematics. In this section we define metric spaces and give some examples. Then we define continuity on such spaces and proceed to show how this definition of continuity can be stated without mentioning the metrics. The material presented in this section will motivate the definition of a topology in the next section.

The term, Cartesian product, is used in this section. The Cartesian product of two sets is defined in Appendix B, and arbitrary Cartesian products are defined in Appendix D.

Definition. A metric on a set X is a function $d: X \times X \to \mathbb{R}$ that satisfies the following conditions:

(a) $d(x, y) \geq 0$ for all $x, y \in X$.

(b) $d(x, y) = 0$ if and only if $x = y$.

(c) $d(x, y) = d(y, x)$ for all $x, y \in X$.

(d) $d(x, z) \leq d(x, y) + d(y, z)$ for all $x, y, z \in X$.

If d is a metric on a set X, the ordered pair (X, d) is called a **metric space,** and if $x, y \in X$, then $d(x, y)$ is the **distance from x to y.** ■

Property **(d)** of a metric is an abstraction of the fact that the length of one side of a triangle is less than or equal to the sum of the lengths of the other two sides. Thus it is called the **triangle inequality.** (See Figure 1.1.)

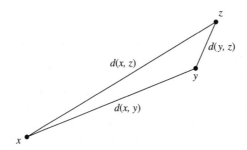

Figure 1.1

Notice that a metric space is simply a set together with a "distance function" on the set. The distance function must satisfy certain properties, and these properties parallel the familiar properties of the usual distance functions on $\mathbb{R}$ and $\mathbb{R}^2$.

EXAMPLE 1. Note that the function $d: \mathbb{R} \times \mathbb{R} \to \mathbb{R}$ defined by $d(x, y) = |x - y|$ satisfies the four conditions of the definition and hence this function is a metric on $\mathbb{R}$. It is called the **usual metric on** $\mathbb{R}$. Also, the function $d: \mathbb{R}^2 \times \mathbb{R}^2 \to \mathbb{R}$ defined by $d((x_1, y_1), (x_2, y_2)) = \sqrt{(x_1 - x_2)^2 + (y_1 - y_2)^2}$ satisfies the four conditions of the definition (see Exercise 4). It is called the **usual metric on** $\mathbb{R}^2$.

EXAMPLE 2. Let X be a set and define $d: X \times X \to \mathbb{R}$ by $d(x, x) = 0$ for all $x \in X$ and $d(x, y) = 1$ if $x, y \in X$ and $x \neq y$. Then d is a metric of X (see Exercise 1), and it is called the **discrete metric.**

EXAMPLE 3. The usual metrics for $\mathbb{R}$ and $\mathbb{R}^2$ are special cases of an important class of metric spaces, the **Euclidean spaces** $\mathbb{R}^n$, where $n \in \mathbb{N}$. Recall that if $n \in \mathbb{N}$, then $\mathbb{R}^n = \{x = (x_1, x_2, ..., x_n): x_i \in \mathbb{R}$ for each $i = 1, 2, ..., n\}$. The function $d: \mathbb{R}^n \times \mathbb{R}^n \to \mathbb{R}$ defined by $d((x, y)) = (\sum_{i=1}^{n}(x_i - y_i)^2)^{1/2}$, where $x = (x_1, x_2, ..., x_n)$ and $y = (y_1, y_2, ..., y_n)$ is a metric for $\mathbb{R}^n$ (see Exercise 4). It is called the **usual** or **Euclidean metric** for $\mathbb{R}^n$.

If we take X to be $\mathbb{R}$ in Example 2, then we have two metrics on $\mathbb{R}$, the usual metric and the discrete metric. In the following example, we give a second metric on

$\mathbb{R}^n$, where n is an arbitrary natural number. Thus we will have defined three metrics on $\mathbb{R}(=\mathbb{R}^1)$.

EXAMPLE 4. Let $n \in \mathbb{N}$ and define a function $d: \mathbb{R}^n \times \mathbb{R}^n \to \mathbb{R}$ by $d((x, y)) = \sum_{i=1}^{n}|x_i - y_i|$, where $x = (x_1, x_2, ..., x_n)$ and $y = (y_1, y_2, ..., y_n)$. Then d is a metric (see Exercise 2) on $\mathbb{R}^n$, and it is called the **taxicab metric**.

The metric in Example 4 is called the taxicab metric because, in the plane, the distance from one point to another is the sum of the lengths of a horizontal line segment and a vertical line segment.

The following example provides yet another metric on $\mathbb{R}^n$.

EXAMPLE 5. Let $n \in \mathbb{N}$ and define a function $\rho: \mathbb{R}^n \times \mathbb{R}^n \to \mathbb{R}$ by $\rho(x, y) = \max\{|x_1 - y_1|, |x_2 - y_2|, ..., |x_n - y_n|\}$, where $x = (x_1, x_2, ..., x_n)$ and $y = (y_1, y_2, ..., y_n)$. Then ρ is a metric on $\mathbb{R}^n$, and it is called the **square metric**.

Analysis. It is clear that $\rho(x, y) \geq 0$ for all $x, y \in \mathbb{R}^n$, $\rho(x, y) = 0$ if and only if $x = y$ and $\rho(x, y) = \rho(y, x)$. Let $x, y, z \in \mathbb{R}^n$. For each $i = 1, 2, ..., n$, $|x_i - z_i| \leq |x_i - y_i| + |y_i - z_i|$. Therefore, by definition of ρ, $|x_i - z_i| \leq \rho(x, y) + \rho(y, z)$ for each $i = 1, 2, ..., n$. Hence $\rho(x, z) \leq \rho(x, y) + \rho(y, z)$. Therefore ρ is a metric on $\mathbb{R}^n$. ∎

Note that for $\mathbb{R}^1$, the Euclidean metric, the taxicab metric, and the square metric are precisely the same.

Thus far, the discrete metric is the only metric we have defined on a set that is possibly different from $\mathbb{R}^n$. Now we give an example of a metric on a subset of $\mathbb{R}^\omega$ ($\mathbb{R}^\omega$ is defined in Appendix D), but first we introduce some notation, prove the Cauchy-Schwartz Inequality, and state the Minkowski Inequality.

Let $x = (x_1, x_2, ..., x_n)$ and $y = (y_1, y_2, ..., y_n)$ be members of $\mathbb{R}^n$, and let $c \in \mathbb{R}$. The **sum** $x + y$ and **difference** $x - y$ are defined by $x + y = (x_1 + y_1, x_2 + y_2, ..., x_n + y_n)$ and $x - y = (x_1 - y_1, x_2 - y_2, ..., x_n - y_n)$, and cx is defined to be $(cx_1, cx_2, ..., cx_n)$. The **dot product** $x \cdot y$ is defined by $x \cdot y = \sum_{i=1}^{n} x_i y_i$. The **norm** $\|x\|$ of x is defined by $\|x\| = (\sum_{i=1}^{n}(x_i)^2)^{1/2}$. Note that the norm of x is the distance between x and the origin.

THEOREM 1.1 (The Cauchy-Schwartz Inequality). Let $x = (x_1, x_2, ..., x_n)$ and $y = (y_1, y_2, ..., y_n)$ be members of $\mathbb{R}^n$. Then $|x \cdot y| \leq \|x\|\|y\|$.

Proof. If $x = (0, 0, ..., 0)$ or $y = (0, 0, ..., 0)$, both sides of $|x \cdot y| \leq \|x\|\|y\|$ are zero. Therefore we may assume that $x \neq 0$ and $y \neq 0$, and therefore that $\|x\|$ and $\|y\|$ are both positive numbers. For each $i = 1, 2, ..., n$, $0 \leq ((|x_i|/\|x\|) - (|y_i|/\|y\|))^2$. Hence

$$\frac{2|x_i y_i|}{\|x\|\|y\|} \leq \frac{(x_i)^2}{\|x\|^2} + \frac{(y_i)^2}{\|y\|^2}.$$

Therefore

$$\sum_{i=1}^{n} \frac{2|x_i y_i|}{\|x\|\|y\|} \le \sum_{i=1}^{n} \left(\frac{(x_i)^2}{\|x\|^2} + \frac{(y_i)^2}{\|y\|^2} \right).$$

Thus

$$\frac{2}{\|x\|\|y\|} \sum_{i=1}^{n} |x_i y_i| \le \frac{1}{\|x\|^2} \sum_{i=1}^{n} (x_i)^2 + \frac{1}{\|y\|^2} \sum_{i=1}^{n} (y_i)^2 = \frac{\|x\|^2}{\|x\|^2} + \frac{\|y\|^2}{\|y\|^2} = 2.$$

So

$$\frac{1}{\|x\|\|y\|} \sum_{i=1}^{n} |x_i y_i| \le 1 \text{ or } \sum_{i=1}^{n} |x_i y_i| \le \|x\|\|y\|.$$

Therefore, since $|x \cdot y| = |\sum_{i=1}^{n} x_i y_i| \le \sum_{i=1}^{n} |x_i y_i|$, $|x \cdot y| \le \|x\|\|y\|$. ∎

THEOREM 1.2 (The Minkowski Inequality). Let $x = (x_1, x_2, ..., x_n)$ and $y = (y_1, y_2, ..., y_n)$ be members of $\mathbb{R}^n$. Then $\|x + y\| \le \|x\| + \|y\|$.

Proof. See Exercise 3. ∎

The Cauchy-Schwartz Inequality, as we have stated it, is due to Augustin L. Cauchy (1784–1857), and the Minkowski Inequality was proved, in 1909, by Hermann Minkowski (1864–1909).

Notice that a metric is a function whose range is a subset of $\mathbb{R}$. Sometimes a metric is defined in such a way that it is not obvious that $d(x, y)$ is a real number for all x and y. In this case we must prove that $d(x, y)$ is a real number for all x and y, and once this is done we say that d is **well-defined**. The following example illustrates this notion.

EXAMPLE 6. Let $\ell^2 = \{x = (x_1, x_2, ...) \in \mathbb{R}^\omega : \sum_{i=1}^{\infty}(x_i)^2 \text{ converges}\}$. For each $x \in \ell^2$, $\|x\| = (\sum_{i=1}^{\infty}(x_i)^2)^{1/2}$ is called the **norm** of x. For $x = (x_1, x_2, ...)$ and $y = (y_1, y_2, ...)$ in ℓ^2, define $d(x, y) = (\sum_{i=1}^{\infty}(x_i - y_i)^2)^{1/2}$. Then d is a well-defined metric on ℓ^2, and it is called the ℓ^2-**metric**.

Analysis. See Exercise 7 and 8. ∎

The fact that ℓ^2 consists only of those points in $\mathbb{R}^\omega$ such that $\sum_{i=1}^{\infty}(x_i)^2$ converges is sufficient to ensure that $(\sum_{i=1}^{\infty}(x_i - y_i)^2)^{1/2}$ is a real number for all $(x_1, x_2, ...), (y_1, y_2, ...) \in \ell^2$.

We now give an example of a metric on a set of functions.

EXAMPLE 7. Let $a, b \in \mathbb{R}$ with $a < b$, and let X denote the set of all continuous functions that map $[a, b]$ into $\mathbb{R}$. Define a function $d: X \times X \to \mathbb{R}$ by $d(f, g) = \int_a^b |f(x) - g(x)|\, dx$. Then (X, d) is a metric space.

Analysis. See Exercise 14. ∎

The following example will be used in Section 1.5.

EXAMPLE 8. Let (X, d) be a metric space, and define a function $d': X \times X \to \mathbb{R}$ by $d'(x, y) = d(x, y)/1 + d(x, y))$. Then (X, d') is a metric space.

Analysis. See Exercise 16. ∎

Let (X, d) be a metric space and let $A \subseteq X$. Then $d|_{(A \times A)}$ is a metric on A (see Exercise 9), and it is called the **subspace metric**. We adopt the convention that if no metric is specified on a subset A of a metric space, then we assume that A has the subspace metric.

Metrics give us a way to define continuity in an abstract setting. The following definition becomes familiar if we replace X and Y by $\mathbb{R}$ and d and ρ by the usual distance (absolute value of the difference of two real numbers) on $\mathbb{R}$.

Definition. Let (X, d) and (Y, ρ) be metric spaces. A function $f: X \to Y$ is **continuous at a point** a in X provided that for each positive number ε there is a positive number δ such that if $x \in X$ and $d(a, x) < \delta$, then $\rho(f(a), f(x)) < \varepsilon$. A function $f: X \to Y$ is **continuous** provided it is continuous at each point of X. ∎

EXAMPLE 9. Let (X, d) and (Y, ρ) be metric spaces and let $y \in Y$. Then the functions $i: X \to X$ defined by $i(x) = x$ and $c: X \to Y$ defined by $c(x) = y$ for all $x \in X$ are continuous.

Analysis. Let $a \in X$ and let $\varepsilon > 0$. Choose $\delta = \varepsilon$. Then if $x \in X$ and $d(a, x) < \delta$, $d(i(a), i(x)) = d(a, x) < \varepsilon$. Therefore i is continuous.

Let $a \in X$ and let $\varepsilon > 0$. Let δ be any positive number. Then if $x \in X$ and $d(a, x) < \delta$, $\rho(c(a), c(x)) = \rho(y, y) = 0 < \varepsilon$. Therefore c is continuous. ∎

In the introduction to this section we mentioned that we would state the definition of continuity without mentioning the metrics. The following notation is a first step in that direction.

Let (X, d) be a metric space, let ε be a positive number and let $x \in X$. The set $B_d(x, \varepsilon) = \{y \in X: d(x, y) < \varepsilon\}$ is called the **ε-ball with center at x**. When no confusion will arise, we omit d from the notation and write $B(x, \varepsilon)$. We also refer to $B(x, \varepsilon)$ as the **open ball centered at x with radius ε**. Note that if d is the usual metric on $\mathbb{R}$, $\varepsilon > 0$, and $x \in \mathbb{R}$, then $B(x, \varepsilon)$ is the open interval $(x - \varepsilon, x + \varepsilon)$. Also if d is the usual metric on $\mathbb{R}^2$, $\varepsilon > 0$ and $(x, y) \in \mathbb{R}^2$, then $B((x, y), \varepsilon)$ is an open disk with center at (x, y) and radius ε.

The definition of continuity in metric spaces can be stated in terms of open balls.

THEOREM 1.3. Let (X, d) and (Y, ρ) be metric spaces. A function $f: X \to Y$ is continuous at a point a in X if and only if for each open ball, $B_\rho(f(a), \varepsilon)$, there is an open ball, $B_d(a, \delta)$, such that $f(B_d(a, \delta)) \subseteq B_\rho(f(a), \varepsilon)$.

Proof. See Exercise 10. ∎

Figure 1.2 illustrates this restatement of continuity.

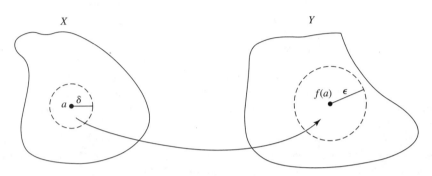

Figure 1.2

This restatement, together with the next definition, makes it possible to define continuity without mentioning the metrics.

Definition. subset U of a metric space (X, d) is **open** if for each $x \in U$, there is an open ball $B_d(x, \varepsilon)$ such that $B_d(x, \varepsilon) \subseteq U$. (See Figure 1.3.) ∎

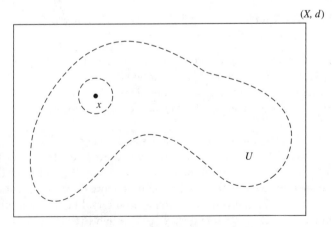

The set U is open in the metric space (X, d).

Figure 1.3

The following theorem provides the motivation for the definition of a topology on a set, which will be given in Section 1.2.

THEOREM 1.4. The open subsets of a metric space (X, d) have the following properties.

(a) X and $\emptyset$ are open sets.

(b) The union of any collection of open sets is open.

(c) The intersection of any finite collection of open sets is open.

Proof.

(a) Let $x \in X$, and let ε be any positive number. Then $B_d(x, \varepsilon) \subseteq X$ and hence X is open. The empty set is open because there is no $x \in \emptyset$.

(b) Let $\{U_\alpha : \alpha \in \Lambda\}$ be a collection of open sets and let $x \in \bigcup_{\alpha \in \Lambda} U_\alpha$. Then there exists $\beta \in \Lambda$ such that $x \in U_\beta$. Since U_β is open, there exists an open ball $B_d(x, \varepsilon)$ such that $B_d(x, \varepsilon) \subseteq U_\beta$. Since $U_\beta \subseteq \bigcup_{\alpha \in \Lambda} U_\alpha, B_d(x, \varepsilon) \subseteq \bigcup_{\alpha \in \Lambda} U_\alpha$. Therefore $\bigcup_{\alpha \in \Lambda} U_\alpha$ is open.

(c) Let $n \in \mathbb{N}$. For each $i \in I_n$ (I_n is defined in Appendix D) let U_i be an open set, and let $x \in \bigcap_{i=1}^{n} U_i$. Then $x \in U_i$ for each $i \in I_n$, and hence for each $i \in I_n$ there exists a positive number ε_i such that $B_d(x, \varepsilon_i) \subseteq U_i$. Let $\varepsilon = \min \{\varepsilon_i : i \in I_n\}$. Then $B_d(x, \varepsilon) \subseteq \bigcap_{i=1}^{n} B_d(x, \varepsilon_i) \subseteq \bigcap_{i=1}^{n} U_i$ and hence $\bigcap_{i=1}^{n} U_i$ is open. ∎

EXAMPLE 10. An open nonempty subset of $\mathbb{R}$ (with the usual metric) is any set that can be expressed as the union of open intervals.

Analysis. Let $U \subseteq \mathbb{R}$ such that $U = \bigcup_{\alpha \in \Lambda}(a_\alpha, b_\alpha)$ and let $x \in U$. Then there exists $\beta \in \Lambda$ such that $x \in (a_\beta, b_\beta)$. Let $\varepsilon = \min\{x - a_\beta, b_\beta - x\}$. Then $x \in (x - \varepsilon, x + \varepsilon) \subseteq (a_\beta, b_\beta) \subseteq U$. Thus each subset of $\mathbb{R}$ that can be expressed as the union of open intervals is open.

Now suppose U is an open subset of $\mathbb{R}$, and let $x \in U$. Then there exists a positive number ε such that $B_d(x, \varepsilon) \subseteq U$. But $B_d(x, \varepsilon) = (x - \varepsilon, x + \varepsilon)$. Therefore U can be expressed as the union of open intervals. ∎

The preceding analysis also shows that every "open interval" in $\mathbb{R}$ of the form (a, b) is an open set. We leave as Exercise 11 the proof that every open ball in a metric space is an open set.

Notice that arbitrary intersections of open sets need not be open. In particular, for each $n \in \mathbb{N}$, let $U_n = (-\frac{1}{n}, \frac{1}{n})$. Then for each $n \in \mathbb{N}$, U_n is an open subset of $\mathbb{R}$ (with the usual metric), but $\bigcap_{n \in \mathbb{N}} U_n = \{0\}$. Since, by Example 10, $\{0\}$ is not open in $\mathbb{R}$, $\bigcap_{n \in \mathbb{N}} U_n$ is not open.

EXAMPLE 11. Let X be a nonempty set and let d be the discrete metric on X. Then for each $x \in X$ and each ε with $0 < \varepsilon \leq 1$, $B_d(x, \varepsilon) = \{x\}$ and hence for each $x \in X$, $\{x\}$ is an open subset of (X, d).

Thus we have seen that while $\{0\}$ is not open in $\mathbb{R}$ with the usual metric, it is open in $\mathbb{R}$ with respect to another metric. In fact most sets can support more than one metric.

The following theorem provides the motivation for the definition of continuous functions on arbitrary topological spaces (which is given in Section 1.7).

THEOREM 1.5. Let (X, d) and (Y, ρ) be metric spaces, let $f: X \to Y$ be a function, and let $a \in X$. Then f is continuous at a, if and only if, for each open subset V of Y such that $f(a) \in V$, there is an open subset U of X such that $a \in U$ and $f(U) \subseteq V$.

Proof. Suppose f is continuous at a, and let V be an open subset of Y such that $f(a) \in V$. Then there is a positive number ε such that $B_\rho(f(a), \varepsilon) \subseteq V$. Since f is continuous, there is a positive number δ such that $f(B_d(a, \delta)) \subseteq B_\rho(f(a), \varepsilon)$. Since $B_d(a, \delta)$ is an open subset of (X, d), the proof is complete.

Suppose that for each open subset V of Y such that $f(a) \in V$, there is an open subset U of X such that $a \in U$ and $f(U) \subseteq V$. Let ε be a positive number. Then $V = B_\rho(f(a), \varepsilon)$ is an open set such that $f(a) \in V$. Thus there is an open subset U of X such that $a \in U$ and $f(U) \subseteq V$. Since U is open and $a \in U$, there is a positive number δ such that $B_d(a, \delta) \subseteq U$. Hence $f(B_d(a, \delta)) \subseteq B_\rho(f(a), \varepsilon)$, and f is continuous. ∎

Now that we have given an introduction to metric spaces to motivate the study of topological spaces, we begin an axiomatic development of topological spaces in the next section. Although topological spaces are defined in the next section, we do not define continuity in topological spaces until Section 1.7. There you will see that the definition of continuity is just a restatement of Theorem 1.5.

EXERCISES 1.1

1. Let X be a set and define $d: X \times X \to \mathbb{R}$ by $d(x, x) = 0$ for all $x \in X$ and $d(x, y) = 1$ if $x, y \in X$ and $x \neq y$. Prove that d is a metric on X.

2. Prove that the taxicab metric is a metric on $\mathbb{R}^n$.

3. Prove the Minkowski Inequality. *Hint:* Use the Cauchy-Schwartz Inequality and prove that if $x, y \in \mathbb{R}^n$, then $\|x + y\|^2 \leq (\|x\| + \|y\|)^2$.

4. Prove that the Euclidean metric d for $\mathbb{R}^n$ is indeed a metric. *Hint:* Note that if $x, y \in \mathbb{R}^n$, then $d(x, y) = \|x - y\|$. Use the Minkowski Inequality.

5. Show that if $x = (x_1, x_2, \ldots)$ and $y = (y_1, y_2, \ldots)$ are members of ℓ^2, then $\sum_{i=1}^{\infty} |x_i y_i|$ converges.

6. Let $x, y \in \ell^2$ and let $c \in \mathbb{R}$. Prove that $x + y$ and cx are members of ℓ^2.

7. Show that the function d defined in Example 6 is well-defined; that is, show that $\sum_{i=1}^{\infty}(x_i - y_i)^2$ converges whenever $x = (x_1, x_2,...)$ and $y = (y_1, y_2,...)$ are members of ℓ^2.

8. Show that the function d defined in Example 6 is a metric.

9. Let (X, d) be a metric space and let $A \subseteq X$. Prove that $d|_{(A \times A)}$ is a metric on A.

10. Let (X, d) and (Y, ρ) be metric spaces, let $a \in X$, and let $f : X \to Y$ be a function. Prove that f is continuous at a if and only if for each open ball, $B_\rho(f(a), \varepsilon)$, there is an open ball, $B_d(a, \delta)$, such that $f(B_d(a, \delta)) \subseteq B_\rho(f(a), \varepsilon)$.

11. Let (X, d) be a metric space, let $x \in X$, and let ε be a positive number. Prove that $B_d(x, \varepsilon)$ is an open set.

12. Let (X, d_1) and (Y, d_2) be metric spaces. Define a function $d: (X \times Y) \times (X \times Y) \to \mathbb{R}$ by $d((x_1, y_1), (x_2, y_2)) = \max\{d_1(x_1, x_2), d_2(y_1, y_2)\}$. Prove that d is a metric on $X \times Y$.

13. Let (X, d_1) and (Y, d_2) be metric spaces. Define a function $d: (X \times Y) \times (X \times Y) \to \mathbb{R}$ by $d((x_1, y_1), (x_2, y_2)) = \min\{d_1(x_1, x_2), d_2(y_1, y_2)\}$. Is d a metric on $X \times Y$? Explain.

14. Let $a, b \in \mathbb{R}$ with $a < b$, and let X denote the set of all continuous functions that map $[a, b]$ into $\mathbb{R}$. Define a function $d: X \times X \to \mathbb{R}$ by $d(f, g) = \int_a^b |f(x) - g(x)| dx$. Prove that (X, d) is a metric space.

15. Let (X, d) be the metric space in Exercise 14, and let ρ be the usual metric on $\mathbb{R}$. Define a function $F: (X, d) \to (\mathbb{R}, \rho)$ by $F(f) = \int_a^b f(x) dx$. Prove that F is continuous.

16. Let (X, d) be a metric space. Define a function $d': X \times X \to \mathbb{R}$ by
$$d'(x, y) = \frac{d(x, y)}{1 + d(x, y)}.$$
Prove that (X, d') is a metric space.

17. Let (X, d) be a metric space and let r be a positive real number. Define a function $d_r: X \times X \to \mathbb{R}$ by $d_r(x, y) = r \cdot d(x, y)$.

    (a) Prove that d_r is a metric on X.

    (b) Prove that the function $i: (X, d) \to (X, d_r)$ defined by $i(x) = x$ is continuous.

    (c) Prove that the function $i: (X, d_r) \to (X, d)$ defined by $i(x) = x$ is continuous.

18. Let (X, d) and (Y, ρ) be metric spaces. An **isometry** is a one-to-one function f that maps X onto Y so that $d(x, y) = \rho(f(x), f(y))$ for all $x, y \in X$. Prove that if $f: X \to Y$ is an isometry, then f and f^{-1} are continuous.

19. Let (X, d) be a metric space and let A be a nonempty subset of X. Define $f: X \to \mathbb{R}$ by $f(x) = \inf\{d(x, a) : a \in A\}$. Prove that f is continuous.

20. Let (X, d_1) and (Y, d_2) be metric spaces, and let $f: (X, d_1) \to (X, d_2)$ be a function. Prove that f is continuous, if and only if, for each open subset U of (Y, d_2), $f^{-1}(U)$ is an open subset of (X, d_1).

21. Let $X = \{1, 2, 3, 4\}$. List the different metrics (different in the sense that they yield a different collection of open sets) that can be defined on X.

22. Let (X_1, d_1), (X_2, d_2), and (X_3, d_3) be metric spaces, and let $f: X_1 \to X_2$ and $g: X_2 \to X_3$ be continuous functions. Prove that $g \circ f: X_1 \to X_3$ is continuous.

23. Define $f: \mathbb{R} \to \mathbb{R}$ by $f(x) = |x|$. Prove that f is continuous.

1.2 Topological Spaces: the Definition and Examples

In Section 1.1, and axiomatic framework for generalizing the notion of distance in Euclidean space was developed. In this section we introduce an axiomatic framework in which "nearness" does not refer to a metric, and thus we obtain a greater degree of generality. This framework is possible because many of the concepts in the first section can be expressed in terms of open sets. This passage to a higher level of abstraction simplifies many concrete situations. For example, it is often the case that the easiest proof of a fact about a metric space is obtained by considering the metric space to be a topological space. In addition, some ideas, such as pointwise convergence of a sequence of functions, cannot be expressed in terms of a metric on the set of functions.

The open sets in a metric space have the properties listed in Theorem 1.4, and these properties are used to define a topology on a set.

Definition. A **topology** on a set X is a collection $\mathcal{T}$ of subsets of X having the following properties:

(a) $\varnothing \in \mathcal{T}$ and $X \in \mathcal{T}$.

(b) If $U \in \mathcal{T}$ and $V \in \mathcal{T}$, then $U \cap V \in \mathcal{T}$.

(c) If $U_\alpha \in \mathcal{T}$ for each α in an index set Λ, then $\bigcup\{U_\alpha : \alpha \in \Lambda\} \in \mathcal{T}$.

The members of $\mathcal{T}$ are called **open sets**. ∎

A topology is defined so that the intersection of any two open sets is open. Theorem 1.4, however, states that in a metric space the intersection of any finite number of open sets is open. We use mathematical induction to prove that the intersection of a finite number of members of $\mathcal{T}$ is a member of $\mathcal{T}$.

THEOREM 1.6. Let $\mathcal{T}$ be a topology on a set X and let Λ be a finite set. If $U_\alpha \in \mathcal{T}$ for each $\alpha \in \Lambda$, then $\bigcap \{U_\alpha : \alpha \in \Lambda\} \in \mathcal{T}$.

Proof. Let $S = \{n \in \mathbb{N}:$ the intersection of any n members of $\mathcal{T}$ is a member of $\mathcal{T}\}$. It is clear that 1 and 2 belong to S. Suppose $n \geq 2$ and $\{1, 2, ..., n\} \subseteq S$. Let $\{U_i : i \in I_{n+1}\}$ be a collection of members of $\mathcal{T}$. Since $n \in S$, $\bigcap_{i=1}^n U_i \in \mathcal{T}$. Therefore, since $2 \in S$, $\bigcap_{i=1}^{n+1} U_i = (\bigcap_{i=1}^n U_i) \bigcap U_{n+1} \in \mathcal{T}$. Thus $n + 1 \in S$, and by the Second Principle of Mathematical Induction, $S = \mathbb{N}$. ∎

Definition. A **topological space** is an ordered pair $(X, \mathcal{T})$, where X is a set and $\mathcal{T}$ is a topology on X. ∎

Note that if $(X, \mathcal{T})$ is a topological space, then the union of any collection of open sets is an open set and the intersection of any finite collection of open sets is an open set.

EXAMPLE 12. Let (X, d) be a metric space. By Theorem 1.4, the open subsets of (X, d) form a topology on X. This topology is called the **topology on X induced by d** or the **topology on X generated by d.** Thus every metric space is a topological space, and whenever we speak of a metric space (X, d), we assume that the topology on X is the one induced by d.

Definition. A topological space $(X, \mathcal{T})$ is **metrizable** if there is a metric d on X such that the topology induced by d is $\mathcal{T}$. ∎

Note the distinction between a metric space and a metrizable space. A metrizable space is a topological space whose topology is generated by some metric, and a metric space is a set with a metric on it. Of course the metric on a set X generates a topology on X, and thus a metric space (X, d) determines a topological space $(X, \mathcal{T})$. Given a metrizable space $(X, \mathcal{T})$, there is a metric d on X such that the topology induced by d is $\mathcal{T}$. As we shall see in Example 18, the metric d is not unique. In fact there are many metrics that generate $\mathcal{T}$.

EXAMPLE 13. Let X be a nonempty set, and let $\mathcal{T} = \{\emptyset, X\}$. Then $\mathcal{T}$ is a topology on X, and it is called the **trivial** or **indiscrete topology** on X.

The following example shows that there are topological spaces that are not metrizable. Other examples will be given throughout the text.

EXAMPLE 14. Let X be a set with at least two members, and let $\mathcal{T}$ be the trivial topology on X. Then $(X, \mathcal{T})$ is not metrizable.

Analysis. Let d be a metric on X and let $\mathcal{U}$ be the topology generated by d. We show that $(X, \mathcal{T})$ is not metrizable by showing that $\mathcal{U} \neq \mathcal{T}$. Let a and b denote distinct members of X. By the definition of a metric, there is a positive number r such that

$d(a, b) = r$. Therefore $a \in B(a, r)$ but $b \notin B(a, r)$. Hence $B(a, r) \in \mathcal{U}$, $B(a, r) \neq X$, and $B(a, r) \neq \emptyset$. But the only members of $\mathcal{T}$ are $\emptyset$ and X. Thus $\mathcal{U} \neq \mathcal{T}$. ∎

EXAMPLE 15. Let X be a nonempty set, and let $\mathcal{T}$ denote the collection of all subsets of X (that is, $\mathcal{T} = \mathcal{P}(X)$). Then $\mathcal{T}$ is a topology on X, and it is called the **discrete topology** on X.

We leave as Exercise 1 the proof that the discrete metric on a nonempty set X generate the discrete topology on X.

EXAMPLE 16. Let $\mathcal{T} = \{U \in \mathcal{P}(\mathbb{R})\colon \text{for each } x \in U, \text{ there is a positive number } \delta_x$ such that the open interval $(x - \delta_x, x + \delta_x) \subseteq U\}$. Then it follows immediately from the definition of open subsets of a metric space that $\mathcal{T}$ is the topology on $\mathbb{R}$ induced by the usual metric (see Example 1) on $\mathbb{R}$.

The topology $\mathcal{T}$ on $\mathbb{R}$ in Example 16 is called the **usual** or **Euclidean topology** on $\mathbb{R}$. Note that a subset of $\mathbb{R}$ is open if and only if it is the union of a collection of open intervals. The **usual** or **Euclidean topology** on the Euclidean plane is given in the following example.

EXAMPLE 17. Let $\mathcal{T} = \{U \in \mathcal{P}(\mathbb{R}^2)\colon \text{for each } (a, b) \in U, \text{ there is a positive number } \delta \text{ such that } \{(x, y) \in \mathbb{R}^2\colon (x - a)^2 + (y - b)^2 < \delta^2\} \subseteq U\}$. Then $\mathcal{T}$ is the topology on $\mathbb{R}^2$ induced by the usual metric (see Example 1) on $\mathbb{R}^2$.

Note that if δ is a positive number, then $\{(x, y) \in \mathbb{R}^2\colon (x - a)^2 + (y - b)^2 < \delta^2\}$ is the set of points inside the circle with center (a, b) and radius δ. This set of points is the open ball (or disk) with center (a, b) and radius δ.

Throughout the remainder of this book, unless otherwise stated, we assume that $\mathbb{R}$ and $\mathbb{R}^2$ have the usual topology.

Two metrics d and ρ on a set X induce the same topology on X provided that a subset of X is open in (X, d) if and only if it is open in (X, ρ). We use this fact in the following examples.

EXAMPLE 18. The topology generated by the square metric ρ (see Example 5) on $\mathbb{R}^2$ is the usual topology.

Before we show that the topology induced by ρ is the topology induced by the usual metric d, let us see what $B_\rho(x, \varepsilon)$ looks like. Let $x = (x_1, x_2) \in \mathbb{R}^2$ and let $\varepsilon > 0$. Then, as shown in Figure 1.4, $B_\rho(x, \varepsilon)$ is the inside of a square.

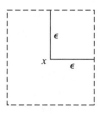

Figure 1.4

Analysis. Let d be the usual metric on $\mathbb{R}^2$. Let U be an open subset of $(\mathbb{R}^2, \rho)$, and let $x \in U$. Then there exists a positive number ε such that $B_\rho(x, \varepsilon) \subseteq U$. Since $\rho(a, b) \leq d(a, b)$ for all $a, b \in \mathbb{R}^2$ (see Exercise 2), $B_d(x, \varepsilon) \subseteq B_\rho(x, \varepsilon)$. Therefore $B_d(x, \varepsilon) \subseteq U$ and hence U is an open subset of $(\mathbb{R}^2, d)$.

Now let U be an open subset of $(\mathbb{R}^2, d)$, and let $x \in U$. Then there exists a positive number ε such that $B_d(x, \varepsilon) \subseteq U$. Since $d(a, b) \leq \sqrt{2} \cdot \rho(a, b)$ for all $a, b \in \mathbb{R}^2$ (see Exercise 2), $B_\rho\left(x, \frac{\sqrt{2}}{2}\varepsilon\right) \subseteq B_d(x, \varepsilon)$. Therefore $B_\rho\left(x, \frac{\sqrt{2}}{2}\varepsilon\right) \subseteq U$, and hence U is an open subset of $(\mathbb{R}^2, \rho)$. ∎

EXAMPLE 19. Let $X = \{1, 2, 3\}$. What are some of the topologies on X?

Analysis. Of course, we have the trivial topology and the discrete topology. We list some other topologies.

$\mathcal{T}_1 = \{\{1\}, \emptyset, X\}$ $\mathcal{T}_2 = \{\{1, 2\}, \emptyset, X\}$
$\mathcal{T}_3 = \{\{1\}, \{2\}, \{1, 2\}, \emptyset, X\}$ $\mathcal{T}_4 = \{\{1\}, \{2, 3\}, \emptyset, X\}$
$\mathcal{T}_5 = \{\{1, 2\}, \{2, 3\}, \{2\}, \emptyset, X\}$ $\mathcal{T}_6 = \{\{1\}, \{1, 2\}, \emptyset, X\}$
$\mathcal{T}_7 = \{\{1\}, \{2\}, \{1, 2\}, \{2, 3\}, \emptyset, X\}$

The nonempty members of $\mathcal{T}_5$ are shown in Figure 1.5. ∎

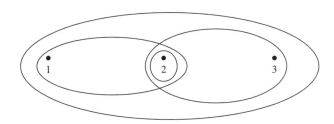

Figure 1.5

There are collections of subsets of $X = \{1, 2, 3\}$ that are not topologies on X. For example, $\{\{1, 2\}, \{2, 3\}, \emptyset, X\}$ and $\{\{1\}, \{2\}, \emptyset, X\}$ are not topologies on X.

As Example 19 illustrates, there are many topologies on a set. If $\mathcal{T}_1$ and $\mathcal{T}_2$ are topologies on a set X and $\mathcal{T}_1 \subseteq \mathcal{T}_2$, then we say that $\mathcal{T}_2$ is **finer (stronger, larger)** than $\mathcal{T}_1$ and $\mathcal{T}_1$ is **coarser (weaker, smaller)** than $\mathcal{T}_2$. If $\mathcal{T}_1$ is a proper subset of $\mathcal{T}_2$, then we say that $\mathcal{T}_2$ is **strictly finer (strictly stronger, strictly larger)** than $\mathcal{T}_1$ and $\mathcal{T}_1$ is **strictly coarser (strictly weaker, strictly smaller)** than $\mathcal{T}_2$. In Example 19, $\mathcal{T}_3$ is strictly coarser than $\mathcal{T}_7$ and $\mathcal{T}_7$ is strictly finer than $\mathcal{T}_3$. In the same example, $\mathcal{T}_3$ and $\mathcal{T}_4$ are not comparable. Notice that if a topology $\mathcal{T}$ on a set X is finer than a topology $\mathcal{U}$ on X, then there are more open sets in $(X, \mathcal{T})$ than in $(X, \mathcal{U})$.

EXAMPLE 20. Let X be a set and let $\mathcal{T} = \{U \in \mathcal{P}(X): U = \emptyset \text{ or } X - U \text{ is finite}\}$. Then $\mathcal{T}$ is a topology on X (see Exercise 4), and it is called the **finite complement topology**. This topology is of interest only when X is an infinite set because if X is finite, the finite complement topology is the discrete topology (see Exercise 6).

EXAMPLE 21. Let X be a set and let $\mathcal{T} = \{U \in \mathcal{P}(X): U = \emptyset \text{ or } X - U \text{ is countable}\}$. Then $\mathcal{T}$ is a topology on X (see Exercise 7), and it is called the **countable complement topology** on X.

Notice that if X is a finite set, then the finite complement topology, the countable complement topology, and the discrete topology are the same topology. The countable complement topology on $\mathbb{N}$ is the discrete topology on $\mathbb{N}$. However $\{1\}$ is not a member of the finite complement topology on $\mathbb{N}$. Also $\{1\}$ is not a member of either the finite complement topology or the countable complement topology on $\mathbb{R}$. Observe that the countable complement topology is finer than the finite complement topology on a set X.

The following theorem provides a useful criterion for proving that a subset of a topological space is open.

THEOREM 1.7. Let $(X, \mathcal{T})$ be a topological space, and let A be a subset of X with the property that for each $a \in A$ there is a member U_a of $\mathcal{T}$ such that $a \in U_a$ and $U_a \subseteq A$. Then $A \in \mathcal{T}$.

Proof. For each $a \in A$, let $U_a \in \mathcal{T}$ such that $a \in U_a$ and $U_a \subseteq A$. Then $A = \bigcup_{a \in A} U_a$, and, since the union of open sets is open, A is open. ∎

EXERCISES 1.2

1. Let d be the discrete metric on a nonempty set X. Prove that the topology on X generated by d is the discrete topology.

2. Let d be the usual metric and let ρ be the square metric on $\mathbb{R}^2$. Prove that:
   (a) $\rho(a, b) \leq d(a, b)$ for all $a, b \in \mathbb{R}^2$.
   (b) $d(a, b) \leq \sqrt{2} \cdot \rho(a, b)$ for all $a, b \in \mathbb{R}^2$.

3. For $i = 1, 2, ..., 7$, let $\mathcal{T}_i$ denote the topology on X given in Example 19. Are any of the topological spaces $(X, \mathcal{T}_i)$ metrizable? Explain your answer.

4. Let X be an infinite set and let $\mathcal{T} = \{U \in \mathcal{P}(X): U = \emptyset \text{ or } X - U \text{ is finite}\}$. Prove that $\mathcal{T}$ is a topology on X.

5. Let $(X, \mathcal{T})$ be a topological space. Prove that $\mathcal{T}$ is the discrete topology on X if and only if for each $x \in X$, $\{x\} \in \mathcal{T}$.

6. Let X be a set and let $\mathcal{T}$ be the finite complement topology on X. Prove that $\mathcal{T}$ is the discrete topology on X if and only if X is a finite set.

7. Let X be an infinite set and let $\mathcal{T} = \{U \in \mathcal{P}(X): U = \varnothing$ or $X - U$ is countable$\}$. Prove that $\mathcal{T}$ is a topology on X.

8. Let X be a set and let $\mathcal{T}$ be the countable complement topology on X. Let Λ be a countable set and for each $\alpha \in \Lambda$, let $U_\alpha \in \mathcal{T}$. Prove that $\bigcap\{U_\alpha: \alpha \in \Lambda\} \in \mathcal{T}$.

9. Let (X, d) be a metric space, let $\mathcal{T}$ be the topology induced by d, and let $\mathcal{U}$ be the finite complement topology on X. Prove that $\mathcal{T}$ is finer than $\mathcal{U}$.

10. Let $\{\mathcal{T}_\alpha: \alpha \in \Lambda\}$ be a collection of topologies on a set X. Prove that $\bigcap\{\mathcal{T}_\alpha: \alpha \in \Lambda\}$ is a topology on X.

11. Give an example of a set X and topologies $\mathcal{T}_1$ and $\mathcal{T}_2$ on X such that $\mathcal{T}_1 \cup \mathcal{T}_2$ is not a topology on X.

12. Let $\{\mathcal{T}_\alpha: \alpha \in \Lambda\}$ be a collection of topologies on a set X. Prove that there is a unique topology $\mathcal{T}$ on X such that: (1) for each $\alpha \in \Lambda$, $\mathcal{T}$ is finer than $\mathcal{T}_\alpha$, and (2) if $\mathcal{T}'$ is a topology on X that is finer than $\mathcal{T}_\alpha$ for each $\alpha \in \Lambda$, then $\mathcal{T}$ is coarser than $\mathcal{T}'$.

13. Let $\{\mathcal{T}_\alpha: \alpha \in \Lambda\}$ be a collection of topologies on a set X. Prove that there is a unique topology $\mathcal{T}$ on X such that: (1) for each $\alpha \in \Lambda$, $\mathcal{T}$ is coarser than $\mathcal{T}_\alpha$, and (2) if $\mathcal{T}'$ is a topology on X that is coarser than for each $\alpha \in \Lambda$, then $\mathcal{T}$ is finer than $\mathcal{T}'$.

1.3 Basis for a Topology

A topology on a set can be a complicated collection of subsets of a set, and it can be difficult to describe the entire collection. In most cases one describes a subcollection that "generates" the topology. One such collection is called a basis and another is called a subbasis. In this section we study these two concepts.

Definition. Let $(X, \mathcal{T})$ be a topological space. **A basis** for $\mathcal{T}$ is a subcollection $\mathcal{B}$ of $\mathcal{T}$ with the property that if $U \in \mathcal{T}$ then $U = \varnothing$ or there is a subcollection $\mathcal{B}'$ of $\mathcal{B}$ such that $U = \bigcup\{B: B \in \mathcal{B}'\}$. ∎

EXAMPLE 22. The collection $\mathcal{B}$ of all open intervals is a basis for the usual topology on $\mathbb{R}$, and the collection $\mathcal{B}$ of all open disks is a basis for the usual topology on the plane.

EXAMPLE 23. If X is a set, then $\mathcal{B} = \{\{x\}: x \in X\}$ is a basis for the discrete topology on X.

EXAMPLE 24. Let (X, d) be a metric space. Then the family $\mathscr{B} = \{B(x, \varepsilon): x \in X \text{ and } \varepsilon > 0\}$ is a basis for the topology generated by d.

The following example shows that there is a subset of $\mathscr{P}(X)$ that is not a basis for a topology on X.

EXAMPLE 25. Let $X = \{1, 2, 3\}$ and $\mathscr{B} = \{\{1, 2\}, \{2, 3\}, X\}$. Then $\mathscr{B}$ is not a basis for a topology on X.

Analysis. Suppose $\mathscr{B}$ is a basis for a topology $\mathscr{T}$ on X. Then by the definition of basis, $\mathscr{B} \subseteq \mathscr{T}$. Hence $\{1, 2\}, \{2, 3\} \in \mathscr{T}$ and so $\{1, 2\} \cap \{2, 3\} = \{2\} \in \mathscr{T}$. But $\{2\} \neq \emptyset$ and there is no subcollection $\mathscr{B}'$ of $\mathscr{B}$ such that $\{2\} = \bigcup\{B: B \in \mathscr{B}'\}$. Hence $\mathscr{B}$ is not a basis for $\mathscr{T}$. ∎

The following theorem provides a necessary and sufficient condition for a subset of $\mathscr{P}(X)$ to be a basis for a topology on X.

THEOREM 1.8. A family $\mathscr{B}$ of subsets of a set X is a basis for some topology on X if and only if: **(1)** $X = \bigcup\{B: B \in \mathscr{B}\}$, and **(2)** if $B_1, B_2 \in \mathscr{B}$ and $x \in B_1 \cap B_2$, then there exists $B \in \mathscr{B}$ such that $x \in B$ and $B \subseteq B_1 \cap B_2$.

Proof. Let $\mathscr{B}$ be a family or subsets of X such that $\mathscr{B}$ is a basis for a topology $\mathscr{T}$ on X. Since $X \in \mathscr{T}$, $X = \bigcup\{B: B \in \mathscr{B}\}$. Suppose $B_1, B_2 \in \mathscr{B}$ and $x \in B_1 \cap B_2$. Then $B_1, B_2 \in \mathscr{T}$, so $B_1 \cap B_2 \in \mathscr{T}$. Since $B_1 \cap B_2 \in \mathscr{T}$ and $\mathscr{B}$ is a basis for $\mathscr{T}$, there is a subcollection $\mathscr{B}'$ of $\mathscr{B}$ such that $B_1 \cap B_2 = \bigcup\{B: B \in \mathscr{B}'\}$. Therefore there exists $B \in \mathscr{B}$ such that $x \in B$ and $B \subseteq B_1 \cap B_2$.

Let $\mathscr{B}$ be a family of subsets of X satisfying conditions **(1)** and **(2)**. Let $\mathscr{T}$ be the collection of subsets of X consisting of $\emptyset$ and those sets that are unions of members of $\mathscr{B}$. Then $\emptyset \in \mathscr{T}$ and $X \in \mathscr{T}$. Since $\mathscr{T}$ consists of all unions of members of $\mathscr{B}$, the union of any collection of members of $\mathscr{T}$ is also in $\mathscr{T}$. Suppose $U_1, U_2 \in \mathscr{T}$ and $x \in U_1 \cap U_2$. Then there exists $B_1, B_2 \in \mathscr{B}$ such that $x \in B_1$, $B_1 \subseteq U_1$, $x \in B_2$, and $B_2 \subseteq U_2$. By conditions **(2)**, there exists $B \in \mathscr{B}$ such that $x \in B$ and $B \subseteq B_1 \cap B_2$. Therefore $U_1 \cap U_2$ can be expressed as the union of members of a subcollection of $\mathscr{B}$ and hence $U_1 \cap U_2 \in \mathscr{T}$. ∎

Notice that if $X = \{1, 2, 3\}$ and $\mathscr{B} = \{\{2\}, \{1, 2\}, \{2, 3\}\}$, then $\mathscr{B}$ satisfies both conditions of Theorem 1.8 and therefore it is a basis for a topology $\mathscr{T}$ on X. The topology $\mathscr{T}$ is $\{\emptyset, \{2\}, \{1, 2\}, \{2, 3\}, X\}$.

EXAMPLE 26. Let $\mathscr{B}$ be the collection of all regions in the plane that are interiors of rectangles, where the rectangles have sides that are parallel to the coordinate axes. Then $\mathscr{B}$ satisfies both conditions of Theorem 1.8. The second condition is illustrated in Figure 1.6.

1.3 Basis for a Topology

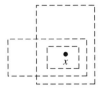

Figure 1.6

In Example 26, the intersection of any two members of $\mathcal{B}$ is either $\varnothing$ or a member of $\mathcal{B}$. Thus it is not necessary to draw a smaller rectangle that contains x since the intersection of the two rectangles gives the desired rectangle. As observed in Example 22, the collection of all open disks in the plane is a basis for a topology on $\mathbb{R}^2$. Notice that the intersection of two open disks need not be an open disk. As indicated in Figure 1.7, however, if x belongs to the intersection of two open disks, there is an open disk that contains x and is contained in the intersection.

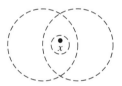

Figure 1.7

The proof of Theorem 1.8 tells us how to obtain a topology from a collection of subsets of a set that satisfies conditions **(1)** and **(2)** of Theorem 1.8. In particular, if $\mathcal{B}$ is a collection of subsets of a set X that satisfies conditions **(1)** and **(2)**, then we obtain a topology $\mathcal{T}$ on X by saying that $\varnothing \in \mathcal{T}$ and any set that can be expressed as the union of the members of some subcollection of $\mathcal{B}$ is a member of $\mathcal{T}$.

Observe that if we start with a topological space $(X, \mathcal{T})$, then a basis for $\mathcal{T}$ is a subcollection of $\mathcal{T}$ that has the property described in the definition of a basis for $\mathcal{T}$. If we start with a set X and a collection $\mathcal{B}$ of subsets of X that satisfy conditions **(1)** and **(2)** of Theorem 1.8, then we obtain a topology on X according to the manner described in the proof of Theorem 1.8.

Definition. If $\mathcal{B}$ is a basis for a topology $\mathcal{T}$ on a set X, then $\mathcal{T}$ is the **topology generated by** $\mathcal{B}$. ∎

EXAMPLE 27. Let $\mathcal{B} = \{B \in \mathcal{P}(\mathbb{R}) : B \text{ is an interval of the form } [a, b)\}$. Then $\mathcal{B}$ satisfies conditions **(1)** and **(2)** of Theorem 1.8 and hence $\mathcal{B}$ is a basis for a topology on X. This topology is called the **lower-limit topology** on $\mathbb{R}$.

Definition. Let $\mathcal{B}_1$ and $\mathcal{B}_2$ be bases for topologies $\mathcal{T}_1$ and $\mathcal{T}_2$ on a set X. The $\mathcal{B}_1$ and $\mathcal{B}_2$ are **equivalent** provided $\mathcal{T}_1 = \mathcal{T}_2$. ∎

The following result is an immediate consequence of Example 18.

EXAMPLE 28. The collection of open disks and the collection of open squares are equivalent bases for topologies in the plane. In each case the topology generated by the basis is the usual topology.

The following theorem gives a characterization for a topology $\mathcal{T}'$ to be finer than a topology $\mathcal{T}$ in terms of bases for $\mathcal{T}$ and $\mathcal{T}'$.

THEOREM 1.9. Let $\mathcal{T}$ and $\mathcal{T}'$ be topologies on a set X and let $\mathcal{B}$ and $\mathcal{B}'$ be bases for $\mathcal{T}$ and $\mathcal{T}'$ respectively. Then the following are equivalent:

(a) $\mathcal{T}'$ is finer than $\mathcal{T}$.

(b) For each $x \in X$ and each $B \in \mathcal{B}$ such that $x \in B$, there is a member B' of $\mathcal{B}'$ such that $x \in B'$ and $B' \subseteq B$.

Proof. (a) $\Rightarrow$ (b). Suppose $\mathcal{T}'$ is finer than $\mathcal{T}$, let $x \in X$, and let $B \in \mathcal{B}$ such that $x \in B$. Since $B \in \mathcal{T}$ and $\mathcal{T}'$ is finer than $\mathcal{T}$, $B \in \mathcal{T}'$. Since $\mathcal{T}'$ is generated by $\mathcal{B}'$, there is a member B' of $\mathcal{B}'$ such that $x \in B'$ and $B' \subseteq B$.

(b) $\Rightarrow$ (a). Let $U \in \mathcal{T}$, and let $x \in U$. Since $\mathcal{T}$ is generated by $\mathcal{B}$ there is a member B of $\mathcal{B}$ such that $x \in B$ and $B \subseteq U$. By condition (b), there is a member B' of $\mathcal{B}'$ such that $x \in B'$ and $B' \subseteq B$. Since $B' \subseteq B$ and $B \subseteq U$, $B' \subseteq U$. Therefore U is the union of the members of a subcollection of $\mathcal{B}'$ and hence $U \in \mathcal{T}'$. ∎

The following corollary is an immediate consequence of Theorem 1.9.

Corollary 1.10. Bases $\mathcal{B}_1$ and $\mathcal{B}_2$ for topologies on a set X are equivalent if and only if: **(1)** for each $B_1 \in \mathcal{B}_1$ and each $x \in B_1$, there exists $B_2 \in \mathcal{B}_2$ such that $x \in B_2$ and $B_2 \subseteq B_1$, and **(2)** for each $B_2 \in \mathcal{B}_2$ and each $x \in B_2$ there exists $B_1 \in \mathcal{B}_1$ such that $x \in B_1$ and $B_1 \subseteq B_2$. ∎

EXAMPLE 29. Let $\mathcal{B}_1$ be the collection of regions in the plane that are interiors of rectangles, and let $\mathcal{B}_2$ be the collection of all open disks in the plane. Then Figure 1.8 illustrates how Corollary 1.10 can be used to show that $\mathcal{B}_1$ and $\mathcal{B}_2$ are equivalent.

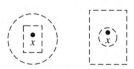

Figure 1.8

1.3 Basis for a Topology

EXAMPLE 30. Use corollary 1.10 to show that the lower-limit topology on $\mathbb{R}$ is not the usual topology on $\mathbb{R}$.

Analysis. As seen in Example 27, $\mathcal{B}_1 = \{[a, b): a, b \in \mathbb{R} \text{ and } a < b\}$ is a basis for the lower-limit topology. As seen in Example 22, $\mathcal{B}_2 = \{(a, b): a, b \in \mathbb{R} \text{ and } a < b\}$ is a basis for the usual topology. Now $[1, 2) \in \mathcal{B}_1$ and $1 \in [1, 2)$. However if $(a, b) \in \mathcal{B}_2$ and $1 \in (a, b)$, then $a < 1$ and hence $(a, b) \not\subseteq [1, 2)$. ∎

Notice that every topology has a basis since the topology itself forms a basis. The following theorem provides a condition for one to check to see if a subcollection of a topology $\mathcal{T}$ is a basis for $\mathcal{T}$.

THEOREM 1.11. Let $(X, \mathcal{T})$ be a topological space, and suppose $\mathcal{B}$ is a subcollection of $\mathcal{T}$ such that for each $x \in X$ and each member U of $\mathcal{T}$ such that $x \in U$, there is an element B of $\mathcal{B}$ such that $x \in B$ and $B \subseteq U$. Then $\mathcal{B}$ is a basis for $\mathcal{T}$.

Proof. First we show that $\mathcal{B}$ is a basis for some topology on X. Let $x \in X$. Since $X \in \mathcal{T}$, there is an element B of $\mathcal{B}$ such that $x \in B$ and $B \subseteq X$. Therefore $X = \bigcup\{B: B \in \mathcal{B}\}$.

Suppose $B_1, B_2 \in \mathcal{B}$ and $x \in B_1 \cap B_2$. Since $B_1, B_2 \in \mathcal{T}$, $B_1 \cap B_2 \in \mathcal{T}$. Thus there is an element B of $\mathcal{B}$ such that $x \in B$ and $B \subseteq B_1 \cap B_2$.

By Theorem 1.8, $\mathcal{B}$ is a basis for a topology $\mathcal{T}'$ on X. By Theorem 1.9, $\mathcal{T}'$ is finer than $\mathcal{T}$. Since $\mathcal{B} \subseteq \mathcal{T}$, arbitrary unions of members of $\mathcal{B}$ are members of $\mathcal{T}$. Therefore $\mathcal{T}' \subseteq \mathcal{T}$ and hence $\mathcal{T} = \mathcal{T}'$. ∎

As we have illustrated, there are advantages to specifying a basis rather than a topology. It is also sometimes advantageous to specify an even smaller collection.

Definition. Let $(X, \mathcal{T})$ be a topological space. A subcollection $\mathcal{S}$ of $\mathcal{T}$ is a **subbasis** for $\mathcal{T}$ provided the family of all finite intersections of members of $\mathcal{S}$ is a basis for $\mathcal{T}$. ∎

EXAMPLE 31. If $\mathcal{S}$ is the collection of all intervals of the form (a, ∞) and $(-\infty, b)$, then $\mathcal{S}$ is a subbasis for the usual topology on $\mathbb{R}$.

One advantage of the concept of subbasis is that we can define a topology on a set X by simply choosing an arbitrary collection of subsets of X whose union is X.

THEOREM 1.12. Let X be a set and let $\mathcal{S}$ be a collection of subsets of X such that $X = \bigcup\{S: S \in \mathcal{S}\}$. Then there is a unique topology $\mathcal{T}$ on X such that $\mathcal{S}$ is a subbasis for $\mathcal{T}$.

Proof. $\mathcal{B} = \{B \in \mathcal{P}(X): B \text{ is the intersection of a finite number of members of } \mathcal{S}\}$. Let $\mathcal{T} = \{U \in \mathcal{P}(X): U = \emptyset \text{ or there is a subcollection } \mathcal{B}' \text{ of } \mathcal{B} \text{ such that } U = \bigcup\{B: B \in \mathcal{B}'\}\}$. It suffices to prove that $\mathcal{T}$ is a topology on X. Note that $\emptyset \in \mathcal{T}$ by definition. Since $X = \bigcup\{S: S \in \mathcal{S}\}$, $X \in \mathcal{T}$. Let $U_\alpha \in \mathcal{T}$ for each α in an index set Λ. Then there is a subcollection $\mathcal{B}_\alpha$ of $\mathcal{B}$ such that $U_\alpha = \bigcup\{B\colon B \in \mathcal{B}_\alpha\}$. Hence $\bigcup\{U_\alpha\colon \alpha \in \Lambda\} = \bigcup_{\alpha \in \Lambda}(\bigcup_{B \in \mathcal{B}_\alpha} B)$ and so $\bigcup\{U_\alpha\colon \alpha \in \Lambda\} \in \mathcal{T}$. Suppose U_1, $U_2 \in \mathcal{T}$ and $x \in U_1 \cap U_2$. Then there exist $B_1, B_2 \in \mathcal{B}$ such that $x \in B_1 \cap B_2$, $B_1 \subseteq U_1$, and $B_2 \subseteq U_2$. Since each of B_1 and B_2 is the intersection of a finite number of members of $\mathcal{S}$, $B_1 \cap B_2 \in \mathcal{B}$. Therefore there is a subcollection $\mathcal{B}'$ of $\mathcal{B}$ such that $U_1 \cap U_2 = \bigcup\{B\colon B \in \mathcal{B}'\}$ and hence $U_1 \cap U_2 \in \mathcal{T}$. Therefore $\mathcal{T}$ is a topology, and it is clear that $\mathcal{T}$ is the unique topology that has $\mathcal{S}$ as a subbasis. ∎

The proof of Theorem 1.12 describes how to obtain a topology on a set X from a collection $\mathcal{S}$ of subsets of X whose union is X. This topology is called the **topology generated by** $\mathcal{S}$.

EXAMPLE 32. Let $X = \{1, 2, 3, 4, 5\}$. Then by Theorem 1.12, $\mathcal{S} = \{\{1\}, \{1, 2, 3\}, \{2, 3, 4\}, \{3, 5\}\}$ is a subbasis for a topology $\mathcal{T}$ on X. What is $\mathcal{T}$?

Analysis. The collection $\mathcal{B}$ of all finite intersections of members of $\mathcal{S}$ is a basis for $\mathcal{T}$. Note that $\mathcal{B} = \{\{1\}, \{2, 3\}, \{3\}, \{1, 2, 3\}, \{2, 3, 4\}, \{3, 5\}\}$. Then $\mathcal{T}$ consists of $\varnothing$ and all subsets of X that can be expressed as the union of members of $\mathcal{B}$, so $\mathcal{T} = \{\varnothing, \{1\}, \{3\}, \{1, 3\}, \{2, 3\}, \{3, 5\}, \{1, 2, 3\}, \{1, 3, 5\}, \{2, 3, 4\}, \{2, 3, 5\}, \{1, 2, 3, 4\}, \{1, 2, 3, 5\}, \{2, 3, 4, 5\}, X\}$. ∎

Notice the difference between Theorems 1.8 and 1.12. Theorem 1.12 says that if $\mathcal{S}$ is a collection of subsets of a set X such that each member of X belongs to some member of $\mathcal{S}$, then $\mathcal{S}$ is a subbasis for some topology $\mathcal{T}$ on X. The topology $\mathcal{T}$ is called the topology generated by $\mathcal{S}$. Furthermore, the collection of all finite intersections of members of $\mathcal{S}$ is a basis for $\mathcal{T}$, but $\mathcal{S}$ is not necessarily a basis for any topology. (Note that the collection $\mathcal{S}$ in Example 32 is not a basis for a topology because $2 \in \{1, 2, 3\} \cap \{2, 3, 4\}$, but there is no member S of $\mathcal{S}$ such that $2 \in S \subseteq \{1, 2, 3\} \cap \{2, 3, 4\}$). If, however, $\mathcal{B}$ is basis for a topology $\mathcal{T}$ on a set X, then $\mathcal{B}$ is a subbasis for $\mathcal{T}$. Notice also that if $\mathcal{B}$ is a basis for a topology on a set X, then a subset U of X is a member of the topology generated by $\mathcal{B}$ if and only if U is the union of members of $\mathcal{B}$. In contrast, if $\mathcal{S}$ is a subbasis for a topology on a set X, then a subset U of X is a member of the topology generated by $\mathcal{S}$ if and only if U is the union of finite intersections of members of $\mathcal{S}$. (Notice that the collection $\mathcal{B}$ in the analysis of Example 32 is the collection of finite intersections of members of $\mathcal{S}$ and each nonempty member of $\mathcal{T}$ is the union of members of $\mathcal{B}$.)

The following theorem describes sufficient conditions for one to check to see if two subcollections of a set X (such that the union of the members of each is X) generate the same topology on X. Note that the converse of the theorem is false (see Exercise 20).

THEOREM 1.13. Let $\mathcal{S}_1$ and $\mathcal{S}_2$ be collections of subsets of a set X such that $X = \bigcup\{S\colon S \in \mathcal{S}_1\} = \bigcup\{S\colon S \in \mathcal{S}_2\}$. Moreover, suppose that:

(1) for each $S_1 \in \mathcal{S}_1$ and each $x \in S_1$ there exists $S_2 \in \mathcal{S}_2$ such that $x \in S_2$ and $S_2 \subseteq S_1$, and

(2) for each $S_2 \in \mathscr{S}_2$ and each $x \in S_2$ there exists $S_1 \in \mathscr{S}_1$ such that $x \in S_1$ and $S_1 \subseteq S_2$.

Then $\mathscr{S}_1$ and $\mathscr{S}_2$ are subbases for the same topology on X.

Proof. See Exercise 1. ∎

Recall from Example 24 that if (X, d) is a metric space, then the family $\mathscr{B} = \{B(x, \varepsilon) : x \in X$ and $\varepsilon > 0\}$ is a basis for a topology on X. The subcollection $\mathscr{B}' = \left\{B\left(x, \dfrac{1}{n}\right) : x \in X \text{ and } n \in \mathbb{N}\right\}$ of $\mathscr{B}$ is also a basis for a topology on X. Furthermore $\mathscr{B}$ and $\mathscr{B}'$ are equivalent. Thus for each $x \in X$, $\left\{B\left(x, \dfrac{1}{n}\right) : n \in \mathbb{N}\right\}$ is a countable collection of open subsets of X such that if U is any open set and $x \in U$, then there exists $n \in \mathbb{N}$ such that $B\left(x, \dfrac{1}{n}\right) \subseteq U$. This leads to the following two definitions.

Definition. Let $(X, \mathscr{T})$ be a topological space and let $a \in X$. A **local basis** at a is a collection $\mathscr{B}_a$ of $\mathscr{T}$ such that: (1) if $B \in \mathscr{B}_a$ then $a \in B$, and (2) if $U \in \mathscr{T}$ and $a \in U$, then there exists $B \in \mathscr{B}_a$ such that $B \subseteq U$. ∎

EXAMPLE 33. If $\mathscr{T}$ is the usual topology on $\mathbb{R}$, then $\left\{\left(-\dfrac{1}{n}, \dfrac{1}{n}\right) : n \in \mathbb{N}\right\}$ is a local basis at 0.

Definition. A topological space $(X, \mathscr{T})$ is **first countable** or **satisfies the first axiom of countability** provided there is countable local basis at each point of X.

Based on the preceding discussion, we have the following theorem:

THEOREM 1.14. *Every metric space is first countable.* ∎

The importance of first countability will be illustrated in Section 1.6. An even more important property of topological spaces is the following.

Definition. A topological space $(X, \mathscr{T})$ is **second countable** or **satisfies the second axiom of countability** provided there is a countable basis for $\mathscr{T}$. ∎

The first and second axioms of countability were introduced by Hausdorff.

EXAMPLE 34. The set $\mathbb{R}$ with the usual topology $\mathscr{T}$ is second countable because $\mathscr{B} = \{(a, d) : a < b$ and a and b are rational$\}$ is a countable basis for $\mathscr{T}$, while $\mathbb{R}$ with the lower-limit topology is not second countable.

Analysis. See Exercises 2 and 3. ∎

The relation between first countable and second countable is given by the following theorem and Example 36.

THEOREM 1.15. Every second countable space is first countable.

Proof. Let $(X, \mathcal{T})$ be a second countable space, let $\mathcal{B}$ be a countable basis for $\mathcal{T}$, and let $p \in X$. Then the subcollection of $\mathcal{B}$ consisting of all those members of $\mathcal{B}$ that contain p is a countable local basis at p. Therefore $(X, \mathcal{T})$ is first countable. ∎

It is customary to shorten the phrase "U is an open set such that $x \in U$" to the phrase "U is a **neighborhood** of x."

Example 35 provided an example of a topological space that is not first countable, and Example 36 provides an example of a first countable space that is not second countable.

EXAMPLE 35. Let $\mathcal{T}$ be the finite complement topology on $\mathbb{R}$. Then $(\mathbb{R}, \mathcal{T})$ is not first countable.

Analysis. Let $x \in \mathbb{R}$ and suppose $\mathcal{B} = \{B_n : n \in \mathbb{N}\}$ is countable local basis at x. For each $y \in \mathbb{R}$ such that $y \neq x$, $\mathbb{R} - \{y\}$ is a neighborhood of x. Therefore for each $y \in \mathbb{R}$ such that $y \neq x$, there exists $n \in \mathbb{N}$ such that $y \notin B_n$. Hence $\bigcap \{B_n : n \in \mathbb{N}\} = \{x\}$. Thus $\mathbb{R} - \{x\} = \mathbb{R} - \bigcap \{B_n : n \in \mathbb{N}\} = \bigcup \{\mathbb{R} - B_n : n \in \mathbb{N}\}$. For each $n \in \mathbb{N}$, $\mathbb{R} - B_n$ is finite, so $\bigcup \{\mathbb{R} - B_n : n \in \mathbb{N}\}$ is countable. Therefore $\mathbb{R} - \{x\}$ is countable. This is a contradiction. ∎

EXAMPLE 36. Let $\mathcal{T}$ be the lower-limit topology on $\mathbb{R}$. Then $(\mathbb{R}, \mathcal{T})$ is first countable but not second countable.

Analysis. See Exercise 3. ∎

The following theorem provides a basis for a topology on the product of two topological spaces.

THEOREM 1.16. Let $(X, \mathcal{T}_1)$ and $(Y, \mathcal{T}_2)$ be topological spaces, and let $\mathcal{B} = \{U \times V : U \in \mathcal{T}_1 \text{ and } V \in \mathcal{T}_2\}$. Then $\mathcal{B}$ is a basis for a topology on $X \times Y$.

Proof. See Exercise 9. ∎

The following theorem describes a property of a first countable space that will be used in Section 1.6.

THEOREM 1.17. Let $(X, \mathcal{T})$ be a first countable space and let $x \in X$. Then there is a countable local basis $\mathcal{B} = \{B_n : n \in \mathbb{N}\}$ at x, such that $B_{n+1} \subseteq B_n$ for each $n \in \mathbb{N}$.

We conclude this section with a brief discussion of the standard topology on a linearly ordered set (see Appendix E). This discussion provides additional examples of bases and subbases.

Let $(X, \leq)$ be a linearly ordered set. Throughout the remainder of this section we assume that X has at least two members. If $a, b \in X$, then we write $a < b$ or $b > a$ to indicate that $a \leq b$ and $a \neq b$. If $a, b \in X$ with $a < b$, then there are four subsets, called **intervals**, of X determined by a and b. They are:

$$(a, b) = \{x \in X: a < x < b\} \qquad [a, b) = \{x \in X: a \leq x < b\}$$
$$(a, b] = \{x \in X: a < x \leq b\} \qquad [a, b] = \{x \in X: a \leq x \leq b\}$$

Notice that the notation is the familiar notation that is used to denote intervals on the real line. The set (a, b) is the called an **open interval** while the set $[a, b]$ is called a **closed interval**. The sets $[a, b)$ and $(a, b]$ are called **half-open intervals**. This terminology suggests a definition of a topology on $(X, \leq)$.

Definition. Let $(X, \leq)$ be a linearly ordered set, and let $\mathscr{S}$ be the collection of all subsets of X of the form $\{x \in X: x < a\}$ and $\{x \in X: x > a\}$, where $a \in X$. Then the topology that has $\mathscr{T}$ as a subbasis is called the **ordered topology** on X. ∎

In order to see that $\mathscr{T}$ is a subbasis for a topology on X, it is sufficient (see Theorem 1.12) to show that each member of X belongs to some member of $\mathscr{S}$. Let a, b be distinct members of X. Since any two members of X are comparable, $a < b$ or $b < a$. We may assume without loss of generality that $a < b$. Then $a \in \{x \in X: x < b\}$ and $b \in \{x \in X: x > a\}$. The collection $\mathscr{B}$ of all finite intersections of members of $\mathscr{T}$ is a basis for the order topology on X. Notice that $\mathscr{B}$ is the collection of all subsets of X of the following types:

(a) all open intervals.

(b) all half-open intervals of the form $[a_0, b)$, where a_0 is the smallest member of X (provided X has a smallest member).

(c) all half-open intervals of the form $(a, b_0]$, where b_0 is the largest member of X (provided X has a largest member).

By Theorem 10 in Appendix H, there is an uncountable well-ordered set Ω of ordinal numbers with maximal element ω_1 having the property that if $x \in \Omega$ and $x \neq \omega_1$, then $\{y \in \Omega: y \leq x\}$ is countable. Observe that in Ω,

$$(a, b] = \{x \in X: x > a\} \cap \{x \in X: x < b + 1\}$$

is a member of the basis $\mathscr{B}$ for the order topology. So a basis for the order topology on Ω consists of all sets of the form $(a, b]$, where $a < b$, and all sets of the form $[1, a)$, where $1 < a$.

Notice that if $\mathscr{T}$ is the usual topology on $\mathbb{R}$ and $\leq$ is the usual order on $\mathbb{R}$, then the order topology on $\mathbb{R}$ determined by $\leq$ is $\mathscr{T}$.

Definition. Let $(X, \leq_X)$ and $(Y, \leq_Y)$ be linearly ordered sets. Define a relation $\leq$ on $X \times Y$ by defining $(x_1, y_1) < (x_2, y_2)$ provided $x_1 <_X x_2$, or $x_1 = x_2$ and $y_1 <_Y y_2$. The relation $\leq$ is called the **dictionary order relation** or **lexicographic order relation** on $X \times Y$. ∎

Notice that $(X \times Y, \leq)$ is a linearly ordered set.

EXAMPLE 37. Let $\leq_\mathbb{R}$ denote the usual order on $\mathbb{R}$ and let $\leq$ denote the dictionary order on $\mathbb{R}^2$ determined by $\leq_\mathbb{R}$. Then $(\mathbb{R}^2, \leq)$ has neither a smallest nor a largest member. Therefore the members of the basis that we previously described for the order topology are open intervals. The two types of open intervals are shown in Figure 1.9.

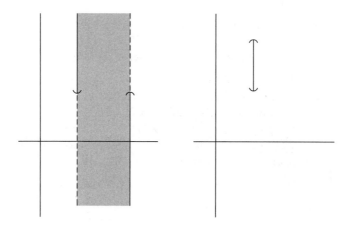

Figure 1.9

Exercises 4, 5, and 6 provide more examples of topologies on $\mathbb{R}$. Exercise 9 gives an example of a topology on the Cartesian product of two topological spaces, while Exercise 8 provides an example of a topology on a set of functions. Exercises 12 and 13 provide examples of topologies on subsets of $\mathbb{R}^2$ and Exercise 14 gives an example of a topology on $\mathbb{R}^2$. Many of these examples will be encountered later in the book.

EXERCISES 1.3

1. Prove Theorem 1.13.

2. Let $\mathcal{T}$ be the usual topology on $\mathbb{R}$. Prove that $\mathcal{B} = \{(a, b): a < b$ and a and b are rational$\}$ is a countable basis for $\mathcal{T}$.

3. Let $\mathcal{T}$ be the lower-limit topology on $\mathbb{R}$. Prove that $(\mathbb{R}, \mathcal{T})$ is first countable but not second countable.

1.3 Basis for a Topology

4. Let $\mathcal{B}$ be the collection of all intervals of the form (a, ∞). Show that $\mathcal{B}$ is a basis for a topology on $\mathbb{R}$.

5. Let $\mathcal{B}$ be the collection of all intervals of the form $[a, b)$, where $a < b$ and a and b are rational. Prove that $\mathcal{B}$ is a basis for a topology $\mathcal{T}$ on $\mathbb{R}$. Is $\mathcal{T}$ the lower-limit topology on $\mathbb{R}$?

6. Let $A = \left\{\dfrac{1}{n} : n \in \mathbb{N}\right\}$ and let $\mathcal{B} = \{(a, b) : a < b\} \cup \{(a, b) - A : a < b\}$. Prove that $\mathcal{B}$ is a basis for a topology $\mathcal{T}$ on $\mathbb{R}$. Is $\mathcal{T}$ the usual topology on $\mathbb{R}$?

7. Let $n \in \mathbb{N}$, let X be a set with exactly n members, and let $\mathcal{T}$ be the discrete topology on X. What is the least number of members that a basis for $\mathcal{T}$ can have?

8. Let X be the set of all functions that map $[0, 1]$ into $[0, 1]$. For each subset A of $[0, 1]$, let $B_A = \{f \in X : f(x) = 0 \text{ for all } x \in A\}$. Prove that $\mathcal{B} = \{B_A : A \subseteq [0, 1]\}$ is a basis for a topology on X.

9. Let $(X, \mathcal{T}_1)$ and $(Y, \mathcal{T}_2)$ be topological spaces. Let $\mathcal{B} = \{U \times V : U \in \mathcal{T}_1 \text{ and } V \in \mathcal{T}_2\}$. Prove that $\mathcal{B}$ is a basis for a topology on $X \times Y$.

10. Prove that the lower-limit topology on $\mathbb{R}$ is strictly finer than the usual topology on $\mathbb{R}$.

11. Let $\mathcal{B}$ be a basis for a topology on a set X. Show that the topology generated by $\mathcal{B}$ is the intersection of all topologies on X that contain $\mathcal{B}$.

12. Let $A = \{(x, y) \in \mathbb{R}^2 : y \geq 0\}$. For each $z = (x, y) \in A$ with $y > 0$ and each ε with $0 < \varepsilon < y$, let $D(z, \varepsilon) = \{(a, b) \in A : (x - a)^2 + (y - b)^2 < \varepsilon^2\}$. For each $(x, 0) \in A$ and each $\varepsilon > 0$, let $E((x, 0), \varepsilon) = \{(x, 0)\} \cup \{(a, b) \in A : b > 0 \text{ and } (x - a)^2 + b^2 < \varepsilon^2\}$. Prove that $\mathcal{B} = \{D(z, \varepsilon)\} \cup \{E((x, 0), \varepsilon)\}$ is a basis for a topology on A. *Hint:* The two types of members of $\mathcal{B}$ are the open disks that lie above the horizontal axis and sets of the form:

13. Let $A = \{(x, y) \in \mathbb{R}^2 : y \geq 0\}$. For each $z = (x, y) \in A$ with $y > 0$ and each ε with $0 < \varepsilon < y$, let $D(z, \varepsilon) = \{(a, b) \in A : (x - a)^2 + (y - b)^2 < \varepsilon^2\}$. For each $(x, 0) \in A$ and each $\varepsilon > 0$, let $E((x, 0), \varepsilon) = \{(x, 0)\} \cup \{(a, b) \in A : (x - a)^2 + (\varepsilon - b)^2 < \varepsilon^2\}$.

   (a) Prove that $\mathcal{B} = \{D(z, \varepsilon)\} \cup \{E((x, 0), \varepsilon)\}$ is a basis for a topology on A. *Hint:* The two types of members of $\mathcal{B}$ are the open disks that lie above the horizontal axis and sets of the form:

(b) Use Corollary 1.10 to show that the basis given in this exercise is not equivalent to the basis given Exercise 12.

14. For each $z = (x, y) \in \mathbb{R}^2$ such that $x^2 + y^2 \neq 0$ and each positive number ε with $\varepsilon^2 < x^2 + y^2$, let $D(z, \varepsilon) = \{(a, b) \in \mathbb{R}^2 : (x - a)^2 + (y - b)^2 < \varepsilon^2\}$. For each positive number ε, let $E(\varepsilon) = \{(a, b) \in \mathbb{R}^2 : a^2 + b^2 < \varepsilon^2\} - \{(a, b) \in \mathbb{R}^2 : a > 0 \text{ and } b = 0\}$. Prove that $\mathcal{B} = \{D(z, \varepsilon)\} \cup \{E(\varepsilon)\}$ is a basis for a topology on $\mathbb{R}^2$.

15. Let $X = \{1, 2, 3, 4, 5, 6\}$, and let $\mathcal{S} = \{\{1, 2\}, \{2, 3\}, \{3, 4\}, \{4, 5, 6\}\}$. Then by Theorem 1.12, $\mathcal{S}$ is a subbasis for a topology $\mathcal{T}$ on X. What is $\mathcal{T}$?

16. Let $X = \{1, 2, 3, 4, 5, 6\}$, and let $\mathcal{S} = \{\{1\}, \{1, 2\}, \{2, 3\}, \{3, 4\}, \{4, 5\}, \{5, 6\}, \{6\}\}$. Then by Theorem 1.12, $\mathcal{S}$ is a subbasis for a topology $\mathcal{T}$ on X. What is $\mathcal{T}$?

17. Describe a collection $\mathcal{S}$ of subsets of $\mathbb{R}^2$ such that $\mathcal{S}$ is a subbasis for a topology on $\mathbb{R}^2$ but $\mathcal{S}$ is not a basis for a topology on $\mathbb{R}^2$.

18. For each prime p let $S_p = \{n \in \mathbb{N} : n \text{ is a multiple of } p\}$. By Theorem 1.12, $\mathcal{S} = \{1\} \cup \{S_p : p \text{ is a prime}\}$ is a subbasis for a topology $\mathcal{T}$ on $\mathbb{N}$. Describe $\mathcal{T}$.

19. Let $(X, \mathcal{T})$ be a first countable space and let $x \in X$. Show that there is a countable local basis $\mathcal{B} = \{B_n : n \in \mathbb{N}\}$ at x such that $B_{n+1} \subseteq B_n$ for each $n \in \mathbb{N}$.

20. Find an example of a set X and subbases $\mathcal{S}_1$ and $\mathcal{S}_2$ that generate the same topology for X such that at least one of the two conditions in Theorem 1.13 is false.

21. Let $X = \{1, 2, 3, 4, 5, 6, 7\}$.

    a) List several different (different in the sense that they generate different topologies) bases for topologies on X.

    b) List several different (different in the sense that they generate different topologies) subbases (that are not bases) for topologies on X.

22. Give an example of a subbasis $\mathcal{S}$ for a topology $\mathcal{T}$ on $\mathbb{R}$ such that $\mathcal{T}$ is not the usual topology and $\mathcal{S}$ is not a basis for a topology.

23. Let $(X, \leq)$ be a linearly ordered set, and let $\mathcal{T}$ denote the order topology on X. Prove that if $a, b \in X$ and $a < b$, then there is a neighborhood U of a and a neighborhood V of b such that $x < y$ whenever $x \in U$ and $y \in V$.

1.4 Closed Sets, Closures, and Interiors of Sets

Recall that an open set in a topological space $(X, \mathcal{T})$ is a subset of X that is a member of $\mathcal{T}$. We have introduced a topological space in terms of the properties that the open sets must satisfy. In this section we study closed sets, the closure of a set, a limit point of a set, and the interior of a set. In addition, we encounter separable spaces that arise naturally from set closures.

1.4 Closed Sets, Closures, and Interiors of Sets 27

Definition. A subset A of a topological space $(X, \mathcal{T})$ is **closed** provided its complement, $X - A$, is open. ∎

EXAMPLE 38. A closed interval $[a, b]$, $a < b$, in $\mathbb{R}$ is closed with respect to the usual topology on $\mathbb{R}$ because its complement, $\mathbb{R} - [a, b]$, is $(-\infty, a) \cup (b, \infty)$. If $a \in \mathbb{R}$, then $\{a\}$ is closed with respect to the usual topology on $\mathbb{R}$ because $\mathbb{R} - \{a\} = (-\infty, a) \cup (a, \infty)$. With respect to the finite complement topology on a set X, a proper subset of X is closed if and only if it is a finite set. With respect to the countable complement topology on a set X, a proper subset of X is closet if and only if it is a countable set.

If (X, d) is a metric space, $x \in X$, and $\varepsilon > 0$, then $B(x, \varepsilon)$ is an open set. The following example illustrates some closed sets in a metric space.

EXAMPLE 39. Let (X, d) be a metric space, let $x \in X$, and let $\varepsilon > 0$. Then $A = \{y \in X: d(x, y) \leq \varepsilon\}$ is a closed subset of X.

Analysis. To show that A is closed, we show that $X - A$ is open. Let $y \in X - A$, and let $r_y = d(x, y)$. Then $r_y > \varepsilon$, so $\delta_y = r_y - \varepsilon$ is a positive number. Thus $B(y, \delta_y)$ is an open set, and by Theorem 1.7 the proof will be complete when we show that $B(y, \delta_y) \subseteq X - A$. Suppose $z \in B(y, \delta_y)$. Then $d(y, z) < \delta_y$. Now $d(x, z) + d(z, y) \geq d(x, y)$, so $d(x, z) \geq d(x, y) - d(z, y)$. Since $d(y, z) < \delta_y$, $d(x, y) - d(z, y) > d(x, y) - \delta_y$. But $d(x, y) = r_y$ and $\varepsilon = r_y - \delta_y$. Then $d(x, z) > \varepsilon$ and hence $z \notin A$. Therefore $B(y, \delta_y) \subseteq X - A$. ∎

As the following theorem indicates, the collection of closed subsets of a topological space has properties that are similar to those satisfied by the collection of open subsets.

THEOREM 1.18. Let $(X, \mathcal{T})$ be a topological space. Then the following conditions hold:

(a) X and $\emptyset$ are closed sets.

(b) If $\mathcal{A}$ is a finite family of closed subsets of X, then $\bigcup \{A: A \in \mathcal{A}\}$ is closed set.

(c) If $\mathcal{A}$ is a family of closed subsets of X, then $\bigcap \{A: A \in \mathcal{A}\}$ is a closed set.

Proof. See Exercises 1, 2, and 3. ∎

Instead of using open sets, we could define a topology on a set in terms of a collection of sets satisfying the three conditions stated in Theorem 1.18. In this case, each member of the collection would be a closed set.

If $(X, \mathcal{T})$ is a topological space, then X and $\emptyset$ are both open and closed. The following example illustrates that there can be proper subsets of a topological space that are both open and closed and there can be proper subsets that are neither open nor closed.

28   Chapter 1   ■   Topological Spaces

EXAMPLE 40. With respect to the usual topology on $\mathbb{R}$, if $a, b \in \mathbb{R}$ and $a < b$, then $[a, b)$ is neither open nor closed. With respect to the lower-limit topology on $\mathbb{R}$, if $a, b \in \mathbb{R}$ and $a < b$, then $[a, b)$ is both and closed.

Analysis. See Exercises 4 and 5. ■

Definition. The closure $\overline{A}$ of a subset A of a topological space $(X, \mathcal{T})$ is the intersection of the closed sets that contain A. ■

Note that $\overline{A}$ is the "smallest" closed set that contains A; that is, if B is any closed set such that $A \subseteq B$, then $\overline{A} \subseteq B$. By Theorem 1.18, the intersection of any number of closed sets is closed. Therefore the closure of a subset of a topological space is a closed set.

The following theorem provides a useful method of determining whether a point belongs to the closure of a set.

THEOREM 1.19. Let A be a subset of a topological space $(X, \mathcal{T})$, and let $x \in X$. Then $x \in \overline{A}$ if and only if every neighborhood of x has a nonempty intersection with A.

Proof. Suppose $x \notin \overline{A}$. Then there is a closed set C such that $A \subseteq C$ and $x \notin C$. So $X - C$ is a neighborhood of x such that $(X - C) \cap A = \emptyset$.

Suppose there is a neighborhood U of x such that $U \cap A = \emptyset$. Then $X - U$ is a closed set such that $A \subseteq X - U$ and $x \notin X - U$. Hence $x \notin \overline{A}$. ■

If (X, d) is a metric space and $A \subseteq X$, then it follows from Theorem 1.19 that a member x of X is in $\overline{A}$ if and only if for each positive number ε, $B(x, \varepsilon) \cap A \neq \emptyset$. Intuitively speaking, this means that there are members of A that are arbitrarily close to x. Thus from an intuitive point of view, we have the notion of "closeness" in an arbitrary topological space.

We give a characterization of the closure of a set in terms of "limit points."

Definition. Let A be a subset of a topological space $(X, \mathcal{T})$. A point x in X is a **limit point** of A provided that every neighborhood of x contains a point of A that is different from x. We let A' denote the set of all limit points of A. ■

Notice that if x is a limit point of A and U is any neighborhood of x, then $U \cap (A - \{x\}) \neq \emptyset$. (See Figure 1.10.)

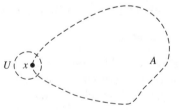

The set U is an arbitrary neighborhood of x, so x is a limit of point A.

Figure 1.10

1.4 Closed Sets, Closures, and Interiors of Sets

EXAMPLE 41. With respect to the usual topology on $\mathbb{R}$, if $a, b \in \mathbb{R}$ and $a < b$, then a is a limit point of each of $[a, b]$, $[a, b)$ and (a, b). With respect to the lower-limit topology on $\mathbb{R}$, if $a, b \in \mathbb{R}$ and $a < b$, then b is not a limit point of $[a, b)$.

EXAMPLE 42. Let (X, d) be a metric space, let $A \subseteq X$, and let $x \in X$. Then x is a limit point of A if and only if for each positive number ε there exists $y \in x$ such that $y \neq x$ and $d(x, y) < \varepsilon$. Thus from an intuitive point of view, x is a limit point of A if and only if there are members of A, different from x, that are arbitrarily close to x. This idea is pursed further in Section 1.6 (see Theorem 1.41).

THEOREM 1.20. Let A be a subset of a topological space $(X, \mathcal{T})$. Then $\overline{A} = A \cup A'$.

Proof. Let $x \in \overline{A}$. If U is any neighborhood of x, then by Theorem 1.19, $U \cap A \neq \emptyset$. If every neighborhood of x contains a point of A different from x, then $x \in A'$. If there is a neighborhood U of x such that $U \cap A = \{x\}$, then $x \in A$. In either case $x \in A \cup A'$, so $\overline{A} \subseteq A \cup A'$.

Suppose $x \in A \cup A'$. If $x \in A$, then x belongs to every closed set that contains A and hence $x \in \overline{A}$. If $x \in A'$, then every neighborhood of x contains a point of A that is different from x. Therefore, by Theorem 1.19, $x \in \overline{A}$. Thus $A \cup A' \subseteq \overline{A}$. ∎

THEOREM 1.21. Let A be a subset of a topological space $(X, \mathcal{T})$. Then A is closed if and only if $A' \subseteq A$.

Proof. Suppose A is closed, and $x \notin A$. Then $X - A$ is a neighborhood of x whose intersection with A is empty. Therefore $x \notin A'$. Hence $A' \subseteq A$.

Suppose $A' \subseteq A$. Then for each $x \in X - A$, there is a neighborhood U_x of x such that $U_x \cap A = \emptyset$. Since $\bigcup \{U_x : x \in X - A\}$ is an open set and $A = X - \bigcup \{U_x : x \in X - A\}$, A is closed. ∎

THEOREM 1.22. Let A and B be subset of a topological space $(X, \mathcal{T})$. Then:

(a) A is closed if and only if $A = \overline{A}$.
(b) $\overline{\overline{A}} = \overline{A}$.
(c) $\overline{\emptyset} = \emptyset$.
(d) $\overline{A} \subseteq \overline{B}$ whenever $A \subseteq B$.
(e) $\overline{A \cup B} = \overline{A} \cup \overline{B}$.
(f) $\overline{A \cap B} \subseteq \overline{A} \cap \overline{B}$.

Proof. Note that (c) follows immediately from (a). We prove (a) and (e) and leave the proofs of the remaining parts as Exercise 7.

(a) Suppose A is closed. Then A is a closed set that contains A, so the intersection of all the closed sets contain A is A. Hence $\overline{A} = A$. Suppose $\overline{A} = A$. Since $\overline{A}$ is a closed set, A is closed.

(e) Let $x \in \overline{A \cup B}$. Then $x \in \overline{A}$ or $x \in \overline{B}$. Suppose $x \in \overline{A}$. Let U be any neighborhood of x. Then $U \cap A \neq \emptyset$. Since $A \subseteq A \cup B$, $U \cap (A \cup B) \neq \emptyset$. Hence $x \in \overline{A \cup B}$. The proof that if $x \in \overline{B}$ then $x \in \overline{A \cup B}$ is similar. Therefore $\overline{A} \cup \overline{B} \subseteq \overline{A \cup B}$. Now suppose $x \notin \overline{A} \cup \overline{B}$. Then $x \notin \overline{A}$ and $x \notin \overline{B}$. Therefore there exist neighborhood U and V of x such that $U \cap A = \emptyset$ and $V \cap B = \emptyset$. Now $U \cap V$ is a neighborhood of x and $(U \cap V) \cap (A \cup B) = \emptyset$. Therefore $x \notin \overline{A \cup B}$, so $\overline{A \cup B} \subseteq \overline{A} \cup \overline{B}$. ∎

As the following example illustrates, there are subsets A and B of a topological space $(X, \mathcal{T})$ such that $\overline{A \cap B} \neq \overline{A} \cap \overline{B}$.

EXAMPLE 43. Let $\mathcal{T}$ be the usual topology on $\mathbb{R}$, let $A = (0, 1)$ and $B = (1, 2)$. Then $A \cap B = \emptyset$, so $\overline{A \cap B} = \emptyset$. However $\overline{A} = [0, 1]$ and $\overline{B} = [1, 2]$, so $\overline{A} \cap \overline{B} = \{1\}$.

Definition. Let A be a subset of a topological space $(X, \mathcal{T})$. A point $x \in X$ is a **boundary point** of A provided $x \in \overline{A} \cap \overline{(X - A)}$. The **boundary** of A, bd(A), is the set of all boundary points of A. ∎

Notice that if x is a boundary point of a subset of A of a topological space, then $x \in \overline{A}$ and $x \in \overline{X - A}$. Thus $x \in A$ or x is a limit point of A, and $x \in X - A$ or x is a limit point of $X - A$. However x cannot belong to both A and $X - A$. Therefore if x is a boundary point of A, then x is a limit point of A or x is limit point of $X - A$. (See Figure 1.11.)

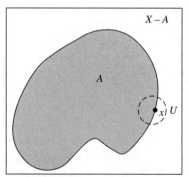

The set U is an arbitrary neighborhood of x, so x is a boundary point of A.

Figure 1.11

A consequence of the second axiom of countability that will be useful to us later is given by Theorem 1.23. First we give two definitions and an example.

1.4 Closed Sets, Closures, and Interiors of Sets

Definition. A subset A of a topological space $(X, \mathcal{T})$ is **dense** in X provided $\overline{A} = X$. If X has a countable dense subset, then $(X, \mathcal{T})$ is a **separable space.** ∎

Cantor defined open sets, closed sets, limit points, and dense sets for the Euclidean spaces during the period 1872–1890. Separability, for metric spaces, was introduced by Fréchet in 1906.

EXAMPLE 44. Let $\mathcal{T}$ be the usual topology on $\mathbb{R}$. Then the set of rational numbers is a countable dense subset and hence $(\mathbb{R}, \mathcal{T})$ is a separable space.

Analysis. Let $x \in \mathbb{R}$ and let U be a neighborhood of x. Then there exists $a, b \in \mathbb{R}$ such that $x \in (a, b) \subseteq U$. Since (a, b) contains a rational number, x belongs to the closure of the set of rational numbers. Therefore the set of rational numbers is a countable dense subset of $\mathbb{R}$. ∎

THEOREM 1.23. Every second countable space is separable.

Proof. Let $(X, \mathcal{T})$ be a second countable space, and let $\mathcal{B}$ be a countable basis for $\mathcal{T}$. For each nonempty member B_i of $\mathcal{B}$, choose a point $x_i \in B_i$ and let $A = \{x_i : B_i \in \mathcal{B}\}$. Then A is a countable subset of X and we need to prove that $\overline{A} = X$. Let $x \in X$ and let U be a neighborhood of x. Then there exists $B_j \in \mathcal{B}$ such that $x \in B_j$ and $B_j \subseteq U$. Since $x_j \in A \cap B_j$, $x \in \overline{A}$. ∎

Example 45 shows that there is a separable space that is not second countable, and Example 46 shows that there is a separable space that is not first countable.

EXAMPLE 45. Let $\mathcal{T}$ be the lower-limit topology on $\mathbb{R}$. By Example 36 of Section 1.3, $(\mathbb{R}, \mathcal{T})$ is not second countable. However by Exercise 9, the set of rational numbers is a countable dense subset and hence $(\mathbb{R}, \mathcal{T})$ is a separable space.

In Section 1.5 we prove that every separable metric space is second countable. Thus it follows that if $\mathcal{T}$ is the lower limit topology on $\mathbb{R}$, then $(\mathbb{R}, \mathcal{T})$ is not metrizable. This illustrates a method of showing that a given topological space is not metrizable. That is, if we can find two properties that are equivalent in a metric space and show that the given topological space has one of the properties but not the other, then we will know that the topological space is not metrizable.

EXAMPLE 46. Let $\mathcal{T}$ be the finite complement topology on $\mathbb{R}$. Then every infinite subset is dense in $\mathbb{R}$ and hence $(\mathbb{R}, \mathcal{T})$ is separable. By Example 35, however, $(\mathbb{R}, \mathcal{T})$ is not first countable.

Analysis. Let A be an infinite subset of $\mathbb{R}$. We prove that $(\mathbb{R}, \mathcal{T})$ is separable by showing that A is dense in $\mathbb{R}$. (If every infinite subset of $\mathbb{R}$ is dense in $\mathbb{R}$, then it is certainly true that set of integers is a countable dense subset of $\mathbb{R}$.) Let $x \in \mathbb{R}$ and let U be a neighborhood of x. Since the $\mathbb{R} - U$ is finite, U must contain all but a finite numbers of members of A. In particular, $A \cap U \neq \emptyset$. Therefore $x \in \overline{A}$ and hence $\overline{A} = \mathbb{R}$. ∎

EXAMPLE 47. Let $\mathcal{T}$ be the discrete topology on $\mathbb{R}$. Since for each $x \in \mathbb{R}$, $\{x\}$ is a countable local basis at x, $(\mathbb{R}, \mathcal{T})$ is first countable. However, since every subset of $\mathbb{R}$ is closed, the closure of any countable subset A of $\mathbb{R}$ is A. Thus no countable subset of $\mathbb{R}$ is dense in $\mathbb{R}$ and hence $(\mathbb{R}, \mathcal{T})$ is not separable.

Definition. The **interior**, int(A), of a subset A of a topological space $(X, \mathcal{T})$ is the union of the open sets that are subsets of A. ■

By definition, the interior of a subset of a topological space is open.

Definition. Let A be a subset of a topological space $(X, \mathcal{T})$ and let $x \in A$. Then x is an **interior point** if A provided there is a neighborhood U of x such that $U \subseteq A$. (See Figure 1.12.) ■

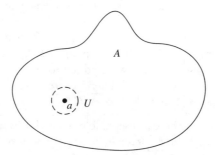

The set U is a neighborhood of a, so a is an interior point of A.

Figure 1.12

Note that if A is a subset of a topological space $(X, \mathcal{T})$ and $B = \{x \in X : x$ is an interior point of $A\}$, then $B = \text{int}(A)$. Also if (X, d) is a metric space, $A \subseteq X$, and $x \in A$, then $x \in \text{int}(A)$ if and only if there exists a positive number ε such that $B(x, \varepsilon) \subseteq A$.

THEOREM 1.24. Let A and B be subsets of a topological space $(X, \mathcal{T})$. Then:

(a) A is open if and only if $A = \text{int}(A)$.
(b) $\text{int}(A) \subseteq \text{int}(B)$ whenever $A \subseteq B$.
(c) $\text{int}(A \cap B) = \text{int}(A) \cap \text{int}(B)$.
(d) $\text{int}(A) \cup \text{int}(B) \subseteq \text{int}(A \cup B)$.

Proof. We prove parts (b) and (d) and leave the proofs of the remaining parts as Exercise 10.

(b) Suppose $A \subseteq B$, and let $x \in \text{int}(A)$. Then there exists a neighborhood U of x such that $U \subseteq A$. Since $A \subseteq B$, $U \subseteq B$. Therefore $x \in \text{int}(B)$ and hence $\text{int}(A) \subseteq \text{int}(B)$.

(d) Let $x \in \text{int}(A) \cup \text{int}(B)$. Then $x \in \text{int}(A)$ or $x \in \text{int}(B)$. If $x \in \text{int}(A)$, there exists a neighborhood U of x such that $U \subseteq A$. If $x \in \text{int}(B)$, there exists a neighborhood U of x such that $U \subseteq B$. In either case, $U \subseteq A \cup B$. Therefore $x \in \text{int}(A \cup B)$, and so $\text{int}(A) \cup \text{int}(B) \subseteq \text{int}(A \cup B)$. ∎

As the following example shows, there are subsets A and B of a topological space $(X, \mathcal{T})$ such that $\text{int}(A) \cup \text{int}(B) \neq \text{int}(A \cup B)$.

EXAMPLE 48. Let $\mathcal{T}$ be the usual topology on $\mathbb{R}$, let $A = [0, 1]$ and $B = [1, 2]$. Then $\text{int}(A \cup B) = (0, 2)$. However, $\text{int}(A) = (0, 1)$ and $\text{int}(B) = (1, 2)$, so $\text{int}(A) \cup \text{int}(B) = (0, 1) \cup (1, 2)$.

Let A be a subset of a topological space $(X, \mathcal{T})$. If $x \in A$, then x is an interior point of A or x is a boundary point of A. There may also be point of $X - A$ that are boundary points of A. In Exercise 21, you are asked to prove that $\overline{A} = \text{int}(A) \cup \text{bd}(A)$. You can also observe that X can be expressed as the union of three pairwise disjoint sets: (1) $\text{int}(A)$, (2) $\text{int}(X - A)$, and (3) $\text{bd}(A)$.

We conclude this section with a definition that will be used in Section 1.6.

Definition. A subset A of a topological space is **nowhere dense** provided $\text{int}(\overline{A}) = \emptyset$. ∎

Paul du Bois-Reymond (1831–1889) introduced nowhere dense sets in his study of Euclidean spaces.

Notice that if A is dense in X, then $\text{int}(\overline{A}) = \text{int}(X) = X$. Thus nowhere dense and dense are opposite extremes. We give some characterizations of nowhere dense in Exercise 23. Exercise 28 provides a condition that ensures that a subset of a topological space be a nowhere dense subset of the space. The following definition is used.

Definition. A subset A of a topological space $(X, \mathcal{T})$ is **relatively discrete** provided that for each $a \in A$, there exists $U \in \mathcal{T}$ such that $U \cap A = \{a\}$. ∎

The following two theorems will be used throughout the text.

THEOREM 1.25. Let $(X, \mathcal{T})$ be a topological space, let C be a closed subset of X, and let U be an open subset of X. Then $C - U$ is closed and $U - C$ is open.

Proof. See Exercise 29. ∎

THEOREM 1.26. Let (X, d) be a metric space, let $x \in X$, and let ε and δ be positive numbers such that $\delta < \varepsilon$. Then $\overline{B(x, \delta)} \subseteq B(x, \varepsilon)$.

Proof. See Exercise 30. ∎

Two definitions are needed in order to state the last theorem of the section.

Definition. A subset A of a topological space $(X, \mathcal{T})$ is a **perfect set** if $A = A'$. ∎

Definition. A point x of a subset A of a topological space $(X, \mathcal{T})$ is an **isolated point** if there is a neighborhood of x that contains no point of A different from x. ∎

THEOREM 1.27. A subset A of a topological space $(X, \mathcal{T})$ is a perfect set if and only if it is closed and has no isolated points.

Proof. See Exercise 24. ∎

EXERCISES 1.4

1. Let $(X, \mathcal{T})$ be a topological space. Prove that X and $\varnothing$ are closed sets.

2. Let $(X, \mathcal{T})$ be a topological space and let $\mathcal{A}$ be a finite family of closed subsets of X. Prove that $\bigcup \{A : A \in \mathcal{A}\}$ is a closed set.

3. Let $(X, \mathcal{T})$ be a topological space and let $\mathcal{A}$ be a family of closed subsets of X. Prove that $\bigcap \{A : A \in \mathcal{A}\}$ is a closed set.

4. Let $\mathcal{T}$ be the usual topology on $\mathbb{R}$ and let $a, b \in \mathbb{R}$ with $a < b$. Prove that $[a, b)$ is neither open nor closed.

5. Let $\mathcal{T}$ be the lower-limit topology on $\mathbb{R}$ and let $a, b \in \mathbb{R}$ with $a < b$. Prove that $[a, b)$ is both open and closed.

6. Let $X = \{1, 2, 3\}$ and let $\mathcal{T} = \{X, \varnothing, \{1\}, \{1, 2\}, \{1, 3\}\}$. Then $\mathcal{T}$ is a topology on X.

    (a) List the closed subsets of $(X, \mathcal{T})$.

    (b) Find $\overline{\{1\}}$.

    (c) Find $\overline{\{2\}}$.

    (d) Find $\text{int}(\{2, 3\})$.

    (e) Find $\text{bd}(\{2, 3\})$.

7. Let A and B be subsets of a topological space $(X, \mathcal{T})$. Prove that:

    (a) $\overline{\overline{A}} = \overline{A}$

    (b) if $A \subseteq B$ then $\overline{A} \subseteq \overline{B}$

    (c) $\overline{A \cap B} \subseteq \overline{A} \cap \overline{B}$.

8. Let $A = \{(x, y) \in \mathbb{R}^2 : y \geq 0\}$ and let $\mathcal{T}$ be the topology on A defined in Exercise 12 of Section 1.3.

    (a) Find the closure of $\{(x, y) \in A : x \text{ is rational and } y = 0\}$.

    (b) Find the closure of $\{(x, y) \in A : x \text{ is irrational and } y = 0\}$.

1.4 Closed Sets, Closures, and Interiors of Sets

9. Let $\mathcal{T}$ be the lower-limit topology on $\mathbb{R}$. Prove that the set of rational numbers is a countable dense subset of $(X, \mathcal{T})$.

10. Let A and B be subsets of a topological space $(X, \mathcal{T})$. Prove that:

    (a) A is open if and only if $A = \text{int}(A)$

    (b) $\text{int}(A \cap B) = \text{int}(A) \cap \text{int}(B)$.

11. Let $(X, \mathcal{T})$ be a topological space and let $U \in \mathcal{T}$. Is it necessarily true that $\overline{\text{int}(U)} = U$? Either prove that it is or give a counterexample to show that it is not.

12. Let A be a closed subset of a topological space $(X, \mathcal{T})$. Is it necessarily true that $\overline{\text{int}(A)} = A$? Either prove that it is or give a counterexample to show that it is not.

13. Let A and B be subsets of a topological space $(X, \mathcal{T})$. Prove that $X - \overline{A \cup B} = (X - \overline{A}) \cap (X - \overline{B})$.

14. Let A and B be subsets of a topological space $(X, \mathcal{T})$. Is it necessarily true that $X - \overline{A \cap B} = (X - \overline{A}) \cup (X - \overline{B})$? Either prove that it is or give a counterexample to show that it is not.

15. Let A be a subset of a topological space $(X, \mathcal{T})$. Prove that:

    (a) $\overline{X - A} = X - \text{int}(A)$.

    (b) $\text{int}(X - A) = X - \overline{A}$.

16. Let $\mathcal{T}$ denote the usual topology on $\mathbb{R}^3$, and let $A = \{(x_1, x_2, x_3) \in \mathbb{R}^3 : x_3 = 0\}$. Find $\text{int}(A)$ and prove your answer.

17. Let $A = \{(x, y)\} \in \mathbb{R}^2 : x^2 + y^2 \geq 1\} \cup (0, 0)\}$, and let d be the usual metric on A. Find the closure of $B_d((0, 0), 1)$ and int $\{(0, 0)\}$. Prove your answers.

18. Let $\mathcal{T}$ be the finite complement topology on $\mathbb{R}$, and let $A = [0, 1]$. Find $\overline{A}$ and $\text{int}(A)$ and prove your answers.

19. Let A be a nonempty proper subset of a set X, and let $\mathcal{T} = \{\emptyset\} \cup \{U \in \mathcal{P}(X) : A \subseteq U\}$.

    (a) Prove that $\mathcal{T}$ is a topology on X.

    (b) Find the interior of A and the closure of A and prove your answers.

    (c) Suppose B is a nonempty proper subset of X such that A is a nonempty proper subset of B. Find the interior of B and the closure of B and prove your answers.

20. Let $\mathcal{T} = \{U \in \mathcal{P}(\mathbb{R}): 0 \notin U \text{ or } U = \mathbb{R}\}$.

    (a) Prove that $\mathcal{T}$ is a topology on $\mathbb{R}$.

    (b) Describe the closed subsets of $(\mathbb{R}, \mathcal{T})$.

    (c) Find $\overline{\{1\}}$.

21. Let A be a subset of a topological space $(X, \mathcal{T})$. Prove that:

    (a) $\text{int}(A)$, $\text{int}(X - A)$ and $\text{bd}(A)$ are pairwise disjoint sets whose union is X.

    (b) $\text{bd}(A)$ is a closed set.

    (c) $\overline{A} = \text{int}(A) \cup \text{bd}(A)$.

    (d) $\text{bd}(A) = \emptyset$ if and only if A is both open and closed.

22. Let A be a subset of a topological space $(X, \mathcal{T})$. Prove that A is dense in X if and only if $\text{int}(X - A) = \emptyset$.

23. Let A be a subset of a topological space $(X, \mathcal{T})$. Prove that the following statements are equivalent:

    (a) A is nowhere dense.

    (b) $X - \overline{A}$ is dense in X.

    (c) $X - \overline{(X - \overline{A})} = \emptyset$.

    (d) $A \subseteq \overline{(X - \overline{A})}$.

24. Let A be a subset of a topological space $(X, \mathcal{T})$. Prove that A is a perfect set if and only if it is closed and has no isolated points.

25. Let A be a subset of a topological space $(X, \mathcal{T})$. Prove that $\text{int}(X - A) = X - \overline{A}$.

26. Let X be a set, and let cl: $\mathcal{P}(X) \to \mathcal{P}(X)$ be a function such that the following conditions hold:

    (a) For each $A \in \mathcal{P}(X)$, $A \subseteq \text{cl}(A)$.

    (b) For each $A \in \mathcal{P}(X)$, $\text{cl}(\text{cl}(A)) = \text{cl}(A)$.

    (c) $\text{cl}(\emptyset) = \emptyset$.

    (d) If $A, B \in \mathcal{P}(X)$, then $\text{cl}(A \cup B) = \text{cl}(A) \cup \text{cl}(B)$.

    Let $\mathcal{T} = \{U \in \mathcal{P}(X):$ there is a subset C of X such that $\text{cl}(C) = C$ and $U = X - C\}$. Prove that $\mathcal{T}$ is a topology on X. Properties (a)–(d) are called the **Kuratowski Closure Properties** in honor of K. Kuratowski (1896–1980).

27. Let X be a set and let $D \subseteq X$. Define a function $f: \mathcal{P}(X) \to \mathcal{P}(X)$ by $f(A) = A \cup D$ for each $A \in \mathcal{P}(X)$.

    (a) Prove that f satisfies the Kuratowski Closure Properties.

    (b) Describe the members of $\mathcal{T}$, where $\mathcal{T}$ is the topology defined as in Exercise 19.

    (c) What is the topology $\mathcal{T}$ when $D = \emptyset$?

    (d) What is the topology $\mathcal{T}$ when $D = X$?

28. Let (X, d) be a metric space with no isolated points, and let A be a relatively discrete subset of X. Prove that A is nowhere dense in X.

29. Let $(X, \mathcal{T})$ be a topological space, let C be a closed subset of X, and let U be an open subset of X. Prove that $C - U$ is closed and $U - C$ is open.

30. Let (X, d) be a metric space, let $x \in X$, and let ε and δ be positive numbers such that $\delta < \varepsilon$. Prove that $\overline{B(x, \delta)} \subseteq B(x, \varepsilon)$.

31. Let U be an open subset of a topological space $(X, \mathcal{T})$. Prove that $\overline{X - \overline{X - U}} = \overline{U}$.

32. This problem is often called the Kuratowski 14-set problem. Let A be a subset of a topological space $(X, \mathcal{T})$.

    (a) Prove that there are at most 14 subsets of X that can be obtained from A by applying the operations of closures and complements successively.

    (b) Find a subset A of $\mathbb{R}$ (with the usual topology) such that 14 subsets of $\mathbb{R}$ can be obtained from A by applying the operations of closure and complements successively.

    (c) Is there a finite set X, a topology $\mathcal{T}$ on X, and a subset A of X such that 14 subsets of X can be obtained from A by applying the operations of closure and complements successively?

1.5 Metric Spaces Revisited

We introduced metric spaces in Section 1.1 to motivate the definition of a topological space in Section 1.2. In this section we study products of metric spaces and investigate "bounded" metrics and some properties of topological spaces that are possessed by metric spaces.

Let $n \in \mathbb{N}$ and for each i, $1 \leq i \leq n$, let (X_i, d_i) be a metric space. The **product metric** on $X = \prod_{i=1}^{n} X_i$ is given by $d(x, y) = (\sum_{i=1}^{n}(d_i(x_i, y_i))^2)^{1/2}$, where $x = (x_1, x_2, ..., x_n)$ and $y = (y_1, y_2, ..., y_n)$. This metric is obviously a generalization of the usual metric on $\mathbb{R}^n$ defined in Section 1.1. Two other choices of a metric on X are given by $d'(x, y) = \max\{d_i(x_i, y_i): i = 1, 2, ..., n\}$ and $d''(x, y) = \sum_{i=1}^{n} d_i(x_i, y_i)$.

THEOREM 1.28. The functions d, d', and d'' are metrics on X.

Proof. See Exercise 1. ■

THEOREM 1.29. The metrics d, d', and d'' generate the same topology on X.

Proof. See Exercise 2. ■

The metric space (X, d) is called the **product metric space** of $\{(X_i, d_i): i = 1, 2, ..., n\}$. Since d' and d'' generate the same topology as d, we can use any one of these metrics as the "product metric."

Definition. Let (X, d) be a metric space. A subset A of X is **bounded** if there exists a real number M such that $d(a, b) \leq M$ for all $a, b \in A$. If A is bounded, the **diameter** of A is lub$\{d(a, b): a, b \in A\}$. ■

The following example shows that boundedness depends upon the metric.

EXAMPLE 49. Let (X, d) be a metric space. In Example 8 of Section 1.1 we defined a metric d' on X by $d'(x, y) = d(x, y)/(1 + d(x, y))$. Thus, for example, if X is $\mathbb{R}$ and d is the usual metric on $\mathbb{R}$, then $\mathbb{R}$ is not bounded (with respect to d). It is the case, however, that $\mathbb{R}$ is a bounded subset of $(\mathbb{R}, d')$. In particular, if $x, y \in \mathbb{R}$, then $d'(x, y) \leq 1$. In Exercise 3, you are asked to show that for any metric space (X, d) the topology on X induced by d is the same as the topology on X induced by d'. As a consequence, given any metric space (X, d), there is another metric d' such that the topology induced by d' is the same as the topology induced by d and every subset of the metric space (X, d') is bounded. There are other metrics that don't change the topology but that bound each set. In particular the metric $\bar{d}$ given by the following theorem is called the **standard bounded metric** corresponding to d.

THEOREM 1.30. Let (X, d) be a metric space, and define a function $\bar{d}: X \times X \to \mathbb{R}$ by $\bar{d}(x, y) = \min\{d(x, y), 1\}$. Then $\bar{d}$ is a metric on X, and the topology induced by $\bar{d}$ is the topology induced by d.

Proof. It is clear that $\bar{d}(x, y) \geq 0$ for all $x, y \in X$, $\bar{d}(x, y) = 0$ if and only if $x = y$, and $\bar{d}(x, y) = \bar{d}(y, x)$. Let $x, y, z \in X$. If $d(x, y) \geq 1$ or $d(y, z) \geq 1$, then $\bar{d}(x, y) + \bar{d}(y, z) \geq 1$ whereas $\bar{d}(x, z) \leq 1$. Hence in this case, $\bar{d}(x, z) \leq \bar{d}(x, y) + \bar{d}(y, z)$. Suppose $d(x, y) < 1$ and $d(y, z) < 1$. Then $\bar{d}(x, y) + \bar{d}(y, z) = d(x, y) + d(y, z) \geq d(x, z) \geq \bar{d}(x, z)$. Therefore $\bar{d}$ is a metric on X.

We use Corollary 1.10 to prove that the topology induced by $\bar{d}$ is the topology induced by d. Let $y \in B_d(x, \varepsilon)$ and let $\delta = \min\{\varepsilon - d(x, y), 1\}$ (see Figure 1.13(a)). Then $B_d(y, \delta) \subseteq B_d(x, \varepsilon)$. Since $B_{\bar{d}}(y, \delta) \subseteq B_d(y, \delta)$, $y \in B_{\bar{d}}(y, \delta) \subseteq B_d(x, \varepsilon)$. Now let $y \in B_{\bar{d}}(x, \varepsilon)$, and let $\delta = \varepsilon - \bar{d}(x, y)$ (see Figure 1.13(b)). Then $B_{\bar{d}}(y, \delta) \subseteq B_{\bar{d}}(x, \varepsilon)$. Since $B_d(y, \delta) \subseteq B_{\bar{d}}(y, \delta)$, $y \in B_d(y, \delta) \subseteq B_{\bar{d}}(x, \varepsilon)$. ■

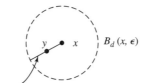

(a) δ is the smaller of this distance (with respect to d) and 1.

Figure 1.13a

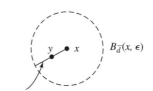

(b) δ is this distance (with respect to $\bar{d}$).

Figure 1.13b

We have seen that every second countable space is separable (Theorem 1.23) and that there is a separable space that is not second countable (Example 45). We prove that in a metric space these concepts are equivalent.

THEOREM 1.31. Let (X, d) be a separable metric space. Then (X, d) is second countable.

Proof. Let A be a countable dense subset of (X, d). Then $\mathcal{B} = \{B(a, \frac{1}{n}) : a \in A$ and $n \in \mathbb{N}\}$ is a countable collection. We show that $\mathcal{B}$ is a basis for the topology $\mathcal{T}$ generated by d. Let $U \in \mathcal{T}$ and let $x \in U$. Then there is a positive number ε such that $B(x, \varepsilon) \subseteq U$. Let n be a natural number such that $\frac{1}{n} < \frac{\varepsilon}{2}$. Since A is dense in X, there exists $a \in A \cap B(x, \frac{1}{n})$. Now $B(a, \frac{1}{n}) \in \mathcal{B}$ and $x \in B(a, \frac{1}{n})$. We complete the proof by showing that $B(a, \frac{1}{n}) \subseteq B(x, \varepsilon)$. Let $y \in B(a, \frac{1}{n})$. Then $d(x, y) \leq d(x, a) + d(a, y) < \frac{1}{n} + \frac{1}{n} = \frac{2}{n} < \varepsilon$, so $y \in B(x, \varepsilon)$. ∎

EXAMPLE 50. Let d be the usual metric for $\mathbb{R}^n$. Then $A = \{(x_1, x_2, ..., x_n) \in \mathbb{R}^n :$ for each $i = 1, 2, ..., n$, x_i is rational$\}$ is a countable dense subset of $\mathbb{R}^n$.

Analysis. The set of rationals is a countable subset of $\mathbb{R}$. By Theorem F.8, a finite Cartesian product of countable sets is countable. Therefore A is a countable subset of

$\mathbb{R}^n$. Let $x = (x_1, x_2, ..., x_n) \in \mathbb{R}^n$, and let U be a neighborhood of x. Then there exists a positive number ε such that $B(x, \varepsilon) \subseteq U$. For each $i = 1, 2, ..., n$, let a_i be a rational number such that $x_i - \frac{\varepsilon}{\sqrt{n}} < a_i < x_i + \frac{\varepsilon}{\sqrt{n}}$. Let $a = (a_1, a_2, ..., a_n)$. Then

$$d(x, a) = \left(\sum_{i=1}^{n}(x_i - a_i)^2\right)^{1/2} < \left(\sum_{i=1}^{n}\left(\frac{\varepsilon}{\sqrt{n}}\right)^2\right)^{1/2}$$

$$= \left(\sum_{i=1}^{n}\frac{\varepsilon^2}{n}\right)^{1/2} = \left(\frac{n\varepsilon^2}{n}\right)^{1/2} = (\varepsilon^2)^{1/2} = \varepsilon.$$

Therefore $a \in A \cap B(x, \varepsilon) \subseteq A \cap U$ and hence $x \in \overline{A}$. Thus A is a dense subset of $\mathbb{R}^n$. ∎

Corollary 1.32. If $n \in \mathbb{N}$ and $\mathcal{T}$ is the usual topology on $\mathbb{R}^n$, then $(\mathbb{R}^n, \mathcal{T})$ is second countable.

Proof. By Example 50, $(\mathbb{R}^n, \mathcal{T})$ is separable. Therefore by Theorem 1.31, $(\mathbb{R}^n, \mathcal{T})$ is second countable. ∎

The real line with the usual order $\leq$ is a partially ordered set (see Appendix E). Therefore the definitions of upper bound, least upper bound, lower bound, and greatest lower bound in a partially ordered set given in Appendix E are applicable to the real line with the usual order $\leq$. We use the notion of greatest lower bound on the real line with the usual order $\leq$ to define the distance between a point and a set and the distance between two sets in a metric space. But first we recall some properties of the real line.

THEOREM 1.33. Let d be the usual metric on $\mathbb{R}$. A subset A of $(\mathbb{R}, d)$ is bounded if any only if it has both a lower bound and an upper bound.

Proof. See Exercise 7. ∎

We accept the following property of $\mathbb{R}$ as an axiom.

Least-Upper-Bound Axiom. Every nonempty subset of $\mathbb{R}$ that has an upper bound has a least upper bound. ∎

We have as Exercise 8 the proof that the following "axiom" is a consequence of the Least-Upper-Bound Axiom.

Greatest-Lower-Bound Axiom. Every nonempty subset of $\mathbb{R}$ that has a lower bound has a greatest lower bound. ∎

Let (X, d) be a metric space, let A and B be subsets of X, and let $x \in X$. We define $d(x, A) = \text{glb}\{d(x, a): a \in A\}$ and $d(A, B) = \text{glb}\{d(a, b): a \in A \text{ and } b \in B\}$.

THEOREM 1.34. Let (X, d) be a metric space. A subset A of X is closed if and only if for each $x \in X - A$, $d(x, A) \neq 0$.

Proof. See Exercise 9. ∎

The following theorem gives a characterization of continuity in metric spaces.

THEOREM 1.35. Let (X, d) and (Y, ρ) be metric spaces and let $f : X \to Y$ be a function. Then f is continuous if and only if for each closed subset C of Y, $f^{-1}(C)$ is closed in X.

Proof. See Exercise 10. ∎

The following four theorems describe topological properties of a metric space. We prove one of them and leave the proofs of the remaining three as exercises.

THEOREM 1.36. Let (X, d) be a metric space and let x and y be distinct members of X. Then there exists a pair of disjoint open sets U and V in X such that $x \in U$ and $y \in V$.

Proof. See Exercise 11. ∎

THEOREM 1.37. Let (X, d) be a metric space, let A be a closed subset of X, and let $x \in X - A$. Then there exists a pair of disjoint open sets U and V in X such that $A \subseteq U$ and $x \in V$.

Proof. By Theorem 1.34, there is a positive number ε such that $d(x, A) = \varepsilon$. Let $U = \bigcup_{a \in A} B\left(a, \frac{\varepsilon}{2}\right)$ and $V = B\left(x, \frac{\varepsilon}{2}\right)$. Then U and V are open sets, $A \subseteq U$, and $x \in V$. Suppose $U \cap V \neq \emptyset$. Then there exists $y \in U \cap V$. Since $y \in U$, there is an $a \in A$ such that $y \in B\left(a, \frac{\varepsilon}{2}\right)$. So $d(a, y) < \frac{\varepsilon}{2}$. Since $y \in V$, $d(x, y) < \frac{\varepsilon}{2}$. Therefore $d(a, x) \leq d(a, y) + d(y, x) < \frac{\varepsilon}{2} + \frac{\varepsilon}{2} = \varepsilon$. This is a contradiction because $d(a, x) \geq d(x, A) = \varepsilon$. Therefore $U \cap V = \emptyset$. ∎

THEOREM 1.38. Let (X, d) be a metric space and let A and B be disjoint closed subsets of X. Then there exists a pair of disjoint open sets U and V in X such that $A \subseteq U$ and $B \subseteq V$.

Proof. See Exercise 12. ∎

Definition. A subset of a topological space is a G_δ-set provided it can be expressed as the intersection of a countable number of open sets. A subset of a topological space is an F_σ-set provided it can be expressed as the union of a countable number of closed sets. ∎

THEOREM 1.39.

(a) If A is a G_δ-set in a topological space $(X, \mathcal{T})$, then $X - A$ is an F_σ-set.

(b) If A is an F_σ-set in a topological space $(X, \mathcal{T})$, then $X - A$ is a G_δ-set.

(c) If A is an F_σ-set in a topological space $(X, \mathcal{T})$, then there is an increasing sequence $C_1 \subseteq C_2 \subseteq C_3 \subseteq \cdots$ of closed subsets of X such that $A = \bigcup_{i=1}^{\infty} C_i$.

(d) Every closed subset of a metric space is a G_δ-set.

Proof. See Exercise 13. ∎

EXERCISES 1.5

1. Prove Theorem 1.28.

2. Prove Theorem 1.29.

3. Let (X, d) be a metric space and define a metric d' on X by $d'(x, y) = d(x, y)/(1 + d(x, y))$. Prove that the topology on X, induced by d is the same as the topology on X induced by d'.

4. Let (X, d) be a metric space and let $x \in X$. Prove that $\{x\}$ is closed.

5. Let (X, d) be a metric space, let A and B be subsets of X and let $x \in X$. Prove that $d(x, A) = d(\{x\}, A)$, $d(A, B) = \text{glb}\{d(a, B): a \in A\} = \text{glb}\{d(b, A): b \in B\}$.

6. Give an example of a metric space (X, d) and subsets A and B of X such that $A \cap B = \emptyset$ but $d(A, B) = 0$.

7. Let d be the usual metric on $\mathbb{R}$. Prove that a subset A of $(\mathbb{R}, d)$ is bounded if and only if it has both a lower bound and an upper bound.

8. Prove the Greatest-Lower-Bound Axiom.

9. Let (X, d) be a metric space. Prove that a subset A of X is closed if and only if for each $x \in X - A$, $d(x, A) \neq 0$.

10. Prove Theorem 1.35.

11. Let (X, d) be a metric space and let x and y be distinct members of X. Prove that there exists a pair of disjoint open sets U and V in X such that $x \in U$ and $y \in V$.

12. Let (X, d) be a metric space and let A and B be disjoint closed subset of X. Prove that there exists a pair of disjoint open sets U and V in X such that $A \subseteq U$ and $B \subseteq V$.

13. (a) Let A be a G_δ-set in a topological space $(X, \mathcal{T})$. Prove that $X - A$ is an F_σ-set.

(b) Let A be an F_σ-set in a topological space $(X, \mathcal{T})$. Prove that $X - A$ is a G_δ-set.

(c) Let A be an F_σ-set in a topological space $(X, \mathcal{T})$. Prove that there is an increasing sequence $C_1 \subseteq C_2 \subseteq C_3 \subseteq \cdots$ of closed subsets of X such that $A = \bigcup_{i=1}^{\infty} C_i$.

(d) Prove that every closed subset of a metric space is a G_δ-set.

14. Let $X = \mathbb{R}^2 - \{(0, 0)\}$. The taxicab metric on X is appropriate for a traveler who walks only along lines that are parallel to the x-axis or along lines that are parallel to the y-axis. Define a metric on X that is appropriate for a traveler who walks only along rays that originate at $(0, 0)$ or along arcs of circles with centers at $(0, 0)$.

15. Let X denote the set of all continuous functions that map the closed interval $[0, 1]$ into $\mathbb{R}$, and let $d: X \times X \to \mathbb{R}$ be the metric (see Example 7 in Section 1.1) defined by $d(f, g) = \int_0^1 |f(x) - g(x)| dx$. Describe each of the following open balls.

(a) $B(f, 1)$, where f is the function defined by $f(x) = 2$ for all $x \in [0, 1]$.

(b) $B(g, 1)$, where g is the function defined by $g(x) = x$ for all $x \in [0, 1]$.

(c) $B(h, 4)$, where h is the function defined by $h(x) = x^2$ for all $x \in [0, 1]$.

1.6 Convergence

You may recall from calculus that a function $f: \mathbb{R} \to \mathbb{R}$ is continuous at a point x in $\mathbb{R}$ if and only if whenever $\langle x_n \rangle$ is a sequence in $\mathbb{R}$ that converges to x then $\langle f(x_n) \rangle$ converges to $f(x)$. We introduced topologies to provide a general setting for the study of continuous functions. Thus it is natural to ask whether we can define convergence of sequences in arbitrary topological spaces and, if so, does convergence describe continuous functions. An answer to the first question is provided in this section, and a partial answer to the second question is provided in the next section.

Recall that a sequence $\mathbf{x} = \langle x_n \rangle$ in a set X (see Appendix C) is a function $\mathbf{x}: \mathbb{N} \to X$.

Definition. A sequence $\langle x_n \rangle$ of points in a topological space $(X, \mathcal{T})$ **converges to a point** x **in** X provided that for each neighborhood U of x there is a natural number N such that $x_i \in U$ for all $i \geq N$. ∎

If $\langle x_n \rangle$ converges to x, we write $\langle x_n \rangle \to x$. Intuitively, a sequence should converge to at most one point. The following example shows that this is not true.

EXAMPLE 51. Let $\mathcal{T}$ be the trivial topology on a set X, let $\langle x_n \rangle$ be a sequence in X, and let x be any member of X. Then $\langle x_n \rangle \to x$.

Analysis. Let $\langle x_n \rangle$ be a sequence in X, let $x \in X$, and let U be a neighborhood of x. Since $\mathcal{T}$ is the trivial topology and $x \in U$, $U = X$. Therefore $x_n \in U$ for every $n \in \mathbb{N}$. ∎

The property that is needed in order to assure convergence to a unique point is a property that is possessed by every metric space (see Theorem 1.36 in Section 1.5).

Definition. A topological space $(X, \mathcal{T})$ is a **Hausdorff space** provided that if x and y are distinct members of X then there exist disjoint open sets U and V such that $x \in U$ and $y \in V$. ∎

Hausdorff spaces are named in honor of Felix Hausdorff. They will be discussed later. At this time, we simply want to establish the following theorem.

THEOREM 1.40. Let $(X, \mathcal{T})$ be a Hausdorff space, let $\langle x_n \rangle$ be a sequence in X that converges to a point x of X, and let $y \in X$ such that $y \neq x$. Then $\langle x_n \rangle$ does not converge to y.

Proof. Let U and V be disjoint open sets such that $x \in U$ and $y \in V$. Since $\langle x_n \rangle \to x$, there is a natural number N such that $x_i \in U$ for all $i \geq N$. Therefore if $i \geq N$, then $x_i \notin V$ and hence $\langle x_n \rangle$ does not converge to y. ∎

Intuitively if x is in the closure of a subset A of a topological space $(X, \mathcal{T})$, then there should be a sequence in A that converges to x. The following example shows that this is not true. However the next theorem establishes that it is true in every first countable space and hence in every metric space (see Theorem 1.14).

EXAMPLE 52. Let $\mathcal{T}$ be the countable complement topology on $\mathbb{R}$, and let A be the set of irrational numbers. Then $0 \in \overline{A}$, but there is no sequence in A that converges to 0.

Analysis. Let U be a neighborhood of 0. Then $\mathbb{R} - U$ is countable. Thus, since A is uncountable, $A \not\subseteq \mathbb{R} - U$. Therefore $U \cap A \neq \emptyset$, and so $0 \in \overline{A}$. Let $\langle x_n \rangle$ be any sequence in A. Then $\mathbb{R} - \{x_n : n \in \mathbb{N}\}$ is a neighborhood of 0 that does not contain any point of $\langle x_n \rangle$. Therefore $\langle x_n \rangle$ does not converge to 0. ∎

THEOREM 1.41. Let $(X, \mathcal{T})$ be a topological space, let $A \subseteq X$, and let $x \in X$.

(a) If there is a sequence of points of A that converges to x, then $x \in \overline{A}$.

(b) If $(X, \mathcal{T})$ is first countable and $x \in \overline{A}$, then there is a sequence of points of A that converges to x.

Proof.

(a) Suppose $\langle x_n \rangle$ is a sequence of points of A that converges to x. Then every neighborhood of x has a nonempty intersection with A. Therefore, by Theorem 1.19, $x \in \overline{A}$.

(b) Let $x \in \overline{A}$ and let $\mathscr{B} = \{B_n : n \in \mathbb{N}\}$ be a countable local basis at x. For each $n \in \mathbb{N}$, let $V_n = \bigcap_{i=1}^n B_i$. Then $\mathscr{V} = \{V_n : n \in \mathbb{N}\}$ is a countable local basis at x such that if $m > n$, then $V_m \subseteq V_n$. For each $n \in \mathbb{N}$, let $x_n \in V_n$. Let U be a neighborhood of x. Then there exists $p \in \mathbb{N}$ such that $V_p \subseteq U$. Since $V_m \subseteq V_p$ for all $m \geq p$, $x_m \in U$ for all $m \geq p$. Therefore $\langle x_n \rangle \to x$. ∎

Let $\langle x_n \rangle$ be a sequence of points in a topological space $(X, \mathscr{T})$. If there is a point x of X such that $\langle x_n \rangle \to x$, we say that $\langle x_n \rangle$ is a **convergent sequence** and that $\langle x_n \rangle$ **converges**. It is useful to have some method to determine that a sequence converges without having to find the point to which it converges. The following necessary condition for a sequence in a metric space to converge is named in honor of the French mathematician, Augustin Cauchy.

Definition. A sequence $\langle x_n \rangle$ in a metric space (X, d) is a **Cauchy sequence** provided that for each $\varepsilon > 0$ there is a natural number N such that if m and n are greater than or equal to N then $d(x_m, x_n) < \varepsilon$. ∎

THEOREM 1.42. Let (X, d) be a metric space and let $\langle x_n \rangle$ be a convergent sequence in X. Then $\langle x_n \rangle$ is a Cauchy sequence.

Proof. Since $\langle x_n \rangle$ converges, there is a point x of X such that $\langle x_n \rangle \to x$. Let $\varepsilon > 0$. Since $\langle x_n \rangle \to x$, there is a natural number N such that $x_i \in B\left(x, \frac{\varepsilon}{2}\right)$ for all $i \geq N$. Suppose m and n are greater than or equal to N. Then $d(x_m, x) < \frac{\varepsilon}{2}$ and $d(x_n, x) < \frac{\varepsilon}{2}$. Therefore $d(x_m, x_n) \leq d(x_m, x) + d(x, x_n) < \frac{\varepsilon}{2} + \frac{\varepsilon}{2} = \varepsilon$. ∎

As the following example indicates, it is not always true that a Cauchy sequence is convergent.

EXAMPLE 53. Let d be the usual metric on $\mathbb{R}$, let $X = \mathbb{R} - \{0\}$, and let ρ be the subspace metric on X. For each natural number n, let $x_n = \frac{1}{2^n}$. Then $\langle x_n \rangle$ is a Cauchy sequence in (X, ρ), and yet it does not converge (see Exercise 1).

Definition. A metric space (X, d) is **complete** provided that every Cauchy sequence in X converges (to a point of X). ∎

THEOREM 1.43. Let (X, d) be a metric space such that every Cauchy sequence in X has a convergent subsequence. Then (X, d) is complete.

Proof. Let $\langle x_n \rangle$ be a Cauchy sequence and let $\langle x_{i_n} \rangle$ be a subsequence of $\langle x_n \rangle$ that converges to x. We prove that $\langle x_n \rangle \to x$. Let $\varepsilon > 0$. There is a natural number N such that if $m, n \geq N$, then $d(x_m, x_n) < \frac{\varepsilon}{2}$. Since $\langle x_{i_n} \rangle \to x$, there is a natural number i_p

such that $i_p > N$ and $d(x_{i_p}, x) < \frac{\varepsilon}{2}$. Then if $n \geq N$, $d(x_n, x) \leq d(x_n, x_{i_p}) + d(x_{i_p}, x) < \varepsilon$. Therefore $\langle x_n \rangle \to x$ and hence (X, d) is complete. ∎

In order to show that $\mathbb{R}$ with the usual metric is complete, we prove the following results.

THEOREM 1.44. Let (X, d) be a metric space, let $\langle x_n \rangle$ be a Cauchy sequence in X, and let $A = \{x_n : n \in \mathbb{N}\}$. Then A is bounded.

Proof. Since $\langle x_n \rangle$ is a Cauchy sequence, there is a natural number N such that if m and n are greater that or equal to N, then $d(x_m, x_n) < 1$. In particular, for each $n \geq N$, $d(x_N, x_n) < 1$. Since $S = \{2d(x_1, x_N), 2d(x_2, x_N), ..., 2d(x_{N-1}, x_N), 2\}$ is a finite set, it has a largest member M. Let $x_i, x_j \in A$. Then $d(x_i, x_j) \leq d(x_i, x_N) + d(x_N, x_j)$. We may assume without loss of generality that $d(x_i, x_N) \geq d(x_N, x_j)$. Then $d(x_i, x_j) \leq 2d(x_i, x_N)$. If $i \geq N$, then $d(x_i, x_N) < 1$. If $i < N$, then $2d(x_i, x_N) \in S$. Thus in either case, $d(x_i, x_j) \leq M$. ∎

THEOREM 1.45. Let (X, d) be a metric space, and let $\langle x_n \rangle$ be a Cauchy sequence in X whose range is finite. Then $\langle x_n \rangle$ is convergent.

Proof. Since the range of $\langle x_n \rangle$ is finite, there is an x such that $x_n = x$ for infinitely many $n \in \mathbb{N}$. If x is the only member of the range of $\langle x_n \rangle$, then $\langle x_n \rangle \to x$. Suppose the range of $\langle x_n \rangle$ has a member other than x. Let $\delta = \min\{d(a, b) : a \text{ and } b \text{ belong to the range of } \langle x_n \rangle \text{ and } a \neq b\}$. Then $\delta > 0$, and since $\langle x_n \rangle$ is a Cauchy sequence, there is a natural number N such that $d(x_m, x_n) < \delta$ for all m and n greater than or equal to N. Thus $x_m = x$ for all $m \geq N$. Therefore for each $\varepsilon > 0$, $x_n \in B(x, \varepsilon)$ for all $n \geq N$, and hence $\langle x_n \rangle \to x$. ∎

THEOREM 1.46. If d is the usual metric on $\mathbb{R}$, then $(\mathbb{R}, d)$ is complete.

Proof. It is convenient here to use the following notation for a sequence. Let $f : \mathbb{N} \to \mathbb{R}$ be a Cauchy sequence in $\mathbb{R}$ and let $x_n = f(n)$ for each $n \in \mathbb{N}$. In light of Theorem 1.45, we may assume that $f(\mathbb{N})$ is infinite. Let $A = \{x \in \mathbb{R} : f(\mathbb{N}) \cap (-\infty, x) \text{ is infinite}\}$. By Theorem 1.44, A is bounded. By Theorem 1.33, A has a lower bound. By the Greatest-Lower-Bound Axiom, A has a greatest lower bound a. We show that $\langle x_n \rangle \to a$. Let $\varepsilon > 0$. Since $\langle x_n \rangle$ is a Cauchy sequence, there is a natural number N such that $d(x_N, x_n) < \varepsilon/4$ for all $n \geq N$. Since $a - \varepsilon/4 < a$, $a - \varepsilon/4 \notin A$. Therefore $f(\mathbb{N}) \cap (-\infty, a - \varepsilon/4)$ is finite. Since $a + \varepsilon/4 > a$ and a is the greatest lower bound of A, $a + \varepsilon/4$ is not a lower bound of A. Hence there is a $b \in A$ such that $a \leq b < a + \varepsilon/4$. Since $b \in A$, $f(\mathbb{N}) \cap (-\infty, b)$ is infinite. Therefore $f(\mathbb{N}) \cap (a - \varepsilon/4, a + \varepsilon/4)$ is infinite. Thus there is an $n \in \mathbb{N}$ such that $n \geq N$ and $x_n \in (a - \varepsilon/4, a + \varepsilon/4)$. So $d(x_N, a) \leq d(x_N, x_n) + d(x_n, a) < \varepsilon/4 + \varepsilon/4 = \varepsilon/2$. Therefore if $n \geq N$, then $d(x_n, a) \leq d(x_n, x_N) + d(x_N, a) < \varepsilon/4 + \varepsilon/2 < \varepsilon$. Hence $\langle x_n \rangle \to a$. ∎

1.6 Convergence

We have used the Greatest-Lower-Bound Axiom to prove completeness of $\mathbb{R}$. Completeness is sometimes used as an axiom in place of the Greatest-Lower-Bound Axiom.

THEOREM 1.47. Let (X, d) be a complete metric space and let A be a subset of X with the subspace metric $\rho = d|_{(A \times A)}$. Then (A, ρ) is complete if and only if A is a closed subset of X.

Proof. Suppose A is a closed subset of X and let $\langle x_n \rangle$ be a Cauchy sequence in A. Then $\langle x_n \rangle$ is a Cauchy sequence in X and hence there is a point x in X such that $\langle x_n \rangle \to x$. By Theorem 1.41 *(a)*, $x \in \overline{A} = A$. Thus $\langle x_n \rangle$ converges in A and hence (A, ρ) is complete.

Suppose (A, ρ) is complete and A is not closed. Let $x \in \overline{A} - A$. By Theorem 1.41 *(b)*, there is a sequence $\langle x_n \rangle$ of points in A that converges to x. By Theorem 1.42, $\langle x_n \rangle$ is a Cauchy sequence. Thus for each $\varepsilon > 0$, there is a natural number N such that if m and n are greater than or equal to N, then $d(x_m, x_n) < \varepsilon$. Since $\langle x_n \rangle$ is a sequence in A, $\rho(x_m, x_n) = d(x_m, x_n)$. Therefore $\langle x_n \rangle$ is a Cauchy sequence in A. Since (A, ρ) is complete, there is a $y \in A$ such that $\langle x_n \rangle \to y$. That is, for each $\varepsilon > 0$, there is a natural number N such that if $n \geq N$, then $\rho(y, x_n) < \varepsilon$. But since $\langle x_n \rangle$ is a sequence in A and $y \in A$, $d(y, x_n) = \rho(y, x_n)$. Hence $\langle x_n \rangle$ converges in (X, d) to both x and y. Since $x \notin A$ and $y \in A$, $x \neq y$. This contradicts Theorem 1.40. ∎

THEOREM 1.48. Let (X, d) be a metric space and let $\bar{d}$ be the standard bounded metric corresponding to d. Then (X, d) is complete if and only if $(X, \bar{d})$ is complete.

Proof. See Exercise 16. ∎

Definition. A topological space that is the union of a countable family of nowhere dense sets is said to be of the **first category**. A topological space that is not of the first category is said to be of the **second category**. ∎

EXAMPLE 54. Let d be the usual metric on $\mathbb{R}$, let $\mathbb{Q} = \{x \in \mathbb{R} : x \text{ is rational}\}$, and let ρ be the subspace metric on $\mathbb{Q}$. Then $\bigcup \{\{x\} : x \in \mathbb{Q}\}$ is a countable union of nowhere dense sets and hence $(\mathbb{Q}, \rho)$ is of the first category.

THEOREM 1.49. Let A be a subset of a metric space (X, d). Then the following statements are equivalent:

(a) A is nowhere dense.

(b) If U is a nonempty open subset of X, then there exists a nonempty open set V such that $V \subseteq U$ and $V \cap \overline{A} = \emptyset$.

(c) Every nonempty open set in X contains an open ball whose closure is disjoint from A.

Proof. (a) → (b). Suppose there exists a nonempty open subset U of X such that if V is any nonempty open subset of U then $V \cap \overline{A} \neq \emptyset$. Let $x \in U$, and let W be a neighborhood of x. Then $W \cap U \subseteq U$, so $W \cap U \cap \overline{A} \neq \emptyset$. Therefore $W \cap A \neq \emptyset$. Thus $x \in \overline{A} = \overline{A}$, so $U \subseteq \overline{A}$. Hence $\text{int}(\overline{A}) \neq \emptyset$, so A is not nowhere dense.

(b) → (c). Let U be a nonempty open subset of X. Then there exists a nonempty open set V such that $V \subseteq U$ and $V \cap \overline{A} = \emptyset$. Let $x \in V$. Then there exists a positive number ε such that $B(x, \varepsilon) \subseteq V$. By Theorem 1.26, $\overline{B\left(x, \dfrac{\varepsilon}{2}\right)} \subseteq B(x, \varepsilon)$. Thus $B\left(x, \dfrac{\varepsilon}{2}\right) \subseteq U$ and $\overline{B\left(x, \dfrac{\varepsilon}{2}\right)} \cap A = \emptyset$.

(c) → (a). The proof is by contradiction. Suppose A is not nowhere dense. Then $\text{int}(\overline{A}) \neq \emptyset$, so there exists a nonempty open set U such that $U \subseteq \overline{A}$. Let $B(x, \varepsilon)$ be an open ball such that $B(x, \varepsilon) \subseteq U$. Then $\overline{B(x, \varepsilon)} \cap A \neq \emptyset$. ∎

The following theorem is named in honor of René Baire (1874–1932), who proved it for the real line in 1889.

THEOREM 1.50. (The Baire Category Theorem). *Every complete metric space is of the second category.*

Proof. The proof is by contradiction. Suppose there is a complete metric space (X, d) that is not of the second category. Then there is a countable family $\{A_n : n \in \mathbb{N}\}$ of nowhere dense sets such that $X = \bigcup \{\overline{A_n} : n \in \mathbb{N}\}$. By Theorem 1.49, there is an open ball $B_1(x_1, \varepsilon_1)$ such that $\overline{B_1(x_1, \varepsilon_1)} \cap A_1 = \emptyset$, and we may assume that $\varepsilon_1 < 1$. Since $B_1(x_1, \varepsilon_1)$ is a nonempty open set, by Theorem 1.49, there is an open ball $B_2(x_2, \varepsilon_2)$ such that $\overline{B_2(x_2, \varepsilon_2)} \subseteq B_1(x_1, \varepsilon_1)$ and $\overline{B_2(x_2, \varepsilon_2)} \cap A_2 = \emptyset$, and we may assume that $\varepsilon_2 < \dfrac{1}{2}$. Proceeding inductively, for each $n \in \mathbb{N}$, there is an open ball $B_n(x_n, \varepsilon_n)$ such that $\overline{B_n(x_n, \varepsilon_n)} \subseteq B_{n-1}(x_{n-1}, \varepsilon_{n-1})$ and $\overline{B_n(x_n, \varepsilon_n)} \cap A_n = \emptyset$, and we may assume that $\varepsilon_n < \dfrac{1}{n}$. Then $\langle x_n \rangle$ is a Cauchy sequence, and, since (X, d) is complete, there is an $x \in X$ such that $\langle x_n \rangle \to x$. Now $x \in \overline{B_n(x_n, \varepsilon_n)}$ for each $n \in \mathbb{N}$. Therefore $x \notin \bigcup \{A_n : n \in \mathbb{N}\}$.. This contradicts the fact that $X = \bigcup \{\overline{A_n} : n \in \mathbb{N}\}$. Therefore every complete metric space is of the second category. ∎

Some important results in functional analysis are consequences of the Baire Category Theorem. The proof of Theorem 1.51 illustrates the typical way in which the Baire Category Theorem is used to prove an existence theorem. That is, we show that some member of a topological space $(X, \mathcal{T})$ must have a given property P by showing that $(X, \mathcal{T})$ is of the second category and that the set of members of X that do not have P is of the first category. But first we give an example that will be used in the proof of Theorem 1.51.

Definition. Let X be a set. A function $f: X \to \mathbb{R}$ is **bounded** provided $f(X)$ is a bounded subset of $\mathbb{R}$. ∎

EXAMPLE 55. Let (X, d) be a metric space and let $C^*(X) = \{f: X \to \mathbb{R}: f \text{ is continuous and bounded}\}$. Define $\rho: C^*(X) \times C^*(X) \to \mathbb{R}$ by $\rho(f, g) = \text{lup } \{|f(x) - g(x)|: x \in X\}$. In Exercise 7, you are asked to prove that ρ is a metric on $C^*(X)$. The metric ρ is called the **uniform metric** for $C^*(X)$. To see that $(C^*(X), \rho)$ is complete, let $\langle f_n \rangle$ be a Cauchy sequence in $C^*(X)$. Then, for each $x \in X$, $\langle f_n(x) \rangle$ is a Cauchy sequence in $(\mathbb{R}, d)$ (see Exercise 8). Therefore by Theorem 1.46, $\langle f_n(x) \rangle$ converges. Define a function $f: X \to \mathbb{R}$ by letting $f(x)$ be the member of $\mathbb{R}$ to which $\langle f_n(x) \rangle$ converges. We finish the proof that $(C^*(X), \rho)$ is complete by showing that $f \in C^*(X)$ (see Exercise 9).

In the proof of the following theorem, we use the fact that the product of two metric spaces is a metric space (see Theorem 1.28). Since, by Theorem 1.29, the three metrics that we have defined on the product of two metric spaces generate the same topology, we can use whichever metric is most convenient.

THEOREM 1.51. There is a continuous function $f: [0, 1] \to \mathbb{R}$ that does not have a derivative at any point.

Proof. By a theorem in calculus, every real-valued continuous function on $[0, 1]$ is bounded. (We prove this fact in Chapter 6.) Thus $C^*([0, 1]) = \{f: [0, 1] \to \mathbb{R}: f \text{ is continuous}\}$. Therefore by Example 55, $(C^*([0, 1]), \rho)$ is a complete metric space. By the Baire Category Theorem (Theorem 1.50), $(C^*([0, 1]), \rho)$ is of the second category. We complete the proof of the theorem by showing that $D = \{f \in C^*([0, 1]): f \text{ has a derivate somewhere}\}$ is of the first category. For each $n \in \mathbb{N}$, let $D_n = \{f \in C^*([0, 1]): \text{ there exists } x \in \left[0, 1 - \frac{1}{n}\right]$ such that if $h \in \left(0, \frac{1}{n}\right]$, then $|(f(x + h) - f(x))/h| \leq n\}$. In Exercise 10, you are asked to show that $D \subseteq \bigcup_{n \in \mathbb{N}} D_n$. We show that for each $n \in \mathbb{N}$, D_n is nowhere dense by showing that D_n is closed and that int $(D_n) = \emptyset$.

Let $n \in \mathbb{N}$. We show that D_n is closed by showing that it is the intersection of closed sets. Define a function $e: C^*([0, 1]) \times [0, 1] \to \mathbb{R}$ by $e(f, x) = f(x)$. Then e is continuous (see Exercise 11). Let $h \in \left(0, \frac{1}{n}\right]$ and define $e_h: C^*([0, 1]) \times \left[0, 1 - \frac{1}{n}\right] \to \mathbb{R}$ by $e_h(f, x) = |(f(x + h) - f(x)/h|$. In Exercise 12, you are asked to prove that e_h is continuous. Thus, by Exercise 13, $e_h^{-1}([0, n])$ is a closed subset of $C^*([0, 1]) \times \left[0, 1 - \frac{1}{n}\right]$. Let $E_h = \{f \in C^*([0, 1]): \text{ there exists } x \in \left[0, 1 - \frac{1}{n}\right]$ such that $(f, x) \in e_h^{-1}([0, n])\}$. Then E_h is a closed subset of $C^*([0, 1])$ (see Exercise 14) and $D_n = \bigcap \left\{E_h: h \in \left(0, \frac{1}{n}\right]\right\}$ (see Exercise 15).

Let $n \in \mathbb{N}$, let $f \in D_n$, and let $\varepsilon > 0$. We prove that $\mathrm{int}(D_n) = \varnothing$ by showing that there is a member g of $C^*([0, 1])$ such that $\rho(f, g) < \varepsilon$ but $g \notin D_n$. By the Weierstrass Approximation Theorem (which we use without proof—see the comment following the completion of this proof), there exists a polynomial p on $[0, 1]$ such that $\rho(f, p) < \frac{\varepsilon}{2}$. Since the derivative p' of p is continuous on $[0, 1]$, p' is bounded. Therefore p has a maximum slope N. Let r be the continuous function on $[0, 1]$ whose graph is drawn below, where the slope of each line segment is either $N + n + 1$ or $-(N + n + 1)$, the maximum is $\frac{\varepsilon}{2}$, and the minimum is $-\frac{\varepsilon}{2}$.

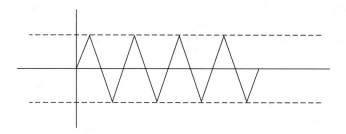

(For example, if $\varepsilon = \frac{1}{10}$, $n = 4$, and $N = 5$, then $r(x) = 10x$, if $0 \leq x \leq \frac{1}{200}$, $r(x) = -10x + \frac{1}{10}$, if $\frac{1}{200} \leq x \leq \frac{3}{200}$, etc.) Then $r \in C^*([0, 1])$. Define $g\colon [0, 1] \to \mathbb{R}$ by $g(x) = p(x) + r(x)$. Then $g \in C^*([0, 1])$ and $\rho(p, g) \leq \frac{\varepsilon}{2}$, so $\rho(f, g) \leq \rho(f, p) + \rho(p, g) < \frac{\varepsilon}{2} + \frac{\varepsilon}{2} = \varepsilon$. But

$$\left|\frac{g(x+h) - g(x)}{h}\right| = \left|\frac{p(x+h) + r(x+h) - p(x) - r(x)}{h}\right|$$

$$\geq \left|\left|\frac{r(x+h) - r(x)}{h}\right| - \left|\frac{p(x+h) - p(x)}{h}\right|\right|$$

Since for each $x \in \left[0, 1 - \frac{1}{n}\right]$, there is an $h \in \left(0, \frac{1}{n}\right]$ such that

$$\left|\left|\frac{r(x+h) - r(x)}{h}\right| - \left|\frac{p(x+h) - p(x)}{h}\right|\right| \geq (N + n + 1) - N = n + 1,$$

$g \notin D_n$. ∎

1.6 Convergence

The Weierstrass Approximation Theorem is named in honor of Karl Weierstrass (1815–1897), one of the great mathematicians of the nineteenth century. This theorem says that if $\varepsilon > 0$ and f is a continuous, real-valued function on $[0, 1]$, then there exists a polynomial p on $[0, 1]$ such that $\rho(f(x), p(x)) < \varepsilon$ for all $x \in [0, 1]$.

We conclude this section by introducing the notion of clustering, which is weaker than convergence.

Definition. A sequence $\langle x_n \rangle$ of points in a topological space $(X, \mathcal{T})$ is **frequently in** a set A provided that for each $i \in \mathbb{N}$, there exists a $j \in \mathbb{N}$ such that $j \geq i$ and $x_j \in A$. A sequence $\langle x_n \rangle$ of points in a topological space $(X, \mathcal{T})$ **clusters at a point** x in X provided that for each neighborhood U of x, $\langle x_n \rangle$ is frequently in U, and $\langle x_n \rangle$ **clusters in** X provided it clusters at come point of X. A point x in X is a **cluster point** if there is a sequence in X that clusters at x. ∎

Although in a Hausdorff space a sequence cannot converge to more than one point, it can cluster at more than one point.

EXAMPLE 56. Let $\mathcal{T}$ be the usual topology on $\mathbb{R}$ and define a sequence $\langle x_n \rangle$ in $\mathbb{R}$ by $x_n = 0$ if n is even and $x_n = 1 - \dfrac{1}{n}$ if n is odd. Then $\langle x_n \rangle$ clusters at both 0 and 1.

If a sequence in a topological space converges to a point then it clusters at that point. Example 56 provides a counterexample to the converse of this statement. It is true, however, that in a first countable space a sequence clusters at a point if and only if some subsequence converges to the point.

THEOREM 1.52. Let $(X, \mathcal{T})$ be a first countable space, let $\langle x_n \rangle$ be a sequence in X, and let $x \in X$. Then $\langle x_n \rangle$ clusters at x if and only if there is a subsequence of $\langle x_n \rangle$ that converges to x.

Proof. Suppose there is a subsequence $\langle x_{i_n} \rangle$ of $\langle x_n \rangle$ that converges to x. Let U be a neighborhood of x and let $p \in \mathbb{N}$. Since $\langle x_{i_n} \rangle \to x$, there is a $q \in \mathbb{N}$ such that if $r \geq q$, then $x_{i_r} \in U$. Suppose $i_j \geq \max\{p, i_q\}$. Then $i_j \geq p$ and $x_{i_j} \in U$, so $\langle x_n \rangle$ clusters at x.

Suppose $\langle x_n \rangle$ clusters at x. By Theorem 1.17, there exists a countable local basis $\mathcal{B} = \{B_n : n \in \mathbb{N}\}$ at x such that $B_{n+1} \subseteq B_n$ for each $n \in \mathbb{N}$. Using the fact that $\langle x_n \rangle$ clusters at x, we construct a subsequence $\langle x_{i_n} \rangle$ of $\langle x_n \rangle$ as follows: Let $i_1 \in \mathbb{N}$ such that $i_1 \geq 1$ and $x_{i_1} \in B_1$. Let $i_2 \in \mathbb{N}$ such that $i_2 \geq i_1$ and $x_{i_2} \in B_2$. Suppose $i_1, i_2, ..., i_{k-1}$ have been chosen so that $i_j \geq i_{j-1}$ and $x_{i_j} \in B_j$ for each $j = 2, 3, ..., k - 1$. Let $i_k \in \mathbb{N}$ such that $i_k \geq i_{k-1}$ and $x_{i_k} \in B_k$. Now we prove that $\langle x_{i_n} \rangle \to x$. Let U be a neighborhood of x. Then there exists $n \in \mathbb{N}$ such that $B_n \subseteq U$. Since $B_{k+1} \subseteq B_k$ for each $k \in \mathbb{N}$, $B_k \subseteq B_n$ for all $k \geq n$. Therefore $x_{i_k} \in B_n$ for all $i_k \geq n$ and hence $\langle x_{i_n} \rangle \to x$. ∎

EXERCISES 1.6

1. Let d be the usual metric on $\mathbb{R}$, let $X = \mathbb{R} - \{0\}$, and let ρ be the subspace metric on X. For each natural number n, let $x_n = 1/2^n$. Prove that $\langle x_n \rangle$ is a Cauchy sequence in (X, ρ) that does not converge.

2. Let X be a set and let d be the discrete metric on X. Prove that (X, d) is complete.

3. Let A be a bounded subset of a metric space (X, d). Prove that $\overline{A}$ is bounded.

4. Let d be the usual metric on $\mathbb{R}^n$. Prove that $(\mathbb{R}^n, d)$ is complete. *Hint:* Let $\langle x_m \rangle$ be a Cauchy sequence in $\mathbb{R}^n$. Prove that for each $i = 1, 2, \ldots, n$, the sequence consisting of the ith coordinate of the points of $\langle x_m \rangle$ is a Cauchy sequence in $\mathbb{R}$ and hence converges to a point y_i of $\mathbb{R}$. Prove that $\langle x_m \rangle$ converges to $(y_1, y_2, \ldots, y_n)$.

5. Give an example of a set X and metrics d and ρ on X such that the topology induced by d is the same as the topology induced by ρ, but (X, d) is complete while (X, ρ) is not.

6. Let (X, d) be a metric space and let A be a dense subset of X such that every Cauchy sequence in A converges in X. Prove that (X, d) is complete.

7. Prove that the function ρ defined in Example 55 is a metric on $C^*(X)$.

8. Prove that if $\langle f_n \rangle$ is a Cauchy sequence in $(C^*(X), \rho)$ and $x \in X$, then $\langle f_n(x) \rangle$ is a Cauchy sequence in $(\mathbb{R}, d)$.

9. Prove that the function f defined in Example 55 is a member of $C^*(X)$.

10. Let D and $\{D_n : n \in \mathbb{N}\}$ be as defined in the proof of Theorem 1.51. Prove that $D \subseteq \bigcup_{n \in \mathbb{N}} D_n$.

11. Prove that the function e defined in the proof of Theorem 1.51 is continuous.

12. Prove that the function e_h defined in the proof of Theorem 1.51 is continuous. *Hint:* Use Exercise 11.

13. Let (X, d) and (Y, ρ) be metric spaces and let $f : X \to Y$ be a function. Prove that the following statements are equivalent:

    (a) f is continuous.

    (b) If V is an open subset of (Y, ρ), then $f^{-1}(V)$ is an open subset of (X, d).

    (c) If C is closed subset of (Y, ρ), then $f^{-1}(V)$ is a closed subset of (X, d).

14. Prove that the set E_h defined in the proof of Theorem 1.51 is closed.

15. Prove that $D_n = \bigcap \{E_h : h \in \left(0, \frac{1}{n}\right]\}$, where D_n and E_h are as defined in the proof of Theorem 1.51.

16. Let (X, d) be a metric space and let $\bar{d}$ be the standard bounded metric corresponding to d. Prove that (X, d) is complete if and only if $(X, \bar{d})$ is complete.

17. Let $(X, \leq)$ be a linearly ordered set, and let $\mathcal{T}$ denote the order topology on X. Prove that $(X, \mathcal{T})$ is a Hausdorff space. *Note:* This exercise is used in Chapters 5, 6 and 7.

1.7 Continuous Functions and Homeomorphisms

Theorem 1.5 provides a definition of continuity of a function from one metric space into another that does not mention the metrics. We adopt this theorem as the definition of continuity of a function from one topological space into another.

Definition. Let $(X, \mathcal{T})$ and $(Y, \mathcal{U})$ be topological spaces. A function $f: X \to Y$ is **continuous at a point** a in X provided that for each neighborhood V of $f(a)$ there is a neighborhood U of a such that $f(U) \subseteq V$. A function $f: X \to Y$ is **continuous** provided it is continuous at each point of x. ∎

Continuity is of fundamental importance in topology. Indeed it is basic to much of mathematics. It is convenient to have a variety of equivalent formulations of continuity because: *(1)* one form may be easier to use than another to establish continuity of a particular function, or *(2)* it may be desirable to draw certain conclusions based on the continuity of a particular function. We were able to conclude, for example, in the proof of Theorem 1.51, that the inverse image of a closed subset of a metric space was a closed subset of a metric space because we knew that the function e_h was continuous. This section is primarily devoted to proving the equivalence of certain conditions, and the first theorem generalizes Exercise 13 of Section 1.6 to arbitrary topological spaces.

THEOREM 1.53. Let $(X, \mathcal{T})$ and $(Y, \mathcal{U})$ be topological spaces and let $f: X \to Y$. The following statements are equivalent:

(a) f is continuous.

(b) If $V \in \mathcal{U}$, then $f^{-1}(V) \in \mathcal{T}$.

(c) If C is a closed subset of $(Y, \mathcal{U})$, then $f^{-1}(C)$ is a closed subset of $(X, \mathcal{T})$.

Proof. (a) → (b). Let $x \in f^{-1}(V)$. Then $f(x) \in V$. Since f is continuous, there is a neighborhood U_x of x such that $f(U_x) \subseteq V$. Since $f(U_x) \subseteq V$, $U_x \subseteq f^{-1}(V)$. Let $U = \bigcup \{U_x : x \in f^{-1}(V)\}$. Then $U = f^{-1}(V)$ and hence $f^{-1}(V) \in \mathcal{T}$.

(b) → (c). Let C be a closed subset of $(Y, \mathcal{U})$. Then $Y - C \in \mathcal{U}$. By (b), $f^{-1}(Y - C) \in \mathcal{T}$. By Theorem C.1(h), $f^{-1}(C) = X - f^{-1}(Y - C)$. Therefore $f^{-1}(C)$ is a closed subset of $(X, \mathcal{T})$.

(c) → (a). Let $x \in X$ and let V be a neighborhood of $f(x)$. Then $Y - V$ is a closed subset of $(Y, \mathcal{U})$ and hence $f^{-1}(Y - V)$ is a closed subset of $(X, \mathcal{T})$. Let $U = X - f^{-1}(Y - V)$. Then U is a neighborhood of x and, by Theorem C.1(i), $f(U) \subseteq V$. ∎

There are other equivalent formulations of continuity that are useful at various times, and the following theorem establishes some of these conditions.

THEOREM 1.54. Let $(X, \mathcal{T})$ and $(Y, \mathcal{U})$ be topological spaces and let $f : X \to Y$. Then the following statements are equivalent:

(a) f is continuous.

(b) If $A \subseteq X$, then $f(\overline{A}) \subseteq \overline{f(A)}$.

(c) If $A \subseteq X$ and p is a limit point of A, then $f(P) \in \overline{f(A)}$.

(d) If $B \subseteq Y$, then $\overline{f^{-1}(B)} \subseteq f^{-1}(\overline{B})$.

(e) If $\mathcal{B}$ is a basis for $\mathcal{U}$, then $f^{-1}(B) \in \mathcal{T}$ for each $B \in \mathcal{B}$.

(f) If $\mathcal{S}$ is a subbasis for $\mathcal{U}$, then $f^{-1}(S) \in \mathcal{T}$ for each $S \in \mathcal{S}$.

Proof. (a) → (b). Let $y \in f(\overline{A})$. Then there exists $x \in \overline{A}$ such that $f(x) = y$. Let V be a neighborhood of y. Since f is continuous, there is a neighborhood U of x such that $f(U) \subseteq V$. Since $x \in \overline{A}$, there is a point $z \in U \cap A$. Then $f(z) \in f(U \cap A)$. By Theorem C.1(e), $f(U \cap A) \subseteq f(U) \cap f(A)$. Since $f(U) \subseteq V, V \cap f(A) \neq \emptyset$. Theorem $y \in \overline{f(A)}$.

(b) → (c). Let $A \subseteq X$ and let p be a limit point of A. Then $p \in \overline{A}$ and hence $f(p) \in \overline{f(A)}$.

(c) → (d). Let $x \in \overline{f^{-1}(B)}$. Then by Theorem 1.20, $x \in f^{-1}(B)$ or x is a limit point of $f^{-1}(B)$. If $x \in f^{-1}(B)$, then $x \in f^{-1}(\overline{B})$. Suppose x is a limit point of $f^{-1}(B)$. Then $f(x) \in \overline{f(f^{-1}(B))}$. By Theorem C.6 (c) and 1.22 (d), $f(x) \in \overline{B}$. Hence $x \in f^{-1}(\overline{B})$.

(d) → (a). We assume (d) and prove (a) by showing that condition (c) of Theorem 1.53 is satisfied. Let C be a closed subset of $(Y, \mathcal{U})$. Then $\overline{f^{-1}(C)} \subseteq f^{-1}(\overline{C})$. Since C is closed, $f^{-1}(\overline{C}) = f^{-1}(C)$. Since $\overline{f^{-1}(C)} \subseteq f^{-1}(C), f^{-1}(C)$ is a closed subset of $(X, \mathcal{T})$.

(a) → (e). Since each member of $\mathcal{B}$ is a member of $\mathcal{U}$, by part (b) of Theorem 1.53, (a) → (e).

(a) → (f). Since each member of $\mathcal{S}$ is a member of $\mathcal{U}$, by part (b) of Theorem 1.53 (a) → (h).

(e) → (a). We assume (e) and prove (a) by showing that condition (b) of Theorem 1.53 is satisfied. Let $V \in \mathcal{U}$. Then there is a collection $\{B_\alpha : \alpha \in \Lambda\}$ of members of $\mathcal{B}$ such that $V = \bigcup \{B_\alpha : \alpha \in \Lambda\}$. By Exercise 7 (d) of Appendix D, $f^{-1}(V) = \bigcup \{f^{-1}(B_\alpha) : \alpha \in \Lambda\}$. Since $f^{-1}(B_\alpha) \in \mathcal{T}$ for each $\alpha \in \Lambda, f^{-1}(V) \in \mathcal{T}$.

1.7 Continuous Functions and Homeomorphisms

(f) → (a). We assume *(f)* and prove *(a)* by showing that condition *(b)* of Theorem 1.53 is satisfied. Let $\mathcal{B}'$ be the family of all finite intersections of members of $\mathcal{S}$, and let $V \in \mathcal{U}$. Then there is a collection $\{B_\alpha : \alpha \in \Lambda\}$ of members of $\mathcal{B}'$ such that $V = \bigcup \{B_\alpha : \alpha \in \Lambda\}$. For each α, there is finite collection $\{S_{\alpha,i_\alpha} : i_\alpha \in \Gamma_\alpha\}$ of members of $\mathcal{S}$ such that $B_\alpha = \bigcap \{S_{\alpha,i_\alpha} : i_\alpha \in \Gamma_\alpha\}$. Therefore $V = \bigcup_{\alpha \in \Lambda}(\bigcap_{i_\alpha \in \Gamma_\alpha} S_{\alpha,i_\alpha})$. By Exercise 7 *(d)* and *(c)* of Appendix D, $f^{-1}(V) = \bigcup_{\alpha \in \Lambda}(\bigcap_{i_\alpha \in \Gamma_\alpha} f^{-1}(S_{\alpha,i_\alpha}))$. Since $f^{-1}(S_{\alpha,i_\alpha}) \in \mathcal{T}$ for each $\alpha \in \Lambda$ and each $i_\alpha \in \Gamma_\alpha$, $\bigcap \{f^{-1}(S_{\alpha,i_\alpha}) : i_\alpha \in \Gamma_\alpha\} \in \mathcal{T}$ for each $\alpha \in \Lambda$. Therefore $f^{-1}(V) \in \mathcal{T}$. ∎

Condition *(e)* of Theorem 1.54 is useful because it permits us to deal with the members of a basis for the topology $\mathcal{U}$ rather than with an arbitrary member of $\mathcal{U}$. This is illustrated in the following example.

EXAMPLE 57. Let $\mathcal{T}$ denote the usual topology on $\mathbb{R}$ and define $f : (\mathbb{R}, \mathcal{T}) \to (\mathbb{R}, \mathcal{T})$ by $f(x) = x^3$. The collection $\mathcal{B}$ of all open intervals is a basis for $\mathcal{T}$. Thus, by Theorem 1.54 part *(e)*, in order to show that f is continuous, it is sufficient to show that the inverse image of each open interval is open. Let (a, b) be an open interval. Then $f^{-1}((a, b)) = (\sqrt[3]{a}, \sqrt[3]{b})$, so f is continuous.

Many applications of Theorem 1.53 and 1.54 will be found throughout the text.

We emphasize that continuity of functions is expressed in terms of inverse images of open sets. This is not to be confused with the image of open sets being open.

Definition. Let $(X, \mathcal{T})$ and $(Y, \mathcal{U})$ be topological spaces. A function $f : X \to Y$ is **open** provided that if $U \in \mathcal{T}$ then $f(U) \in \mathcal{U}$. ∎

The following examples show that there is a continuous function that is not open and an open function that is not continuous.

EXAMPLE 58. Let $\mathcal{T}$ denote the discrete topology on $\mathbb{R}$ and let $\mathcal{U}$ denote the usual topology on $\mathbb{R}$. Then $f : (\mathbb{R}, \mathcal{T}) \to (\mathbb{R}, \mathcal{U})$ defined by $f(x) = x$ for each $x \in \mathbb{R}$ is continuous because if $U \in \mathcal{U}$ then $f^{-1}(U) \in \mathcal{T}$ [$\mathcal{T} = \mathcal{P}(\mathbb{R})$]. However, f is not open because $\{1\} \in \mathcal{T}$ whereas $\{1\} \notin \mathcal{U}$. The function $g : (\mathbb{R}, \mathcal{U}) \to (\mathbb{R}, \mathcal{T})$ defined by $g(x) = x$ for all $x \in \mathbb{R}$ is open but not continuous.

Notice that the continuity of functions is also expressed in terms of inverse images of closed sets. Once again this is not to be confused with the image of closed sets being closed.

Definition. Let $(X, \mathcal{T})$ and $(Y, \mathcal{U})$ be topological spaces. A function $f : X \to Y$ is **closed** provided that if C is a closed subset of $(X, \mathcal{T})$ then $f(C)$ is a closed subset of $(Y, \mathcal{U})$. ∎

The function in Example 58 that is open is also closed, and the function in Example 58 that is not open is also not closed. In Exercise 7 you are asked to prove that f^{-1} is continuous if and only if f is open, where f is a one-to-one function that maps one topological space onto another.

Definition. Let $(X, \mathcal{T})$ and $(Y, \mathcal{U})$ be topological spaces. A bijection $f: X \to Y$ is **homeomorphism** provided that f and f^{-1} are continuous. Topological spaces $(X, \mathcal{T})$ and $(Y, \mathcal{U})$ are said to be **homeomorphic** if there is a homeomorphism $f: X \to Y$. In this case Y is said to be **homeomorphic image** of X. ■

Note that another way to define a homeomorphism is to say that it is a bijection $f: X \to Y$ such that $f(U)$ is open if and only if U is open. (See Figure 1.14.)

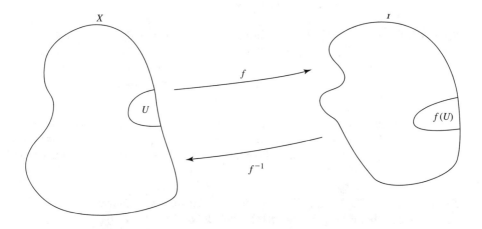

Figure 1.14

Thus a homeomorphism is not only a bijection between X and Y but also between the topologies on X and Y. Therefore any property of X that can be expressed exclusively in terms of the topology on X yields, via the homeomorphism, the same property for Y.

The first systematic treatment of continuity and homeomorhpism was given by Hausdorff.

If you have studied group theory, then you know that an isomorphism between groups is a bijection that preserves the algebraic structure (operation). Thus an isomorphism of groups is essentially a renaming of the elements of the group. A homeomorphism is the analogous concept in topology. It is a bijection that preserves the topological structure (open sets). So a homeomorphism of topological spaces is a renaming of the members of the topological space. There is also an analogy between a homeomorphism and a continuous function; namely, a homeomorphism preserves the algebraic structure, while the inverse image of a continuous function preserves open sets.

We can now give a precise definition of the branch of mathematics known as topology. Topology, as a branch of mathematics, was for many years known as **analysis** situs, and the literal meaning of the two Greek words from which topology is derived is "the science of position." **A topological property** is a property that, if possessed by a topological space X, is also possessed by every homeomorphic image of X. **Topology** is the study of topological properties of topological spaces.

In topology we seek invariants to determine if a function is a homeomorphism. One of the goals of the topologist is to classify spaces and to determine invariants to detect if a given space falls into a given class. This goal has an analogy in algebra because one of the goals of the algebraist is to classify groups and to determine invariants to detect if a given groups falls into a given class.

The following theorem shows that sequences suffice to describe continuity in first countable spaces.

THEOREM 1.55. Let $(X, \mathcal{T})$ and $(Y, \mathcal{U})$ be topological spaces, and let $f : X \to Y$.

(a) If f is continuous and $\langle x_n \rangle$ is a sequence in X that converges to x, then the sequence $\langle f(x_n) \rangle$ converges to $f(x)$.

(b) Suppose $(X, \mathcal{T})$ is first countable and for each $x \in X$ and each sequence $\langle x_n \rangle$ such that $\langle x_n \rangle \to x$, the sequence $\langle f(x_n) \rangle$ converges to $f(x)$. Then f is continuous.

Proof.

(a) Suppose f is continuous. Let $x \in X$ and let $\langle x_n \rangle$ be a sequence such that $\langle x_n \rangle \to x$. Let V be a neighborhood of $f(x)$. Then $f^{-1}(V)$ is a neighborhood of x, and hence there is a natural number N such that $x_i \in f^{-1}(V)$ for all $i \geq N$. Thus $f(x_i) \in V$ for all $i \geq N$ and hence $\langle f(x_n) \rangle \to f(x)$.

(b) Suppose that $(X, \mathcal{T})$ is first countable and for each $x \in X$ and each sequence $\langle x_n \rangle$ that converges to x, $\langle f(x_n) \rangle$ converges to $f(x)$. We prove that f is continuous by showing that if $A \subseteq X$, then $f(\overline{A}) \subseteq \overline{f(A)}$. Let $A \subseteq X$ and let $y \in f(\overline{A})$. Then there exists $x \in \overline{A}$ such that $f(x) = y$. By Theorem 1.41 **(b)**, there is a sequence $\langle x_n \rangle$ of points of A that converges to x. Therefore $\langle f(x_n) \rangle$ converges to $f(x)$. Since $f(x_n) \in f(A)$ for each natural number n, by Theorem 1.41 **(a)**, $y = f(x) \in \overline{f(A)}$. ∎

THEOREM 1.56. Let $(X, \mathcal{T})$, $(Y, \mathcal{U})$ and $(Z, \mathcal{V})$ be topological spaces. If $f : X \to Y$ and $g : Y \to Z$ are continuous, then $g \circ f : X \to Z$ is continuous.

Proof. See Exercise 1. ∎

THEOREM 1.57. Metrizability is a topological property.

Proof. Let (X, d) be a metric space, let $(Y, \mathcal{T})$ be a topological space, and suppose $f : X \to Y$ is a homeomorphism. Define $\rho : Y \times Y \to \mathbb{R}$ by $\rho((y_1, y_2)) = d(f^{-1}(y_1),$

$f^{-1}(y_2)$). We leave as Exercise 2 the proof that ρ is a metric on Y and that the topology induced by ρ is $\mathcal{T}$. ∎

We now define convergence of a sequence of functions.

Definition. Let X be a set, let (Y, d) be a metric space, let $f: X \to Y$ be a function and, for each $n \in \mathbb{N}$, let $f_n: X \to Y$ be a function. Then the sequence $\langle f_n \rangle$ **converges uniformly** to f provided that for each $\varepsilon > 0$ there is a natural number N such that if $n \geq N$, then $d(f_n(x), f(x)) < \varepsilon$ for all $x \in X$. ∎

Notice that uniform convergence depends upon the metric on Y.

THEOREM 1.58. Let $(X, \mathcal{T})$ be a topological space, let (Y, d) be a metric space, let $f: X \to Y$ be a function, and, for each $n \in \mathbb{N}$, let $f_n: X \to Y$ be a continuous function such that the sequence $\langle f_n \rangle$ converges uniformly to f. Then f is continuous.

Proof. Let $x_0 \in X$ and let V be a neighborhood of $f(x_0)$. Then there exists $\varepsilon > 0$ such that $B_d(f(x_0), \varepsilon) \subseteq V$. Since $\langle f_n \rangle$ converges uniformly to f, there exists $N \in \mathbb{N}$ such that if $n \geq N$ and $x \in X$, then $d(f_n(x), f(x)) < \frac{\varepsilon}{4}$. Since f_N is continuous, there exists a neighborhood U of x_0 such that $f_N(U) \subseteq B_d\left(f_N(x_0), \frac{\varepsilon}{2}\right)$. We claim that $f(U) \subseteq V$. Let $x \in U$. Then $d(f(x), f_N(x)) < \frac{\varepsilon}{4}$, $d(f_N(x), f_N(x_0)) < \frac{\varepsilon}{2}$, and $d(f_N(x_0), f(x_0)) < \frac{\varepsilon}{4}$. Therefore by the triangle inequality, $d(f(x), f(x_0)) < \varepsilon$. Thus $f(x) \in V$ and hence f is continuous. ∎

We now state three theorems that involve topological properties.

THEOREM 1.59. Let $(X, \mathcal{T})$ be a separable space, and let $(Y, \mathcal{U})$ be a topological space. If $f: X \to Y$ is a continuous function that maps X onto Y, then $(Y, \mathcal{U})$ is separable.

Proof. See Exercise 3. ∎

THEOREM 1.60. Let $(X, \mathcal{T})$ be a topological space, let $(Y, \mathcal{U})$ be a Hausdorff space, and let A be a nonempty subset of X. If $f, g: X \to Y$ are continuous functions such that $f(a) = g(a)$ for all $a \in A$, then $f(x) = g(x)$ for all $x \in \overline{A}$.

Proof. See Exercise 14. ∎

THEOREM 1.61. Let $(X, \mathcal{T})$ be a topological space, let $(Y, \mathcal{U})$ be a Hausdorff space, and let $f, g: X \to Y$ be continuous functions. If $A = \{x \in X: f(x) = g(x)\}$, then A is closed.

Proof. See Exercise 16. ∎

The final theorem of this section is useful in the study of quotient spaces.

THEOREM 1.62. Let $(X, \mathcal{T})$ be a topological space, let Y be a set, and let $f: X \to Y$ be a function. Define a subset U of Y to be open provided $f^{-1}(U)$ is an open subset of X. Then:

(a) the family of open subsets of Y is a topology $\mathcal{U}$ on Y.

(b) $f: (X, \mathcal{T}) \to (Y, \mathcal{U})$ is continuous.

(c) If $\mathcal{V}$ is a topology on Y such that $f: (X, \mathcal{T}) \to (Y, \mathcal{V})$ is continuous, then $\mathcal{U}$ is finer than $\mathcal{V}$.

Proof. See Exercise 17. ∎

Many of the exercises in this section will be used throughout the text. For example, Exercise 6 and 8 provide characterizations of homeomorphisms and Exercise 7 provides characterizations of the continuity of the inverse of a function.

EXERCISES 1.7

1. Let $(X, \mathcal{T})$, $(Y, \mathcal{U})$ and $(Z, \mathcal{V})$ be topological spaces, and suppose $f: X \to Y$ and $g: Y \to Z$ are continuous. Prove that $g \circ f: X \to Z$ is continuous.

2. Complete the proof of Theorem 1.57.

3. Let $(X, \mathcal{T})$ be separable space, let $(Y, \mathcal{U})$ be a topological space, and let $f: X \to Y$ be a continuous function that maps X onto Y. Prove that $(Y, \mathcal{U})$ is separable.

4. Give examples of topological spaces $(X, \mathcal{T})$ and $(Y, \mathcal{U})$ and a function $f: X \to Y$ such that:

    (a) f is open but not closed.

    (b) f is closed but not open.

5. Let $(X, \mathcal{T})$ be a second countable space, let $(Y, \mathcal{U})$ be a topological space, and let $f: X \to Y$ be an open continuous surjection. Prove that $(Y, \mathcal{U})$ is second countable.

6. Let $(X, \mathcal{T})$ and $(Y, \mathcal{U})$ be topological spaces and let $f: X \to Y$ be a bijection. Prove that the following statements are equivalent:

    (a) f is a homeomorphism.

(b) f and f^{-1} are open.

(c) f and f^{-1} are closed.

7. Let $(X, \mathcal{T})$ and $(Y, \mathcal{U})$ be a topological spaces and let $f: X \to Y$ be a bijection. Prove that the following statements are equivalent:

    (a) f^{-1} is continuous.

    (b) f is an open mapping.

    (c) f is a closed mapping.

8. Let $(X, \mathcal{T})$ and $(Y, \mathcal{U})$ be topological spaces and let $f: X \to Y$ be a bijection. Prove that the following statements are equivalent:

    (a) f is a homeomorphism.

    (b) f is open and continuous.

    (c) f is closed and continuous.

9. Prove that first countability is a topological property.

10. Prove that Hausdorff is a topological property.

11. Is completeness a topological property of metric spaces? Either prove that it is or give an example to show that it is not.

12. For topological spaces $(X, \mathcal{T})$ and $(Y, \mathcal{U})$, define $X \sim Y$ to mean that there is a homeomorphism $f: X \to Y$. Prove that:

    (a) if $(X, \mathcal{T})$ is a topological space, then $X \sim X$.

    (b) $\sim$ is symmetric.

    (c) $\sim$ is transitive.

13. Let $f: \mathbb{R} \to \mathbb{R}$ be a continuous function, and define $g: \mathbb{R} \to \mathbb{R}^2$ by $g(x) = (x, f(x))$. Prove that g is continuous.

14. Let $(X, \mathcal{T})$ be a topological space, let $(Y, \mathcal{U})$ be a Hausdorff space, let A be a nonempty subset of X, and suppose $f, g: X \to Y$ are continuous functions such that $f(a) = g(a)$ for all $a \in A$. Prove that $f(x) = g(x)$ for all $x \in \overline{A}$.

15. Let X be a set, let d be the discrete metric on X, let $(Y, \mathcal{T})$ be a topological space, and let $f: X \to Y$ be a function. Prove that f is continuous.

16. Let $(X, \mathcal{T})$ be a topological space, let $(Y, \mathcal{U})$ be a Hausdorff space, let $f, g: X \to Y$ be continuous functions, and let $A = \{x \in X : f(x) = g(x)\}$. Prove that A is closed.

17. Prove Theorem 1.62.

18. Let X be a set, let $(Y, \mathcal{U})$ be a topological space, and let $f : X \to Y$ be a function. Define a subset U of X to be open provided there is an open subset V of Y such that $U = f^{-1}(V)$.

    (a) Prove that the family of open subsets of X is a topology $\mathcal{T}$ on X.

    (b) Prove that $f : (X, \mathcal{T}) \to (Y, \mathcal{U})$ is continuous.

    (c) Prove that if $\mathcal{V}$ is a topology on X such that $f : (X, \mathcal{V}) \to (Y, \mathcal{U})$ is continuous, then $\mathcal{T}$ is coarser than $\mathcal{V}$.

19. Let $(X, \mathcal{T})$ be a topological space, and let $f, g : X \to \mathbb{R}$ be continuous functions.

    (a) Prove that the function $h : X \to \mathbb{R}$ defined by $h(x) = f(x) + g(x)$ is continuous.

    (b) Prove that the function $h : X \to \mathbb{R}$ defined by $h(x) = f(x)g(x)$ is continuous.

    (c) Let $A = \{x \in X : g(x) = 0\}$. Prove that the function $h : (X - A) \to \mathbb{R}$ defined by $h(x) = f(x)/g(x)$ is continuous.

    (d) Prove that the function $h : X \to \mathbb{R}$ defined by $h(x) = \min\{f(x), g(x)\}$ is continuous.

    (e) Prove that the function $h : X \to \mathbb{R}$ defined by $h(x) = \max\{f(x), g(x)\}$ is continuous.

20. Let $\mathcal{T}$ denote the lower limit topology on $\mathbb{R}$, and let $\mathcal{U}$ denote the usual topology. Define $f : (\mathbb{R}, \mathcal{T}) \to (\mathbb{R}, \mathcal{U})$ by $f(x) = x$ for each $x \in \mathbb{R}$.

    (a) Prove that f is continuous.

    (b) Find a subset A of $\mathbb{R}$ such that $f(\overline{A}) \neq \overline{f(A)}$.

    (c) Find a subset B or $\mathbb{R}$ such that $\overline{f^{-1}(B)} \neq f^{-1}(\overline{B})$.

21. Let $X = \{1, 2, 3\}$, and let $\mathcal{T}_1, \ldots, \mathcal{T}_7$ be as defined in Example 19 (Section 1.2).

    (a) Find i and $j (i \neq j)$ and a continuous bijection $f : (X, \mathcal{T}_i) \to (X, \mathcal{T}_j)$.

    (b) Does there exist i and j $(i \neq j)$ such that $(X, \mathcal{T}_i)$ is homeomorphic to $(X, \mathcal{T}_j)$? Explain your answer.

22. Let $(Y, \leq)$ be a linearly ordered set, and let $\mathcal{U}$ denote the order topology on Y. Furthermore, let $(X, \mathcal{T})$ be a topological space and let $f, g : X \to Y$ be continuous functions.

    (a) Prove that $\{x \in X : f(x) \leq g(x)\}$ is a closed subset of X.

    (b) Define a function $h : X \to Y$ by $h(x) = \min\{f(x), g(x)\}$ for all $x \in X$. Prove that h is continuous.

Chapter 2
New Spaces from Old Ones

In this chapter we study the basic properties of subspaces, product spaces, quotient spaces, and the order topology on a linearly ordered set.

2.1 Subspaces

In Section 1.1, we defined the subspace metric on a subset of a metric space. The following theorem suggests a natural way of making a subset of a topological space into a topological space.

THEOREM 2.1. Let (X, d) be a metric space and let A be a subset of X with the subspace metric $\rho = d|_{(A \times A)}$. Then a subset U of (A, ρ) is open if and only if there is an open subset V of (X, d) such that $U = A \cap V$.

Proof. Let U be an open subset of (A, ρ). Then for each $x \in U$, there is a positive number ε_x such that $B_\rho(x, \varepsilon_x) \subseteq U$ (note that $B_\rho(x, \varepsilon_x)$ is an open ball in A). So $U = \bigcup_{x \in U} B_\rho(x, \varepsilon_x)$. For each $x \in U$, $B_\rho(x, \varepsilon_x) = A \cap B_d(x, \varepsilon_x)$. Let $V = \bigcup_{x \in U} B_d(x, \varepsilon_x)$. Then V is an open subset of (X, d) such that $U = \bigcup_{x \in U}(A \cap B_d(x, \varepsilon_x)) = A \cap V$.

Suppose U is a subset of (A, ρ) with the property that there is an open subset V of (X, d) such that $U = A \cap V$. Then for each $x \in V$, there is a positive number ε_x such that $B_d(x, \varepsilon_x) \subseteq V$. So $V = \bigcup_{x \in V} B_d(x, \varepsilon_x)$. For each $x \in V$, $B_d(x, \varepsilon_x) \cap A = B_\rho(x, \varepsilon_x)$. Thus $U = A \cap (\bigcup_{x \in V} B_d(x, \varepsilon_x)) = \bigcup_{x \in V} B_\rho(x, \varepsilon_x)$, and hence U is open in (A, ρ). ∎

Definition. Let A be a subset of a topological space $(X, \mathcal{T})$. The **subspace topology** or **relative topology** on A determined by $\mathcal{T}$ is the collection $\mathcal{T}_A = \{U \cap A : U \in \mathcal{T}\}$. ∎

The subspace topology was also introduced by Hausdorff.

EXAMPLE 1. Let $\mathcal{T}$ be the usual topology on $\mathbb{R}$, and let A be the closed interval $[0, 1]$. Then a subset U of $[0, 1]$ that does not contain either 0 or 1 is a member of $\mathcal{T}_A$

if and only if it is member of $\mathcal{T}$. Suppose $0 < b < 1$. Since $(-1, b) \cap [0, 1] = [0, b)$, $[0, b)$ is a member of $\mathcal{T}_A$. Since $(b, 2) \cap [0, 1] = (b, 1]$, $(b, 1]$ is a member of $\mathcal{T}_A$. Thus $\mathcal{B} = \{(a, b): 0 < a < b < 1\} \cup \{[0,b): 0 < b < 1\} \cup \{(b, 1]: 0 < b < 1\}$ is a basis for $\mathcal{T}_A$.

EXAMPLE 2. Let $\mathcal{T} = \{U \in \mathcal{P}(\mathbb{R}): 0 \in U \text{ or } U = \varnothing\}$. (This topology was introduced in Exercise 19 of Section 1.4), and let A be the closed interval $[1, 2]$. For each $x \in A$, $\{0, x\} \in \mathcal{T}$. Since $\{0, x\} \cap A = \{x\}$, $\{x\} \in \mathcal{T}_A$. Therefore $\mathcal{T}_A$ is the discrete topology.

THEOREM 2.2. Let A be a subset of a topological space $(X, \mathcal{T})$. Then $\mathcal{T}_A$ is a topology on A.

Proof. Since $X, \varnothing \in \mathcal{T}$, $A = X \cap A$ and $\varnothing = \varnothing \cap A$ are members of $\mathcal{T}_A$.

Suppose $U_1, U_2 \in \mathcal{T}_A$. Then for each $i = 1, 2$, there exists $V_i \in \mathcal{T}$ such that $U_i = V_i \cap A$. So $U_1 \cap U_2 = (V_1 \cap A) \cap (V_2 \cap A) = (V_1 \cap V_2) \cap A$. Since $V_1 \cap V_2 \in \mathcal{T}$, $U_1 \cap U_2 \in \mathcal{T}_A$.

Let $\{U_\alpha: \alpha \in \Lambda\}$ be a collection of members of $\mathcal{T}_A$. Then for each $\alpha \in \Lambda$, there is a member V_α of $\mathcal{T}$ such that $U_\alpha = V_\alpha \cap A$. By Theorem D.2 (a), $\bigcup_{\alpha \in \Lambda} U_\alpha = \bigcup_{\alpha \in \Lambda}(V_\alpha \cap A) = (\bigcup_{\alpha \in \Lambda} V_\alpha) \cap A$. Since $\bigcup_{\alpha \in \Lambda} V_\alpha \in \mathcal{T}$, $\bigcup_{\alpha \in \Lambda} U_\alpha \in \mathcal{T}_A$. ∎

Definition. The members of $\mathcal{T}_A$ are called **relatively open sets** or simply **open sets** in A. ∎

Open sets in A are those sets that can be written in the form $U \cap A$, where U is open in X. In Figure 2.1, the shaded region represents an open set in A.

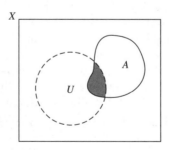

Figure 2.1

For convenience, we sometimes omit mention of the topologies and $\mathcal{T}$ and $\mathcal{T}_A$ and refer to A as a subspace of X. Throughtout the remainder of this book, unless otherwise stated, we assume that if $(X, \mathcal{T})$ is a topological space and $A \subseteq X$, then A has the subspace topology.

EXAMPLE 3.

(a) If $\mathcal{T}$ is the usual topology on $\mathbb{R}^2$, them the subspace topology on $\mathbb{R}$ (regarded as a subset of $\mathbb{R}^2$) is the usual topology on $\mathbb{R}$.

(b) If $\mathcal{T}$ is the usual topology on $\mathbb{R}$, then the subspace topology on the integers is the discrete topology.

(c) If $\mathcal{T}$ is the discrete topology on a set X and $A \subseteq X$, then $\mathcal{T}_A$ is the discrete topology on A.

(d) If $\mathcal{T}$ is the trivial topology on a set X and $A \subseteq X$, then $\mathcal{T}_A$ is the trivial topology on A.

(e) If $(X, \mathcal{T})$ is a topological space and $A \subseteq B \subseteq X$, then the subspace topology on A determined by $\mathcal{T}$ is the same as the subspace topology on A determined by $\mathcal{T}_B$ (see Exercise 21).

In Section 1.4, we introduced the notion of relatively discrete. Notice that a subset A of a topological space $(X, \mathcal{T})$ is relatively discrete if and only if $\mathcal{T}_A$ is the discrete topology. Thus the topological space $(A, \mathcal{T}_A)$ in Example 2 is relatively discrete.

THEOREM 2.3. Let $(X, \mathcal{T})$ be a topological space, let A be a subset of X, and let $\mathcal{B}$ be a basis for $\mathcal{T}$. Then $\mathcal{B}_A = \{B \cap A : B \in \mathcal{B}\}$ is a basis for the subspace topology on A.

Proof. Let $U \in \mathcal{T}$ and let $x \in U \cap A$. There is a member B of $\mathcal{B}$ such that $x \in B$ and $B \subseteq U$. Therefore $x \in B \cap A \subseteq U \cap A$ and hence, by Theorem 1.11, $\mathcal{B}_A$ is a basis for the subspace topology on A. ∎

Notice that in Example 1, $[0, 1) \in \mathcal{T}_A$ but $[0, 1) \notin \mathcal{T}$. The following theorem gives a special situation in which every member of the subspace topology is also a member of the topology on X.

THEOREM 2.4. Let $(X, \mathcal{T})$ be a topological space and let A be an open subset of X. If $U \in \mathcal{T}_A$, then $U \in \mathcal{T}$.

Proof. Let $U \in \mathcal{T}_A$. Then there is an open subset V of X such that $U = A \cap V$. Since A and V are open in X, so is $A \cap V = U$. ∎

THEOREM 2.5. Let $(A, \mathcal{T}_A)$ be a subspace of a topological space $(X, \mathcal{T})$. A subset C of A is closed in $(A, \mathcal{T}_A)$ if and only if there is a closed subset D of $(X, \mathcal{T})$ such that $C = A \cap D$.

Proof. Suppose C is a closed subset of $(A, \mathcal{T}_A)$. Then there is an open subset U of $(X, \mathcal{T})$ such that $A - C = A \cap U$. Therefore $C = A - (A - C) = A - (A \cap U) = A \cap (A - U) = A \cap (X - U)$. Since $X - U$ is a closed subset of $(X, \mathcal{T})$, the proof is complete.

Suppose C is a subset of A and there is a closed subset D of $(X, \mathcal{T})$ such that $C = A \cap D$. Then $X - D$ is open in X and $A - C = A - (A \cap D) = A \cap (A - D) = A \cap (X - D)$. Therefore $A - C$ is open in $(A, \mathcal{T}_A)$ and hence C is a closed subset of $(A, \mathcal{T}_A)$. ∎

There is also a criterion for a closed subset of a subspace to be closed in the topological space. The proof is left as Exercise 2.

THEOREM 2.6. Let A be a closed subset of a topological space $(X, \mathcal{T})$. If C is closed in $(A, \mathcal{T}_A)$, then C is closed in $(X, \mathcal{T})$. ∎

When dealing with subspaces of a topological space, one needs to exercise care in taking closures of a set because the closure in the subspace may be quite different from the closure in the space. The following theorem establishes a criterion for dealing with this situation.

THEOREM 2.7. Let $(A, \mathcal{T}_A)$ be subspace of a topological space $(X, \mathcal{T})$, and let B be a subset of A. Then the closure of B in $(A, \mathcal{T}_A)$ is $A \cap \overline{B}$, where $\overline{B}$ is the closure of B in X.

Proof. Since $\overline{B}$ is closed in $(X, \mathcal{T})$, by Theorem 2.5, $A \cap \overline{B}$ is closed in $(A, \mathcal{T}_A)$. Since $A \cap \overline{B}$ contains B and the closure of B in $(A, \mathcal{T}_A)$ is the intersection of all closed subsets of A that contain B, the closure of B in $(A, \mathcal{T}_A)$ is a subset of $A \cap \overline{B}$.

Since the closure of B in $(A, \mathcal{T}_A)$ is closed in A, by Theorem 2.5, there is a closed subset C of $(X, \mathcal{T})$ such that the closure of B in $(A, \mathcal{T}_A)$ is $A \cap C$. Thus C is a closed subset of $(X, \mathcal{T})$ that contains B. Since $\overline{B}$ is the intersection of all such closed sets, $\overline{B} \subseteq C$. Therefore $A \cap \overline{B} \subseteq A \cap C$. Since $A \cap C$ is the closure of B in $(A, \mathcal{T}_A)$, the proof is complete. ∎

The following example shows that the interior operator and boundary operator on subspaces behave differently from the closure operator.

EXAMPLE 4. Let $\mathcal{T}$ be the usual topology on $\mathbb{R}^2$, and let $B = A = \{(x, y) \in \mathbb{R}^2 : y = 0\}$. Then the interior of B, considered as a subset of A, is B while the interior of B, considered as a subset of $\mathbb{R}^2$, is $\varnothing$. Thus the former cannot be obtained by intersecting the latter with A. The boundary of B, considered as a subset of A, is $\varnothing$, while the boundary of B, considered as a subset of $\mathbb{R}^2$, is B. Thus the former cannot be obtained by intersecting the latter with A. In Exercise 13, you are asked to show that if $(X, \mathcal{T})$ is a topological space and $B \subseteq A \subseteq X$, then the interior of B, considered as subset of X, is a subset of the interior of B, considered as a subset of A. In Exercise 14, you are asked to show that if $(X, \mathcal{T})$ is a topological space and $B \subseteq A \subseteq X$, then the boundary of B, considered as a subset of A, is a subset of the boundary of B, considered as a subset of X, intersected with A.

THEOREM 2.8. Let $(A, \mathcal{T}_A)$ be a subspace of a topological space $(X, \mathcal{T})$, let $U \subseteq A$, and let $a \in A$. Then U is a neighborhood of a with respect to $\mathcal{T}_A$ if and only if there is a neighborhood V of a with respect to $\mathcal{T}$ such that $U = A \cup V$.

Proof. See Exercise 3. ∎

THEOREM 2.9. Let A be a subset of a metric space (X, d), let ρ be the subspace metric on A, let $\mathcal{T}$ be the topology on X generated by d, let $\mathcal{T}_A$ be the subspace topology determined by $\mathcal{T}$, and let $\mathcal{T}_\rho$ be the topology on A generated by ρ. Then $\mathcal{T}_A = \mathcal{T}_\rho$.

Proof. Let $U \in \mathcal{T}_A$. Then there exists a member V of $\mathcal{T}$ such that $U = A \cap V$. Now $V = \bigcup_{x \in V} B_d(x, \varepsilon_x)$ and hence $U = A \cap V = A \cap (\bigcup_{x \in V} B_d(x, \varepsilon_x)) = \bigcup_{x \in A \cap V}(A \cap B_d(x, \varepsilon_x)) = \bigcup_{x \in U} B_\rho(x, \varepsilon_x)$. Therefore $U \in \mathcal{T}_\rho$.
Now let $U \in \mathcal{T}_\rho$. Then $U = \bigcup_{x \in U} B_\rho(x, \varepsilon_x) = \bigcup_{x \in U}(A \cap B_d(x, \varepsilon_x)) = A \cap (\bigcup_{x \in U} B_d(x, \varepsilon_x))$. Since $\bigcup_{x \in U} B_d(x, \varepsilon_x) \in \mathcal{T}$, $U \in \mathcal{T}_A$. ∎

If a topological space has a certain topological property, it is natural to ask whether every subspace also has the property.

Definition. A topological property P is **hereditary** provided every subspace of a topological space with P also has P. ∎

THEOREM 2.10. The topological properties—Hausdorff, first countable, second countable, and metrizable—are hereditary.

Proof. Theorem 2.9 establishes that metrizability is hereditary. We prove that second countability is hereditary and leave the proofs that Hausdorff and first countability are hereditary as Exercises 4 and 5.

Let A be a subset of a second countable space $(X, \mathcal{T})$. Then there is a countable basis $\mathcal{B}$ for $\mathcal{T}$. For each $B \in \mathcal{B}$, let $A_B = A \cap B$. Then $\mathcal{B}_A = \{A_B : B \in \mathcal{B}\}$ is a countable subcollection of $\mathcal{T}_A$, and, by Theorem 2.3, it is a basis for $\mathcal{T}_A$. ∎

The following example shows that separability is not hereditary.

EXAMPLE 5. Let $A = \{(x, y) \in \mathbb{R}^2 : y = 0\}$ and let $X = A \cup \{(0, 1)\}$. Let $\mathcal{T} = \{U \in \mathcal{P}(X) : U = \emptyset \text{ or } (0, 1) \in U\}$. Then $\mathcal{T}$ is a topology on X. Furthermore, $(X, \mathcal{T})$ is separable because $\{(0, 1)\}$ is a dense subset of X. However, if $(a, 0) \in A$, then $\{(a, 0), (0, 1)\} \in \mathcal{T}$, so $\{(a, 0)\} \in \mathcal{T}_A$. Therefore $\mathcal{T}_A$ is the discrete topology on A, and hence $(A, \mathcal{T}_A)$ is not separable.

We do, however, have the following two results. The proof of the first is left as Exercise 15.

THEOREM 2.11. Let $(X, \mathcal{T})$ be a separable space and let A be an open subset of X. Then $(A, \mathcal{T}_A)$ is separable. ∎

THEOREM 2.12. Every subspace of a separable metric space is separable.

Proof. Let (X, d) be a separable metric space, let A be a subset of X, and let $D = \{x_j : j \in \Lambda\}$ be a countable dense subset of X. For each $j \in \Lambda$ and each $n \in \mathbb{N}$ such that $B(x_j, \frac{1}{n}) \cap A \neq \emptyset$, let $y_{jn} \in B(x_j, \frac{1}{n}) \cap A$. By Theorem F.7 and Exercise 7 in Appendix F, $B = \{y_{jn} : j \in \Lambda \text{ and } n \in \mathbb{N}\}$ is countable. We complete the proof by showing that B is dense in A.

Let $a \in A$ and let $\varepsilon > 0$. Let $n \in \mathbb{N}$ such that $\frac{2}{n} < \varepsilon$. Since D is dense in X, there exists $x_j \in D$ such that $d(x_j, a) < \frac{1}{n}$. Then $B(x_j, \frac{1}{n}) \cap A \neq \emptyset$, so there exists $y_{jn} \in B$. Thus $d(a, y_{jn}) \leq d(a, x_j) + d(x_j, y_{jn}) < \frac{2}{n} < \varepsilon$. Hence B is dense in A. ∎

Definition. Let $(X, \mathcal{T})$ and $(Y, \mathcal{U})$ be topological spaces. If $(A, \mathcal{U}_A)$ is a subspace of $(Y, \mathcal{U})$ and $f : (X, \mathcal{T}) \to (A, \mathcal{U}_A)$ is a homeomorphism, then X is said to be **embedded** in Y and f is called an **embedding** of X in Y. ∎

EXAMPLE 6. The function $f : \mathbb{R} \to \mathbb{R}^2$ defined by $f(x) = (x, 0)$ for each $x \in \mathbb{R}$ is an embedding of $\mathbb{R}$ in $\mathbb{R}^2$.

Analysis. It is clear that f is one-to-one. Let $\mathcal{T}$ denote the usual topology on $\mathbb{R}^2$ and let $A = \{(x, y) \in \mathbb{R}^2 : y = 0\}$. Then f maps $\mathbb{R}$ onto A. Since every metric space is first countable (Theorem 1.14), in order to show that f is continuous, it is sufficient (by Theorem 1.55) to show that if $\langle x_n \rangle$ is a sequence in $\mathbb{R}$ that converges to x in $\mathbb{R}$, then $\langle f(x_n) \rangle$ converges to $f(x)$ in A. But this is obvious. Also by Theorems 1.14 and 1.55, f^{-1} is continuous. So if $\mathcal{U}$ denotes the usual topology on $\mathbb{R}$, then $f : (\mathbb{R}, \mathcal{U}) \to (A, \mathcal{T}_A)$ is a homeomorphism. ∎

THEOREM 2.13. Let $(X, \mathcal{T})$ and $(Y, \mathcal{U})$ be topological spaces, let $(A, \mathcal{T}_A)$ be a subspace of $(X, \mathcal{T})$, and let $(B, \mathcal{U}_B)$ be a subspace of $(Y, \mathcal{U})$.

(a) The inclusion map $i: A \to X$ defined by $i(a) = a$ for each $a \in A$ is continuous.

(b) If $f: X \to Y$ is continuous, then $f|_A: A \to Y$ is continuous.

(c) If $f: X \to B$ is continuous, then the function $g: X \to Y$ defined by $g(x) = f(x)$ for each $x \in X$ is continuous.

(d) If $f: X \to Y$ is continuous and $f(X) \subseteq B$, then the function $g: X \to B$ defined by $g(x) = f(x)$ for each $x \in X$ is continuous.

Proof. We prove part **(d)** and leave the remaining parts as Exercises 16, 17, and 18. Suppose $f: X \to Y$ is continuous and $f(X) \subseteq B$. Let $U \in \mathcal{U}_B$. Then there exists $V \in \mathcal{U}$ such that $U = B \cap V$. Since $f: X \to Y$ is continuous, $f^{-1}(V) \in \mathcal{T}$. But $g^{-1}(U) = g^{-1}(B) \cap g^{-1}(V) = X \cap f^{-1}(V) = f^{-1}(V)$. Therefore $g: X \to B$ is continuous. ∎

THEOREM 2.14. Let $(X, \mathcal{T})$ and $(Y, \mathcal{U})$ be topological spaces, let $f: X \to Y$ be a function, and let $\{U_\alpha : \alpha \in \Lambda\}$ be a collection of open subsets of X such that $X = \bigcup_{\alpha \in \Lambda} U_\alpha$ and $f|_{U_\alpha}: U_\alpha \to Y$ is continuous for each $\alpha \in \Lambda$. Then f is continuous.

Proof. Let $U \in \mathcal{U}$. Then for each $\alpha \in \Lambda$, $(f|_{U_\alpha})^{-1}(U)$ is open in U_α. Since U_α is open in X, by Theorem 2.4 $(f|_{U_\alpha})^{-1}(U)$ is open in X. But $f^{-1}(U) = \bigcup_{\alpha \in \Lambda}(f|_{U_\alpha})^{-1}(U)$, so f is continuous. ∎

The following theorem is sometimes called the *pasting lemma*.

THEOREM 2.15. Let $(X, \mathcal{T})$ and $(Y, \mathcal{U})$ be topological spaces, let A and B be closed subsets of X such that $X = A \cup B$, and let $f: (A, \mathcal{T}_A) \to (Y, \mathcal{U})$ and $g: (B, \mathcal{T}_B) \to (Y, \mathcal{U})$ be continuous functions such that $f(x) = g(x)$ for all $x \in A \cap B$. Then the function $h: (X, \mathcal{T}) \to (Y, \mathcal{U})$ defined by $h(x) = f(x)$ for all $x \in A$ and $h(x) = g(x)$ for all $x \in B$ is continuous.

Proof. Let C be a closed subset of $(Y, \mathcal{U})$. Since f is continuous, by Theorem 1.53, $f^{-1}(C)$ is closed in $(A, \mathcal{T}_A)$. Since g is continuous, by Theorem 1.53, $g^{-1}(C)$ is closed in $(B, \mathcal{T}_B)$. By Theorem 2.6, $f^{-1}(C)$ and $g^{-1}(C)$ are closed in $(X, \mathcal{T})$. Therefore $h^{-1}(C) = f^{-1}(C) \cup g^{-1}(C)$ is closed in $(X, \mathcal{T})$. Hence by Theorem 1.53, h is continuous. ∎

EXERCISES 2.1

1. Let $X = \{1, 2, 3\}$ and let $A = \{2, 3\}$.

    (a) If $\mathcal{T} = \{\emptyset, \{1\}, X\}$, what is $\mathcal{T}_A$?

    (b) If $\mathcal{T} = \{\emptyset, \{1, 2\}, X\}$, what is $\mathcal{T}_A$?

2. Let A be a closed subset of a topological space $(X, \mathcal{T})$ and let C be a closed subset of $(A, \mathcal{T}_A)$. Prove that C is closed in $(X, \mathcal{T})$.

3. Prove Theorem 2.8.

4. Prove that Hausdorff is hereditary.

5. Prove that the first axiom of countability is a hereditary property.

6. Give an example of a topological space $(X, \mathcal{T})$, a subspace $(A, \mathcal{T}_A)$ of $(X, \mathcal{T})$, and an open set in $(A, \mathcal{T}_A)$ that is not open in $(X, \mathcal{T})$.

7. Give an example of a topological space $(X, \mathcal{T})$, a subspace $(A, \mathcal{T}_A)$ of $(X, \mathcal{T})$, and a closed set in $(A, \mathcal{T}_A)$ that is not closed in $(X, \mathcal{T})$.

8. Let $(A, \mathcal{T}_A)$ be a subspace of a topological space $(X, \mathcal{T})$ and let B be a subset of A. Prove that a point $a \in A$ is a limit point of B with respect to $\mathcal{T}_A$ if and only if a is a limit point of B with respect to $\mathcal{T}$.

9. Give an example of a subspace $(A, \mathcal{T}_A)$ of a topological space $(X, \mathcal{T})$ and a subset B of X such that the closure of $A \cap B$ in $(A, \mathcal{T}_A)$ is not $A \cap \overline{B}$.

10. Let $(X, \mathcal{T})$, $(Y, \mathcal{U})$, and $(Z, \mathcal{V})$ be topological spaces such that there is an embedding of X in Y and an embedding of Y in Z. Prove that there is an embedding of X in Z.

11. Give an example of topological spaces $(X, \mathcal{T})$ and $(Y, \mathcal{U})$ such that there is an embedding of X in Y and an embedding on Y in X but $(X, \mathcal{T})$ and $(Y, \mathcal{U})$ are not homeomorphic.

12. Give an example of a separable Hausdorff space $(X, \mathcal{T})$ that has a subspace $(A, \mathcal{T}_A)$ that is not separable.

13. Let $(X, \mathcal{T})$ be a topological space, and let $B \subseteq A \subseteq X$. Show that the interior of B, considered as a subset of X, is a subset of the interior of B, considered as a subset of A.

14. Let $(X, \mathcal{T})$ be a topological space, and let $B \subseteq A \subseteq X$. Show that the boundary of B, considered as a subset of A, is a subset of the boundary of B, considered as a subset of X, intersected with A.

15. Let A be an open subset of a separable space $(X, \mathcal{T})$. Prove that $(A, \mathcal{T}_A)$ is separable.

16. Let $(A, \mathcal{T}_A)$ be a subspace of a topological space $(X, \mathcal{T})$. Prove that the inclusion map $i: A \to X$ defined by $i(a) = a$ for each $a \in A$ is continuous.

17. Let $(X, \mathcal{T})$ and $(Y, \mathcal{U})$ be topological spaces, let $(A, \mathcal{T}_A)$ be a subspace of $(X, \mathcal{T})$, and let $f: X \to Y$ be a continuous function. Prove that $f|_A: A \to Y$ is continuous.

18. Let $(X, \mathcal{T})$ and $(Y, \mathcal{U})$ be topological spaces, let $(B, \mathcal{U}_B)$ be a subspace of $(Y, \mathcal{U})$, and let $f: X \to B$ be a continuous function. Prove that the function $g: X \to Y$ defined by $g(x) = f(x)$ for each $x \in X$ is continuous.

19. Let $\mathcal{T} = \{U \in \mathcal{P}(\mathbb{R}^2): U = \emptyset$ or for each $(x, y) \in U$ there is an open line segment in each direction about (x, y) that is contained in $U\}$.

    (a) Show that $\mathcal{T}$ is a topology on $\mathbb{R}^2$.

    (b) Compare $\mathcal{T}$ with the usual topology $\mathcal{U}$ on $\mathbb{R}^2$; that is, is it weaker, stronger, the same, or none of these.

    (c) Let A denote a straight line in $\mathbb{R}^2$. Compare $\mathcal{T}_A$ and $\mathcal{U}_A$.

    (d) Let A denote a circle in $\mathbb{R}^2$. Describe $\mathcal{T}_A$.

20. Let $\mathcal{B}' = \{B \in \mathcal{P}(\mathbb{R}^2): B$ is an open disk with a finite number of straight lines through the center removed$\}$, and let $\mathcal{B} = \{B \cup \{c\}: B \in \mathcal{B}'$ and c is the center of $B\}$.

    (a) Show that $\mathcal{B}$ is a basis for a topology $\mathcal{T}$ on $\mathbb{R}^2$.

    (b) Compare $\mathcal{T}$ with the usual topology $\mathcal{U}$ on $\mathbb{R}^2$.

    (c) Let A denote a straight line in $\mathbb{R}^2$. Describe $\mathcal{T}_A$.

    (d) Let A denote a circle in $\mathbb{R}^2$. Compare $\mathcal{T}_A$ and $\mathcal{U}_A$.

21. Let $(X, \mathcal{T})$ be a topological space and let $A \subseteq B \subseteq X$. Prove that the subspace topology on A determined by $\mathcal{T}$ is the same as the subspace topology on A determined by $\mathcal{T}_B$.

22. Let $\mathcal{T}$ denote the subspace topology on $[0, 1)$ determined by the usual topology on $\mathbb{R}$, and let $\mathcal{U}$ denote the subspace topology on $S^1 = \{(x, y) \in \mathbb{R}^2: x^2 + y^2 = 1\}$ determined by the usual topology on $\mathbb{R}^2$. Define $f: [0, 1) \to (S^1, \mathcal{U})$ by $f(x) = (\cos 2\pi x, \sin 2\pi x)$.

    (a) Prove that f is a bijection.

    (b) Prove that f is continuous.

    (c) Prove that f^{-1} is not continuous.

23. Give an example of a linearly ordered set $(X, \leq)$ and a subset A of X such that the order topology on A determined by $\leq$ restricted to $A \times A$ is not the same as the relative topology on A determined by the order topology on X.

2.2 The Product Topology on $X \times Y$

In Section 1.5 the product metric space of a finite collection of metric spaces was defined. The purpose of this section is to define a topology on the product of two topological spaces. This topology has the property that it is the same as the topology generated by the product metric on the product of two metric spaces. For simplicity, we work with the product of two spaces, but we emphasize that the same construction holds for a finite product. As we shall see, if $(X_1, \mathcal{T}_1)$ and $(X_2, \mathcal{T}_2)$ are topological spaces, then the product topology on $X_1 \times X_2$ is the smallest topology such that both projections $\pi_1: X_1 \times X_2 \to X_1$ and $\pi_2: X_1 \times X_2 \to X_2$ are continuous. (The projections are defined by $\pi_1((x_1, x_2)) = x_1$ and $\pi_2((x_1, x_2)) = x_2$ (see Appendix C).)

Definition. Let $(X, \mathcal{T}_1)$ and $(Y, \mathcal{T}_2)$ be topological spaces, and let $\mathcal{B} = \{U \times V: U \in \mathcal{T}_1 \text{ and } V \in \mathcal{T}_2\}$. The **product topology** on $X \times Y$ is the topology $\mathcal{T}$ that has $\mathcal{B}$ as a basis, and $(X \times Y, \mathcal{T})$ is called the **product space** of X and Y. ∎

By Exercise 9 of Section 1.3, $\mathcal{B}$ is a basis for a topology on $X \times Y$. In 1910, Fréchet introduced finite products of abstract spaces. Some credit Ernst Steinitz (1871–1928) because he had considered a special case in 1908.

EXAMPLE 7. Let $\mathcal{T}$ denote the usual topology on $\mathbb{R}$, let $U = \{x \in \mathbb{R}: 1 < x < 2 \text{ or } 3 < x < 4\}$, and let $V = \{x \in \mathbb{R}: 2 < x < 3, 3.5 < x < 4, \text{ or } 5 < x < 6\}$. Then $U, V \in \mathcal{T}$ and hence $U \times V$ is the member of the product topology shown in Figure 2.2.

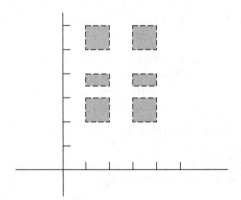

Figure 2.2

Notice that the union of two members of the basis $\mathcal{B}$ (where $\mathcal{B}$ is as defined in the definition of the product topology) need not be a member of $\mathcal{B}$. Let $U_1 = \{x \in \mathbb{R}: 1 < x < 2\}$, $V_1 = \{x \in \mathbb{R}: 1 < x < 2\}$, $U_2 = \{x \in \mathbb{R}: 2 < x < 3\}$, and $V_2 = \{x \in \mathbb{R}: 3 < x < 4\}$. Then $(U_1 \times V_1) \cup (U_2 \times V_2)$ is shown in Figure 2.3.

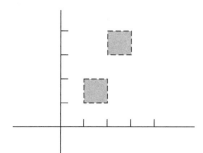

Figure 2.3

THEOREM 2.16. Let (X_1, d_1) and (X_2, d_2) be metric spaces, for each $i = 1, 2$ let $\mathcal{T}_i$ be the topology on X_i generated by d_i, and let $\mathcal{T}$ denote the product topology on $X = X_1 \times X_2$. Furthermore let $\mathcal{U}$ denote the topology on X generated by the product metric d. Then $\mathcal{T} = \mathcal{U}$.

Proof. There is a great deal of similarity between this proof and the analysis of Example 18 in Chapter 1. Let $V \in \mathcal{T}$ and let $(a_1, a_2) \in V$. Then for each $i = 1, 2$ there exists $U_i \in \mathcal{T}_i$ such that $(a_1, a_2) \in U_1 \times U_2 \subseteq V$. Then, since $\mathcal{T}_i$ is the topology generated by d_i, for each $i = 1, 2$ there exists $\varepsilon_i > 0$ such that $B_{d_i}(a_i, \varepsilon_i) \subseteq U_i$. Let $\varepsilon = \min\{\varepsilon_1, \varepsilon_2\}$. Then if $(x_1, x_2) \in X$ and $d((a_1, a_2), (x_1, x_2)) < \varepsilon$, $d_i(a_i, x_i) < \varepsilon \leq \varepsilon_i$ for each $i = 1, 2$. Therefore $B_d((a_1, a_2), \varepsilon) \subseteq B_{d_1}(a_1, \varepsilon_1) \times B_{d_2}(a_2, \varepsilon_2)$. Since $B_d((a_1, a_2), \varepsilon) \in \mathcal{U}$ and $(a_1, a_2) \in B_d((a_1, a_2), \varepsilon) \subseteq V$, $V \in \mathcal{U}$. Hence $\mathcal{T} \subseteq \mathcal{U}$.

Let $V \in \mathcal{U}$ and let $(a_1, a_2) \in V$. Then there exists $\varepsilon > 0$ such that $B_d((a_1, a_2), \varepsilon) \subseteq V$. Now $U = B_{d_1}(a_1, \frac{\sqrt{2}}{2}\varepsilon) \times B_{d_2}(a_2, \frac{\sqrt{2}}{2}\varepsilon) \in \mathcal{T}$ and $U \subseteq B_d((a_1, a_2), \varepsilon)$. Therefore $V \in \mathcal{T}$ and so $\mathcal{U} \subseteq \mathcal{T}$. ∎

The following theorem provides a description of a basis for the product topology on $X \times Y$ in terms of bases for the topologies on X and Y.

THEOREM 2.17. Let $(X_1, \mathcal{T}_1)$ and $(X_2, \mathcal{T}_2)$ be topological spaces, and for each $i = 1, 2$ let $\mathcal{B}_i$ be a basis for $\mathcal{T}_i$. Then $\mathcal{B} = \{U \times V : U \in \mathcal{B}_1 \text{ and } V \in \mathcal{B}_2\}$ is a basis for the product topology $\mathcal{T}$ on $X_1 \times X_2$.

Proof. Let $W \in \mathcal{T}$ and let $(x_1, x_2) \in W$. By the definition of the product topology, for each $i = 1, 2$ there exists $U_i \in \mathcal{T}_i$ such that $(x_1, x_2) \in U_1 \times U_2$ and $U_1 \times U_2 \subseteq W$. Then for each $i = 1, 2$ there exists $B_i \in \mathcal{B}_i$ such that and $x_i \in B_i$ and $B_i \subseteq U_i$. Therefore $(x_1, x_2) \in B_1 \times B_2$ and $B_1 \times B_2 \subseteq W$. Hence by Theorem 1.11, $\mathcal{B}$ is a basis for $\mathcal{T}$. ∎

Note that the collection of open rectangles is a basis for the usual topology on $\mathbb{R}^2$.

EXAMPLE 8. Let $\mathcal{T}$ denote the usual topology on $\mathbb{R}$, and let $\mathcal{U}$ denote the lower limit topology on $\mathbb{R}$. Then all sets of the form $\{(x, y) \in \mathbb{R}^2: a < x < b$ and $c \leq y < d\}$ (illustrated in Figure 2.4) is a basis for the product topology on $\mathbb{R}^2$.

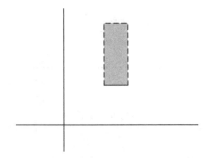

Figure 2.4

THEOREM 2.18. Let $(X_1, \mathcal{T}_1)$ and $(X_2, \mathcal{T}_2)$ be topological spaces, and let $(X_1 \times X_2, \mathcal{T})$ be the product space. Then the projections $\pi_1: X_1 \times X_2 \to X_1$ and $\pi_2: X_1 \times X_2 \to X_2$ are continuous. Moreover, the product topology is the smallest topology for which both projections are continuous.

Proof. Let $U_i \in \mathcal{T}_i$. Then $\pi_1^{-1}(U_1) = U_1 \times X_2$ and $\pi_2^{-1}(U_2) = X_1 \times U_2$. Since $U_1 \times X_2$ and $X_1 \times U_2$ are members of $\mathcal{T}$, by Theorem 1.53, π_1 and π_2 are continuous.

Let $\mathcal{U}$ be a topology on $X_1 \times X_2$ such that both projections are continuous. We show that $\mathcal{T}$ is smaller than $\mathcal{U}$. Let B be a member of the basis $\mathcal{B}$ (given by the definition of the product topology) for $\mathcal{T}$. Then there are members U_1 of $\mathcal{T}_1$ and U_2 of $\mathcal{T}_2$ such that $B = U_1 \times U_2$. Since π_1 and π_2 are continuous with respect to $\mathcal{U}$, $\pi_1^{-1}(U_1)$ and $\pi_2^{-1}(U_2)$ are members of $\mathcal{U}$. Since $B = U_1 \times U_2 = \pi_1^{-1}(U_1) \cap \pi_2^{-1}(U_2)$, $B \in \mathcal{U}$. Therefore $\mathcal{B} \subseteq \mathcal{U}$, and since the union of members of a topology is a member of the topology, $\mathcal{T} \subseteq \mathcal{U}$. Hence $\mathcal{T}$ is smaller than $\mathcal{U}$. ∎

THEOREM 2.19. Let $(X_1, \mathcal{T}_1)$ and $(X_2, \mathcal{T}_2)$ be topological spaces and let π_1 and π_2 denote the projection maps. Then $\mathcal{S} = \{\pi_1^{-1}(U): U \in \mathcal{T}_1\} \cup \{\pi_2^{-1}(V): V \in \mathcal{T}_2\}$ is a subbasis for the product topology on $X_1 \times X_2$.

Proof. It is clear that $X_1 \times X_2 = \bigcup \{S: S \in \mathcal{S}\}$. Hence by Theorem 1.12, there is a unique topology $\mathcal{T}'$ on $X_1 \times X_2$ such that $\mathcal{S}$ is a subbasis for $\mathcal{T}'$. Let $\mathcal{T}$ denote the product topology on $X_1 \times X_2$. Since every member of $\mathcal{S}$ belongs to $\mathcal{T}$, arbitrary unions of finite intersections of members of $\mathcal{S}$ also belong to $\mathcal{T}$. Therefore $\mathcal{T}' \subseteq \mathcal{T}$. Let $\mathcal{B}$ be the bais for $\mathcal{T}$ given by the definition of the product topology, and let $U \times V \in \mathcal{B}$. Since $U \times V = \pi_1^{-1}(U) \cap \pi_2^{-1}(V)$, $U \times V$ is the intersection of two members of $\mathcal{S}$. Hence $U \times V \in \mathcal{T}'$. It follows that $\mathcal{T} \subseteq \mathcal{T}'$. ∎

The following theorem shows that there is no ambiguity involving subspaces of product spaces.

2.2 The Product Topology on X × Y

THEOREM 2.20. Let $(C, \mathcal{T}_C)$ be a subspace of the topological space $(X, \mathcal{T})$, and let $(D, \mathcal{U}_D)$ be a subspace of the topological space $(Y, \mathcal{U})$. Then the product topology on $C \times D$ determined by $\mathcal{T}_C$ and $\mathcal{U}_D$ is the same as the subspace topology on $C \times D$ determined by the product topology on $X \times Y$.

Proof. Let $\mathcal{B}$ be the basis for the product topology on $X \times Y$ given by the definition of the product topology, let $\mathcal{B}_{(C \times D)}$ be the basis for the subspace topology on $C \times D$ given by Theorem 2.3, and let $\mathcal{B}'$ be the basis for the product topology on $C \times D$ given by the definition of the product topology. Note that $\mathcal{B}_{(C \times D)}$ is determined by $\mathcal{B}$, and $\mathcal{B}'$ determined by $\mathcal{T}_C$ and $\mathcal{U}_D$. We prove that $\mathcal{B}_{(C \times D)} = \mathcal{B}'$. Let $M \in \mathcal{B}_{(C \times D)}$. Then there is a member N of $\mathcal{B}$ such that $M = (C \times D) \cap N$. There exists $U \in \mathcal{T}$ and $V \in \mathcal{U}$ such that $N = U \times V$. Since $C \cap U \in \mathcal{T}_C$, $D \cap V \in \mathcal{U}_D$ and $(C \times D) \cap (U \times V) = (C \cap U) \times (D \cap V)$, $M \in \mathcal{B}'$. Therefore $\mathcal{B}_{(C \times D)} \subseteq \mathcal{B}'$. Now let $M \in \mathcal{B}'$. Then there exists $U_C \in \mathcal{T}_C$ and $V_D \in \mathcal{U}_D$ such that $M = U_C \times V_D$. Now there exists $U \in \mathcal{T}$ and $V \in \mathcal{U}$ such that $U_C = C \cap U$ and $V_D = D \cap V$. Since $U \times V \in \mathcal{B}$ and $(C \cap U) \times (D \cap V) = (C \times D) \cap (U \times V)$, $M \in \mathcal{B}_{(C \times D)}$. Hence $\mathcal{B}' \subseteq \mathcal{B}_{(C \times D)}$. ∎

We use Theorem 2.19 to provide a characterization of continuity for functions for which the range is a product space.

THEOREM 2.21. Let $(X, \mathcal{T})$, $(Y_1, \mathcal{U}_1)$, and $(Y_2, \mathcal{U}_2)$ be topological spaces and let $f: X \to Y_1 \times Y_2$ be a function. Then f is continuous if and only if $\pi_i \circ f$ is continuous for each $i = 1, 2$.

Proof. Suppose f is continuous. By Theorem 2.18, for each $i = 1, 2$, π_i is continuous. Therefore by Theorem 1.56, $\pi_i \circ f$ is continuous for each $i = 1, 2$.

Suppose $\pi_i \circ f$ is continuous for each $i = 1, 2$. Let $\mathcal{S}$ be the subbasis for the product topology on $Y_1 \times Y_2$ given by Theorem 2.19, and let $\pi_i^{-1}(U_i) \in \mathcal{S}$. Then $f^{-1}(\pi_i^{-1}(U_i)) = (\pi_i \circ f)^{-1}(U_i)$. Since $\pi_i \circ f$ is continuous, $(\pi_i \circ f)^{-1}(U_i) \in \mathcal{T}$. Therefore by Theorem 1.54, f is continuous. ∎

The following result is an immediate consequence of Theorem 2.21.

Corollary 2.22. Let $(X, \mathcal{T})$, $(Y_1, \mathcal{U}_1)$ and $(Y_2, \mathcal{U}_2)$ be topological spaces, let $f_1: X \to Y_1$ and $f_2: X \to Y_2$ be functions, and define $f: X \to Y_1 \times Y_2$ by $f(x) = (f_1(x), f_2(x))$. Then f is continuous if and only if f_1 and f_2 are continuous. ∎

EXAMPLE 9. Let $f, g: \mathbb{R} \to \mathbb{R}$ be defined by $f(x) = x^2 + 4$ and $g(x) = x^3 - 2x + 6$. Then f and g are polynomials, so they are continuous functions. By Corollary 2.22, the function $h: \mathbb{R} \to \mathbb{R} \times \mathbb{R}$ defined by $h(x) = (x^2 + 4, x^3 - 2x + 6)$ is continuous.

If two topological spaces have a certain topological property, it is natural to ask whether the product space also has the property. We already know that metrizability is such a property (Theorem 2.16). The following four theorems, three of whose proofs are left as exercises, establish that Hausdorff, separable, first countable, and second countable are properties that are enjoyed by the product of two spaces having these properties.

THEOREM 2.23. Let $(X_1, \mathcal{T}_1)$ and $(X_2, \mathcal{T}_2)$ be Hausdorff spaces, and let $\mathcal{T}$ denote the product topology on $X = X_1 \times X_2$. Then $(X, \mathcal{T})$ is a Hausdorff space.

Proof. Let (a_1, a_2) and (b_1, b_2) be distinct members of X. Without loss of generality we may assume that $a_1 \neq b_1$. Since $(X_1, \mathcal{T}_1)$ is Hausdorff, there exist disjoint open subsets U and V of X_1 such that $a_1 \in U$ and $b_1 \in V$. So $\pi_1^{-1}(U)$ and $\pi_1^{-1}(V)$ are disjoint open subsets of X such that $(a_1, a_2) \in \pi_1^{-1}(U)$ and $(b_1, b_2) \in \pi_1^{-1}(V)$. Therefore $(X, \mathcal{T})$ is a Hausdorff space. ∎

THEOREM 2.24. Let $(X_1, \mathcal{T}_1)$ and $(X_2, \mathcal{T}_2)$ be separable spaces, and let $\mathcal{T}$ denote the product topology on $X = X_1 \times X_2$. Then $(X, \mathcal{T})$ is a separable space.

Proof. See Exercise 6. ∎

THEOREM 2.25. Let $(X_1, \mathcal{T}_1)$ and $(X_2, \mathcal{T}_2)$ be first countable spaces, and let $\mathcal{T}$ denote the product topology on $X = X_1 \times X_2$. Then $(X, \mathcal{T})$ is a first countable space.

Proof. See Exercise 7. ∎

THEOREM 2.26. Let $(X_1, \mathcal{T}_1)$ and $(X_2, \mathcal{T}_2)$ be second countable spaces, and let $\mathcal{T}$ denote the product topology on $X = X_1 \times X_2$. Then $(X, \mathcal{T})$ is a second countable space.

Proof. See Exercise 8. ∎

The final result in this section permits us to "combine" continuous functions in order to obtain a continuous function on a product space.

THEOREM 2.27. Let $(X_1, \mathcal{T}_1)$, $(X_2, \mathcal{T}_2)$, $(Y_1, \mathcal{U}_1)$, $(Y_2, \mathcal{U}_2)$ and $(Z, \mathcal{V})$ be topological spaces, and let $f: X_1 \to Y_1$, $g: X_2 \to Y_2$ and $F: Y_1 \times Y_2 \to Z$ be continuous functions. Then the function $G: X_1 \times X_2 \to Z$ defined by $G((x_1, x_2)) = F(f(x_1), g(x_2))$ is continuous.

Proof. See Exercise 16. ∎

EXERCISES 2.2

1. Let $X = \{1, 2, 3\}$, $\mathcal{T} = \{\emptyset, \{1\}, \{1, 2\}, X\}$, $Y = \{4, 5\}$, and $\mathcal{U} = \{\emptyset, \{4\}, Y\}$. Find the basis $\mathcal{B}$ for the product topology on $X \times Y$ described in Theorem 2.17.

2. Let X and Y be infinite sets, let $\mathcal{T}$ be the discrete topology on X, and let $\mathcal{U}$ be the trivial topology on Y. Describe the product topology on X × Y.

3. Let $\mathcal{T} = \{U \in \mathcal{P}(\mathbb{R}): 0 \in U \text{ or } U = \varnothing\}$, and let $\mathcal{U} = \{U \in \mathcal{P}(\mathbb{R}): 1 \in U \text{ or } U = \varnothing\}$. Describe the product topology on $\mathbb{R} \times \mathbb{R}$ determined by $\mathcal{T}$ and $\mathcal{U}$.

4. Let $\mathcal{T}$ be the lower limit topology on $\mathbb{R}$, and let $\mathcal{U}$ be the product topology on $(\mathbb{R}, \mathcal{T}) \times (\mathbb{R}, \mathcal{T})$. Identify the relative topology on the line $L = \{(x, -x): x \in \mathbb{R}\}$.

5. Let $(X, \mathcal{T})$ and $(Y, \mathcal{U})$ be topological spaces, let $a \in X$, and let $b \in Y$. Prove that the functions $f: X \to X \times Y$ and $g: Y \to X \times Y$ defined by $f(x) = (x, b)$ and $g(y) = (a, y)$ are embeddings.

6. Let $(X_1, \mathcal{T}_1)$ and $(X_2, \mathcal{T}_2)$ be separable spaces, and let $\mathcal{T}$ denote the product topology on $X = X_1 \times X_2$. Prove that $(X, \mathcal{T})$ is a separable space.

7. Let $(X_1, \mathcal{T}_1)$ and $(X_2, \mathcal{T}_2)$ be first countable spaces, and let $\mathcal{T}$ denote the product topology on $X = X_1 \times X_2$. Prove that $(X, \mathcal{T})$ is a first countable space.

8. Let $(X_1, \mathcal{T}_1)$ and $(X_2, \mathcal{T}_2)$ be second countable spaces, and let $\mathcal{T}$ denote the product topology on $X = X_1 \times X_2$. Prove that $(X, \mathcal{T})$ is a second countable space.

9. Let $(X_1, \mathcal{T}_1)$ and $(X_2, \mathcal{T}_2)$ be topological spaces, and let $\mathcal{T}$ be the product topology on $X_1 \times X_2$, and let $\mathcal{U}$ be the product topology on $X_2 \times X_1$. Prove that $(X_1 \times X_2, \mathcal{T})$ is homeomorphic to $(X_2 \times X_1, \mathcal{U})$.

10. Let $(X, \mathcal{T})$ be a topological space, let $\mathcal{U}$ denote the product topology on $X \times X$, let $\Delta = \{(x, y) \in X \times X: x = y\}$, and let $\mathcal{U}_\Delta$ be the subspace topology on Δ determined by $\mathcal{U}$. Prove that $(X, \mathcal{T})$ is homeomorphic to $(\Delta, \mathcal{U}_\Delta)$. (The set Δ is called the **diagonal**.)

11. Let $(X, \mathcal{T})$ be a topological space and let $\mathcal{U}$ denote the product topology on $X \times X$. Prove that $(X, \mathcal{T})$ is Hausdorff if and only if the diagonal is a closed subset of $(X \times X, \mathcal{U})$.

12. Let $(X_1, \mathcal{T}_1)$, $(X_2, \mathcal{T}_2)$ and $(X_3, \mathcal{T}_3)$ be topological spaces, let $\mathcal{T}$ be the product topology on $(X_1 \times X_2) \times X_3$ (use $\mathcal{T}_1$ and $\mathcal{T}_2$ to obtain the product topology $\mathcal{T}'$ on $X_1 \times X_2$, then use $\mathcal{T}'$ and $\mathcal{T}_3$ to obtain $\mathcal{T}$), and let $\mathcal{U}$ be the product topology on $X_1 \times (X_2 \times X_3)$. Prove that $((X_1 \times X_2) \times X_3, \mathcal{T})$ is homeomorphic to $(X_1 \times (X_2 \times X_3), \mathcal{U})$.

13. Let $(X_1, \mathcal{T}_1)$ and $(X_2, \mathcal{T}_2)$ be topological spaces and suppose $X_1 \times X_2$ has the product topology. For each $i = 1, 2$ let A_i be a subset of X_i. Prove that $\overline{A_1 \times A_2} = \overline{A_1} \times \overline{A_2}$.

14. Let $(X_1, \mathcal{T}_1)$ and $(X_2, \mathcal{T}_2)$ be topological spaces and suppose $X_1 \times X_2$ has the product topology. For each $i = 1, 2$ let A_i be a subset of X_i. Prove that $\text{int}(A_1 \times A_2) = (\text{int}(A_1)) \times (\text{int}(A_2))$.

15. Let X, Y_1, and Y_2 be sets, for each and $i = 1, 2$, let $U_i \subseteq Y_i$ and let $f_i: X \to Y_i$ be a function, and define $f: X \to Y_1 \times Y_2$ by $f(x) = (f_1(x), f_2(x))$. Prove that $f^{-1}(U_1 \times U_2) = f_1^{-1}(U_1) \cap f_2^{-1}(U_2)$.

16. Prove Theorem 2.27.

17. Prove that he function $f: \mathbb{R} \to \mathbb{R}^2$ defined by $f(x) = (x^2 - 5, 1/(x^2 + 1))$ is continuous.

18. Let $\leq_R$ denote the usual order on $\mathbb{R}$, let $\leq$ denote the dictionary order on $\mathbb{R}^2$ determined by $\leq_R$, and let $\mathcal{U}$ denote the order topology on $\mathbb{R}^2$ determined by $\leq$. Furthermore let $\mathcal{T}_1$ denote the discrete topology on $\mathbb{R}$, let $\mathcal{T}_2$ denote the usual topology on $\mathbb{R}$, and let $\mathcal{T}$ denote the product topology on $\mathbb{R}^2$ determined by $\mathcal{T}_1$ and $\mathcal{T}_2$. Prove that $\mathcal{U} = \mathcal{T}$.

2.3 The Product Topology

In the previous section, we defined the product topology on the product of two topological spaces. We observed that this definition, and the results we established, can be extended in the obvious way to the product of any finite number of topological spaces. In this section we generalize the definition to the product of an arbitrary collection of topological spaces. We present two ways of generalizing the definition, and the one that will prove to be more important is called the **product topology**.

Definition. Let $\{(X_\alpha, \mathcal{T}_\alpha): \alpha \in \Lambda\}$ be an indexed family of topological spaces. Let $\mathcal{B}$ be the collection of all sets of the form $\prod_{\alpha \in \Lambda} U_\alpha$, where $U_\alpha \in \mathcal{T}_\alpha$ for each $\alpha \in \Lambda$. Then $\mathcal{B}$ is a basis for a topology on $\prod_{\alpha \in \Lambda} X_\alpha$, and this topology is called the **box topology**. ∎

We leave as Exercise 1 the proof that $\mathcal{B}$ is a basis for a topology on X. In order to give the second definition, we generalize the notion of a projection. Recall that a member $\langle x_\alpha \rangle_{\alpha \in \Lambda}$ of $\prod_{\alpha \in \Lambda} X_\alpha$ is a function that maps Λ into $\bigcup \{X_\alpha: \alpha \in \Lambda\}$ and has the property that $x_\alpha \in X_\alpha$ for each $\alpha \in \Lambda$.

Definition. Let $\{X_\alpha: \alpha \in \Lambda\}$ be an indexed family of sets and let $\beta \in \Lambda$. The **projection mapping** associated with β is the function $\pi_\beta: \prod_{\alpha \in \Lambda} X_\alpha \to X_\beta$ defined by $\pi_\beta(\langle x_\alpha \rangle_{\alpha \in \Lambda}) = x_\beta$. ∎

Suppose $\Lambda = \{1, 2\}$ and $X_1 = X_2 = \mathbb{R}$. Then the projection map $\pi_1: \mathbb{R} \times \mathbb{R} \to \mathbb{R}$ is the function defined by $\pi_1((x, y)) = x$ (see Appendix C). If A is the open interval $(1, 2)$, then $\pi_1^{-1}(A) = A \times \mathbb{R}$ (see Figure 2.5).

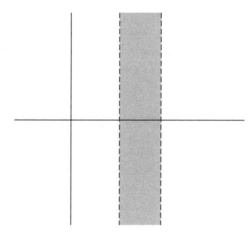

Figure 2.5

Now let $\Lambda = \{1, 2, 3\}$ and $X_1 = X_2 = X_3 = \mathbb{R}$. Then the projection map $\pi_1: \mathbb{R} \times \mathbb{R} \times \mathbb{R} \to \mathbb{R}$ is the function defined by $\pi_1((x_1, x_2, x_3)) = x_1$. If A is the closed interval $[3, 4]$ then $\pi_1^{-1}(A)$ is the set $A \times \mathbb{R} \times \mathbb{R}$. This set can be drawn in 3-space (see Figure 2.6). Intuitively it is a solid infinite sheet.

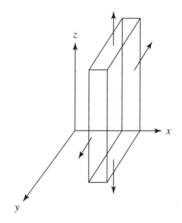

Figure 2.6

In general if $A_\beta \subseteq X_\beta$, then $\pi_\beta^{-1}(A_\beta) = \prod_{\alpha \in \Lambda} B_\alpha$, where $B_\beta = A_\beta$ and $B_\alpha = X_\alpha$ for all $\alpha \neq \beta$. We leave the proof of this fact as Exercise 2. It is clear that $\pi_\beta^{-1}(X_\beta) = \prod_{\alpha \in \Lambda} X_\alpha$.

The definition of the product topology is motivated by the fact that it is the smallest topology on $\prod_{\alpha \in \Lambda} X_\alpha$ such that each projection map is continuous (Theorem 2.33), and it is given in terms of a subbasis.

Definition. Let $\{(X_\alpha, \mathcal{T}_\alpha): \alpha \in \Lambda\}$ be an indexed family of topological spaces. For each $\alpha \in \Lambda$, let $\mathcal{S}_\alpha = \{\pi_\alpha^{-1}(U_\alpha): U_\alpha \in \mathcal{T}_\alpha\}$ and let $\mathcal{S} = \bigcup_{\alpha \in \Lambda} \mathcal{S}_\alpha$. Then $\mathcal{S}$ is a subbasis for a topology $\mathcal{T}$ on $\prod_{\alpha \in \Lambda} X_\alpha$, and $\mathcal{T}$ is called the **product topology**. The topological space $(\prod_{\alpha \in \Lambda} X_\alpha, \mathcal{T})$ is called the **product space**. ∎

It is clear that $\mathcal{S}$ is a subbasis for a topology on $\prod_{\alpha \in \Lambda} X_\alpha$ because if $x \in \prod_{\alpha \in \Lambda} X_\alpha$ and $\alpha \in \Lambda$, then $X_\alpha \in \mathcal{T}_\alpha$ and $x \in \pi_\alpha^{-1}(X_\alpha)$. The general definition of product space was given by A.N. Tychonoff (1906–1993) in 1930. Others, however, had studied countably infinite products in the 1920s.

What is the difference between the box topology and the product topology? In order to answer this question, let us look at the basis $\mathcal{B}'$ determined by $\mathcal{S}$. The collection $\mathcal{B}'$ consists of all finite intersections of members of $\mathcal{S}$. Since for each $\alpha \in \Lambda$, $\pi_\alpha^{-1}(U_\alpha) \cap \pi_\alpha^{-1}(V_\alpha) = \pi_\alpha^{-1}(U_\alpha \cap V_\alpha)$, the intersection of a finite number of members of $\mathcal{S}_\alpha$ is a member of $\mathcal{S}_\alpha$. Therefore each member of $\mathcal{B}'$ that is not a member of $\mathcal{S}$ is obtained by intersecting members of different $\mathcal{S}_\alpha$. That is, we can describe a member of $\mathcal{B}'$ that is not a member of $\mathcal{S}$ as follows: Let $\beta_1, \beta_2, \ldots, \beta_n$ be a finite set of distinct members of Λ, and for each $i = 1, 2, \ldots, n$, let $U_{\beta_i} \in \mathcal{T}_{\beta_i}$. Then $B = \bigcap_{i=1}^n \pi_{\beta_i}^{-1}(U_{\beta_i})$ is a member of $\mathcal{B}'$. We leave as Exercise 3 the proof that $B = \prod_{\alpha \in \Lambda} B_\alpha$, where $B_{\beta_i} = U_{\beta_i}$ for each $i = 1, 2, \ldots, n$ and $B_\alpha = X_\alpha$ for all α different from $\beta_1, \beta_2, \ldots, \beta_n$. We summarize this discussion in the following theorem.

THEOREM 2.28. Let $\{(X_\alpha, \mathcal{T}_\alpha): \alpha \in \Lambda\}$ be an indexed family of topological spaces. The collection of all sets of the form $\prod_{\alpha \in \Lambda} U_\alpha$, where $U_\alpha \in \mathcal{T}_\alpha$ for each $\alpha \in \Lambda$ and $U_\alpha = X_\alpha$ for all but a finite number of members of Λ, is a basis for the product topology on $\prod_{\alpha \in \Lambda} X_\alpha$. ∎

Of course if Λ is a finite set, then the box topology and the product topology are the same. In general, the box topology is finer than the product topology. The following example shows that the two topologies may be different.

EXAMPLE 10. For each $n \in \mathbb{N}$, let $X_n = \{1, 2\}$ and let $\mathcal{T}_n$ be the discrete topology on X_n. Let $X = \prod_{n \in \mathbb{N}} X_n$, let $\mathcal{T}$ be the product topology on X, and let $\mathcal{U}$ be the box topology on X. Then $\mathcal{T} \neq \mathcal{U}$.

Analysis. Let x be the member of X that has property that $x_n = 1$ for each $n \in \mathbb{N}$. Then $\{x\} \in \mathcal{U}$. Let $U \in \mathcal{T}$ such that $x \in U$. Then there is a member B of the basis for $\mathcal{T}$ described in Theorem 2.28 such that $x \in B$ and $B \subseteq U$. Since $B = \prod_{n \in \mathbb{N}} U_n$, where $U_n = X_n$ for all but a finite number of members of $\mathbb{N}$, there exists a natural number N such that if $n > N$, then $U_n = X_n$. Let y be the member of X that has the property that $y_n = 1$ for each $n \leq N$ and $y_n = 2$ for each $n > N$. Then $y \in B$. Therefore $\{x\} \notin \mathcal{T}$. ∎

The reason the product topology is more important than the box topology is not immediately clear. It will become clear as we progress. We shall find that some theorems about finite products will hold for arbitrary products if we use the product topology but not if we use the box topology. Therefore, unless we specify otherwise, whenever we speak of the product space, we shall assume that the Cartesian product has the product topology.

The following two theorems generalize Theorem 2.17. We prove Theorem 2.30 and leave the proof of Theorem 2.29 as Exercise 7.

THEOREM 2.29. Let $\{(X_\alpha, \mathcal{T}_\alpha): \alpha \in \Lambda\}$ be an indexed family of topological spaces, and for each $\alpha \in \Lambda$, let $\mathcal{B}_\alpha$ be a basis for $\mathcal{T}_\alpha$. Then the collection of all sets of the form $\prod_{\alpha \in \Lambda} B_\alpha$, where $B_\alpha \in \mathcal{B}_\alpha$ for each $\alpha \in \Lambda$, is a basis for the box topology on $\prod_{\alpha \in \Lambda} X_\alpha$. ∎

THEOREM 2.30. Let $\{(X_\alpha, \mathcal{T}_\alpha): \alpha \in \Lambda\}$ be an indexed family of topological spaces, and for each $\alpha \in \Lambda$, let $\mathcal{B}_\alpha$ be a basis for $\mathcal{T}_\alpha$. Then the collection $\mathcal{B}$ of all sets of the form $\prod_{\alpha \in \Lambda} B_\alpha$, where $B_\alpha = X_\alpha$ for all but a finite number of members, $\beta_1, \beta_2, ..., \beta_n$ of Λ and $B_{\beta_i} \in \mathcal{B}_{\beta_i}$ for each $i = 1, 2, ..., n$ is a basis for the product topology $\mathcal{T}$ on $\prod_{\alpha \in \Lambda} X_\alpha$.

Proof. It is clear that $\mathcal{B}$ is a subcollection of $\mathcal{T}$. We complete the proof by showing that if $U \in \mathcal{T}$ and $x = \langle x_\alpha \rangle \in U$, then there exists $B \in \mathcal{B}$ such that $x \in B \subseteq U$. Let $U \in \mathcal{T}$ and let $x \in U$. By Theorem 2.28, there exists $B' = \prod_{\alpha \in \Lambda} U_\alpha$, where $U_\alpha = X_\alpha$ for all but a finite number of members $\beta_1, \beta_2, ..., \beta_n$ of Λ, $U_{\beta_i} \in \mathcal{T}_{\beta_i}$ for each $i = 1, 2, ..., n$, and $x \in B' \subseteq U$. For each $i = 1, 2, ..., n$, let $B_{\beta_i} \in \mathcal{B}_{\beta_i}$ such that $x_{\beta_i} \in B_{\beta_i} \subseteq U_{\beta_i}$. Then $B = \prod_{\alpha \in \Lambda} V_\alpha$, where $V_\alpha = X_\alpha$ whenever $\alpha \neq \beta_i$ for any $i = 1, 2, ..., n$ and $V_{\beta_i} = B_{\beta_i}$ for each $i = 1, 2, ..., n$, is a member of $\mathcal{B}$ such that $x \in B \subseteq U$. ∎

The following theorem, whose proof follows immediately from the definition of the product topology, is a generalization of Theorem 2.18.

THEOREM 2.31. Let $\{(X_\alpha, \mathcal{T}_\alpha): \alpha \in \Lambda\}$ be an indexed family of topological spaces, let $\mathcal{T}$ denote the product topology on $\prod_{\alpha \in \Lambda} X_\alpha$, and let $\beta \in \Lambda$. Then the projection map $\pi_\beta: \prod_{\alpha \in \Lambda} X_\alpha \to X_\beta$ is continuous. ∎

Since the product topology is a subset of the box topology, we immediately obtain the following corollary.

Corollary 2.32. Let $\{(X_\alpha, \mathcal{T}_\alpha): \alpha \in \Lambda\}$ be an indexed family of topological spaces, let $\mathcal{T}$ denote the box topology on $\prod_{\alpha \in \Lambda} X_\alpha$, and let $\beta \in \Lambda$. Then the projection map $\pi_\beta: \prod_{\alpha \in \Lambda} X_\alpha \to X_\beta$ is continuous. ∎

THEOREM 2.33. Let $\{(X_\alpha, \mathcal{T}_\alpha): \alpha \in \Lambda\}$ be an indexed family of topological spaces. The product topology is the weakest topology on $\prod_{\alpha \in \Lambda} X_\alpha$ for which each projection map $\pi_\beta: \prod_{\alpha \in \Lambda} X_\alpha \to X_\beta$ is continuous.

Proof. Let $\mathscr{T}$ be the product topology on $X = \prod_{\alpha \in \Lambda} X_\alpha$, and let $\mathscr{U}$ be the weakest topology on X such that each projection map is continuous. We show that $\mathscr{T} \subseteq \mathscr{U}$ by showing that each member of the subbase $\mathscr{S}$ (given in the definition of the product topology) of $\mathscr{T}$ is a member of $\mathscr{U}$. Let $S \in \mathscr{S}$. Then there exists $\beta \in \Lambda$ and a member U_β of $\mathscr{T}_\beta$ such that $S = \pi_\beta^{-1}(U_\beta)$. Since $\mathscr{U}$ is a topology on X such that π_β is continuous, $\pi_\beta^{-1}(U_\beta) \in \mathscr{U}$. Therefore $\mathscr{T} \subseteq \mathscr{U}$.

By Theorem 2.31 each projection map is continuous with respect to $\mathscr{T}$. Therefore since $\mathscr{U}$ is the weakest topology such that each projection map is continuous, $\mathscr{U} \subseteq \mathscr{T}$. ∎

The following theorem, whose proof is left as Exercise 8, shows that if $\mathscr{T}$ is the box topology on $X = \prod_{\alpha \in \Lambda} X_\alpha$, then each projection map is open. Since the product topology is a subset of the box topology, this theorem also shows that if $\mathscr{T}$ is the product topology on X, then each projection map is open.

THEOREM 2.34. Let $\{(X_\alpha, \mathscr{T}_\alpha): \alpha \in \Lambda\}$ be an indexed family of topological spaces, let $X = \prod_{\alpha \in \Lambda} X_\alpha$, and let $\mathscr{T}$ be the box topology on X. Then for each $\beta \in \Lambda$, the projection map $\pi_\beta: X \to X_\beta$ is open. ∎

The following two theorems generalize Theorem 2.20. Since the proof of each is similar to the proof of Theorem 2.20, we prove Theorem 2.35 and leave the proof of Theorem 2.36 as Exercise 9.

THEOREM 2.35. Let $\{(X_\alpha, \mathscr{T}_\alpha): \alpha \in \Lambda\}$ be an indexed family of topological spaces, and for each $\alpha \in \Lambda$, let $(A_\alpha, \mathscr{T}_{A_\alpha})$ be a subspace of $(X_\alpha, \mathscr{T}_\alpha)$. Then the product topology on $\prod_{\alpha \in \Lambda} A_\alpha$ is the same as the subspace topology on $\prod_{\alpha \in \Lambda} A_\alpha$ determined by the product topology on $\prod_{\alpha \in \Lambda} X_\alpha$.

Proof. Let $\mathscr{T}$ be the product topology on $A = \prod_{\alpha \in \Lambda} A_\alpha$, and let $\mathscr{U}_A$ be the subspace topology on A determined by the product topology $\mathscr{U}$ on $X = \prod_{\alpha \in \Lambda} X_\alpha$. Let $\mathscr{B}$ and $\mathscr{B}^*$ be bases (given by the definition of the product topology) for $\mathscr{T}$ and $\mathscr{U}$ respectively. Then by Theorem 2.3, $\mathscr{B}_{A*} = \{A \cap B: B \in \mathscr{B}*\}$ is a basis for $\mathscr{U}_A$. We show that $\mathscr{B} = \mathscr{B}_{A*}$ by showing that each is a subset of the other.

Let $B \in \mathscr{B}$. Then $B = \prod_{\alpha \in \Lambda} U_\alpha$, where $U_\alpha = A_\alpha$ for all but a finite number of members $\beta_1, \beta_2, \ldots, \beta_n$ of Λ and $U_{\beta_i} \in \mathscr{T}_{A_{\beta_i}}$ for each $i = 1, 2, \ldots, n$. Then for each $i = 1, 2, \ldots, n$, there exists $V_{\beta_i} \in \mathscr{T}_{\beta_i}$ such that $U_{\beta_i} = A_{\beta_i} \cap V_{\beta_i}$. Thus $C = \prod_{\alpha \in \Lambda} C_\alpha$, where $C_\alpha = X_\alpha$ if $\alpha \neq \beta_i$ for any $i = 1, 2, \ldots, n$, and $C_{\beta_i} = V_{\beta_i}$ for each $i = 1, 2, \ldots, n$, is a member of $\mathscr{B}^*$. Therefore $C \cap A \in \mathscr{B}_{A*}$. We show that $\mathscr{B} \in \mathscr{B}_{A*}$ by proving that $B = C \cap A$. Let $x \in B$. Then $x \in A$ since $\mathscr{B}$ is a basis for a topology on A and $x_{\beta_i} \in U_{\beta_i}$ for each $i = 1, 2, \ldots, n$. Hence $x_{\beta_i} \in A_{\beta_i} \cap V_{\beta_i}$ for each $i = 1, 2, \ldots, n$ and therefore $x \in C$. This completes the proof that $B \subseteq A \cap C$. Now let $x \in A \cap C$. Then $x \in A$ and $x_{\beta_i} \in V_{\beta_i}$ for each $i = 1, 2, \ldots, n$. Therefore $x_{\beta_i} \in U_{\beta_i}$ for each $i = 1, 2, \ldots, n$, and so $x \in B$. Now we have $A \cap C \subseteq B$, so that $B = A \cap C$. It follows that $\mathscr{B} \subseteq \mathscr{B}_{A*}$.

Now let $C \in \mathcal{B}_{A^*}$. Then there exists $B \in \mathcal{B}^*$ such that $C = A \cap B$. Since $B \in \mathcal{B}^*$, $B = \prod_{\alpha \in \Lambda} D_\alpha$, where $D_\alpha = X_\alpha$ for all but a finite number of members $\beta_1, \beta_2, ..., \beta_n$ of Λ and $D_{\beta_i} \in \mathcal{T}_{\beta_i}$ for each $i = 1, 2, ..., n$. For each $i = 1, 2, ..., n$, let $C_{\beta_i} = D_{\beta_i} \cap A_{\beta_i}$. Then $C_{\beta_i} \in \mathcal{T}_{A_{\beta_i}}$ for each $i = 1, 2, ..., n$. Therefore $E = \prod_{\alpha \in \Lambda} E_\alpha$, where $E_\alpha = A_\alpha$ if $\alpha \neq \beta_i$ for any $i = 1, 2, ..., n$ and $E_{\beta i} = C_{\beta i}$ for each $i = 1, 2, ..., n$ is a member of $\mathcal{B}$. We show that $C \in \mathcal{B}$ by showing that $E = C$. Let $x \in E$. Then $x \in A$ and $x_{\beta_i} \in C_{\beta_i}$ for each $i = 1, 2, ..., n$. Therefore $x_{\beta_i} \in D_{\beta_i}$ for each $i = 1, 2, ..., n$ and hence $x \in B$. Thus $x \in C$, so $E \subseteq C$. Suppose $x \in C$. Then $x \in A$ and $x \in B$, so $x_{\beta_i} \in D_{\beta_i}$ for each $i = 1, 2, ..., n$. Therefore $x_{\beta_i} \in C_{\beta_i}$ for each $i = 1, 2, ..., n$, and hence $x \in E$. This completes the proof that $E = C$ and it follows that $\mathcal{B}_{A^*} \subseteq \mathcal{B}$.

Since $\mathcal{B}$ is a basis for both $\mathcal{T}$ and $\mathcal{U}_A$, $\mathcal{T} = \mathcal{U}_A$. ∎

Notice that Theorem 2.35 is a form of commutativity. It says we get the same topology on the subspace A of X, regardless of whether we take the product topology and then the subspace topology or we take the subspace topology and then the product topology.

THEOREM 2.36. Let $\{(X_\alpha, \mathcal{T}_\alpha) : \alpha \in \Lambda\}$ be an indexed family of topological spaces, and for each $\alpha \in \Lambda$, let $(A_\alpha, \mathcal{T}_{A_\alpha})$ be a subspace of $(X_\alpha, \mathcal{T}_\alpha)$. Then the box topology on $\prod_{\alpha \in \Lambda} A_\alpha$ is the same as the subspace topology on $\prod_{\alpha \in \Lambda} A_\alpha$ determined by the box topology on $\prod_{\alpha \in \Lambda} X_\alpha$. ∎

The following theorem is a generalization of Theorem 2.21.

THEOREM 2.37. Let $\{(Y_\alpha, \mathcal{U}_\alpha) : \alpha \in \Lambda\}$ be an indexed family of topological spaces, let $\mathcal{U}$ be the product topology on $Y = \prod_{\alpha \in \Lambda} Y_\alpha$, let $(X, \mathcal{T})$ be a topological space, and let $f : X \to Y$ be a function. Then f is continuous if and only if $\pi_\alpha \circ f$ is continuous for each $\alpha \in \Lambda$.

Proof. Suppose f is continuous. By Theorem 2.31, each projection map is continuous. Therefore by Theorem 1.56, the composition $\pi_\alpha \circ f$ is continuous for each $\alpha \in \Lambda$.

Suppose $\pi_\alpha \circ f$ is continuous for each $\alpha \in \Lambda$. Let $\mathcal{S}$ be the subbasis for the product topology on Y given by the definition. By Theorem 1.54 part (f), it is sufficient to show that the inverse image under f of each member $\mathcal{S}$ is a member of $\mathcal{T}$. Let $\pi_\alpha^{-1}(U) \in \mathcal{S}$. Then $f^{-1}(\pi_\alpha^{-1}(U)) = (\pi_\alpha \circ f)^{-1}(U)$, and since $\pi_\alpha \circ f$ is continuous, $(\pi_\alpha \circ f)^{-1}(U) \in \mathcal{T}$. ∎

As the following example shows, the analogue of Theorem 2.37 does not hold when Y has the box topology.

EXAMPLE 11. For each $i \in \mathbb{N}$, let $Y_i = \mathbb{R}$. Recall that $\mathbb{R}^\omega$ denotes the Cartesian product $\prod_{i \in \mathbb{N}} Y_i$. Let $\mathcal{T}$ denote the usual topology on $\mathbb{R}$, and let $\mathcal{U}$ denote the box topology on $\mathbb{R}^\omega$. Define a function $f : \mathbb{R} \to \mathbb{R}^\omega$ as follows: For each $x \in \mathbb{R}$ and each $n \in \mathbb{N}$, let $f(x)$ be the sequence whose value at n is x. Then for each $i \in \mathbb{N}$, $\pi_i \circ f$ is continuous but f is not continuous.

Analysis. Let $i \in \mathbb{N}$. Then $\pi_i \circ f : \mathbb{R} \to \mathbb{R}$ is the identity function and hence it is continuous.

In order to show that f is not continuous, we exhibit a member B of $\mathcal{U}$ such that $f^{-1}(B) \notin \mathcal{T}$. For each $i \in \mathbb{N}$, let $B_i = \left(-\frac{1}{i}, \frac{1}{i}\right)$ and let $B = \prod_{i \in \mathbb{N}} B_i$. Then $B \in \mathcal{U}$. The proof that $f^{-1}(B) \notin \mathcal{T}$ is by contradiction. Suppose $f^{-1}(B) \in \mathcal{T}$. Then, since $0 \in f^{-1}(B)$, there is an open interval (a, b) such that $0 \in (a, b)$ and $(a, b) \subseteq f^{-1}(B)$. Therefore $f((a, b)) \subseteq B$, and thus for each natural number n, $(\pi_n \circ f)((a, b)) = (a, b) \subseteq \left(-\frac{1}{n}, \frac{1}{n}\right)$. This is a contradiction. ∎

Theorems 2.33 and 2.37 provide the justification for choosing the product topology over the box topology. In particular, while each projection map is continuous with respect to the box topology, the product topology is the smallest topology for which each projection map is continuous. Moreover a function from a topological space into a product with the product topology is continuous if and only if its composition with each projection map is continuous, while this result is not true if the product has the box topology. We have also seen that subspaces are "well-behaved" with respect to both the product topology and the box topology.

We conclude this section with the statement of two useful theorems whose proofs are left as Exercises 10 and 12.

THEOREM 2.38. Let $\{(X_\alpha, \mathcal{T}_\alpha) : \alpha \in \Lambda\}$ be an indexed family of topological spaces, and for each $\alpha \in \Lambda$ let $A_\alpha \subseteq X_\alpha$.

(a) If A_α is a closed subset of X_α, then $\prod_{\alpha \in \Lambda} A_\alpha$ is a closed subset of $\prod_{\alpha \in \Lambda} X_\alpha$.

(b) $\overline{\prod_{\alpha \in \Lambda} A_\alpha} = \prod_{\alpha \in \Lambda} \overline{A_\alpha}$. ∎

THEOREM 2.39. Let $\{(X_\alpha, \mathcal{T}_\alpha) : \alpha \in \Lambda\}$ be a collection of topological spaces, and let $\mathcal{T}$ be the product topology on $X = \prod_{\alpha \in \Lambda} X_\alpha$. Let $p \in X$, let $\beta \in \Lambda$, and let $H_{p\beta} = \{x \in X; \text{ if } \alpha \neq \beta, \text{ then } x_\alpha = p_\alpha\}$. Define a function $f : X_\beta \to H_{p\beta}$ as follows: for each $x_\beta \in X_\beta$ let $f(x_\beta)$ be the member of $H_{p\beta}$ defined by $[f(x_\beta)]_\beta = x_\beta$ and for $\alpha \neq \beta$, $[f(x_\beta)]_\alpha = p_\alpha$. Then f is a homeomorphism. ∎

EXERCISES 2.3

1. Let $\{(X_\alpha, \mathcal{T}_\alpha) : \alpha \in \Lambda\}$ be an indexed family of topological spaces, and let $\mathcal{B}$ be the collection of all sets of the form $\prod_{\alpha \in \Lambda} U_\alpha$, where $U_\alpha \in \mathcal{T}_\alpha$ for each $\alpha \in \Lambda$. Prove that $\mathcal{B}$ is a basis for a topology on $\prod_{\alpha \in \Lambda} X_\alpha$.

2. Let $\{X_\alpha : \alpha \in \Lambda\}$ be an indexed family of sets, let $\beta \in \Lambda$, and let $A_\beta \subseteq X_\beta$. Prove that $\pi_\beta^{-1}(A_\beta) = \prod_{\alpha \in \Lambda} B_\alpha$, where $B_\beta = A_\beta$ and $B_\alpha = X_\alpha$ for all $\alpha \neq \beta$.

3. Let $\{(X_\alpha, \mathcal{T}_\alpha): \alpha \in \Lambda\}$ be an indexed family of topological spaces, let $\beta_1, \beta_2, \ldots, \beta_n$ be a finite set of distinct members of Λ, for each $i = 1, 2, \ldots, n$ let $U_{\beta_i} \in \mathcal{T}_{\beta_i}$, and let $B = \bigcap_{i=1}^n \pi_{\beta_i}^{-1}(U_{\beta_i})$. Prove that $B = \prod_{\alpha \in \Lambda} B_\alpha$, where $B_{\beta_i} = U_{\beta_i}$ for each $i = 1, 2, \ldots, n$ and $B_\alpha = X_\alpha$ for all α different from $\beta_1, \beta_2, \ldots, \beta_n$.

4. For each $n \in \mathbb{N}$, let $X_n = \mathbb{R}$ and let $\mathcal{T}_n$ be the discrete topology on X_n. Let $\mathcal{T}$ be the product topology on $X = \prod_{n \in \mathbb{N}} X_n$. Is $\mathcal{T}$ the discrete topology on X? Either prove that it is or show by example that it is not.

5. For each $n \in \mathbb{N}$, let $X_n = \mathbb{R}$ and let $\mathcal{T}_n$ be the lower limit topology on X_n. Let $\mathcal{T}$ be the product topology on $X = \prod_{n \in \mathbb{N}} X_n$. Prove that each of the following sets is a closed subset of $(X, \mathcal{T})$.

    (a) $\{x\}$, where $x_n = 1$ for each $n \in \mathbb{N}$.

    (b) $\{x\}$, where $x_n = n$ for each $n \in \mathbb{N}$.

6. For each $n \in \mathbb{N}$, let $X_n = \mathbb{R}$ and let $\mathcal{T}_n$ be the lower limit topology on X_n. Let $\mathcal{T}$ be the product topology on $X = \prod_{n \in \mathbb{N}} X_n$. Which of the following are open subsets of $(X, \mathcal{T})$? If a set is not open, explain why it is not.

    (a) $\prod_{n \in \mathbb{N}} U_n$, where $U_n = [0, 1)$ for each $n \in \mathbb{N}$.

    (b) $\prod_{n \in \mathbb{N}} U_n$, where $U_n = [0, 1)$ if n is even and $U_n = \mathbb{R}$ if n is odd.

    (c) $\prod_{n \in \mathbb{N}} U_n$, where $U_7 = [0, 1)$ and $U_n = \mathbb{R}$ if $n \neq 7$.

7. Prove Theorem 2.29.

8. Let $\{(X_\alpha, \mathcal{T}_\alpha): \alpha \in \Lambda\}$ be an indexed family of topological spaces, let $X = \prod_{\alpha \in \Lambda} X_\alpha$, let $\mathcal{T}$ be a topology on X, and for each $\beta \in \Lambda$, let $\pi_\beta: X \to X_\beta$ be the projection map.

    (a) Prove that if $\mathcal{T}$ is the box topology on X, then π_β is open.

    (b) Give an example to show that if $\mathcal{T}$ is the product topology on X, then π_β need not be closed.

9. Prove Theorem 2.36.

10. Prove Theorem 2.38.

11. Let $\{(X_\alpha, \mathcal{T}_\alpha): \alpha \in \Lambda\}$ be an indexed family of topological spaces, let $\mathcal{T}$ denote the box topology on $X = \prod_{\alpha \in \Lambda} X_\alpha$.

    (a) For each $\alpha \in \Lambda$, let A_α be a closed subset of X_α. Prove or give a counterexample to the conjecture: $\prod_{\alpha \in \Lambda} A_\alpha$ is a closed subset of $(X, \mathcal{T})$.

    (b) For each $\alpha \in \Lambda$, let A_α be a subset of X_α. Prove or give a counterexample to the conjecture: with respect to $\mathcal{T}$, $\overline{\prod_{\alpha \in \Lambda} A_\alpha} = \prod_{\alpha \in \Lambda} \overline{A_\alpha}$.

12. Prove Theorem 2.39.

13. For each $n \in \mathbb{N}$, let $X_n = \mathbb{R}$ and let $\mathcal{T}_n$ denote the usual topology on X_n. Let $\mathcal{T}$ be the product topology on $X = \prod_{n \in \mathbb{N}} X_n$. For each $n \in \mathbb{N}$, let W_n denote the open interval $(0, 1)$, and let $W = \prod_{n \in \mathbb{N}} W_n$.

    (a) Show that W is not open in $(X, \mathcal{T})$.

    (b) Show that if x is the member of X such that $x_n = 0$ for each $n \in \mathbb{N}$, then x is a limit point of W.

    (c) Show that W is not closed in $(X, \mathcal{T})$.

14. Let Λ be an infinite set, and for each $\alpha \in \Lambda$, let $(X_\alpha, \mathcal{T}_\alpha)$ be a topological space. Either prove the following statement or give an example to show that it is not true: If $A_\alpha \subseteq X_\alpha$ for each $\alpha \in \Lambda$, then $\text{int}(\prod_{\alpha \in \Lambda} A_\alpha) = \prod_{\alpha \in \Lambda} \text{int}(A_\alpha)$.

15. For each $n \in \mathbb{N}$, let $X_n = \mathbb{R}$, and let $\mathcal{T}_n$ denote the usual topology on X_n. Let $\mathcal{T}$ be the product topology on $X = \prod_{n \in \mathbb{N}} X_n$. For each $n \in \mathbb{N}$, let A_n be the closed interval $[0, 1]$. Prove that $\prod_{n \in \mathbb{N}} A_n$ is a closed subset of $(X, \mathcal{T})$.

2.4 The Weak Topology and the Product Topology

This section is a continuation of the previous section. Theorem 2.33 is the motivation for the following definition.

Definition. Let X be a set, let $\{(X_\alpha, \mathcal{T}_\alpha): \alpha \in \Lambda\}$ be an indexed family of topological spaces, and for each $\alpha \in \Lambda$ let $f_\alpha: X \to X_\alpha$ be a function. The **weak topology on X induced by** $\{f_\alpha: \alpha \in \Lambda\}$ is the smallest topology on X for which each f_α is continuous. ∎

By Theorem 2.33 the product topology on $\prod_{\alpha \in \Lambda} X_\alpha$ is the weak topology induced by the collection $\{\pi_\alpha: X \to X_\alpha: \alpha \in \Lambda\}$ of projection maps. The proof of the following theorem is similar to the proof of Theorem 2.21.

THEOREM 2.40. Let $\{(X_\alpha, \mathcal{T}_\alpha): \alpha \in \Lambda\}$ be a collection of topological spaces, let X be a set, and for each $\alpha \in \Lambda$ let $f_\alpha: X \to X_\alpha$ be a function. Furthermore, let $\mathcal{T}$ be the weak topology on X induced by $\{f_\alpha: \alpha \in \Lambda\}$ and let $(Y, \mathcal{U})$ be a topological space. Then a function $f: Y \to X$ is continuous if and only if for each $\alpha \in \Lambda$, $f_\alpha \circ f: Y \to X_\alpha$ is continuous.

Proof. Suppose $f: Y \to X$ is continuous. By the definition of the weak topology, $f_\alpha: X \to X_\alpha$ is continuous for each $\alpha \in \Lambda$. Therefore by Theorem 1.56, the composite $f_\alpha \circ f: Y \to X_\alpha$ is continuous for each $\alpha \in \Lambda$.

Suppose $f_\alpha \circ f: Y \to X_\alpha$ is continuous for each $\alpha \in \Lambda$. Let $\mathcal{S} = \{f_\alpha^{-1}(U_\alpha): \alpha \in \Lambda$ and $U_\alpha \in \mathcal{T}_\alpha\}$. In Exercise 1, you are asked to show that $\mathcal{S}$ is a subbasis for $\mathcal{T}$. In order to show that $f: Y \to X$ is continuous, it is sufficient to show that the inverse image under f of each member of $\mathscr{S}$ is a member of $\mathscr{U}$. Let $f_\alpha^{-1}(U_\alpha) \in \mathscr{S}$. Then $f^{-1}(f_\alpha^{-1}(U_\alpha)) = (f_\alpha \circ f)^{-1}(U_\alpha)$. Since $f_\alpha \circ f$ is continuous, $f^{-1}(f_\alpha^{-1}(U_\alpha)) \in \mathscr{U}$. ∎

The following theorem is also a form of commutativity. It provides two ways of obtaining the same topology on a subset of a topological space.

THEOREM 2.41. Let $\{(X_\alpha, \mathscr{T}_\alpha): \alpha \in \Lambda\}$ be a collection of topological spaces, let X be a set, and for each $\alpha \in \Lambda$, let $f_\alpha: X \to X_\alpha$ be a function. Furthermore let $\mathscr{T}$ be the weak topology on X induced by $\{f_\alpha: \alpha \in \Lambda\}$ and let $A \subseteq X$. Then $\mathscr{T}_A$ is the weak topology on A induced by $\{f_\alpha|_A: \alpha \in \Lambda\}$.

Proof. Let $\mathscr{U}$ be the weak topology on A induced by $\{f_\alpha|_A: \alpha \in \Lambda\}$, and let $\mathscr{S} = \{(f_\alpha|_A)^{-1}(U_\alpha): \alpha \in \Lambda \text{ and } U_\alpha \in \mathscr{T}_\alpha\}$. Then $\mathscr{S}$ is a subbasis for $\mathscr{U}$. In order to show that $\mathscr{U} \subseteq \mathscr{T}_A$, it is sufficient to show that $\mathscr{S} \subseteq \mathscr{T}_A$. Let $(f_\alpha|_A)^{-1}(U_\alpha) \in \mathscr{S}$. Now $(f_\alpha|_A)^{-1}(U_\alpha) = f_\alpha^{-1}(U_\alpha) \cap A$. Since $\mathscr{T}$ is the weak topology on X induced by $\{f_\alpha: \alpha \in \Lambda\}$, $f_\alpha^{-1}(U_\alpha) \in \mathscr{T}$. Therefore $f_\alpha^{-1}(U_\alpha) \cap A \in \mathscr{T}_A$, and hence $\mathscr{S} \subseteq \mathscr{T}_A$.

Let $\mathscr{S}' = \{f_\alpha^{-1}(U_\alpha): \alpha \in \Lambda \text{ and } U_\alpha \in \mathscr{T}_\alpha\}$. Then $\mathscr{S}'$ is a subbasis for $\mathscr{T}$. In order to show that $\mathscr{T}_A \subseteq \mathscr{U}$, it is sufficient to show that if $f_\alpha^{-1}(U_\alpha) \in \mathscr{S}'$, then $f_\alpha^{-1}(U_\alpha) \cap A \in \mathscr{U}$. Let $f_\alpha^{-1}(U_\alpha) \in \mathscr{S}'$. Then $f_\alpha^{-1}(U_\alpha) \cap A = (f_\alpha|_A)^{-1}(U_\alpha)$. Since $(f_\alpha|_A)^{-1}(U_\alpha) \in \mathscr{U}$, the proof is complete. ∎

If $(X, \mathscr{T})$ is a topological space and we want to know whether $\mathscr{T}$ is the weak topology induced by a collection $\{f_\alpha: \alpha \in \Lambda\}$ of continuous functions, where for each $\alpha \in \Lambda$, f_α maps X into a topological space $(X_\alpha, \mathscr{T}_\alpha)$, there is often an alternate way of proving that the collection $\{f_\alpha^{-1}(U_\alpha): \alpha \in \Lambda \text{ and } U_\alpha \in \mathscr{T}_\alpha\}$ is a subbasis for $\mathscr{T}$. In order to describe when this alternate method applies, we define two concepts and then give a condition that is equivalent to $\{f_\alpha^{-1}(U_\alpha): \alpha \in \Lambda \text{ and } U_\alpha \in \mathscr{T}_\alpha\}$ being a basis for $\mathscr{T}$. Since every basis for a topology is also a subbasis for the topology, this accomplishes the stated purpose.

Definition. Let $(X, \mathscr{T})$ be a topological space, let $\{(X_\alpha, \mathscr{T}_\alpha): \alpha \in \Lambda\}$ be a collection of topological spaces, and for each $\alpha \in \Lambda$, let $f_\alpha: X \to X_\alpha$ be a function. The collection $\{f_\alpha: \alpha \in \Lambda\}$ **separates points in** X provided that for each pair x, y of distinct members of X, there is an $\alpha \in \Lambda$ such that $f_\alpha(x) \neq f_\alpha(y)$. The collection $\{f_\alpha: \alpha \in \Lambda\}$ **separates points from closed sets** if, whenever C is a closed subset of X and $p \notin C$, there exists $\beta \in \Lambda$ such that $f_\beta(p) \notin \overline{f_\beta(C)}$. ∎

THEOREM 2.42. Let $(X, \mathscr{T})$ be a topological space, let $\{(X_\alpha, \mathscr{T}_\alpha): \alpha \in \Lambda\}$ be a collection of topological spaces, and for each $\alpha \in \Lambda$, let $f_\alpha: X \to X_\alpha$ be a continuous function. Then the collection $\{f_\alpha^{-1}(U_\alpha): \alpha \in \Lambda \text{ and } U_\alpha \in \mathscr{T}_\alpha\}$ is a basis for $\mathscr{T}$ if and only if $\{f_\alpha: \alpha \in \Lambda\}$ separates points from closed sets.

Proof. Suppose the collection $\{f_\alpha^{-1}(U_\alpha): \alpha \in \Lambda \text{ and } U_\alpha \in \mathscr{T}_\alpha\}$ is a basis for $\mathscr{T}$. Let C be a closed subset of X and let $p \in X - C$. Since $X - C$ is open, there exist

$\beta \in \Lambda$ and $U_\beta \in \mathcal{T}_\beta$ such that $p \in f_\beta^{-1}(U_\beta) \subseteq X - C$. Thus if $x \in C$, then $x \notin f_\beta^{-1}(U_\beta)$. Therefore if $x \in C$, then $f_\beta(x) \notin U_\beta$. Hence $U_\beta \cap f_\beta(C) = \emptyset$. So U_β is a neighborhood of $f_\beta(p)$ that does not intersect $f_\beta(C)$. Hence $f_\beta(p) \notin \overline{f_\beta(C)}$ and therefore $\{f_\alpha: \alpha \in \Lambda\}$ separates points from closed sets.

Suppose $\{f_\alpha: \alpha \in \Lambda\}$ separates points from closed sets. Let $p \in X$ and let U be a neighborhood of p. Since $X - U$ is closed, there exists $\beta \in \Lambda$ such that $f_\beta(p) \notin \overline{f_\beta(X - U)}$. Let U_β be a neighborhood of $f_\beta(p)$ such that $U_\beta \cap \overline{f_\beta(X - U)} = \emptyset$. Then $p \in f_\beta^{-1}(U_\beta)$. We complete the proof that $\{f_\alpha^{-1}(U_\alpha): \alpha \in \Lambda$ and $U_\alpha \in \mathcal{T}_\alpha\}$ is a basis for $\mathcal{T}$ by showing that $f_\beta^{-1}(U_\beta) \subseteq U$. Let $x \in f_\beta^{-1}(U_\beta)$. Then $f_\beta(x) U_\beta$, and so $f_\beta(x) \notin \overline{f_\beta(X - U)}$. Therefore $x \notin f_\beta^{-1}(\overline{f_\beta(X - U)})$ and so $x \notin X - U$. Thus $x \in U$ and hence $f_\beta^{-1}(U_\beta) \subseteq U$. ∎

Corollary 2.43. Let $(X, \mathcal{T})$ be a topological space, let $\{(X_\alpha, \mathcal{T}_\alpha): \alpha \in \Lambda\}$ be a collection of topological spaces, and for each $\alpha \in \Lambda$, let $f_\alpha: X \to X_\alpha$ be a continuous function. If $\{f_\alpha: \alpha \in \Lambda\}$ separates points from closed sets, then $\mathcal{T}$ is the weak topology on X induced by $\{f_\alpha: \alpha \in \Lambda\}$.

Proof. See Exercise 2. ∎

The following theorem generalizes Corollary 2.22.

THEOREM 2.44. Let $\{(Y_\alpha, \mathcal{U}_\alpha): \alpha \in \Lambda\}$ be an indexed family of topological spaces, let $\mathcal{U}$ denote the product topology on $\prod_{\alpha \in \Lambda} Y_\alpha$, let $(X, \mathcal{T})$ be a topological space, and for each $\alpha \in \Lambda$ let $f_\alpha: X \to Y_\alpha$ be a function. Define a function $f: X \to \prod_{\alpha \in \Lambda} Y_\alpha$ as follows: For each $x \in X$ and each $\alpha \in \Lambda$, the value of $f(x)$ at α is $f_\alpha(x)$. Then f is continuous if and only if f_α is continuous for each $\alpha \in \Lambda$. ∎

The commutative diagram in Figure 2.7 may be helpful in the proof of this theorem.

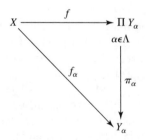

Figure 2.7

Proof. Since, for each $\alpha \in \Lambda$, $f_\alpha = \pi_\alpha \circ f$, this theorem follows immediately from Theorem 2.37. ∎

2.4 The Weak Topology and the Product Topology

Example 11 provides a counterexample to the analogue of Theorem 2.44 when $\prod_{\alpha \in \Lambda} Y_\alpha$ has the box topology. In order to see this, we define a function $f_\alpha: \mathbb{R} \to \mathbb{R}$ by $f_\alpha(x) = x$ for each $\alpha \in \Lambda$ and each $x \in \mathbb{R}$. Then for each $\alpha \in \Lambda$, f_α is continuous, but the function f defined as in Theorem 2.44 is not continuous.

The following theorem generalizes Theorem 2.23, and the proof, which we leave as Exercise 3, is essentially the same as the proof of Theorem 2.23.

THEOREM 2.45. Let $\{(X_\alpha, \mathcal{T}_\alpha): \alpha \in \Lambda\}$ be an indexed family of Hausdorff spaces. If $\mathcal{T}$ is the product topology or box topology on $X = \prod_{\alpha \in \Lambda} X_\alpha$, then $(X, \mathcal{T})$ is a Hausdorff space. ∎

From this point forward, we abandon the systematic study of the box topology, and, unless stated otherwise, our products will always have the product topology.

The following theorem generalizes Theorem 2.24 and characterizes those separable Hausdorff spaces whose products are separable.

THEOREM 2.46. Let $\{(X_\alpha, \mathcal{T}_\alpha): \alpha \in \Lambda\}$ be an indexed family of Hausdorff spaces such that each X_α has at least two points, and let $X = \prod_{\alpha \in \Lambda} X_\alpha$. Then $(X, \mathcal{T})$ is separable if and only if $|\Lambda| \leq c$ and $(X_\alpha, \mathcal{T}_\alpha)$ is separable for each $\alpha \in \Lambda$.

Proof. Suppose $(X, \mathcal{T})$ is separable and let $\beta \in \Lambda$. By Theorem 2.31, the projection map $\pi_\beta: X \to X_\beta$ is continuous. It is clear that π_β maps X onto X_β. Therefore by Theorem 1.59, $(X_\beta, \mathcal{T}_\beta)$ is separable. ∎

In the proof that $|\Lambda| \leq c$, we use the fact that for each $\alpha \in \Lambda$, $(X_\alpha, \mathcal{T}_\alpha)$ is a Hausdorff space with at least two points. For each $\alpha \in \Lambda$, let U_α and V_α be disjoint nonempty members of $\mathcal{T}_\alpha$. Let A be a countable dense subset of $(X, \mathcal{T})$ and for each $\alpha \in \Lambda$, let $A_\alpha = A \cap \pi_\alpha^{-1}(U_\alpha)$. Then for each $\alpha \in \Lambda$, $A_\alpha \neq \emptyset$. Let α and β be distinct members of Λ. Then there exists $x \in \pi_\alpha^{-1}(U_\alpha) \cap \pi_\beta^{-1}(V_\beta) \cap A$, so $x \in A_\alpha$ but $x \notin A_\beta$. Therefore $A_\alpha \neq A_\beta$. Hence the function $f: \Lambda \to \mathcal{P}(A)$ defined by $f(\alpha) = A_\alpha$ is one-to-one. Thus $|\Lambda| \leq |\mathcal{P}(A)| = 2^{\aleph_0} = c$.

Now suppose that $|\Lambda| \leq c$ and $(X_\alpha, \mathcal{T}_\alpha)$ is separable for each $\alpha \in \Lambda$. For each $\alpha \in \Lambda$, let $A_\alpha = \{a_{\alpha i}: i \in \mathbb{N}\}$ be a countable dense subset of X_α. Since $|\Lambda| \leq |I|$ and $|I| = c$, we may assume without loss of generality that $\Lambda \subseteq I$. For each finite sequence $I_1, I_2, \ldots, I_k$ of disjoint closed intervals with rational endpoints and each finite sequence $n_1, n_2, \ldots, n_k$ of natural numbers, define a member f (f depends on $I_1, I_2, \ldots, I_k$ and $n_1, n_2, \ldots, n_k$) of X as follows:

Let $\alpha \in \Lambda$. If there is an i ($i = 1, 2, \ldots, k$) such that $\alpha \in I_i$, then the value of f at α is $a_{\alpha n_i}$. If $\alpha \notin I_i$ for any $i = 1, 2, \ldots, k$, then the value of f at α is $a_{\alpha 1}$.

The subset A of X consisting of all such f is countable. It remains to show that A is dense in X. Let $U \in \mathcal{T}$. We show that $U \cap A \neq \emptyset$. Now U contains a set B of the form $\pi_{\alpha_1}^{-1}(U_{\alpha_1}) \cap \pi_{\alpha_2}^{-1}(U_{\alpha_2}) \cap \ldots \cap \pi_{\alpha_m}^{-1}(U_{\alpha_m})$, where $\alpha_1, \alpha_2, \ldots, \alpha_m$ are members of Λ and for each $i = 1, 2, \ldots, m$, $U_{\alpha_i} \in \mathcal{T}_{\alpha_i}$. Since for each $i = 1, 2, \ldots, m$, A_{α_i} is dense in X_{α_i}, U_{α_i} contains a member $a_{\alpha_i n_i}$ of A_{α_i}. Furthermore there are disjoint closed intervals with rational endpoints $I_1, I_2, \ldots, I_m$ such that for each $i = 1, 2, \ldots, m$, $\alpha_i \in I_i$. The

90   Chapter 2   ■   New Spaces from Old Ones

member f of X defined in terms of $I_1, I_2, \ldots, I_m$ and $n_1, n_2, \ldots, n_m$ belongs to B because for each $i = 1, 2, \ldots, m$, the value of f at α_i is $a_{\alpha_i n_i}$. Therefore $B \cap A \neq \emptyset$ and hence A is dense in X.

The following theorem generalizes Theorem 2.25.

THEOREM 2.47. Let $\{(X_\alpha, \mathcal{T}_\alpha): \alpha \in \Lambda\}$ be an indexed family of first countable spaces, and let $X = \prod_{\alpha \in \Lambda} X_\alpha$. Then $(X, \mathcal{T})$ is first countable if and only if $\mathcal{T}_\alpha$ is the trivial topology for all but a countable number of α.

Proof. Suppose there is a countable subset Γ of Λ such that $\mathcal{T}_\alpha$ is the trivial topology for all $\alpha \in \Lambda - \Gamma$. Let $x \in X$. For each $\alpha \in \Lambda$, let $\mathcal{B}_\alpha$ be a countable local basis at x_α. Then for each $\alpha \in \Lambda - \Gamma$, $\mathcal{B}_a = \{X_a\}$. Let $\mathcal{B}$ denote the family of all finite intersections of sets of the form $\pi_\alpha^{-1}(B_\alpha)$, where $\alpha \in \Lambda$ and $\mathcal{B}_\alpha \in \mathcal{B}_\alpha$. Then $\mathcal{B}$ is countable because $\pi_\alpha^{-1}(B_\alpha) = X$ for all $\alpha \in \Lambda - \Gamma$ and all $B_\alpha \in \mathcal{B}_\alpha$. (There is only one member of $\mathcal{B}_\alpha$ and it is X_α.) It follows from Theorem 2.30 that $\mathcal{B}$ is a local basis at x. Therefore $(X, \mathcal{T})$ is first countable.

The proof that if $(X, \mathcal{T})$ is first countable then $\mathcal{T}_\alpha$ is the trivial topology for all but a countable number of α is by contradiction. Suppose $(X, \mathcal{T})$ is first countable and there is an uncountable subset Γ of Λ such that for each $\alpha \in \Gamma$, $\mathcal{T}_\alpha$ is not the trivial topology. Let $x \in X$ such that for each $\alpha \in \Gamma$ there is a neighborhood U_α of x_α such that $U_\alpha \neq X_\alpha$. Since $(X, \mathcal{T})$ is first countable, there is a countable local basis $\mathcal{B}$ at x. Let $B \in \mathcal{B}$. By Theorem 2.30, for each $\alpha \in \Lambda$ there is a member U_α of $\mathcal{T}_\alpha$ such that $x \in \prod_{\alpha \in \Lambda} U_\alpha$, $\prod_{\alpha \in \Lambda} U_\alpha \subseteq B$, and $U_\alpha = X_\alpha$ for all but a finite number of members of Λ. Therefore $\pi_\alpha(B) = X_\alpha$ for all but a finite number of members α of Λ. Since Γ is uncountable, there is a member β of Γ such that $\pi_\beta(B) = X_\alpha$ for every $B \in \mathcal{B}$. But there is a neighborhood U_β of x_β such that $U_\beta \neq X_\beta$, and, since $\pi_\beta(B) = X_\beta$ for each $B \in \mathcal{B}$, no member of $\mathcal{B}$ is a subset of $\pi_\beta^{-1}(U_\beta)$. This is a contradiction. ∎

The following theorem generalizes Theorem 2.26.

THEOREM 2.48. Let $\{(X_\alpha, \mathcal{T}_\alpha): \alpha \in \Lambda\}$ be an indexed family of topological spaces, and let $X = \prod_{\alpha \in \Lambda} X_\alpha$. Then $(X, \mathcal{T})$ is second countable if and only if $(X_\alpha, \mathcal{T}_\alpha)$ is second countable for all $\alpha \in \Lambda$ and $\mathcal{T}_\alpha$ is the trivial topology for all but a countable number of α.

Proof. Suppose $(X_\alpha, \mathcal{T}_\alpha)$ is second countable for all $\alpha \in \Lambda$ and $\mathcal{T}_\alpha$ is the trivial topology for all but a countable number of members $\beta_1, \beta_2, \ldots$ of Λ. For each $i \in \mathbb{N}$, let $\mathcal{B}_{\beta i}$ be a countable base for $\mathcal{T}_{\beta i}$. Then by Theorem F.7, $\bigcup_{i=1}^{\infty} \mathcal{B}_{\beta_i}$ is a countable collection. Let $\mathcal{S} = \{\pi_{\beta_i}^{-1}(B_{\beta_i}): i \in \mathbb{N} \text{ and } B_{\beta_i} \in \mathcal{B}_{\beta_i}\}$, and let $\mathcal{B}$ denote the collection of all finite intersections of members of $\mathcal{S}$. Since each member of $\mathcal{B}$ is a finite intersection, each member of $\mathcal{B}$ corresponds to a finite subset of the natural numbers. Hence by Exercise 12 of Appendix F, $\mathcal{B}$ is countable.

We show that $\mathcal{B}$ is a base for the product topology. Let U be an open set in X and let $x \in U$. Then there exists a set of the form $\prod_{\alpha \in \Lambda} U_\alpha$, where $U_\alpha \in \mathcal{T}_\alpha$ and

$U_\alpha = X_\alpha$ for all but a finite number of members $\gamma_1, \gamma_2, ..., \gamma_n$ of Λ, such that $x \in \prod_{\alpha \in \Lambda} U_\alpha$ and $\prod_{\alpha \in \Lambda} U_\alpha \subseteq U$. For each $\alpha \in \Lambda$, $x_\alpha \in U_\alpha$. Thus for each $i = 1, 2, ..., n$ there exists $B_{\gamma_i} \in \mathcal{B}_{\gamma_i}$ such that $x_{\gamma_i} \in B_{\gamma_i} \subseteq U_{\gamma_i}$. Then $\bigcap_{i=1}^n \pi_{\gamma_i}^{-1}(B_{\gamma_i}) \in \mathcal{B}$ and $x \in \bigcap_{i=1}^n \pi_{\gamma_i}^{-1}(B_{\gamma_i}) \subseteq \bigcap_{i=1}^n \pi_{\gamma_i}^{-1}(U_{\gamma_i}) \subseteq U$. Hence $\mathcal{B}$ is a basis for the product topology.

Suppose $(X, \mathcal{T})$ is second countable and let $\mathcal{B}$ be a countable base for $\mathcal{T}$. By Theorem 2.34, π_α is open for each $\alpha \in \Lambda$. Therefore for each $\alpha \in \Lambda$ and each $B \in \mathcal{B}$, $\pi_\alpha(B)$ is a member of $\mathcal{T}_\alpha$. For each $\alpha \in \Lambda$, $\mathcal{B}_\alpha = \{\pi_\alpha(B) : B \in \mathcal{B}\}$ is countable. Let $U_\alpha \in \mathcal{T}_\alpha$ and let $x_\alpha \in U_\alpha$. Then $\pi_\alpha^{-1}(U_\alpha) \in \mathcal{T}$. Let $y \in \pi_\alpha^{-1}(U_\alpha)$ such that $y_\alpha = x_\alpha$. There exists $B \in \mathcal{B}$ such that $y \in B \subseteq \pi_\alpha^{-1}(U_\alpha)$. Now $\pi_\alpha(B) \in \mathcal{B}_\alpha$ and $x_\alpha = y_\alpha \in \pi_\alpha(B) \subseteq \pi_\alpha(\pi_\alpha^{-1}(U_\alpha) = U_\alpha)$ Therefore $\mathcal{B}_\alpha$ is a base for $\mathcal{T}_\alpha$ and hence $(X_\alpha, \mathcal{T}_\alpha)$ is second countable. By Theorem 1.15, $(X, \mathcal{T})$ is first countable. Therefore, by Theorem 2.47, $\mathcal{T}_\alpha$ is the trivial topology for all but a countable number of α. ∎

EXERCISES 2.4

1. Prove that the collection $\mathcal{S}$ defined in the proof of Theorem 2.40 is a subbasis for $\mathcal{T}$.

2. Prove Corollary 2.43.

3. Prove Theorem 2.45.

4. Let $\mathcal{U}$ be the usual topology on $\mathbb{R}$. Describe the weak topology $\mathcal{T}$ on $\mathbb{R}$ induced by each of the following families of functions.

    (a) The family of constant functions that map $\mathbb{R}$ into $(\mathbb{R}, \mathcal{U})$.

    (b) The family that consists only of the function $i: \mathbb{R} \to (\mathbb{R}, \mathcal{U})$ defined by $i(x) = x$.

5. Let $\mathcal{U}$ be the usual topology on $\mathbb{R}$ and let $\mathcal{T}$ be the weak topology on $\mathbb{R}$ induced by the family of all bounded real-valued continuous functions that map $(\mathbb{R}, \mathcal{U})$ into $(\mathbb{R}, \mathcal{U})$. Describe $\mathcal{T}$.

6. Let $\{(Y_\alpha, \mathcal{U}_\alpha) : \alpha \in \Lambda\}$ be a collection of topological spaces, let $(X, \mathcal{T})$ be a topological space, and let $\mathcal{A} = \{f : X \to Y_\alpha : \alpha \in \Lambda\}$ be a family of continuous functions. Prove that if $\mathcal{U}$ is the weak topology on X induced by $\mathcal{A}$, then $\mathcal{U} \subseteq \mathcal{T}$.

2.5 The Uniform Metric

We have seen (Exercise 4 of Section 1.6) that $\mathbb{R}^n$ with the usual topology is a complete metric space, and we have also seen (Example 6 of Chapter 1) that there is a subspace ℓ^2 of $\mathbb{R}^\omega$ that is a metric space. But can we define a metric on $\mathbb{R}^\omega$? The natural thing to attempt is to generalize either the Euclidean metric by defining $d(x, y) = (\sum_{i=1}^{\infty}(x_i - y_i)^2)^{1/2}$ or the square metric by defining $p(x, y) = \text{lub}\{|x_i - y_i|\}$ for all $x = (x_1, x_2, ...), y = (y_1, y_2, ...) \in \mathbb{R}^\omega$. These attempts are doomed to failure because $(\sum_{i=1}^{\infty}(x_i - y_i)^2)^{1/2}$ need not converge; that is why we defined a metric on the subspace ℓ^2 of $\mathbb{R}^\omega$ rather than on $\mathbb{R}^\omega$, and the set $\{|x_i - y_i|\}$ may not be bounded. We can, however, avoid the problem with the metric ρ by replacing the metric $|x - y|$ on $\mathbb{R}$ by the standard bounded metric $\bar{d}$ defined by $\bar{d}(x, y) = \min\{|x - y|, 1\}$, and define $\bar{\rho}(x, y) = \text{lub}\{\bar{d}(x_i, y_i)\}$ for all $x = (x_1, x_2, ...), y = (y_1, y_2, ...) \in \mathbb{R}^\omega$. We leave as Exercise 1 the proof that $\bar{\rho}$ is a metric on $\mathbb{R}^\omega$. As we shall see, however, $\bar{\rho}$ does not induce the product topology on $\mathbb{R}^\omega$. We have defined a metric on $\mathbb{R}^\omega$, but there is nothing special about $\mathbb{R}^\omega$. If Λ is any nonempty set, in a similar fashion we can define a metric on $\mathbb{R}^\Lambda$ (see Appendix D for the definition of X^Λ).

Definition. Let Λ be a nonempty set, and let $\bar{d}$ be the standard bounded metric on $\mathbb{R}$. Define a metric $\bar{\rho}$ on $\mathbb{R}^\Lambda$ by $\bar{\rho}(x, y) = \text{lub}\{\bar{d}(x_\alpha, y_\alpha): \alpha \in \Lambda\}$. By Exercise 1, $\bar{\rho}$ is a metric, and it is called the **uniform metric** on $\mathbb{R}^\Lambda$. The topology on $\mathbb{R}^\Lambda$ induced by $\bar{\rho}$ is called the **uniform topology** on $\mathbb{R}^\Lambda$. ∎

In Exercise 2, you are asked to show that if Λ is an infinite set, then the uniform topology on $\mathbb{R}^\Lambda$ is different from the product topology on $\mathbb{R}^\Lambda$.

THEOREM 2.49. If Λ is a nonempty set, the uniform topology on $\mathbb{R}^\Lambda$ is finer than the product topology.

Proof. Let $x \in \mathbb{R}^\Lambda$ and let U be a neighborhood of x with respect to the product topology. Then there exist $\alpha_1, \alpha_2, ..., \alpha_n$ in Λ and open sets $U_{\alpha_1}, U_{\alpha_2}, ..., U_{\alpha_n}$ such that $x \in \bigcap_{i=1}^{n} \pi_{\alpha_i}^{-1}(U_{\alpha_i}) \subseteq U$. For each $i = 1, 2, ..., n$, let $\varepsilon_i > 0$ such that $B_{\bar{d}}(x_{\alpha_i}, \varepsilon_{\alpha_i}) \subseteq U_{\alpha_i}$, let $\varepsilon = \min\{\varepsilon_1, \varepsilon_2, ..., \varepsilon_n\}$, and let $y \in B_{\bar{\rho}}(x, \varepsilon)$. Since $\bar{\rho}(x, y) < \varepsilon$, $\bar{d}(x_{\alpha_i}, y_{\alpha_i}) < \varepsilon_i$ for each $i = 1, 2, ..., n$. Therefore $y \in \bigcap_{i=1}^{n} \pi_{\alpha_i}^{-1}(U_{\alpha_i})$, and hence $B_{\bar{\rho}}(x, \varepsilon) \subseteq U$. So U is a neighborhood of x with respect to the uniform topology on $\mathbb{R}^\Lambda$. ∎

By modifying the uniform metric, we can obtain a metric on $\mathbb{R}^\omega$ that induces the product topology.

THEOREM 2.50. Let $\bar{d}$ be the standard bounded metric on $\mathbb{R}$, and define $D(x, y) = \text{lub}\{\bar{d}(x_i, y_i)/i\}$ for all $x, y \in \mathbb{R}^\omega$. Then D is a metric on $\mathbb{R}^\omega$ that induces the product topology.

Proof. It is clear that $D(x, y) = 0$ if and only if $x = y$ and that $D(x, y) = D(y, x)$ for all $x, y \in \mathbb{R}^\omega$. Let $x, y, z \in \mathbb{R}^\omega$. Note that for each $i \in \mathbb{N}$,

$$\frac{\bar{d}(x_i, z_i)}{i} \leq \frac{\bar{d}(x_i, y_i)}{i} + \frac{\bar{d}(y_i, z_i)}{i} \leq D(x, y) + D(y, z).$$

Thus $D(x, y) = \text{lub}\{\bar{d}(x_i, z_i)/i \leq D(x, y) + D(y, z)$. Therefore D is a metric on $\mathbb{R}^\infty$.

Let $x \in \mathbb{R}^\omega$, and let U be a neighborhood of x with respect to the topology induced by D. Let $\varepsilon > 0$ such that $B_D(x, \varepsilon) \subseteq U$, and let $N \in \mathbb{N}$ such that $1/N < \varepsilon$. Let $V = \prod_{i \in \mathbb{N}} V_i$, where $V_i = (x_i - \varepsilon, x_i + \varepsilon)$ for each $i = 1, 2, \ldots, N$ and $V_i = \mathbb{R}$ for all $i > N$. Then V is a neighborhood of x with respect to the product topology. We claim that $V \subseteq B_D(x, \varepsilon)$. Let $y \in V$. By the definition of the standard bounded metric $\bar{d}(x_i, y_i) \leq 1$, for each $i \in \mathbb{N}$. Therefore for each $i \geq N$, $\bar{d}(x_i, y_i)/i \leq 1/N$. Thus

$$D(x, y) \leq \max\left\{\frac{\bar{d}(x_1, y_1)}{1}, \frac{\bar{d}(x_2, y_2)}{2}, \ldots, \frac{\bar{d}(x_N, y_N)}{N}, \frac{1}{N}\right\}.$$

Since $\bar{d}(x_i, y_i) < \varepsilon$ for each $i = 1, 2, \ldots, N$ and $1/N < \varepsilon$, $D(x, y) < \varepsilon$. Therefore $y \in B_D(x, \varepsilon)$, and hence $V \subseteq B_D(x, \varepsilon) \subseteq U$. Thus U is a neighborhood of x with respect to the product topology.

Now let $x \in \mathbb{R}^\omega$ and let U be a neighborhood of x with respect to the product topology. Then there exist $\alpha_1, \alpha_2, \ldots, \alpha_n$ in $\mathbb{N}$ and open sets $U_{\alpha_1}, U_{\alpha_2}, \ldots, U_{\alpha_n}$ such that $x \in \bigcap_{i=1}^n \pi_{\alpha_i}^{-1}(U_{\alpha_i}) \subseteq U$. For each $i = 1, 2, \ldots, n$, let $\varepsilon_{\alpha_i} > 0$ such that $\varepsilon_{\alpha_i} \leq 1$ and $(x_{\alpha_i} - \varepsilon_{\alpha_i}, x_{\alpha_i} + \varepsilon_{\alpha_i}) \subseteq U_{\alpha_i}$. Let $\varepsilon = \min\{\varepsilon_{\alpha_1}/1, \varepsilon_{\alpha_2}/2, \ldots, \varepsilon_{\alpha_n}/n\}$. We claim that $B_D(x, \varepsilon) \subseteq U$. Let $y \in B_D(x, \varepsilon)$. Then for each $i \in \mathbb{N}$, $\bar{d}(x_i, y_i)/i \leq D(x, y) < \varepsilon$. Also for each $i = 1, 2, \ldots, n$, $\varepsilon \leq \varepsilon_{\alpha_i}/i$, and hence $\bar{d}(x_{\alpha_i}, y_{\alpha_i}) \leq \varepsilon_{\alpha_i} \leq 1$. It follows that for each $i = 1, 2, \ldots, n$, $|x_{\alpha_i} - y_{\alpha_i}| < \varepsilon_{\alpha_i}$. Therefore $y \in \bigcap_{i=1}^n \pi_{\alpha_i}^{-1}(U_{\alpha_i}) \subseteq U$. Hence U is a neighborhood of x with respect to the topology induced by D. ∎

EXAMPLE 12. If $\mathcal{T}$ is the box topology on $\mathbb{R}^\omega$, then $(\mathbb{R}^\omega, \mathcal{T})$ is not metrizable.

Analysis. Since every metric space is first countable (Theorem 1.14) and, in a first countable space, a point belongs to the closure of a set if and only if there is a sequence of points in the set that converges to the point (Theorem 1.41), it is sufficient to define a subset A of $\mathbb{R}^\omega$ such that there exists $a \in \bar{A}$ with the property that no sequence in A converges to a.

Let $A = \{(x_1, x_2, \ldots): x_i > 0 \text{ for all } i \in \mathbb{N}\}$, and let $a = (0, 0, \ldots, 0, \ldots)$. Let $U = (a_1, b_1) \times (a_2, b_2) \times \cdots$ be a basic open set containing a. Then the point $(\frac{1}{2}b_1, \frac{1}{2}b_2, \ldots)$ belongs to $U \cap A$, and hence $a \in \bar{A}$.

We claim that no sequence in A converges to a. Let $\langle a_n \rangle$ be a sequence of points in A. Then, for each $n \in \mathbb{N}$, $a_n = (a_{1n}, a_{2n}, \ldots, a_{in}, \ldots)$, where $a_{in} > 0$ for each $i, n \in \mathbb{N}$. Then $V = (-a_{11}, a_{11}) \times (-a_{22}, a_{22}) \times \cdots$ is a basis open set containing a that does not contain any member of the sequence $\langle a_n \rangle$. Therefore the sequence $\langle a_n \rangle$ does not converges to a. ∎

EXAMPLE 13. Let Λ be an uncountable set, and let $\mathcal{T}$ be the product topology on $\mathbb{R}^\Lambda$. Then $(\mathbb{R}^\Lambda, \mathcal{T})$ is not metrizable.

Analysis. The proof uses the same theorems that were used in the previous example. Let $A = \{x \in \mathbb{R}^\Lambda$: there is a finite subset Γ of Λ such that $x_\alpha = 0$ for all $\alpha \in \Gamma$ and

94 Chapter 2 ■ New Spaces from Old Ones

$x_\alpha = 1$ for all $\alpha \in \Lambda - \Gamma\}$, and let a be the member of $\mathbb{R}^\Lambda$ with the property that $a_\alpha = 0$ for all α in Λ.

We claim that a belongs to $\overline{A}$. Let U be a basic open set such that $a \in U$. Then there exists a finite subset $\{\alpha_1, \alpha_2, ..., \alpha_n\}$ of Λ and a collection $U_{\alpha_1}, U_{\alpha_2}, ..., U_{\alpha_n}$ of open subsets of $\mathbb{R}$ such that $U = \bigcap_{i=1}^n \pi_{\alpha_i}^{-1}(U_{\alpha_i})$. Thus the point x defined by $x_{\alpha_i} = 0$ for each $i = 1, 2, ..., n$ and $x_\alpha = 1$ for all $\alpha \neq \alpha_i$ for any $i = 1, 2, ..., n$ belongs to $U \cap A$. Therefore $a \in \overline{A}$.

We claim that no sequence of points in A converges to a. Let $\langle a_n \rangle$ be a sequence of points in A. For each $n \in \mathbb{N}$, let $\Lambda_n = \{\alpha \in \Lambda : (a_n)_\alpha = 0\}$. Then $\bigcup_{n \in \mathbb{N}} \Lambda_n$ is the countable union of finite sets and therefore countable. Thus there exists $\beta \in \Lambda - \bigcup_{n \in \mathbb{N}} \Lambda_n$. Therefore for each $n \in \mathbb{N}$, $(a_n)_\beta = 1$. Hence the open set $\pi_\beta^{-1}((-1, 1))$ in $\mathbb{R}^\Lambda$ is a neighborhood of a that does not contain any member of the sequence $\langle a_n \rangle$. Thus the sequence $\langle a_n \rangle$ does not converge to a. ■

THEOREM 2.51. Let $\mathcal{T}$ be the product topology on $\mathbb{R}^\omega$, and let D be the metric, defined in Theorem 2.50, which induces $\mathcal{T}$. Then $(\mathbb{R}^\omega, D)$ is complete.

Proof. Let $\langle x_n \rangle$ be a Cauchy sequence in $(\mathbb{R}^\omega, D)$, and let $i \in \mathbb{N}$. Since $\overline{d}(\pi_i(x), \pi_i(y)) \leq iD(x, y)$ for each $x, y \in \mathbb{R}^\omega$, the sequence $\langle \pi_i(x_n) \rangle$ is a Cauchy sequence in $(\mathbb{R}, \overline{d})$. Since $(\mathbb{R}, \overline{d})$ is complete, this sequence converges to a point a_i in $\mathbb{R}$. Let a be the member of $\mathbb{R}^\omega$ whose ith coordinate is a_i. We claim that $\langle x_n \rangle$ converges to a. Let $\bigcap_{i=1}^m \pi_{j_i}^{-1}(U_{j_i})$ be a basic open set containing a. For each $i = 1, 2, ..., m$, let $N_{j_i} \in \mathbb{N}$ such that $\pi_{j_i}(x_n) \in U_{j_i}$ for all $n \geq N_{j_i}$. Let $N = \max\{N_{j_1}, N_{j_2}, ..., N_{j_m}\}$. Then $x_n \in \bigcap_{i=1}^n \pi_{j_i}^{-1}(U_{j_i})$ for all $n \geq N$. Therefore $\langle x_n \rangle$ converges to a. ■

Now we show that if $\overline{\rho}$ is the uniform metric on $\mathbb{R}^\Lambda$, then $(\mathbb{R}^\Lambda, \overline{\rho})$ is complete. First we generalize the definition of the uniform metric. In this definition and throughout the remainder of this section, it is convenient to use functional notation for the members of X^Λ rather than "tuple" notation.

Definition. Let (X, d) be a metric space, and let $\overline{d}$ be the standard bounded metric on X corresponding to d ($\overline{d}(x, y) = \min\{d(x, y), 1\}$). Let Λ be a nonempty set, and define a metric $\overline{\rho}$ on X^Λ by $\overline{\rho}(f, g) = \text{lub}\{\overline{d}(f(\alpha), g(\alpha)) : \alpha \in \Lambda\}$ for all $f, g \in X^\Lambda$. By Exercise 1, $\overline{\rho}$ is a metric, and it is called the **uniform metric** on X^Λ. ■

THEOREM 2.52. If (X, d) is a complete metric space, Λ is a nonempty set, and $\overline{\rho}$ is the uniform metric on X^Λ corresponding to d, then the metric space $(X^\Lambda, \overline{\rho})$ is complete.

Proof. Let (X, d) be a complete metric space. Then, by Theorem 1.48, $(X, \overline{d})$ is complete ($\overline{d}$ is the standard bounded metric corresponding to d). Let $\langle f_n \rangle$ be a Cauchy sequence in $(X, \overline{\rho})$. Let $\alpha \in \Lambda$. Since, for each $m, n \in \mathbb{N}$, $\overline{d}(f_m(\alpha), f_n(\alpha)) \leq \overline{\rho}(f_m, f_n)$, the sequence $\langle f_n(\alpha) \rangle$ is a Cauchy sequence in $(X, \overline{d})$. Hence this sequence converges to a point $x_\alpha \in X$. Define $f: \Lambda \to X$ by $f(\alpha) = x_\alpha$. We claim that the sequence $\langle f_n \rangle$ converges to f in $(X^\Lambda, \bar{\rho})$.

Let $\varepsilon > 0$. Then there exists $N \in \mathbb{N}$ such that $\bar{\rho}(f_m, f_n) < \varepsilon/2$ whenever $m, n \geq N$. Therefore for each $\alpha \in \Lambda$, $\bar{d}(f_m(\alpha), f_n(\alpha)) < \varepsilon/2$ whenever $m, n \geq N$. Therefore $\bar{d}(f_n(\alpha), f(\alpha)) \leq \varepsilon/2$ for all $\alpha \in \Lambda$ and all $n \geq N$. Hence $\bar{\rho}(f_m, f) < \varepsilon$ whenever $m \geq N$. ∎

Now suppose $(X, \mathcal{T})$ is a topological space and (Y, d) is a metric space. We consider the subset $C(X, Y)$ of Y^X consisting of all continuous functions $f: X \to Y$. We already know that if (Y, d) is complete and $\bar{\rho}$ is the uniform metric corresponding to d, then $(Y^X, \bar{\rho})$ is complete. But what about the subspace $(C(X, Y), \bar{\rho})$ (with, of courses, $\bar{\rho}$ restricted to $C(X, Y)$)? It turns out that this subspace is also complete.

THEOREM 2.53. Let $(X, \mathcal{T})$ be a topological space, let (Y, d) be a metric space, and let $\bar{\rho}$ be the uniform metric on Y^X corresponding to d. Then $C(X, Y)$ is a closed subset of $(Y^X, \bar{\rho})$.

Proof. First we show that if a sequence $\langle f_n \rangle$ of members of Y^X converges to $f \in Y^X$ relative to $\bar{\rho}$, then $\langle f_n \rangle$ converges uniformly to f relative to the standard bounded metric $\bar{d}$ that corresponds to d.

Suppose $\langle f_n \rangle$ converges to f relative to $\bar{\rho}$, and let $\varepsilon > 0$. Then there exists $N \in \mathbb{N}$ such that if $n \geq N$, $\bar{\rho}(f_n, f) < \varepsilon$. Therefore for all $n \geq N$ and all $x \in X$, $\bar{d}(f_n(x), f(x)) \leq \bar{\rho}(f_n, f) < \varepsilon$. Hence $\langle f_n \rangle$ converges uniformly to f.

Now we show that $C(X, Y)$ is a closed subset of $(Y^X, \bar{\rho})$. Suppose $f \in Y^X$ is a limit point of $C(X, Y)$. Then there exists a sequence $\langle f_n \rangle$ of members of $C(X, Y)$ that converges to f relative to $\bar{\rho}$. Then $\langle f_n \rangle$ converges uniformly to f relative to $\bar{d}$, and so by Theorem 1.58, f is continuous. Therefore $f \in C(X, Y)$. ∎

Corollary 2.54. Let $(X, \mathcal{T})$ be a topological space, let (Y, d) be a complete metric space, and let $\bar{\rho}$ be the uniform metric on Y^X corresponding to d. Then $(C(X, Y), \bar{\rho})$ is complete.

Proof. By Theorem 2.52, $(Y^X, \bar{\rho})$ is complete. Theorem 2.53, $C(X, Y)$ is a closed subset of Y^X. Therefore, by Theorem 1.47, $(C(X, Y), \bar{\rho})$ is complete. ∎

Definition. Let X be a set, let (Y, d) be a metric space, and suppose $\mathcal{F}$ is a subset of Y^X such that if $f, g \in \mathcal{F}$, then $\{d(f(x), g(x)): x \in X\}$ is bounded. Exercise 3 proves that the function ρ defined by $\rho(f, g) = \text{lub}\{d(f(x), g(x)): x \in X\}$ is a metric on $\mathcal{F}$. It is called the **sup metric**. ∎

How does the sup metric ρ compare with the uniform metric $\bar{\rho}$? We leave as Exercise 4 the proof that if $f, g \in \mathcal{F}$, then $\bar{\rho}(f, g) = \min\{\rho(f, g), 1\}$. Thus on $\mathcal{F}$, $\bar{\rho}$ is the standard bounded metric that corresponds to ρ. Therefore ρ and $\bar{\rho}$ induce the same topology on $\mathcal{F}$, and $(\mathcal{F}, \rho)$ is complete if and only if $(\mathcal{F}, \bar{\rho})$ is complete.

EXERCISES 2.5

1. Let (X, d) be a metric space, and let $\bar{d}$ be the standard bounded metric on X that corresponds to d. Let Λ be a nonempty set, and define a metric $\bar{\rho}$ on X^Λ by $\bar{\rho}(f, g) = \text{lub}\{\bar{d}(f(\alpha), g(\alpha)): \alpha \in \Lambda\}$ for all $f, g \in X^\Lambda$. Prove that $\bar{\rho}$ is a metric.

2. Let Λ be an infinite set. Prove that the uniform topology on $\mathbb{R}^\Lambda$ is different from the product topology on $\mathbb{R}^\Lambda$.

3. Let X be a set, let (Y, d) be a metric space, and suppose $\mathcal{F}$ is a subset of Y^X such that if $f, g \in \mathcal{F}$, then $\{d(f(x), g(x)): x \in X\}$ is bounded. Prove that the function ρ defined by $\rho(f, g) = \text{lub}\{d(f(x), g(x)): x \in X\}$ is a metric on $\mathcal{F}$.

4. Let X be a set, let (Y, d) be a metric space, and suppose $\mathcal{F}$ is a subset of Y^X such that if $f, g \in \mathcal{F}$, then $\{d(f(x), g(x)): x \in X\}$ is bounded. Let ρ be the sup metric on $\mathcal{F}$ and let $\bar{\rho}$ be the uniform metric on Y^X. Prove that if $f, g \in \mathcal{F}$, then $\bar{\rho}(f, g) \min\{\rho(f, g), 1\}$.

5. For each $n \in \mathbb{N}$, let $f_n: [0, 1] \to \mathbb{R}$ be the function defined by $f_n(x) = x^n$.

    (a) Show that for each $x \in [0, 1]$, the sequence $\langle f_n(x) \rangle$ converges.

    (b) Show that the sequence $\langle f_n \rangle$ does not converge uniformly.

6. Let X be a set, let $\bar{\rho}$ be the uniform metric on $\mathbb{R}^X$, let $f: X \to \mathbb{R}$ be a function, and, for each $n \in \mathbb{N}$, let $f_n: X \to \mathbb{R}$ be a function. Show that the sequence $\langle f_n \rangle$ converges uniformly to f if and only if $\langle f_n \rangle$ converges to f in $(\mathbb{R}^X, \bar{\rho})$.

7. Let $(X, \mathcal{T})$ be a topological space, let $\langle x_n \rangle$ be a sequence of members of X that converges to $x \in X$, let (Y, d) be a metric space, let $f: X \to Y$ be a function, and, for each $n \in \mathbb{N}$, let $f_n: X \to \mathbb{R}$ be a continuous function. Prove that if the sequence $\langle f_n \rangle$ converges uniformly to f, then $\langle f_n(x_n) \rangle$ converges to $f(x)$.

8. Let $(X, \mathcal{T})$ be a metric space, let $\mathcal{B}(X, \mathbb{R}) = \{f: X \to \mathbb{R}: f \text{ is bounded}\}$, and let ρ be the sup metric on $\mathcal{B}(X, \mathbb{R})$. Prove that $(\mathcal{B}(X, \mathbb{R}), \rho)$ is complete.

9. Let $X = x \in \mathbb{R}^\omega$: there exists $N \in \mathbb{N}$ such that $x_i = 0$ for all $i \geq N\}$, and let $\bar{\rho}$ be the uniform metric on $\mathbb{R}^\omega$. Is $(X, \bar{\rho})$ complete?

2.6 Quotient Spaces

Unlike many concepts in topology, quotient spaces did not originate from analysis. Instead, their origin is the cut-and-paste techniques of geometry. For example, a circle can be obtained from a closed interval by pasting the end points together. The **torus** can be constructed from a rectangle by pasting the edges together as shown in Figure 2.8.

2.6 Quotient Spaces 97

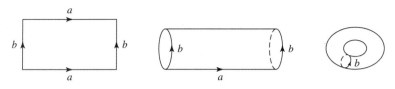

Figure 2.8

Also, the **Möbius strip** (named in honor of the German mathematician A.F. Möbius (1790–1868)) can be constructed by "twisting" a rectangle and pasting the edges together as shown in Figure 2.9.

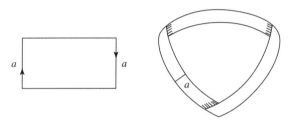

Figure 2.9

The Möbius strip has the remarkable property that it has only one side and one edge (see Exercise 1). This remark is intuitive and depends on how we view the strip in $\mathbb{R}^3$. However if you do Exercise 1 you can see what we mean.

The **Klein bottle** (named in honor of Felix Klein (1849–1925)) can also be constructed from a rectangle. Two edges are pasted together to obtain a cyclinder, but the rectangle is twisted before the other two sides are pasted together. The manner in which the edges are pasted in illustrated in Figure 2.10.

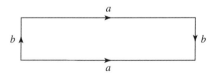

Figure 2.10

The **2-sphere**, $\{(x, y, z) \in \mathbb{R}^3 : x^2 + y^2 + z^2 = 1\}$, can be constructed from a disk, $\{(x, y) \in \mathbb{R}^2 : x^2 + y^2 \leq 1\}$, by collapsing its boundary, $\{(x, y) \in \mathbb{R}^2 : x^2 + y^2 = 1\}$, to a single point (that is, the 2-sphere is homeomorphic to the resulting space), and the projective plane can be constructed from a disk by pasting together diametrically opposite points x and –x as shown in Figure 2.11 (a). The projective plane can also be constructed from a rectangle by pasting together the edges as shown in Figure 2.11 (b).

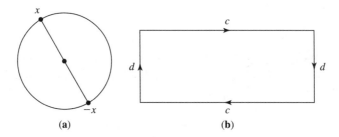

Figure 2.11

It is interesting to note that neither the Klein bottle nor the projective plane can be embedded in $\mathbb{R}^3$.

The construction of quotient spaces developed from the idea of identifying (that is, glueing together) parts of a figure. This basic construction was used by Möbius in 1858 and by Klein in 1882. (Abstract) Quotient spaces appeared in the work of R.L. Moore in 1925 and Paul S. Alexandroff (1896–1982) in 1927. The general concepts of quotient spaces and quotient maps were introduced in 1932 by R.W. Baer and F. Levi.

We can begin with any set X and describe two geometric constructions. The **cone over** X, denoted by CX, is obtained from $X \times [0,1]$ by collapsing $X \times \{1\}$ to a single point (see Figure 2.12). The **suspension over X**, denoted by SX, is obtained from $X \times [-1, 1]$ by collapsing $X \times \{-1\}$ to a single point and by collapsing $X \times \{1\}$ to a single point (see Figure 2.12).

We give three distinct but equivalent ways of formally defining a quotient space.

Definition. Let $(X, \mathcal{T})$ be a topological space, let Y be a set, and let f be a function that maps X onto Y. Then $\mathcal{U} = \{U \in \mathcal{P}(Y) : f^{-1}(U) \in \mathcal{T}\}$ is a topology on Y (see Theorem 1.62), and it is called the **quotient topology on Y induced by** f. The topological space $(Y, \mathcal{U})$ is called a **quotient space** of X and the function f is called a **quotient map**. ■

By Theorem 1.62, the quotient map f is not only continuous, but $\mathcal{U}$ is the finest topology on Y for which f is continuous. It is easy to see that if $(Y, \mathcal{U})$ is a quotient space of $(X, \mathcal{T})$ with quotient map f, then a subset C of Y is closed in Y if and only if $f^{-1}(C)$ is closed in X.

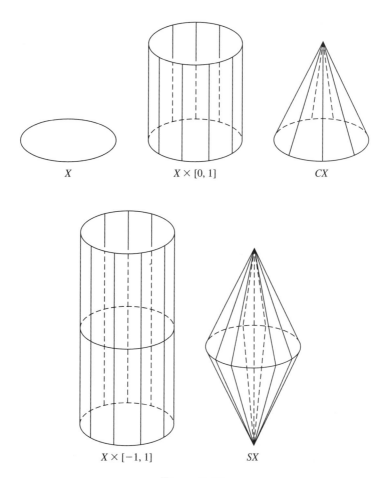

Figure 2.12

Suppose $(X, \mathcal{T})$ and $(Y, \mathcal{V})$ are topological spaces and f is a function that maps X onto Y. The following theorem gives conditions on f that make $\mathcal{V}$ equal to the quotient topology $\mathcal{U}$ on Y induced by f.

THEOREM 2.55. Let $(X, \mathcal{T})$ and $(Y, \mathcal{V})$ be topological spaces, let f be a continuous function that maps X onto Y, and let $\mathcal{U}$ be the quotient topology on Y induced by f. If f is open or closed, then $\mathcal{U} = \mathcal{V}$.

Proof. We prove that if f is open, then $\mathcal{U} = \mathcal{V}$. The proof that if f is closed, then $\mathcal{U} = \mathcal{V}$ is left as Exercise 6.

Suppose f is open. Since $\mathcal{U}$ is the finest topology for which f is continuous (Theorem 1.62), $\mathcal{V} \subseteq \mathcal{U}$. Let $U \in \mathcal{U}$. Then $f^{-1}(U) \in \mathcal{T}$ by the definition of $\mathcal{U}$. Since f is open, $U = f(f^{-1}(U)) \in \mathcal{V}$. Hence $\mathcal{U} \subseteq \mathcal{V}$. ∎

Theorem 2.55 tells us that if $f: (X, \mathcal{T}) \to (Y, \mathcal{V})$ is an open (closed) continuous surjection, then f is a quotient map. However, as the following examples shows, there are quotient maps that are neither open nor closed.

EXAMPLE 14. Let $X = \{(x, y) \in \mathbb{R}^2 : x \geq 0 \text{ or } y = 0\}$, and let $\mathcal{T}$ be the subspace topology on X induced by the usual topology on $\mathbb{R}^2$. Suppose $\mathbb{R}$ has the usual topology, and define $f : X \to \mathbb{R}$ by $f((x, y)) = x$ for all $(x, y) \in X$. Then f is a quotient map, but it is neither open nor closed.

Analysis. See Figure 2.13 and Exercise 5. ∎

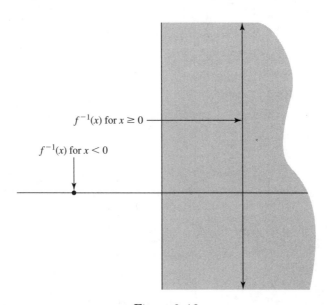

Figure 2.13

THEOREM 2.56. The composition of two quotient maps is a quotient map.

Proof. See Exercise 7. ∎

The following theorem is a fundamental result about quotient spaces.

THEOREM 2.57. Let $(X, \mathcal{T})$ be a topological space, let Y be a set, let f be a function that maps X onto Y, let $\mathcal{U}$ be the quotient topology on Y induced by f, and let $(Z, \mathcal{V})$ be a topological space. Then a function $g: Y \to Z$ is continuous if and only if $g \circ f : X \to Z$ is continuous.

2.6 Quotient Spaces

Proof. Suppose g is continuous. Since the composition of continuous functions is continuous (Theorem 1.56), $g \circ f$ is continuous.

Suppose $g \circ f$ is continuous. Let $V \in \mathcal{V}$. Then $(g \circ f)^{-1}(V) \in \mathcal{T}$, and $(g \circ f)^{-1}(V) = f^{-1}(g^{-1}(V))$. Thus $g^{-1}(V) \in \mathcal{U}$ by the definition of the quotient topology. Therefore g is continuous. ∎

THEOREM 2.58. Let $(X, \mathcal{T})$ and $(Y, \mathcal{U})$ be topological spaces and let $p: X \to Y$ be a continuous surjection. Then p is a quotient map if and only if, for each topological space $(Z, \mathcal{V})$ and each function $g: Y \to Z$, the continuity of $g \circ p$ implies that of g.

Proof. If p is a quotient map, the desired result follows from Theorem 2.57.

Suppose the condition holds. Let $\mathcal{T}'$ be the quotient topology on Y induced by p. Let p' denote p considered as a function from $(X, \mathcal{T})$ into $(Y, \mathcal{T}')$, and let $i: (Y, \mathcal{U}) \to (Y, \mathcal{T}')$ be the identity map. Since $i \circ p = p'$ is continuous, i is continuous. Since $i^{-1} \circ p' = p$ is continuous and p' is a quotient map, the part of the theorem that we have already established shows that i^{-1} is continuous. Thus i is a homeomorphism, so p is a quotient map. ∎

THEOREM 2.59. Let $(X, \mathcal{T})$, $(Y, \mathcal{U})$, and $(Z, \mathcal{V})$ be topological spaces, let $p: X \to Y$ be a quotient map, and let $h: X \to Z$ be continuous. Furthermore, assume that $h \circ p^{-1}$ is single-valued; that is, assume that for each $y \in Y$, h is constant on $p^{-1}(y)$. Then the following conditions hold:

(a) $(h \circ p^{-1}) \circ p = h$, and $h \circ p^{-1}$ is continuous.

(b) $h \circ p^{-1}$ is an open (closed) map if and only if $h(U)$ is open (closed) whenever U is an open (closed) set satisfying $U = p^{-1}(p(U))$.

Proof.

(a)

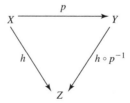

Let $x \in X$. Then $x \in p^{-1}(p(x))$. Since h is constant on $p^{-1}(p(x))$, $h(x) = h(p^{-1}(p(x)))$. But $h(p^{-1}(p(x))) = [(h \circ p^{-1}) \circ p](x)$. Therefore $h = (h \circ p^{-1}) \circ p$. Since h is continuous, the continuity of $h \circ p^{-1}$ follows from Theorem 2.58.

(b) We prove the open part and leave the closed part as an exercise. Suppose $h \circ p^{-1}$ is an open map, and let U be an open subset of X such that

$U = p^{-1}(p(U))$. Since p is a quotient map and $p^{-1}(p(U))$ is open, $p(U)$ is open. Thus, since $h \circ p^{-1}$ is an open map, $(h \circ p^{-1})(p(U))$ is open. Therefore $h(U) = h(p^{-1}(p(U))) = (h \circ p^{-1})(p(U))$ is open.

Now suppose $h(U)$ is open whenever U is an open set such that $U = p^{-1}(p(U))$, and let V be an open subset of Y. Then $U = p^{-1}(V)$ is open in X. Now $p(U) = V$, so $U = p^{-1}(V) = p^{-1}(p(U))$. Thus, by hypothesis, $h(U)$ is open. Therefore $(h \circ p^{-1})(V) = h(p^{-1}(p(U))) = h(U)$ is open, and so $h \circ p^{-1}$ is an open map. ∎

THEOREM 2.60. Let $(X, \mathcal{T})$, $(Y, \mathcal{U})$, and $(Z, \mathcal{V})$ be topological spaces, let $p: X \to Y$ be a quotient map and let $g: Y \to Z$ be a surjection. Then $g \circ p$ is a quotient map if and only if g is a quotient map.

Proof. By Theorem 2.58, $g \circ p$ is continuous if and only if g is continuous. Therefore, for $U \in \mathcal{V}$, $(g \circ p)^{-1}(U)$ is open if and only if $g^{-1}(U)$ is open. ∎

Let $f: X \to Y$ be a continuous function. Define a relation $\sim_f$ on X by $x \sim_f y$ if and only if $f(x) = f(y)$. Then $\sim_f$ is an equivalent relation and we have the identification space (or decomposition space) $X/\sim_f$ and the natural map (or identification map) $p: X \to X/\sim_f$. If $x, y \in p^{-1}(z)$ for some $z \in X/\sim_f$ then $x \sim_f y$ and hence $f(x) = f(y)$. Therefore $f \circ p^{-1}$ is single-valued, and so by Theorem 2.59 **(a)**, $f \circ p^{-1}$ is continuous. Note that p is a surjection because every member of $X/\sim_f$ is an equivalence class and p maps each member of X into the equivalence class that contains it. Suppose $[x_1], [x_2] \in X/\sim_f$ and $(f \circ p^{-1})([x_1]) = (f \circ p^{-1})([x_2])$. Let $y_1 \in p^{-1}([x_1])$ and $y_2 \in p^{-1}([x_2])$. Then $f(y_1) = f(y_2)$, so $y_1 \sim_f y_2$. Therefore $[x_1] = p(y_1) = p(y_2) = [x_2]$, and hence $f \circ p^{-1}$ is an injection. Suppose f is a surjection, and let $y \in Y$. Then there exists $x \in X$ such that $f(x) = y$. Now $[x] \in X/\sim_f$ and $(f \circ p^{-1})([x]) = y$, so $f \circ p^{-1}$ is a surjection. Therefore if f is a surjection, then $f \circ p^{-1}$ is a bijection.

THEOREM 2.61. Let $(X, \mathcal{T})$ and $(Y, \mathcal{U})$ be topological spaces and let $f: X \to Y$ be a continuous surjection. Then $f \circ p^{-1}: X/\sim_f \to Y$ is a homeomorphism if and only if f is a quotient map.

Proof. Suppose f is a quotient map. By Exercise 8 of Section 1.7, it is sufficient to prove that $f \circ p^{-1}$ is an open map. Let U be an open set satisfying $U = p^{-1}(p(U))$. Since $p^{-1}(p(U)) = f^{-1}(f(U))$, $f^{-1}(f(U))$ is open. Since f is a quotient map, $f(U)$ is open. Therefore by Theorem 2.59 **(b)**, $f \circ p^{-1}$ is open.

Suppose $f \circ p^{-1}$ is a homeomorphism. Then it is a quotient map. By Theorem 2.60, $(f \circ p^{-1}) \circ p$ is a quotient map. But $(f \circ p^{-1}) \circ p = f$. ∎

We turn our attention toward the second way of defining a quotient space.

Definition. Let $(X, \mathcal{T})$ be a topological space, and let $\mathcal{D}$ be a partition of X. Define a function $p: X \to \mathcal{D}$ as follows: For each $x \in X$, $p(x)$ is the member of $\mathcal{D}$ that contains x. If $\mathcal{U}$ is the quotient topology on $\mathcal{D}$ induced by p, then $(\mathcal{D}, \mathcal{U})$ is a **quotient**

space of X. The function p is called the **natural map** of X onto $\mathcal{D}$. The set $\mathcal{D}$ is also called a **decomposition** of X and the quotient space $(\mathcal{D}, \mathcal{U})$ is also called a **decomposition space** or an **identification space** of X. ∎

EXAMPLE 15. Let $X = \mathbb{N}$, let $D_1 = \{n \in \mathbb{N}: n \text{ is odd}\}$, let $D_2 = \{n \in \mathbb{N}: n \text{ is even}\}$, and let $\mathcal{D} = \{D_1, D_2\}$. Then $\mathcal{D}$ is a partition of X and the function $p: X \to \mathcal{D}$ is the function defined by $p(n) = D_1$ if n is odd and $p(n) = D_2$ if n is even. Since the usual topology for $\mathbb{N}$ (as a subspace of $\mathbb{R}$) is the discrete topology, let $\mathcal{T}$ be the discrete topology on $\mathbb{N}$. Then, of course, $\mathcal{D}$ has the discrete topology because $p^{-1}(D_1)$ and $p^{-1}(D_2)$ are both open.

EXAMPLE 16. Let $X = \{(x, y) \in \mathbb{R}^2: 0 \le x \le 2 \text{ and } 0 \le y \le 1\}$, and let $\mathcal{T}$ denote the subspace topology on X induced by the usual topology on $\mathbb{R}^2$. For each $(x, y) \in X$ such that $0 < x < 2$ and $0 < y < 1$, let $D_{(x,y)} = \{(x, y)\}$. For each $(x, 0) \in X$, where $0 < x < 2$, let $D_{(x,0)} = \{(x, 0), (x, 1)\}$, for each $(0, y) \in X$, where $0 < y < 1$, let $D_{(0,y)} = \{(0, y), (2, 1 - y)\}$, and let $D_{(0,0)} = \{(0, 0), (2, 0), (0, 1), (2, 0)\}$. Then

$$\mathcal{D} = \{D_{(x,y)}: (x, y) \in X, 0 < x < 2, \text{ and } 0 < y < 1\}$$
$$\cup \{D_{(x,0)}: (x, 0) \in X \text{ and } 0 < x < 2\}$$
$$\cup \{D_{(0,y)}: (0, y) \in X \text{ and } 0 < y < 1\} \cup D_{(0,0)}$$

is a partition of X. Let p be the natural map of X onto $\mathcal{D}$. If $\mathcal{U}$ is the quotient topology on X induced by p, then the quotient space $(\mathcal{D}, \mathcal{U})$ of X is the Klein bottle.

EXAMPLE 17. Let $X = \{(x, y) \in \mathbb{R} \times \mathbb{R}: 0 \le x \le 3 \text{ and } 0 \le y \le 2\}$ (see Figure 2.14), and let $\mathcal{T}$ be the usual topology on X. Let $D_1 = (x, y) \in X: 0 \le x < 1$ and $y = 0\}$,
let $D_2 = \{(x, y) \in X: 0 < x < 1, y < 1 - x, y < x \text{ if } x \ge \frac{1}{4}, \text{ and } y \le x \text{ if } x < \frac{1}{4}\}$,
let $D_3 = \{(x, y) \in X: 0 \le x < \frac{1}{2}, y < 1 - x, y > x \text{ if } x < \frac{1}{4}, \text{ and } y \ge x \text{ if } x \ge \frac{1}{4}\}$,
let $D_4 = \{(x, y) \in X: 0 \le x < 1 \text{ and } 1 \le y \le 2\}$,
let $D_5 = \{(x, y) \in X: 0 < x < 1, 0 < y < 1, \text{ and } y \ge 1 - x\}$,
let $D_6 = \{(x, y) \in X: (x - 2)^2 + (y - 1)^2 < 1\}$,
let $D_7 = \{(x, y) \in X: 1 \le x < 2 \text{ and } 0 \le y < 1\} - D_6$,
let $D_8 = \{(x, y) \in X: 2 \le x \le 3 \text{ and } 0 \le y \le 1\} - D_6$,
let $D_9 = \{(x, y) \in X: 2 < x \le 3 \text{ and } 1 < y \le 2\} - D_6$, and
let $D_{10} = \{(x, y) \in X: 1 \le x \le 2 \text{ and } 1 \le y \le 2\} - D_6$. Then $\mathcal{D} = \{D_i: i = 1, 2, ..., 10\}$ is a partition of X. In Exercise 13, you are asked to prove that the collection

$$\mathcal{B} = \{\{D_6\}, \{D_2, D_3\}, \{D_6, D_9\}, \{D_1, D_2, D_3\}, \{D_2, D_3, D_5\}, \{D_2, D_3, D_4, D_5\},$$
$$\{D_2, D_3, D_4, D_5, D_6, D_7\}, \{D_2, D_3, D_4, D_5, D_6, D_7, D_8, D_9\},$$
$$\{D_2, D_3, D_4, D_5, D_6, D_7, D_9, D_{10}\}\}$$

is a basis for the quotient topology on $\mathcal{D}$ induced by the natural map $p: X \to \mathcal{D}$.

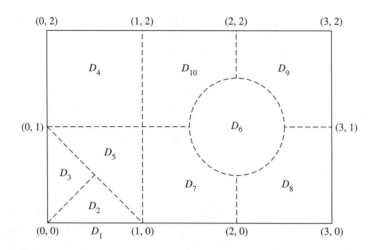

Figure 2.14

THEOREM 2.62. Let $(X, \mathcal{T})$ be a topological space, and let $\mathcal{D}$ be a partition of X. Let $p: X \to \mathcal{D}$ be the natural map, and let $\mathcal{U}$ be the quotient topology on $\mathcal{D}$ induced by p. Then a subset $\mathcal{E}$ of $\mathcal{D}$ is open if and only if $\bigcup \{E : E \in \mathcal{E}\}$ is open in X.

Proof. See Exercise 8. ∎

THEOREM 2.63. Let $(X, \mathcal{T})$ be a topological space, let Y be a set, let f be a function that maps X onto Y, and let $\mathcal{U}$ be the quotient topology on Y induced by f. Let $\mathcal{D} = \{f^{-1}(y): y \in Y\}$. Then there is a homeomorphism $g: Y \to \mathcal{D}$ such that $g \circ f$ is the natural map $p: X \to \mathcal{D}$.

Proof. Define $g: Y \to \mathcal{D}$ by $g(y) = f^{-1}(y)$ for each $y \in Y$. Then g is a bijection. Let $x \in X$. Then $g(f(x)) = f^{-1}(f(x))$ is the member of $\mathcal{D}$ that contains x. Thus it is $p(x)$. Therefore $g \circ f = p$.

Let $\mathcal{E}$ be an open subset of $\mathcal{D}$. Since p is continuous and $g \circ f = p$, $(g \circ f)^{-1}(\mathcal{E}) = f^{-1}(g^{-1}(\mathcal{E}))$ is open in X. By the definition of the quotient topology on Y, $g^{-1}(\mathcal{E})$ is open in Y. Therefore g is continuous.

Let U be an open subset of Y. Then by the definition of the quotient topology, $f^{-1}(U)$ is open in X. But $g(U) = f^{-1}(U)$, and hence g is an open mapping. Therefore g is a homeomorphism. ∎

Theorem 2.63 yields the following commutative diagram, where g is a homeomorphism.

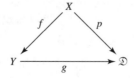

2.6 Quotient Spaces

Theorem 2.37 provides a criterion for determining when a function that maps into a product space is continuous. The following theorem provides a criterion for determining when a function whose domain is a quotient space is continuous.

THEOREM 2.64. Let $(X, \mathcal{T})$ be a topological space, let $(\mathcal{D}, \mathcal{U})$ be a decomposition space of X, and let $X \to \mathcal{D}$ be the natural map. Let $(Z, \mathcal{V})$ be a topological space and let $f: X \to Z$ be a continuous function such that for each $D \in \mathcal{D}$, $f(x) = f(y)$ for all $x, y \in D$. Then there is a continuous function $g: \mathcal{D} \to Z$ such that $g \circ p = f$.

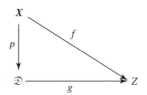

Proof. Let $D \in \mathcal{D}$, and define $g(D)$ to be the only member of $f(D)$. Then for each $x \in X$, $g(p(x)) = f(x)$. Let V be an open set in Z. Then $f^{-1}(V) = (g \circ p)^{-1}(V) = p^{-1}(g^{-1}(V))$ is open in X. Since p is a quotient map, $g^{-1}(V)$ is open in $\mathcal{D}$. Therefore g is continuous. ∎

The following theorem gives a similar result.

THEOREM 2.65. Let $(X, \mathcal{T})$ and $(Z, \mathcal{V})$ be topological spaces, let $f: X \to Z$ be a continuous surjection, let $\mathcal{D} = \{f^{-1}(z): z \in Z\}$, and let $\mathcal{U}$ denote the quotient topology on $\mathcal{D}$ induced by the natural map $p: X \to \mathcal{D}$. The induced continuous function (given by Theorem 2.64) $g: \mathcal{D} \to Z$ is a homeomorphism if and only if f is a quotient map.

Proof. Suppose g is a homeomorphism. Then g and p are quotient maps. Therefore, by Theorem 2.56, $f = g \circ p$ is a quotient map.

Suppose f is a quotient map. By Theorem 2.64, g is continuous. Observe that g is a bijection. Let V be an open set in $\mathcal{D}$. Then $p^{-1}(V)$ is open in X and $f^{-1}(g(V)) = p^{-1}(V)$. Therefore since f is a quotient map, $g(V)$ is open. Hence g is a homeomorphism. ∎

The following theorem gives a criterion for a quotient space to be a Hausdorff space.

THEOREM 2.66. Let $(X, \mathcal{T})$ and $(Z, \mathcal{V})$ be topological spaces, let $f: X \to Z$ be a continuous surjection, let $\mathcal{D} = \{f^{-1}(z): z \in Z\}$, and let $\mathcal{U}$ denote the quotient topology on $\mathcal{D}$ induced by the natural map $p: X \to \mathcal{D}$. If Z is Hausdorff, then so is $\mathcal{D}$.

Proof. By Theorem 2.64, there is a continuous function $g: \mathcal{D} \to Z$. We have observed that g is a bijection. Let D and E be distinct members of $\mathcal{D}$. Then $g(D)$ and $g(E)$ are distinct member of Z. Since Z is Hausdorff, there exist disjoint open sets U and V such that $g(D) \in U$ and $g(E) \in V$. Therefore $g^{-1}(U)$ and $g^{-1}(V)$ are disjoint open subsets of $\mathcal{D}$ such that $D \in g^{-1}(U)$ and $E \in g^{-1}(V)$. ∎

As the following example illustrates, we must proceed with caution when dealing with quotient spaces.

EXAMPLE 18. Let $X = \mathbb{R}^2$, and let $\mathcal{T}$ denote the usual topology on X. Let $\mathcal{D}$ denote the partition of X consisting of all horizontal lines in X. Let $A = \{(x, y) \in X : x = 0 \text{ and } y > 0\} \cup \{(1, 0)\}$. Let $\mathcal{U}$ denote the quotient topology on $\mathcal{D}$ induced by the natural map $p : X \to \mathcal{D}$, and let $\mathcal{U}_A$ denote the subspace topology on A induced by $\mathcal{U}$. Now let $\mathcal{D}|_A = \{D \cap A : D \in \mathcal{D} \text{ and } D \cap A \neq \varnothing\}$, and let $\mathcal{U}|_A$ denote the quotient topology on $\mathcal{D}|_A$ induced by the natural map $p|_A : A \to \mathcal{D}|_A$. Then $\{(1, 0)\} \in \mathcal{U}|_A$, while if U is any member of $\mathcal{U}_A$ that contains $(1, 0)$, then there exists a positive number ε such that $\{(x, y) \in A : x = 0 \text{ and } y < \varepsilon\} \subseteq U$. Therefore $\mathcal{U}_A \neq \mathcal{U}|_A$.

The third way of defining a quotient space is to define it in terms of an equivalence relation.

Definition. Let $(X, \mathcal{T})$ be a topological space, and let $\sim$ be an equivalence relation on X. Then the family of equivalence classes of X with respect to $\sim$ is a partition $\mathcal{D}$ of X. The **identification space** $(X/\sim, \mathcal{U})$ is defined to be the quotient space $(\mathcal{D}, \mathcal{U})$, where $\mathcal{U}$ is the quotient topology induced by the function that maps each x in X into the equivalence class $[x]$ that contains x. ∎

Note that the members of the identification space are equivalence classes.

EXAMPLE 19. Suppose $\mathbb{R}$ has the usual topology. Define an equivalence relation $\sim$ on $\mathbb{R}$ by $a \sim b$ provided there is an even integer k such that $a - b = k\pi$. Let $\mathcal{D}$ be the set of all equivalence classes, and let $\mathcal{U}$ be the quotient topology on $\mathcal{D}$ induced by the function that maps each member of $\mathbb{R}$ into the equivalence class that contains it. Describe the quotient space $(\mathcal{D}, \mathcal{U})$.

Analysis. There is a natural homeomorphism of the quotient space $(\mathcal{D}, \mathcal{U})$ onto the circle (considered as a subspace of $\mathbb{R}^2$ with circumference 2π). The intuitive idea is to continue wrapping the real line around the circle. We may assume that the center of the circle is at the origin and define the homeomorphism by $h([x]) = (\cos y, \sin y)$, where y is the unique member of the interval $[0, 2\pi)$ that belongs to $[x]$. ∎

Many of the important examples of quotient spaces involve compact spaces and hence will be considered in Chapters 4 and 6.

EXERCISES 2.6

1. (a) Construct a Möbius strip.

   (b) Cut the Möbius strip (which you constructed in (a)) along the "equator-line" and explain the result that is obtained in terms of quotient spaces.

2.6 Quotient Spaces

(c) Construct another Möbius strip and cut it along the "$\frac{1}{3}$-equator-line" (that is, cut it along the $\frac{1}{3}$ the distance from the edge of the strip) and explain the result that is obtained in terms of quotient spaces.

(d) Let n be a natural number. Explain, in terms of quotient spaces, the result that is obtained when a Möbius strip is cut along the "$\frac{1}{n}$-equator-line."

2. Define an equivalence relation $\sim$ on $X = \mathbb{R}^2$ by $(x_1, y_1) \sim (x_2, y_2)$ if and only if $y_1 = y_2$. Let $(X/\sim, \mathcal{U})$ be the identification space, and let $\mathcal{T}$ denote the usual topology on $\mathbb{R}$. Prove that $(X/\sim, \mathcal{U})$ is homeomorphic to $(\mathbb{R}, \mathcal{T})$.

3. Let X denote the square $[0, 2\pi] \times [0, 2\pi]$. Let $\mathcal{D}$ be the partition of X that consists of $\{(x, y)\}$ whenever x is neither 0 nor 2π, and $\{(0, y), (2\pi, y)\}$ for each $(0 \leq y \leq 2\pi)$. Let $\mathcal{U}$ be the quotient topology on $\mathcal{D}$ induced by the natural map $p: X \to \mathcal{D}$. Prove that $(\mathcal{D}, \mathcal{U})$ is homeomorphic to $\{(x, y) \in \mathbb{R} \times \mathbb{R}: x^2 + y^2 = 1\} \times [0, 2\pi]$.

4. Let X denote the square $[0, 2\pi] \times [0, 2\pi]$. Let $\mathcal{D}$ be the partition of X that consists of $\{(x, y)\}$ whenever $x \neq 0$, $x \neq 2\pi$, $y \neq 0$, and $y \neq 2\pi$, $\{(0, y), (2\pi, y)\}$ for each y $(0 \leq y \leq 2\pi)$, and $\{(x, 0), (x, 2\pi)\}$ for each x $(0 \leq x \leq 2\pi)$. Let $\mathcal{U}$ be the quotient topology on $\mathcal{D}$ induced by the natural map $p: X \to \mathcal{D}$. Prove that $(\mathcal{D}, \mathcal{U})$ is homeomorphic to $\{(x, y) \in \mathbb{R}^2: x^2 + y^2 = 1\} \times \{(x, y) \in \mathbb{R}^2: x^2 + y^2 = 1\}$.

5. Let $X = \{(x, y) \in \mathbb{R}^2: x \geq 0 \text{ or } y = 0\}$, and let $\mathcal{T}$ be the subspace topology on X induced by the usual topology on $\mathbb{R}^2$, and let $\mathcal{U}$ be the usual topology on $\mathbb{R}$. Define $f: X \to \mathbb{R}$ by $f((x,y)) = x$ for all $(x, y) \in X$. Prove that f is a quotient map and show that it is neither open nor closed.

6. Let $(X, \mathcal{T})$ and $(Y, \mathcal{V})$ be topological spaces, let f be a function that maps X onto Y, and let $\mathcal{U}$ be the quotient topology on Y induced by f. Proved that if f is continuous and closed, then $\mathcal{U} = \mathcal{V}$.

7. Prove that the composition of two quotient maps is a quotient map.

8. Prove Theorem 2.62.

9. Let $Y = \{(x, y) \in \mathbb{R}^2: x = 0 \text{ or } y = 0\}$. Let $\mathcal{T}$ be the usual topology on $\mathbb{R}^2$, and define $f: \mathbb{R}^2 \to Y$ by $f((0, y)) = (0, y)$ and $f((x, y)) = (x, 0)$ if $x \neq 0$. Let $\mathcal{U}$ be the quotient topology on Y induced by f. Show that $(Y, \mathcal{U})$ is not Hausdorff.

10. Let $(X, \mathcal{T})$, $(Y, \mathcal{U})$, and $(Z, \mathcal{V})$ be topological spaces, let $p: X \to Y$ be a quotient map, and let $h: X \to Z$ be continuous. Assume $h \circ p^{-1}$ is single-valued. Prove that $h \circ p^{-1}$ is a closed map if and only if $h(C)$ is closed whenever C is a closed set satisfying $C = p^{-1} p(C)$.

11. Let $Y = \{1, 2, 3\}$ and define $f: \mathbb{R} \to Y$ by $f(x) = 1$ if $x < 0$, $f(x) = 2$ if $x = 0$, and $f(x) = 3$ if $x > 0$. Describe the quotient topology on Y induced by f.

12. Let $X = [0, 1] \cup (2, 3]$, let $Y = [0, 2]$, and suppose X and Y have the usual topologies. Define $f: X \to Y$ by $f(x) = x$ if $x \in [0, 1]$ and $f(x) = x - 1$ if $x \in (2, 3]$. Is f a quotient map? Prove your answer.

13. Prove that the collection $\mathcal{B}$ defined in Example 17 is a basis for the quotient topology on $\mathcal{D}$ induced by the natural map $p: X \to \mathcal{D}$.

14. Let $X = [0, 1] \cup (2, 3]$, and let $\mathcal{T}$ be the subspace topology on X induced by the usual topology on $\mathbb{R}$. Let $Y = [0, 2]$, and let $\mathcal{U}$ be the subspace topology on Y induced by the usual topology on $\mathbb{R}$. Define $p: X \to Y$ by $p(x) = x$ if $x \in [0, 1]$ and $p(x) = x - 1$ if $x \in (2, 3]$. Is p continuous? Prove your answer.

15. Let $X = \bigcup_{n \in \mathbb{N}} (\mathbb{R} \times \{n\})$, and let $Y = \bigcup_{n \in \mathbb{N}} \{(x, y) \in \mathbb{R}^2 : y = nx\}$. Suppose both X and Y have the subspace topology induced by the usual topology on $\mathbb{R}^2$. Define $p: X \to Y$ by $p((x, n)) = (x, nx)$ for each $(x, n) \in X$.

    (a) Show that p maps X onto Y.

    (b) Show that p is not a quotient map.

16. Define a relation $\sim$ on $\mathbb{R}^2$ by $(x_1, y_1) \sim (x_2, y_2)$ provided $x_1 + y_1^2 = x_2 + y_2^2$.

    (a) Show that $\sim$ is an equivalence relation on $\mathbb{R}^2$.

    (b) Describe the identification space $(\mathbb{R}^2/\sim, \mathcal{U})$.

17. Define $p: \mathbb{R}^2 \to [0, \infty)$ by $p((x, y)) = x^2 + y^2$, and suppose $[0, \infty)$ has the subspace topology induced by the usual topology on $\mathbb{R}$. Show that p is a quotient map.

Chapter 3
Connectedness

From an intuitive point of view we say that something is connected if it consists of one piece. In this chapter this notion is made precise. As we shall see, the intermediate-value theorem that is used in calculus can be proved by using theorems that we prove about connectedness. Connectedness can also be used to distinguish, in a topological sense, between a circle and $\mathbb{R}$ and between $\mathbb{R}$ and $\mathbb{R}^2$. Spaces that are not connected are also interesting. One extreme example is the subspace of the real line consisting of the set of rational numbers. Connectedness was defined for bounded, closed subsets of $\mathbb{R}^n$ by Cantor in 1883, but his definition is not suitable for general topological spaces. In 1892, Camille Jordan (1838–1922) gave a different definition of connectedness for bounded, closed subsets of $\mathbb{R}^n$. Then, in 1911, N.J. Lennes extended Jordan's definition to abstract spaces. Hausdroff's Grundzüge der Mengenlehre was the first systematic study of connectedness.

3.1 Connected Spaces

An interval in $\mathbb{R}$ has the property that if it contains two distinct points, then it contains every point between them. Connectedness represents an extension of the idea that an interval is in one piece. The problem of deciding whether a topological space is in one piece is resolved by determining whether it can be broken up into nonempty disjoint open subsets.

Definition. A topological space is **connected** if it cannot be expressed as the union of two nonempty, disjoint open sets. ■

It is convenient to introduce some additional terminology and to prove two theorems that characterize connectedness.

Definition. Let $(X, \mathcal{T})$ be a topological space and let A and B be subsets of X. Then A and B are **separated** in X if $\overline{A} \cap B = A \cap \overline{B} = \emptyset$. If A and B are separated in X and $X = A \cup B$, then A and B are said to form a **separation** of X. ■

THEOREM 3.1. A topological space $(X, \mathcal{T})$ is connected if and only if it cannot be expressed as the union of two nonempty sets that are separated in X.

Proof. Suppose X is not connected. Then there are nonempty, disjoint open sets U and V such that $X = U \cup V$. Then U and V are also closed, so $\overline{U} \cap V = U \cap V = \emptyset$ and $U \cap \overline{V} = U \cap V = \emptyset$. Therefore U and V are separated in X.

Suppose now that there are nonempty subsets A and B such that $X = A \cup B$ and $\overline{A} \cap B = A \cap \overline{B} = \emptyset$. Since $X = A \cup B$ and $\overline{A} \cap B = \emptyset$, $\overline{A} \subseteq A$, so A is closed. The same argument shows that B is closed. Therefore A and B are also open, and hence X is not connected. ∎

Notice that the proof of Theorem 3.1 establishes the following corollary.

Corollary 3.2. If A and B form a separation of the topological space $(X, \mathcal{T})$, then A and B are both open and closed. ∎

THEOREM 3.3. A topological space $(X, \mathcal{T})$ is connected if and only if no nonempty proper subset of X is both open and closed.

Proof. Suppose X is not connected. Then there are nonempty, disjoint open sets U and V such that $X = U \cup V$. Thus U is a nonempty proper subset of X that is both open and closed.

Suppose X has a nonempty proper subset U that is both open and closed. Then U and $X - U$ are nonempty disjoint open sets whose union is X. Therefore X is not connected. ∎

The following theorem provides useful ways of formulating the definition of connectedness for subspaces of a topological space.

THEOREM 3.4. Let $(X, \mathcal{T})$ be a topological space and let $A \subseteq X$. Then the following conditions are equivalent.

(a) The subspace $(A, \mathcal{T}_A)$ is connected.

(b) The set A connot be expressed as the union of two nonempty sets that are separated in X.

(c) There do not exist $U, V \in \mathcal{T}$ such that $U \cap A \neq \emptyset$, $V \cap A \neq \emptyset$, $U \cap V \cap A = \emptyset$, and $A \subseteq U \cup V$.

Proof. We prove three implications. In each case, we prove the contrapositive of the implication.

(a) → (b) Suppose U and V are nonempty sets such that $A = U \cup V$ and $\overline{U} \cap V = U \cap \overline{V} = \emptyset$. Then U and V are separated in A, so, by Theorem 3.1, A is not connected.

(b) → **(c)** Suppose there exist $U, V \in \mathcal{T}$ such that $U \cap A \neq \emptyset$, $V \cap A \neq \emptyset$, $U \cap V \cap A = \emptyset$ and $A \subseteq U \cup V$. Then $U \cap A$ and $V \cap A$ and are nonempty sets that are separated in X and $A = (U \cap A) \cup (V \cap A)$.

(c) → **(a)** Suppose $(A, \mathcal{T}_A)$ is not connected. Then there exist $U', V' \in \mathcal{T}_A$ such that $U' \neq \emptyset \neq V', U' \cap V' = \emptyset$, and $A = U' \cup V'$. Thus there exist $U, V \in \mathcal{T}$ such that $U' = A \cap U$ and $V' = A \cap V$. It is clear that $U \cap A \neq \emptyset$, $V \cap A \neq \emptyset, U \cap V \cap A = \emptyset$, and $A \subseteq U \cup V$. ∎

We give examples of some spaces that are connected and some that aren't.

EXAMPLE 1.

(a) Let X be a set with more than one member, and let $\mathcal{T}$ be the discrete topology on X. Then $(X, \mathcal{T})$ is not connected.

(b) Let X be a nonempty set, and let $\mathcal{T}$ be the trivial topology on X. Then $(X, \mathcal{T})$ is connected.

(c) Let $X = \mathbb{R} - \{0\}$, and let $\mathcal{T}$ be the usual topology on X. Since $X = (-\infty, 0) \cup (0, \infty)$, $(X, \mathcal{T})$ is not connected.

(d) Let X be a set, let p be an object that does not belong to X, and let $Y = X \cup \{p\}$. Let $\mathcal{T} = U \in \mathcal{P}(Y) : U = \emptyset$ or $p \in U\}$. Since any two nonempty open subsets of Y have a nonempty intersection, $(Y, \mathcal{T})$ is connected.

EXAMPLE 2. Let $\mathcal{T}$ be the usual topology on $\mathbb{R}$. Then $(\mathbb{R}, \mathcal{T})$ is connected.

Analysis. The proof is by contradiction. Suppose $(\mathbb{R}, \mathcal{T})$ is not connected. Then there exist disjoint nonempty open sets U and V such that $\mathbb{R} = U \cup V$. Since $U = \mathbb{R} - V$ and $V = \mathbb{R} - U$, U and V are also closed. Let $a \in U$ and let $b \in V$. We may assume without loss of generality that $a < b$. Let $W = U \cap [a, b]$. Since W is bounded, it has a least uppper bound c. Since W is closed, $c \in W$. Since $W \cap V = \emptyset, c \neq b$. Since c is the least upper bound of W, $(c, b] \subseteq V$. Therefore $c \in \overline{V}$. But V is closed, so $c \in V$. Therefore $c \in U \cap V$. This is a contradiction because $U \cap V = \emptyset$. ∎

The following theorem together with Example 2 shows that every continuous image of $\mathbb{R}$ is connected.

THEOREM 3.5. Let $(X, \mathcal{T})$ be a connected space, let $(Y, \mathcal{U})$ be a topological space, and let $f : X \to Y$ be a continuous function that maps X onto Y. Then $(Y, \mathcal{U})$ is connected.

Proof. We prove the contrapositive. Suppose $(Y, \mathcal{U})$ is not connected. Then there are disjoint nonempty open sets U and V such that $Y = U \cup V$. Since f is continuous and maps onto Y, $f^{-1}(U)$ and $f^{-1}(V)$ are nonempty open subsets of X. Since $f^{-1}(U) \cap f^{-1}(V) = \emptyset$ and $X = f^{-1}(U) \cup f^{-1}(V)$, $(X, \mathcal{T})$ is not connected. ∎

An important application of Theorem 3.5 arises when the continuous function f on a connected space is real valued. In that case, if f assumes two distinct values, then it must assume as values all real numbers between these two values. In the proof we also use Exercise 8, which characterizes the connected subsets of $\mathbb{R}$.

Corollary 3.6 (Intermediate-Value Theorem). Let $(X, \mathcal{T})$ be a connected topological space, let $f : X \to \mathbb{R}$ be a continuous function, let $a, b \in X$ such that $f(a) < f(b)$, and let $d \in \mathbb{R}$ such that $f(a) < d < f(b)$. Then there exists $c \in X$ such that $f(c) = d$.

Proof. By Theorem 3.5, $f(X)$ is a connected subset of $\mathbb{R}$. By Exercise 8, $f(X)$ is an interval. Therefore $[f(a), f(b)] \subseteq f(X)$, and so there exists $c \in X$ such that $f(c) = d$. ∎

The following corollary is also an immediate consequence of Theorem 3.5.

Corollary 3.7. Connectedness is a topological property. ∎

The analysis of the following example is now an easy exercise (see Exercise 3).

EXAMPLE 3. Let $a, b \in \mathbb{R}$ with $a < b$. Prove that $[a, b]$ is connected.

The following theorem, which is used to show that the closure of a connected set is connected, establishes that if A and B form a separation of a space X, then every connected subset of X is a subset of either A or B.

THEOREM 3.8. Let $(X, \mathcal{T})$ be a topological space and suppose $X = A \cup B$, where A and B are nonempty subsets that are separated in X. If H is a connected subspace of X, then $H \subseteq A$ or $H \subseteq B$.

Proof. Since A and B are open in X, $A \cap H$ and $B \cap H$ are open in H. If both $A \cap H$ and $B \cap H$ are nonempty, they would form a separation of H. Therefore either $A \cap H = \emptyset$ or $B \cap H = \emptyset$. Thus $H \subseteq A$ or $H \subseteq B$. ∎

THEOREM 3.9. Let $(X, \mathcal{T})$ be a topological space, let $(A, \mathcal{T}_A)$ be a connected subspace of X, and let B be a subset of X such that $A \subseteq B$ and $B \subseteq \overline{A}$. Then $(B, \mathcal{T}_B)$ is connected.

Proof. Suppose $B = C \cup D$ is a separation of B. By Theorem 3.8, $A \subseteq D$ or $A \subseteq C$. We may assume without loss of generality that $A \subseteq C$. Then $\overline{A} \subseteq \overline{C}$. Since $\overline{C} \cap D = \emptyset$, $B \cap D = \emptyset$. This contradicts the fact that D is a nonempty subset of B. ∎

The following theorem is used to show that the product of connected spaces is connected.

3.1 Connected Spaces 113

THEOREM 3.10. Let $\{(A_\alpha, \mathcal{T}_{A_\alpha}): \alpha \in \Lambda\}$ be a collection of connected subspaces of a topological space $(X, \mathcal{T})$, and let $A = \bigcup_{\alpha \in \Lambda} A_\alpha$. Then the following conditions hold:

(a) If $\bigcap_{\alpha \in \Lambda} A_\alpha \neq \emptyset$ then $(A, \mathcal{T}_A)$ is connected.

(b) If for each pair $a, b \in A$ there is a connected subset B of A such that $a, b \in B$, then $(A, \mathcal{T}_A)$ is connected.

(c) If $\Lambda = \mathbb{N}$ and $A_n \cap A_{n+1} \neq \emptyset$ for each $n \in \mathbb{N}$, then $(A, \mathcal{T}_A)$ is connected.

Proof.

(a) Suppose $\bigcap_{\alpha \in \Lambda} A_\alpha \neq \emptyset$ and $(A, \mathcal{T}_A)$ is not connected. Then there are nonempty disjoint open subsets U and V of A such that $A = U \cup V$. Let $\beta \in \Lambda$. Then A_β is connected and hence $A_\beta \subseteq U$ or $A_\beta \subseteq V$. Suppose $A_\beta \subseteq U$. Then for each $\alpha \in \Lambda$, $A_\alpha \cap A_\beta \neq \emptyset$ and hence $A_\alpha \subseteq U$. Therefore $A \subseteq U$ and hence $V = \emptyset$. This is a contradiction.

(b) Let $x \in A$, and for each $y \in A$ let B_{xy} be a connected subset of A that contains x and y. Then $A = \bigcup_{y \in A} B_{xy}$ and $x \in \bigcap_{y \in A} B_{xy}$. Therefore by *(a)*, A is connected.

(c) For each $n \in \mathbb{N}$, let $B_n = \bigcup_{i=1}^n A_i$. The proof that for each $n \in \mathbb{N}$, B_n is connected is by induction. Since $B_1 = A_1$, B_1 is connected. Suppose B_n is connected. Since $A_{n+1} \cap A_n \neq \emptyset$ and $A_{n+1} \cap A_n \subseteq A_{n+1} \cap B_n$, $A_{n+1} \cap B_n \neq \emptyset$. Therefore by *(a)*, $B_{n+1} = A_{n+1} \cup B_n$ is connected. Since $\bigcap_{n \in \mathbb{N}} B_n = A_1$ by *(a)*, $A = \bigcup_{n \in \mathbb{N}} B_n$ is connected. ∎

Definition. Let $(X, \mathcal{T})$ be a topological space and define an equivalence reation $\sim$ on X by $x \sim y$ provided there is a connected subset A of X such that $x, y \in A$. The equivalence classes of X with respect to $\sim$ are called the **components** of X. ∎

We leave as Exercise 1 the proof that $\sim$ is an equivalence relation on X. By Theorem 3.10, each component of a topological space is connected. By Theorem E.1, the collection of components of a topological space $(X, \mathcal{T})$ forms a partition of X. Notice that a topological space is connected if any only if it has exactly one component.

THEOREM 3.11. A nonempty connected subset of a topological space $(X, \mathcal{T})$ is a component of X if and only if it is not a proper subset of a connected subset of X.

Proof. Let A be a nonempty connected subset of X. Suppose there is a connected subset B of X such that A is a proper subset of B. Let $x \in A$ and $y \in B - A$. Then $x, y \in B$, so x and y are in the same component of X. Hence A is not a component.

Suppose A is not a component of X. Since A is connected, it is a subset of some component B of X. Therefore A is a proper subset of B. ∎

THEOREM 3.12. Every component of a topological space $(X, \mathcal{T})$ is a closed subset of X.

Proof. Let A be a component of a topological space $(X, \mathcal{T})$. Then A is connected, and so by Theorem 3.9, $\overline{A}$ is connected. Therefore $\overline{A} \subseteq A$ and hence A is closed. ∎

While a component of a topological space is closed, the following example shows that components need not be open.

EXAMPLE 4. Let $X = \{(x, y) \in \mathbb{R}^2 : x = 0 \text{ or } x = \frac{1}{n} \text{ for some } n \in \mathbb{N}, \text{ and } 0 \leq y \leq 1\}$. We leave as Exercise 7 the proof that $\{0\} \times I$ is a component of X that is not open.

THEOREM 3.13. Let $(X, \mathcal{T})$ be a topological space and let $x \in X$. Then the component of X that contains x is the union of all connected subsets of X that contain x.

Proof. Let A be the component of X that contains x and let $\{C_\alpha : \alpha \in \Lambda\}$ be the collection of all connected sets that contain x. Then for each $\alpha \in \Lambda$, $C_\alpha \subseteq A$, and so $\bigcup_{\alpha \in \Lambda} C_\alpha \subseteq A$. Since A is connected, there exists $\beta \in \Lambda$ such that $A = C_\beta$. Hence $A \subseteq \bigcup_{\alpha \in \Lambda} C_\alpha$. ∎

The proof that the product of connected spaces is connected is somewhat complicated. Thus we have an exercise (Exercise 4), which asks you to prove that the product of two connected spaces is connected.

THEOREM 3.14. Let $\{(X_\alpha, \mathcal{T}_\alpha) : \alpha \in \Lambda\}$ be a collection of topological spaces, and suppose that for each $\alpha \in \Lambda$, $X_\alpha \neq \emptyset$. Let $X = \prod_{\alpha \in \Lambda} X_\alpha$ and let $\mathcal{T}$ be the product topology on X. Then $(X, \mathcal{T})$ is connected if and only if, for each $\alpha \in \Lambda$, $(X_\alpha, \mathcal{T}_\alpha)$ is connected.

Proof. Suppose $(X, \mathcal{T})$ is connected and let $\beta \in \Lambda$. Then π_β is a continuous function that maps X onto X_β, and hence by Theorem 3.5, $(X_\beta, \mathcal{T}_\beta)$ is connected.

Suppose that for each $\alpha \in \Lambda$, $(X_\alpha, \mathcal{T}_\alpha)$ is connected. Suppose $(X, \mathcal{T})$ is not connected, and let K be a component of X. Then there exist $x \in K$ and $y = \langle y_\alpha \rangle \in X - K$. By Theorem 3.12, K is closed. Therefore there is a finite subset $\{\alpha_1, \alpha_2, ..., \alpha_n\}$ of Λ such that for each $i = 1, 2, ..., n$, there exists $U_{\alpha_i} \in \mathcal{T}_{\alpha_i}$ such that $y \in V = \bigcap_{i=1}^n \pi_{\alpha_i}^{-1}(U_{\alpha_i}) \subseteq X - K$. Define a member p of X as follows: For each $i = 1, 2, ..., n$, let $p_{\alpha_i} = y_{\alpha_i}$, and if $\alpha \in \Lambda - \{\alpha_1, \alpha_2, ..., \alpha_n\}$, let $p_\alpha = x_\alpha$. Then $p \in V$. Let $p_0 = x$ and for each $i = 1, 2, ..., n$, define a member p_i of X as follows: $(p_i)_{\alpha_i} = p_{\alpha_i}$ and for each $\alpha \in \Lambda - \{\alpha_i\}$, $(p_i)_\alpha = (p_{i-1})_\alpha$. Then $p_n = p$.

For each $i = 1, 2, ..., n$, let H_i be the set $H_{(p_i)(\alpha_i)} = \{x \in X : \text{if } \alpha \neq \alpha_i \text{ then } x_\alpha = (p_i)_\alpha\}$. By Theorem 2.39, $(H_i, \mathcal{T}_{H_i})$ is homeomorphic to X_{α_i}. Hence H_i is a connected subset of X. Moreover $p_{i-1}, p_i \in H_i$. Let $H = \bigcup_{i=1}^n H_i$. Then $x, p \in H$, and by Theorem 3.10, $(H, \mathcal{T}_H)$ is connected. By Theorem 3.13, $H \subseteq K$. Therefore $p \in K$, and this is a contradiction. ∎

Definition. Let $(X, \mathcal{T})$ be a topological space and let $a, b \in X$. A **simple chain** connecting a and b is a finite set $U_1, U_2, ..., U_n$ of open sets such that $a \in U_1 - U_2$, $b \in U_n - U_{n-1}$, and for each $i, j = 1, 2, ...n$, $U_i \cap U_j \neq \emptyset$ if and only if $|i - j| \leq 1$. ∎

3.1 Connected Spaces

Definition. A collection $\mathcal{A}$ of subsets of a topological space $(X, \mathcal{T})$ is said to **cover** a subset B of X or to be a **covering** of B provided $B \subseteq \bigcup_{A \in \mathcal{A}} A$. A covering $\mathcal{A}$ of B is said to be an **open cover** of B if each member of $\mathcal{A}$ is open. ∎

THEOREM 3.15. Let $(X, \mathcal{T})$ be a connected space, let $\mathcal{O}$ be an open cover of X, and let a and b be distinct points of X. Then there is a simple chain consisting of members of $\mathcal{O}$ that connects a and b.

Proof. Let $A = \{x \in X: \text{there is a simple chain consisting of members of } \mathcal{O} \text{ that connects } a \text{ and } x\}$. Then A is an open set. Let $U \in \mathcal{O}$ such that $a \in U$. Then U is a simple chain connecting a and a. Therefore $a \in A$.

Let $x \in \overline{A}$, and let $U \in \mathcal{O}$ such $x \in U$. Then $U \cap A \neq \emptyset$, so let $y \in U \cap A$. Then there is a simple chain $U_1, U_2, \ldots, U_n$ consisting of members of $\mathcal{O}$ that connects a and y. If there exists i with $1 \leq i \leq n$ such that $x \in U_i$, let k be the smallest such i. Then $U_1, U_2, \ldots, U_k$, is a simple chain consisting of members of $\mathcal{O}$ that connects a and x. If $x \notin U_j$ for any $j = 1, 2, \ldots, n$, let k be the smallest j such that $U_j \cap U \neq \emptyset$. Then, $U_1, U_2, \ldots, U_k, U$ is a simple chain consisting of members of $\mathcal{O}$ that connects a and x. We have proved that $x \in A$. Therefore A is closed.

Since A is a nonempty subset of the connected space $(X, \mathcal{T})$ that is both open and closed, $A = X$. Therefore $b \in A$. ∎

The following example is the classic example in the next section.

EXAMPLE 5. Let $A = \{(x, y) \in \mathbb{R}^2: x = 0 \text{ and } -1 \leq y \leq 1\}$, let $B = \{(x, y) \in \mathbb{R}^2: 0 < x \leq 1 \text{ and } y = \sin\left(\frac{\pi}{x}\right)\}$, and let $X = A \cup B$. Let $\mathcal{T}$ be the subspace topology on X induced by the usual topology on $\mathbb{R}^2$. Then $(X, \mathcal{T})$ is connected.

Analysis. Define $f: (0, 1] \to B$ by $f(x) = (x, \sin\left(\frac{\pi}{x}\right))$. Then f is a continuous function that maps $(0, 1]$ onto B. Therefore, by Theorem 3.5, B is connected. So, by Theorem 3.9, the closure of B in $\mathbb{R}^2$ is connected. But $\overline{B} = X$, so X is connected. ∎

The topological space $(X, \mathcal{T})$ in Example 5 is called the **topologist's sine curve** and is shown in Figure 3.1.

An application of connectedness is the important fixed-point property.

Definition. A topological space $(X, \mathcal{T})$ has the **fixed-point property** if for each continuous function $f: X \to X$ there exists $x \in X$ such that $f(x) = x$. ∎

THEOREM 3.16. The closed unit interval I has the fixed-point property.

Proof. Let $f: I \to I$ be a continuous function. We show that there exists $a \in I$ such that $f(a) = a$. Notice that if $f(0) = 0$ or $f(1) = 1$, the proof is complete. Therefore we may assume that $0 < f(0)$ and $f(1) < 1$. Define $g: I \to \mathbb{R}$ by $g(x) = x - f(x)$. Then g is a continuous function, $g(0) = -f(0) < 0$, and $g(1) = 1 - f(1) > 0$. By

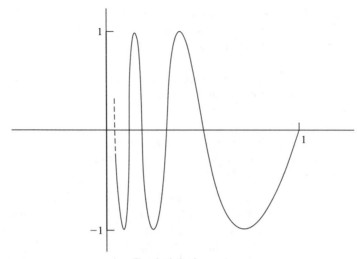

Topologist's sine curve

Figure 3.1

the Intermediate-Value Theorem (Corollary 3.6), there exists $a \in I$ such that $g(a) = 0$. Thus $a - f(a) = 0$, so $f(a) = a$. ∎

The following theorem permits us to conclude that if $(X, \mathcal{T})$ is a topological space that is homeomorphic to a topological space with the fixed-point property, then $(X, \mathcal{T})$ has the fixed-point property.

THEOREM 3.17. The fixed-point property is a topological invariant.

Proof. Let $(X, \mathcal{T})$ be a topological space that has the fixed-point property, and let $(Y, \mathcal{U})$ be a topological space that is homeomorphic to $(X, \mathcal{T})$. Then there is a homeomorphism $h: X \to Y$. We must show that $(Y, \mathcal{U})$ has the fixed-point property. Let $f: Y \to Y$ be a continuous function. Since h is a homeomorphism and the composition of continuous functions is continuous, $h^{-1} \circ f \circ h: X \to X$ is a continuous function. Since $(X, \mathcal{T})$ has the fixed-point property, there exists $a \in X$ such that $(h^{-1} \circ f \circ h)(a) = a$. Since $h \circ h^{-1}$ is the identity, $f \circ h = h \circ h^{-1} \circ f \circ h$. Therefore $f(h(a)) = (h \circ h^{-1} \circ f \circ h)(a) = h \circ (h^{-1} \circ f \circ h)(a) = h(a)$, and hence $(Y, \mathcal{U})$ has the fixed-point property. ∎

We conclude this section with a brief discussion of points in a connected space whose removal leaves the space disconnected, and we use this idea to show that $\mathbb{R}$ is not homeomorphic to $\mathbb{R}^2$ and that a circle is not homeomorphic to a sphere.

Definition. A point p of connected space $(X, \mathcal{T})$ is a **cut point** of X if $X - \{p\}$ is not connected. ∎

If $(X, \mathcal{T})$ is the subset of the plane shown in Figure 3.2, then the point p is a cut point of X.

Figure 3.2

THEOREM 3.18. Let $(X, \mathcal{T})$ and $(Y, \mathcal{U})$ be conected spaces, let p be a cut point of X, and let $h: X \to Y$ be a homeomorphism. Then $h(p)$ is a cut point of Y.

Proof. Since $X - \{p\}$ is not connected, there exist nonempty disjoint open sets U and V such that $X - \{p\} = U \cup V$. Since h is a homeomorphism, $h(U)$ and $h(V)$ are nonempty disjoint open subsets of Y such that $Y - \{h(p)\} = h(U) \cup h(V)$. Therefore $h(p)$ is a cut point of Y. ∎

EXERCISES 3.1

1. Let $(X, \mathcal{T})$ be a topological space and define a relation $\sim$ on X by $x \sim y$ provided there is a connected subset A of X such that $x, y \in X$. Prove that $\sim$ is an equivalence relation on X.

2. Let $Y = \{0, 1\}$ and let $\mathcal{U} = \mathcal{P}(Y)$. Prove that a topological space $(X, \mathcal{T})$ is connected if and only if there is no continuous function that maps X onto Y.

3. Let $a, b \in \mathbb{R}$ with $a < b$. Prove that $[a, b]$ is connected.

4. Prove that the product of two connected topological spaces is connected.

5. Let $\mathcal{T}$ be the lower-limit topology on $\mathbb{R}$. Is $(\mathbb{R}, \mathcal{T})$ connected? Prove your answer.

6. Let $\mathcal{T}$ be the finite complement topology on $\mathbb{R}$. Is $(\mathbb{R}, \mathcal{T})$ connected? Prove your answer.

7. Prove that $\{0\} \times I$ is a component of the space in Example 4, but that $\{0\} \times I$ is not an open set.

8. Let $\mathcal{T}$ be the usual topology on $\mathbb{R}$. Prove that a subset of $(\mathbb{R}, \mathcal{T})$ is connected if and only if it is an interval.

9. Prove that no two of the intervals $[0, 1]$, $(0, 1)$, and $[0, 1)$ are homeomorphic. *Hint:* Use Theorem 3.18.

10. Are any of the following subspaces of $\mathbb{R}^2$ homeomorphic? Prove your answer.

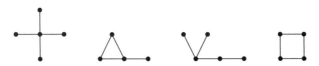

11. Let $A_\alpha: \alpha \in \Lambda\}$ be a collection of connected subsets of X, and let A be a connected subset of X such that $A \cap A_\alpha \neq \emptyset$ for each $\alpha \in \Lambda$. Prove that $A \cup (\bigcup_{\alpha \in \Lambda} A_\alpha)$ is connected.

12. Let $(X, \mathcal{T})$ be a topological space. Prove that the components of X are connected disjoint subsets of X whose union is X and that each connected subset of X intersects only one component.

13. Let p be a cut point of a connected space $(X, \mathcal{T})$ and suppose A and B form a separation of $X - \{p\}$. Prove that $A \cup \{p\}$ is connected.

14. Let $\mathcal{T}$ be the finite complement topology on $\mathbb{R}$. Are there any cut points of $(\mathbb{R}, \mathcal{T})$? Prove your answer.

15. For each natural number $n > 1$, let R_n be the edges of the rectangle in the plane of height $2 - 2/n$ and length n centered at the origin. Let L_1 and L_2 be the horizontal lines $y = 1$ and $y = -1$, and let $X = L_1 \cup L_2 \cup (\bigcup_{n=2}^{\infty} R_n)$. Let $\mathcal{T}$ be the subspace topology on X induced by the usual topology on $\mathbb{R}^2$.

   (a) Prove that the components of X are the sets L_1, L_2 and R_n.

   (b) Prove that there is no separation $X = A \cup B$ of X with $L_1 \subseteq A$ and $L_2 \subseteq B$.

16. Let $(X, \mathcal{T})$ and $(Y, \mathcal{U})$ be topological spaces and let $f: X \to Y$ be a continuous function.

   (a) Prove that the image under f of each component of X must be a subset of a component of Y.

   (b) Under the assumption that f is a homeomorphism, prove that f induces a one-to-one correspondence between the components of X and the components of Y and that corresponding components are homeomorphic.

17. Let $X = \bigcup_{n=0}^{\infty}(3n, 3n+1) \cup \bigcup_{n=0}^{\infty}\{3n+2\}$ and $Y = (0, 1] \cup \bigcup_{n=1}^{\infty}(3n, 3n+1) \cup \bigcup_{n=0}^{\infty}\{3n+2\}$ be subsets of $\mathbb{R}^1$ with the usual subspace topology.

   (a) Prove that there is a continuous bijection $f: X \to Y$.

   (b) Prove that there is a continuous bijection $g: Y \to X$.

   (c) Prove that X and Y are not homeomorphic.

18. Let $S^1 = \{(x, y) \in \mathbb{R} \times \mathbb{R}: x^2 + y^2 = 1\}$. Observe that if $x = (x_1, x_2) \in S^1$, then $-x = (-x_1, -x_2) \in S^1$. Let $f: S^1 \to \mathbb{R}$ be a continuous function.

Prove that there exists $x \in S^1$ such that $f(x) = f(-x)$. *Hint:* Consider the maps $g: S^1 \to \mathbb{R}$ defined by $g(x) = f(x) - f(-x)$ and $p: [0, 1] \to S^1$ defined by $p(t) = (\cos \pi t, \sin \pi t)$.

19. Let $(X, \mathcal{T})$ be a topological space and suppose there exists a covering $\mathcal{A}$ of X consisting of connected subsets of X such that if $A, B \in \mathcal{A}$ then there is a finite subcollection $\{A_1, A_2, ..., A_n\}$ of $\mathcal{A}$ such that $A = A_1$, $B = A_n$, and $A_i \cap A_{i+1} \neq \emptyset$ for each $i = 1, 2, ..., n - 1$. Prove that X is connected.

20. A subset X of $\mathbb{R}^2$ is **convex** provided that for each $x, y \in X$, the line segment $\{(1 - t)x + ty : 0 \leq t \leq 1\}$ is a subset of X. Prove that every convex subset of $\mathbb{R}^2$ is connected.

21. Prove that S^1 is not homeomorphic is $\mathbb{R}$. *Hint:* Prove that if $p \in S^1$, then p is not a cut point of S^1.

22. Prove that $\mathbb{R}^2$ is not homeomorphic to $\mathbb{R}$. *Hint:* Prove that if $p \in \mathbb{R}^2$, then p is not a cut point of $\mathbb{R}^2$.

23. Let $(X, \leq)$ be a linearly ordered set and let $\mathcal{T}$ be the order topology on X. The topological space $(X, \mathcal{T})$ has a **gap** if their exists $x, y \in X$ with $x < y$ for which there is no z in X such that $x < z < y$. The topological space $(X, \mathcal{T})$ is **Dedekind complete** provided each nonempty subset of X that has an upper bound has a least upper bound. Prove that $(X, \mathcal{T})$ is connected if and only if it is Dedekind complete and has no gaps.

24. Let $(X, \mathcal{T})$ be a topological space and define a relation $\approx$ on X by $x \approx y$ provided X cannot be expressed as the union of two disjoint open sets U and V such that $x \in U$ and $y \in V$.

    (a) Prove that $\approx$ is an equivalence relation on X. The equivalence classes of X with respect to $\approx$ are called the **quasi-components** of X.

    (b) Let $p \in X$. Prove that the quasi-component of X that contains p is the intersection of all subsets of X that are both open and closed and contain p.

    (c) Let $p \in X$. Prove that the component of X that contains p is a subset of the quasi-component of X that contains p.

    (d) Let $(X, \mathcal{T})$ be the topological space given in Exercise 15, and let $p = (0,1)$. Prove that the component of X that contains p is a proper subset of the quasi-component of X that contains p.

3.2 Pathwise and Local Connectedness

Connectedness is defined in a negative manner; that is, the definition says that a certain kind of splitting of a space does not exist. A positive approach to a similar but stronger concept is provided by the definition of pathwise connectedness. This concept is particularly useful in studying connectivity from an algebraic point of view.

We begin with some terminology about continuous functions that map the closed unit interval into a topological space.

Definition. A **path** in a topological space $(X, \mathcal{T})$ is a continuous function $f: I \to X$. The path f is also called a **path from** $f(0)$ **to** $f(1)$. The points $f(0)$ and $f(1)$ are called the **endpoints** of f, and f is said to **join the initial point** $f(0)$ **and the terminal point** $f(1)$. Then path $\bar{f}: I \to X$ defined by $\bar{f}(x) = f(1 - x)$ for each $x \in I$ is called the **reverse path** of f. ■

Definition. A topological space $(X, \mathcal{T})$ is **pathwise connected** if for each pair a, b of members of X there is a path f in X such that $f(0) = a$ and $f(1) = b$. ■

The idea of joining points by continuous curves has been used for centuries. Weierstrass was probably the first one to use this idea in a topological sense. He discussed pathwise connectedness of subsets of Euclidean spaces prior to 1890.

THEOREM 3.19. Every pathwise connected space is connected.

Proof. The proof is by contradiction. Suppose there is a pathwise connected space $(X, \mathcal{T})$ that is not connected. Then there are nonempty disjoint open sets U and V such that $X = U \cup V$. Let $a \in U$ and $b \in V$ and let f be a path from a to b. Then $f^{-1}(U)$ and $f^{-1}(V)$ are nonempty disjoint open subsets of I such that $I = f^{-1}(U) \cup f^{-1}(V)$. This is a contradiction because, by Example 3, I is connected. ■

By Example 5 the topologist's sine curve is connected. We prove that it is not pathwise connected. Recall that the topologist's sine curve is the subspace X of $\mathbb{R} \times \mathbb{R}$, where $X = A \cup B$, $A = \{(x, y) \in \mathbb{R} \times \mathbb{R}: x = 0 \text{ and } -1 \le y \le 1\}$, and $B = \{(x, y) \in \mathbb{R} \times \mathbb{R}: 0 < x \le 1 \text{ and } y = \sin(\frac{\pi}{x})\}$.

EXAMPLE 6. The topologist's sine curve is not pathwise connected.

Analysis. The proof is by contradiction. Let $a = (0, a_1) \in A$ and $b = (b_1, b_2) \in B$, and suppose there is a path f from a to b. Let $W = \{x \in I: f([0, x]) \subseteq A\}$, and let w be the least upper bound of W. Then $w \in \overline{W}$, and since A is a closed subset of X and f is continuous, $f(w) \in f(\overline{W}) \subseteq \overline{f(W)} \subseteq \overline{A} = A$. Let d denote the usual metric on $\mathbb{R} \times \mathbb{R}$. Since f is continuous at w, there is a positive number δ such that if $x \in I$ and $|x - w| < \delta$, then $d(f(x), f(w)) < \frac{1}{2}$. Since w is the least upper bound of W, there exists $r \in I$ such that $w < r < w + \delta$ and $f(r) \in B$. Therefore $f([w, r])$ is a connected subset of X with diameter less than 1 such that $f(w) \in A$ and $f(r) \in B$. We complete the proof by showing that the diameter of $f([w, r])$ must be greater than 1. Since $f(r) \in B$, there exists $r_1, r_2 \in \mathbb{R}$ such that $0 < r_1 \le 1, -1 \le r_2 \le 1$, and $f(r) = (r_1, r_2)$. Suppose there exists $s = (s_1, s_2) \in B$ such that $s_1 < r_1$ and $s \notin f([w, r])$. Then $U = \{(x, y) \in \mathbb{R} \times \mathbb{R}: x < s_1\}$ and $V = \{(x, y) \in \mathbb{R} \times \mathbb{R}: x > s_1\}$ are disjoint open subsets of $\mathbb{R} \times \mathbb{R}$ such that $U \cap f([w, r]) \neq \emptyset$, $V \cap f([w, r]) \neq \emptyset$, and $f([w, r]) \subseteq U \cup V$. Therefore by Theorem 3.4, $f([w, r])$ is not connected. Hence, if $s = (s_1, s_2) \in B$ and $s_1 < r_1$, then $s \in f([w, r])$. But there exist

p_1, q_1 such that $p_1 < r_1$, $q_1 < r_1$, $(p_1, 1) \in B$, and $(q_1, -1) \in B$. Therefore the diameter of $f([w, r])$ exceeds 1. ∎

The notion of a path product is useful in the proofs of some theorems, and this concept will be used extensively in Chapter 9.

Definition. Let f_1 and f_2 be paths in a topological space $(X, \mathcal{T})$ such that $f_1(1) = f_2(0)$. The **path product** $f_1 * f_2$ is the path in X defined by $(f_1 * f_2)(x) = f_1(2x)$ if $0 \le x \le \frac{1}{2}$ and $(f_1 * f_2)(x) = f_2(2x - 1)$ if $\frac{1}{2} \le x \le 1$. ∎

It follows from Theorem 2.15 that the path product of two paths is continuous. We use the path product to prove a partial converse to Theorem 3.19, but we need a definition.

Definition. A topological space $(X, \mathcal{T})$ is **locally pathwise connected at a point** p in X if for each neighborhood U of p there is a pathwise connected neighborhood V of p such that $V \subseteq U$. A topological space is **locally pathwise connected** if it is locally pathwise connected at each of its points. ∎

THEOREM 3.20. Let $(X, \mathcal{T})$ be a connected, locally pathwise connected space. Then $(X, \mathcal{T})$ is pathwise connected.

Proof. Let $a \in X$ and let $A = \{x \in X : \text{there is a path from } a \text{ to } x\}$. Since $a \in A$, $A \ne \emptyset$. We prove that $A = X$ by showing that it is open and closed.

Let $b \in A$. Then there is a pathwise connected neighborhood V of b. Let $c \in V$, let f_1 be a path from a to b, and let f_2 be a path from b to c. Then the path product $f_1 * f_2$ is a path from a to c. Thus $c \in A$ and hence $V \subseteq A$. Therefore A is open.

Let $b \in \overline{A}$. There is a pathwise connected neighborhood V of b, and since $b \in \overline{A}$, $V \cap A \ne \emptyset$. Let $c \in V \cap A$, let f_1 be a path from a to c, and let f_2 be a path form c to b. Then $f_1 * f_2$ is a path from a to b, and hence $b \in A$. Therefore A is closed. ∎

Definition. Let $(X, \mathcal{T})$ be a topological space and define an equivalence relation $\sim$ on X by $x \sim y$ provided there is a pathwise connected subset A of X such that $x, y \in A$. The equivalence classes of X with respect to $\sim$ are called the **path components** of X.

We leave as Exercise 1 the proof that $\sim$ is an equivalence relation on X.
The following two theorems are analogues of results obtained in Section 3.1.

THEOREM 3.21. Each path component of a topological space is pathwise connected.

Proof. Let P be a path component of a topological space $(X, \mathcal{T})$, and let $a, b \in P$. Then there is a pathwise connected subset A of X such that $a, b \in A$. Thus there is a path f in A such that $f(0) = a$ and $f(1) = b$. We complete the proof by showing that $A \subseteq P$.

Let $x \in A$. Since A is pathwise connected, there is a path g in A such that $g(0) = x$ and $g(1) = a$. Therefore $x \sim a$ and so $x \in P$. We have proved that $A \subseteq P$, and so f is a path in P. Thus P is pathwise connected. ∎

By Theorem E.1, the collection of path components of a topological space $(X, \mathcal{T})$ forms a partition of X. Notice that a topological space is pathwise connected if and only if it has exactly one path component.

The proof of the following theorem is similar to the proof of Theorem 3.11 and hence it is left as Exercise 2.

THEOREM 3.22. A nonempty pathwise connected subset of a topological space $(X, \mathcal{T})$ is a path component of X if and only if it is not a proper subset of a pathwise connected subset of X. ■

Whereas, by Theorem 3.12, every component of a topological space is a closed subset, we leave as Exercise 3 the proof that one of the path components of the topologist's sine curve is not closed.

Having introduced locally pathwise connected spaces, we now turn to locally connected spaces. Unlike many local properties, there is no relationship between connectedness and local connectedness. Theorem 3.23 illustrates the importance of local connectedness because it tells us that components of a locally connected space are open sets. Example 4 provides an example of a space with a component that is not open. This space is, therefore, not locally connected.

Definition. A topological space $(X, \mathcal{T})$ is **locally connected at a point** p in X if for each neighborhood U of p there is a connected neighborhood V of p such that $V \subseteq U$. A topological space is **locally connected** if it is locally connected at each of its points. ■

Properties similar to local connectedness were studied by Pia Nalli in 1911 and by Stephan Mazurkiewicz (1888–1945) in 1913. Hans Hahn (1879–1934) defined local connectedness in 1914.

By Exercise 4, every locally pathwise connected space is locally connected. The analyses of the following two examples are left as Exercises 5 and 6.

EXAMPLE 7. Let $A = \{\frac{1}{n} : n \in \mathbb{N}\}$ and let $X = (I \times \{0\}) \cup (\{0\} \times I) \cup (A \times I)$. Then X is a subspace of $\mathbb{R} \times \mathbb{R}$ and it is called the **topologist's comb**. The topologist's comb is pathwise connected but it is not locally connected.

The topologist's comb is shown in Figure 3.3.

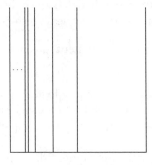

Figure 3.3

EXAMPLE 8. Let $\leq$ denote the dictionary order relation on $I \times I$ determined by less than or equal to on I, and let $\mathcal{T}$ denote the order topology on $I \times I$. Then $(I \times I, \mathcal{T})$ is locally connected but not locally pathwise connected.

THEOREM 3.23. A topological space $(X, \mathcal{T})$ is locally connected if and only if each component of each open set is open.

Proof. Suppose $(X, \mathcal{T})$ is locally connected. Let $U \in \mathcal{T}$, let C be a component of U, and let $p \in C$. There is a connected neighborhood V of p such that $V \subseteq U$. If $V \not\subseteq C$, then C is a proper subset of the connected set $V \cup C$. Therefore $V \subseteq C$ and hence C is open.

Suppose each component of each open set is open. Let $p \in X$ and let U be a neighborhood of p. Then the component V of U that contains p is a connected neighborhood of p such that $V \subseteq U$. Therefore $(X, \mathcal{T})$ is locally connected. ∎

The following corollary is an immediate consequence of Theorem 3.12 and 3.23.

Corollary 3.24. The components of a locally connected space are both open and closed. ∎

The following theorem is an analogue of Theorem 3.10(*a*).

THEOREM 3.25. Let $\{(A_\alpha, \mathcal{T}_\alpha): \alpha \in \Lambda\}$ be a collection of pathwise connected subspaces of a topological space $(X, \mathcal{T})$, and let $A = \bigcup_{\alpha \in \Lambda} A_\alpha$. If $\bigcap_{\alpha \in \Lambda} A_\alpha \neq \emptyset$, then $(A, \mathcal{T}_A)$ is pathwise connected.

Proof. Let $a, b \in A$. Then there exists $\alpha, \beta \in \Lambda$ such that $a \in A_\alpha$ and $b \in A_\beta$. Let $c \in A_\alpha \cap A_\beta$. Then there exists a path f in A_α from a to c and a path g in A_β from c to b. Then the path product $f * g$ is a path in $A_\alpha \cup A_\beta$ from a to b, and therefore $(A, \mathcal{T}_A)$ is pathwise connected. ∎

The following theorem is an analogue of Theorem 3.23.

THEOREM 3.26. A topological space is locally pathwise connected if and only if each path component of each open set is open.

Proof. Let $(X, \mathcal{T})$ be a locally pathwise connected space, let $U \in \mathcal{T}$, let P be a path component of U, and let $p \in P$. There is a pathwise connected neighborhood V of p such that $V \subseteq U$. By Theorem 3.25, $V \cup P$ is pathwise connected. Suppose $V \not\subseteq P$. Then P is a proper subset of the pathwise connected space $V \cup P$. This contradicts Theorem 3.22, so $V \subseteq P$. Hence P is open.

Let $(X, \mathcal{T})$ be a topological space with the property that each path component of each open set is open, let $p \in X$, and let U be a neighborhood of p. Let V be the path component of U that contains p. Then V is a neighborhood of p, and, by Theorem 3.21, V is pathwise connected. Therefore $(X, \mathcal{T})$ is locally pathwise connected. ∎

We now examine the relation between components and path components.

THEOREM 3.27. Each path component of a topological space $(X, \mathcal{T})$ is a subset of a component of X.

Proof. Let P be a path component of a topological space $(X, \mathcal{T})$, let $p \in P$, and let C be the component of X that contains p. We use three theorem to show that $P \subseteq C$. By Theorem 3.21, P is pathwise connected. By Theorem 3.19, P is connected. By Theorem 3.13, $P \subseteq C$. ∎

THEOREM 3.28. Let $(X, \mathcal{T})$ be a locally pathwise connected space. Then the components and path components of X are the same.

Proof. Let $p \in X$, let C be the component of X that contains p, and let P be the path component of X that contains p. Suppose $P \neq C$. By Theorem 3.27, $P \subseteq C$. Therefore there is a path component Q of X such that $Q \neq P$ and $Q \cap C \neq \emptyset$. By Theorem 3.19, Q is connected, so by Theorem 3.13, $Q \subseteq C$. Thus if $R = \bigcup \{Q : Q$ is a path component of X, $Q \neq P$, and $Q \cap C \neq \emptyset\}$, then $C = P \cup R$. By Theorem 3.26, each path component of X is open. Therefore P and R are open subsets of X. Since $P \cap R \neq \emptyset$ and C is connected, we have a contradiction. ∎

The following theorem is nice result about preservation of a local property. It provides a method of proving that the open (closed) continuous image of a locally connected space is locally connected.

THEOREM 3.29. Let $(Y, \mathcal{U})$ be a quotient space of locally connected space $(X, \mathcal{T})$. Then $(Y, \mathcal{U})$ is locally connected.

Proof. Let $f: X \to Y$ be a quotient map. We prove that each component of each open subset of $(Y, \mathcal{U})$ is open. Then by Theorem 3.23, $(Y, \mathcal{U})$ is locally connected.

Let $U \in \mathcal{U}$ and let C be a component of U. For each $x \in f^{-1}(C)$, let C_x be the component of $f^{-1}(U)$ that contains x. Since f is continuous, for each $x \in f^{-1}(C)$, $f(C_x)$ is connected (Theorem 3.5). Since $f(x) \in C \cap f(C_x)$ and C is a component, $f(C_x) \subseteq C$. Therefore $x \in C_x \subseteq f^{-1}(C)$. Since $(X, \mathcal{T})$ is locally connected and C_x is a component of an open set, C_x is open (Theorem 3.23). Therefore $f^{-1}(C)$ is open. Since f is a quotient map, C is open. ∎

The following corollary is an immediate consequence of Theorems 2.55 and 3.29.

Corollary 3.30. The open continuous images and the closed continuous images of locally connected spaces are locally connected. ∎

We conclude this section with a theorem about the product of locally connected spaces.

THEOREM 3.31. Let $\{(X_\alpha, \mathcal{T}_\alpha) : \alpha \in \Lambda\}$ be a colletion of topological spaces, and let $\mathcal{T}$ be the product topology on $X = \prod_{\alpha \in \Lambda} X_\alpha$. Then $(X, \mathcal{T})$ is locally connected if and only if for each $\alpha \in \Lambda$, $(X_\alpha, \mathcal{T}_\alpha)$ is locally connected and for all but a finite number of $\alpha \in \Lambda$, $(X_\alpha, \mathcal{T}_\alpha)$ is connected.

Proof. Suppose $(X, \mathcal{T})$ is locally connected. Let $\alpha \in \Lambda$. Then π_α is an open continuous function, so by Corollary 3.30, $\pi_\alpha(X) = X_\alpha$ is locally connected.

Let $p \in X$, and V be a connected neighborhood of p. Then there exist $\alpha_1, \alpha_2, ..., \alpha_n$ in Λ and open sets $U_{\alpha_1}, U_{\alpha_2}, ..., U_{\alpha_n}$ in $X_{\alpha_1}, X_{\alpha_2}, ..., X_{\alpha_n}$ such that $p \in \bigcap_{i=1}^{n} \pi_{\alpha_i}^{-1}(U_{\alpha_i}) \subseteq V$. Thus if $\alpha \neq \alpha_i$ for any $i = 1, 2, ..., n$, then $\pi_\alpha(V) = X_\alpha$. Therefore, since V is connected and π_α is continuous, X_α is connected for all $\alpha \in \Lambda$ except possibly $\alpha_1, \alpha_2, ..., \alpha_n$.

Suppose that for each $\alpha \in \Lambda$, $(X_\alpha, \mathcal{T}_\alpha)$ is locally connected and that there exists a finite subset Γ of Λ such that if $\alpha \notin \Gamma$, then $(X_\alpha, \mathcal{T}_\alpha)$ is connected. Let $p \in X$ and let W be a neighborhood of p. Then there exist open sets U_{α_i}, $i = 1, 2, ..., n$, such that $p \in \pi_{\alpha_i}^{-1}(U_{\alpha_i}) \subseteq W$. Let $\alpha \in \Lambda$.

(1) If $\alpha \neq \alpha_i$ for any $i = 1, 2, ..., n$, and $\alpha \notin \Gamma$, let $V_\alpha = X_\alpha$.

(2) If $\alpha \neq \alpha_i$ for any $i = 1, 2, ..., n$, and $\alpha \notin \Gamma$, let V_α be a connected open subset of X_α that contains p_α.

(3) If $\alpha = \alpha_i$ for some $i = 1, 2, ..., n$, let V_α be a connected neighborhood of p_α such that $V_\alpha \subseteq U_\alpha$.

Then $\prod_{\alpha \in \Lambda} V_\alpha$ is a connected neighborhood of p that is contained in W. Therefore $(X, \mathcal{T})$ is a locally connected. ∎

EXERCISES 3.2

1. Let $(X, \mathcal{T})$ be a topological space and define a relation $\sim$ on X by $x \sim y$ provided there is a pathwise connected subset A of X such that $x, y \in A$. Prove that $\sim$ is an equivalence relation on X.

2. Prove that a nonempty pathwise connected subset of a topological space $(X, \mathcal{T})$ is a path component of X if and only if it is not a proper subset of a pathwise connected subset of X.

3. Show that one of the path components of the topologist's sine curve is not closed.

4. Prove that every locally pathwise connected space is locally connected.

5. Show that the topologist's comb is pathwise connected but not locally connected.

6. Let $\leq$ denote the dictionary order relation on $I \times I$ determined by less than or equal to on I, and let $\mathcal{T}$ denote the order topology on $I \times I$. Prove that $(I \times I, \mathcal{T})$ is locally connected but not locally pathwise connected.

7. Prove that the continuous image of a pathwise connected space is pathwise connected.

8. Let $N \in \mathbb{N}$, for each $i = 1, 2, ..., N$, let $(X_i, \mathcal{T}_i)$ be a topological space, and let $\mathcal{T}$ be the product topology on $X = \prod_{i=1}^{N} X_i$. Prove that $(X, \mathcal{T})$ is pathwise connected if and only if for each $i = 1, 2, ..., N$, $(X_i, \mathcal{T}_i)$ is pathwise connected.

9. Prove that the path components of a topological space $(X, \mathcal{T})$ are pathwise connected disjoint subsets of X whose union is X and that each pathwise connected subset of X intersects only one path component.

10. Prove that if A is a countable subset $\mathbb{R}^2$, then $\mathbb{R}^2 - A$ is pathwise connected.

11. Prove that if U is an open connected subset of $\mathbb{R}^2$, then U is pathwise connected.

12. The **deleted comb space** is the space obtained from the topologist's comb by deleting the open interval $\{0\} \times (0, 1)$. Prove that the deleted comb space is connected and has two path components.

13. Let $A = \{(0, x) \in I \times I : x \text{ is irrational}\}$, let B be the deleted comb space, and let $X = A \cup B$. Prove that X is connected and has an uncountable number of path components.

14. Let $(X, \mathcal{T})$ be a locally connected space. Prove that the quasi-components of X are the same as the components of X.

15. Let $A = \{(1, x): x \in \mathbb{R}\}$ and let $B = \{(-1, x): x \in \mathbb{R}\}$. For each $n \in \mathbb{N}$, let $R_n = \{(x, y) \in \mathbb{R} \times \mathbb{R} : |x| \leq n/(n + 1) \text{ and } |y| \leq n\}$. Let $X = A \cup B \cup (\bigcup_{n \in \mathbb{N}} \text{bd}(R_n))$ be a subspace of $\mathbb{R} \times \mathbb{R}$.

    (a) Find the component of X that contains $(1, 0)$.

    (b) Find the quasi-component of X that contains $(1, 0)$.

3.3 Totally Disconnected Spaces

Each component of the subspace of the real line consisting of the rational numbers is a set with exactly one member. Another famous example of such a space is the Cantor set, which we define in this section. Surprisingly, this set turns out to be a closed and uncountable subset of $\mathbb{R}$. But first we give a definition.

Definition. A topological space $(X, \mathcal{T})$ is **totally disconnected** provided that each component of X is a set consisting of a single point. ∎

Notice that if $\mathcal{T}$ is the discrete topology on a set X, then $(X, \mathcal{T})$ is totally disconnected. We define the Cantor set.

EXAMPLE 9. Let $C_1 = [0, \frac{1}{3}] \cup [\frac{2}{3}, 1]$ be the subset of I obtained by removing the open middle third $(\frac{1}{3}, \frac{2}{3})$. Let $C_2 = [0, \frac{1}{9}] \cup [\frac{2}{9}, \frac{1}{3}] \cup [\frac{2}{3}, \frac{7}{9}] \cup [\frac{8}{9}, 1]$ be the subset of C_1 obtained by removing the open middle thirds $(\frac{1}{9}, \frac{2}{9})$ and $(\frac{7}{9}, \frac{8}{9})$ of the two components of C_1. (The sets C_1 and C_2 are illustrated in Figure 3.4.) In general for each $n \in \mathbb{N}$, C_{n+1} is the subset of C_n obtained by removing the open middle thirds of the 2^n components of C_n. Then $C = \bigcap_{n=1}^{\infty} C_n$ is the Cantor set. The topology on C is the subspace topology.

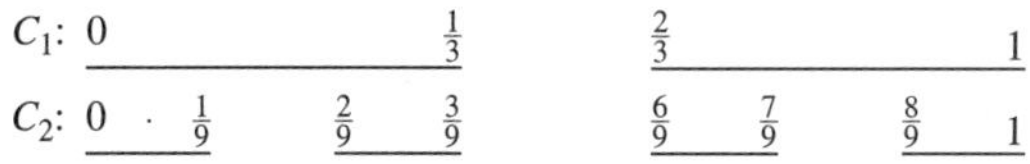

C_1: 0 ─────────── $\frac{1}{3}$ $\frac{2}{3}$ ─────────── 1

C_2: 0 ─ $\frac{1}{9}$ $\frac{2}{9}$ ─ $\frac{3}{9}$ $\frac{6}{9}$ ─ $\frac{7}{9}$ $\frac{8}{9}$ ─ 1

Figure 3.4

Before examining some to the properties of the Cantor set, we give an alternative definition. Note that each $x \in I$ has a **ternary expansion**; that is, each $x \in I$ can be written $x = 0.x_1 x_2 x_3 \ldots = \sum_{n=1}^{\infty} x_n/3^n$, where each x_n is 0, 1, or 2. A ternary expansion of $x \in I$ of is unique except that if $x \neq 1$ and x has a ternary expansion ending in a sequence of 2's, then x has a ternary expansion ending in a sequence of 0's. As examples, note that

$$\frac{1}{3} = \frac{0}{3} + \frac{2}{3^2} + \frac{2}{3^3} + \frac{2}{3^4} + \cdots = \frac{1}{3} + \frac{0}{3^2} + \frac{0}{3^3} + \frac{0}{3^4} + \cdots$$

$$\frac{1}{9} = \frac{0}{3} + \frac{0}{3^2} + \frac{2}{3^3} + \frac{2}{3^4} + \cdots = 0 + \frac{1}{3^2} + \frac{0}{3^3} + \frac{0}{3^4} + \cdots$$

The numbers in I that require 1 in the first place of their ternary expansions are those numbers in the open interval $(\frac{1}{3}, \frac{2}{3})$. The numbers in C_1 that require 1 in the second place of their ternary expansions are those numbers in the open intervals $(\frac{1}{9}, \frac{2}{9})$ and $(\frac{7}{9}, \frac{8}{9})$. In general the numbers in I that require 1 in the nth place of their ternary expansions are those numbers in the open intervals that are removed in order to form C_n from C_{n-1}. Therefore the Cantor set is the set of all $x \in I$ that have a ternary expansion $x = 0.x_1 x_2 x_3 \ldots$, where each x_n is 0 or 2.

For each $n \in \mathbb{N}$, let $X_n = \{0, 2\}$ and let $\mathcal{T}_n$ be the discrete topology on X_n. Let $X = \prod_{n \in \mathbb{N}} X_n$ and let $\mathcal{T}$ be the product topology on X. By Example 1 in Appendix G, X is uncountable. In Exercise 3 you are asked to show that the Cantor set is homeomorphic to $(X, \mathcal{T})$. Thus the Cantor set is uncountable. Since it is the intersection of a collection of closed sets, it is closed. In Exercises 4–6, you are asked to prove that the Cantor set is a perfect, totally disconnected, nowhere dense subset of $\mathbb{R}$.

We now prove that total disconnectedness is preserved by products and subspaces.

THEOREM 3.32. The product of totally disconnected spaces is totally disconnected.

Proof. Let $\{(X_\alpha, \mathcal{T}_\alpha) : \alpha \in \Lambda\}$ be a collection of totally disconnected spaces, let $X = \prod_{\alpha \in \Lambda} X_\alpha$, and let C be a nonempty connected subset of X. Since each projection map is continuous, for each $\alpha \in \Lambda$, $\pi_\alpha(C)$ is connected. Therefore for each

$\alpha \in \Lambda$, $\pi_\alpha(C)$ is a set consisting of a single point. Hence C is a set consisting of a single point, and so X is totally disconnected. ∎

THEOREM 3.33. Every subspace of a totally disconnected space is totally disconnected.

Proof. Let $(X, \mathcal{T})$ be a totally disconnected space, let $A \subseteq X$, and let C be a nonempty connected subset of A. Then C is a nonempty connected subset of X and hence C is a set consisting of a single point. ∎

The following dimensional concept is related to total disconnectedness, but, as Theorem 3.34 and the comment following indicate, it is stronger than total disconnectedness.

Definition. A topological space $(X, \mathcal{T})$ is **0-dimensional** if for each $p \in X$ and each neighborhood U of p there is a closed and open set V such that $p \in V$ and $V \subseteq U$. ∎

An example of a 0-dimensional space is provided in Exercise 8. An additional condition, which is obviously weaker than Hausdorff, is needed in order to obtain the last result of this chapter.

Definition. A topological space $(X, \mathcal{T})$ is a T_0**-space** provided that for each pair of distinct points there is an open set that contains one of the points but not the other. ∎

THEOREM 3.34. Every 0-dimensional T_0-space is totally disconnected.

Proof. Let $(X, \mathcal{T})$ be a 0-dimensional T_0-space, and let C be a subset of X that contains two distinct points p and q. Since X is T_0, we may assume that there exists an open set U such that $p \in U$ and $q \notin U$. Since X is 0-dimensional, there exists a closed and open set A such that $p \in A \subseteq U$. Then $A \cap C$ is a nonempty proper subset of C that is both open and closed in C. By Theorem 3.3, C is not connected. ∎

The Cantor set was studied by Georg Cantor in 1883. It is a good set for examples, counterexamples, intuitiveness, and counter-intuitiveness. Some additional exercises involving the Cantor set will be provided when compactness is introduced.

In a series of articles published in *Mathematische Annalen* in the period between 1870 and 1890, Cantor laid out many properties of $\mathbb{R}$ that are now considered the basic concepts of general topology. Thus he is considered to be one of the originators of general topology. He was born in St. Petersburg in 1845. He moved to Germany in 1856 and studied at the University of Berlin, where he was influenced by Weierstrass, from 1863 to 1869. He taught at the University of Halle from 1869 to 1905, and he died in a mental hospital in 1918.

In Exercise 9, we give an example of a space that is totally disconnected but not 0-dimensional, and in the next chapter we give a condition on a space that ensures these are equivalent concepts.

EXERCISES 3.3

1. Which of the following are in the Cantor set: $\frac{7}{12}, \frac{1}{3}, \frac{1}{4}, \frac{11}{12}$? Give the ternary expansion of each.

2. Prove that each C_n, defined in the construction of the Cantor set, has 2^n components.

3. For each $n \in \mathbb{N}$, let $X_n = \{0, 2\}$ and let $\mathcal{T}_n$ be the discrete topology on X_n. Let $X = \prod_{n \in \mathbb{N}} X_n$, and let $\mathcal{T}$ be the product topology on X. Prove that the Cantor set is homeomorphic to $(X, \mathcal{T})$.

4. Prove that the Cantor set is a nowhere dense subset of $\mathbb{R}$.

5. Prove that the Cantor set is totally disconnected.

6. Prove that the Cantor set is a perfect set.

7. Prove that every subspace of a 0-dimensional space is 0-dimensional.

8. Prove that the real line with the lower limit topology is 0-dimensional.

9. As Example 9 indicates, the Cantor set C is obtained by deleting a countable collection of open intervals from I. Let $A = \{x \in I : x$ is an endpoint of an open interval that is deleted in the construction of the Cantor set$\}$, and let $B = C - A$. Let $p = (\frac{1}{2}, \frac{1}{2}) \in \mathbb{R}^2$, and for each $x \in C$, Let L_x denote the closed line segment joining x and p. For each $x \in A$, let $L_x^* = \{(x_1, x_2) \in L_x : x_2$ is rational,$\}$ and for each $x \in B$, let $L_x^* = (x_1, x_2) \in L_x : x_2$ is irrational. Prove that the subspace $X = \bigcup_{x \in C} L_x^*$ of $\mathbb{R}^2$ is connected, that $X - \{p\}$ is totally disconnected, and that $X - \{p\}$ is not 0-dimensional.

Chapter 4
Compactness

The concept of compactness is a generalization of an important result in classical analysis. The result we have in mind is the Heine-Borel Theorem, named in honor of Eduard Heine (1821–1881) and Emile Borel (1871–1956), which asserts that if a and b are real numbers with $a < b$ and $\mathcal{O}$ is a collection of open intervals such that $[a, b] \subseteq \bigcup \{O: O \in \mathcal{O}\}$, then there is a finite subset $\{O_1, O_2, ..., O_N\}$ of $\mathcal{O}$ such that $[a, b] \subseteq \bigcup_{n=1}^{N} O_n$. The Heine-Borel Theorem guarantees that a continuous real-valued function whose domain is a closed interval $[a, b]$ has several "nice" properties. For example, it guarantees that $f([a, b])$ is closed and bounded. The conclusion of the Heine-Borel Theorem is converted into the definition of a compact topological space.

4.1 Compactness in Metric Spaces

The purpose of this section is to prove some important characterizations of compact metric spaces. We begin by defining compactness in an arbitrary topological space and giving some examples of both compact and non-compact spaces. Recall that a collection $\mathcal{A}$ of subsets of a topological space $(X, \mathcal{T})$ is a **cover** of a subset B of X provided $B \subseteq \bigcup_{A \in \mathcal{A}} A$, and the cover $\mathcal{A}$ of B is an **open cover** of B provided each member of $\mathcal{A}$ is open.

Definition. A subset A of a topological space $(X, \mathcal{T})$ is **compact** provided every open cover of A has a finite subcover. ∎

EXAMPLE 1.

(a) If X is a finite set and $\mathcal{T}$ is any topology on X, then $(X, \mathcal{T})$ is compact.

(b) If $\mathcal{T}$ is the finite complement topology on any set X, then $(X, \mathcal{T})$ is compact.

(c) Let $X = \{0\} \cup \{\frac{1}{n}: n \in \mathbb{N}\}$, and let $\mathcal{T}$ be the subspace topology. If $\mathcal{U}$ is any open cover of X, then there exists $U \in \mathcal{U}$ such that $0 \in U$. Since all but a finite number of members of X are members of U, $(X, \mathcal{T})$ is compact.

(d) Let $\mathcal{U}$ be an open cover of a closed interval $[a, b]$. For each $x \in [a, b]$, let $U_x \in \mathcal{U}$ such that $x \in U_x$. Then there is an open interval I_x such that $x \in I_x \subseteq U_x$. By the Heine-Borel Theorem (see Exercise 2), there is a finite subcollection $\{I_{x_1}, I_{x_2}, ..., I_{x_n}\}$ of $\{I_x : x \in [a, b]\}$ such that $[a, b] \subseteq \bigcup_{i=1}^{n} I_{x_i}$. The collection $\{U_{x_1}, U_{x_2}, ..., U_{x_n}\}$ of members of $\mathcal{U}$ that corresponds to $I_{x_1}, I_{x_2}, ..., I_{x_n}$ is a finite subcollection of $\mathcal{U}$ that covers $[a, b]$. Therefore $[a, b]$ is compact.

EXAMPLE 2.

(a) If X is an infinite set and $\mathcal{T}$ is the discrete topology on X, then $(X, \mathcal{T})$ is not compact.

(b) The open interval $(0, 1)$ is not compact because $\{(\frac{1}{n}, 1) : n \in \mathbb{N}\}$ is an open cover that does not have a finite subcover.

(c) The real line is not compact because $\{(-n, n) : n \in \mathbb{N}\}$ is an open cover that does not have a finite subcover.

(d) The real line with the lower limit topology is not compact because $\{[n, n+1) : n \in \mathbb{Z}\}$ is an open cover that does not have a finite subcover.

One characterization of compact metric spaces involves the following concept.

Definition. Let (X, d) be a metric space and let $\varepsilon > 0$. A finite subset F of X is called an ε**-net for** X if for each $x \in X$ there exists $p \in F$ such that $d(x, p) < \varepsilon$. A metric space (X, d) is **totally bounded** if for each $\varepsilon > 0$, there is an ε-net for X. ∎

We prove that every totally bounded metric space is bounded and give an example to show that the converse is not true.

THEOREM 4.1. Every totally bounded metric space is bounded.

Proof. Let (X, d) be a totally bounded metric space. Let F be a 1-net for X and let $M = \max\{d(p, q) : p, q \in F\}$. Let $x, y \in X$. Then there exists $a, b \in F$ such that $d(x, a) < 1$ and $d(y, b) < 1$. Therefore $d(x, y) \leq d(x, a) + d(a, b) + d(b, y) < M + 2$. Therefore $M + 2$ is a real number such that $d(x, y) < M + 2$ for all $x, y \in X$. Hence (X, d) is bounded. ∎

EXAMPLE 3. For $x, y \in \mathbb{R}$, let $d(x, y) = \min\{|x - y|, 1\}$. Then $(\mathbb{R}, d)$ is a bounded metric space, but it is not totally bounded.

Analysis. By Theorem 1.30, d is a metric on $\mathbb{R}$. Since $d(x, y) \leq 1$ for all $x, y \in \mathbb{R}$, $(\mathbb{R}\ d)$ is bounded.

We show that $(\mathbb{R}, d)$ is not totally bounded by showing that there is no $\frac{1}{2}$-net for $\mathbb{R}$. Let F be any finite subset of $\mathbb{R}$ and let p be the largest member of F. If $q \in F$, then $d(p + 1, q) = 1$, and hence there is no $\frac{1}{2}$-net for $\mathbb{R}$. ∎

4.1 Compactness in Metric Spaces

The following property of a topological space is named in honor of Bernard Bolzano (1781–1848) and Karl Weierstrass. It is a condition that guarantees total boundedness of a metric space.

Definition. A topological space $(X, \mathcal{T})$ has the **Bolzano-Weierstrass property** provided that every infinite subset of X has a limit point. ∎

THEOREM 4.2. Let (X, d) be a metric space that has the Bolzano-Weierstrass property. Then (X, d) is totally bounded.

Proof. We give a contrapositive argument. Suppose (X, d) is a metric space that is not totally bounded. Then there is a positive number ε such that X has no ε-net. We construct an infinite subset of X (a sequence) that has no limit point. Let $x_1 \in X$. Then $\{x_1\}$ is a finite subset of X, so there exists $x_2 \in X$ such that $d(x_1, x_2) \geq \varepsilon$. We proceed inductively. Suppose $n \in \mathbb{N}$ and a subset $\{x_i : i = 1, 2, \ldots, n\}$ of X has been constructed so that $d(x_i, x_j) \geq \varepsilon$ for all $i, j = 1, 2, \ldots, n$ with $i \neq j$. Then $\{x_i : i = 1, 2, \ldots, n\}$ is a finite set and hence there exists $x_{n+1} \in X$ such that $d(x_{n+1}, x_i) \geq \varepsilon$ for each $i = 1, 2, \ldots, n$. Therefore by induction there is an infinite subset $A = \langle x_n \rangle$ of X such that $d(x_i, x_j) \geq \varepsilon$ for all $i, j \in \mathbb{N}$ with $i \neq j$. Hence A has no limit point and therefore (X, d) does not have the Bolzano-Weierstrass property. ∎

Theorem 4.4 is one of the fundamental results about compact metric spaces. In order to prove this theorem, we introduce the following concept and prove Theorem 4.3.

Definition. A topological space $(X, \mathcal{T})$ is a T_1-**space** provided that for each pair x, y of distinct points of X there exist open sets U and V such that $x \in U$, $y \notin U$, $y \in V$, and $x \notin V$. ∎

Notice that every Hausdorff space is obviously a T_1-space.

THEOREM 4.3. Let $(X, \mathcal{T})$ be a T_1-space, let $A \subseteq X$, and let p be a limit point of A. Then every neighborhood of p contains an infinite number of distinct members of A.

Proof. Suppose there is a neighborhood U of p that contains only a finite number of members $x_1, x_2, \ldots, x_n$ of A (distinct from p). For each $i = 1, 2, \ldots, n$, let U_i be a neighborhood of p such that $x_i \notin U_i$. (Since $(X, \mathcal{T})$ is a T_1-space, such a neighborhood exists.) Then $U \cap \bigcap_{i=1}^{n} U_i$ is a neighborhood of p that does not contain a point of A different from p. This is a contradiction since p is a limit point of A. ∎

THEOREM 4.4. Let (X, d) be a metric space that has the Bolzano-Weierstrass property, and let $\mathcal{O}$ be an open cover of X. Then there is a positive number δ such that for each $x \in X$, $B_d(x, \delta)$ is a subset of some member of $\mathcal{O}$.

Proof. The proof is by contradiction. Suppose there is no positive number δ such that for each $x \in X$, $B_d(x, \delta)$ is a subset of some member of $\mathcal{O}$. Then for each natural number n, there exists $x_n \in X$ such that $B_d(x_n, \frac{1}{n})$ is not a subset of any member of $\mathcal{O}$. We show that $A = \{x_n: n \in \mathbb{N}\}$ is an infinite set. Let $y \in X$. Then there exists a member U of $\mathcal{O}$ such that $y \in U$. Since U is open, there is a positive number ε such that $B_d(y, \varepsilon) \subseteq U$. Let $N \in \mathbb{N}$ such that $\frac{1}{N} < \varepsilon$. Then for each $n > N$, $B_d(y, \frac{1}{n}) \subseteq U$. Therefore if $n > N$, $x_n \neq y$, so A is an infinite set.

Since (X, d) has the Bolzano-Weierstrass property, there is a limit point a of A. Let $V \in \mathcal{O}$ such that $a \in V$. Since V is open, there is a positive number r such that $B_d(a, r) \subseteq V$. By Theorem 4.3, $B_d(a, \frac{r}{2})$ contains an infinite number of members of A. Therefore there is a natural number M such that $\frac{1}{M} < \frac{r}{2}$ and $x_M \in B_d(a, \frac{r}{2})$. Let $z \in B_d(x_M, \frac{1}{M})$. Then $d(a, z) \leq d(a, x_M) + d(x_M, z) < \frac{r}{2} + \frac{1}{M} < r$. Hence $B_d(x_M, \frac{1}{M}) \subseteq B_d(a, r) \subseteq V$. This is a contradiction because $B_d(x_M, \frac{1}{M})$ is not a subset of any member of $\mathcal{O}$. ∎

Definition. Let $\mathcal{O}$ be an open cover of a metric space (X, d). A **Lebesgue number** for $\mathcal{O}$ is a positive number ε such that each subset of X of diameter less than ε is a subset of some member of $\mathcal{O}$. ∎

By Theorem 4.4, every open cover of a metric space with the Bolzano-Weierstrass property has a Lebesgue number.

In order to give the first three characterizations of compactness in metric spaces, we introduce some additional concepts and prove a sequence of theorems that do not require a metric setting.

Definition. A topological space $(X, \mathcal{T})$ is **countably compact** provided every countable open cover of X has a finite subcover. ∎

Notice that every compact space is countably compact.

Definition. A family $\mathcal{A}$ of subsets of a set has the **finite intersection property** provided that if $\mathcal{A}'$ is a finite subcollection of $\mathcal{A}$ then $\bigcap \{A: A \in \mathcal{A}'\} \neq \emptyset$. ∎

EXAMPLE 4. Let $\mathcal{A} = \{[a, \infty): a \in \mathbb{R} \text{ and } a > 0\}$. Then $\mathcal{A}$ has the finite intersection property. However, $\bigcap_{A \in \mathcal{A}} A = \emptyset$.

THEOREM 4.5. A topological space $(X, \mathcal{T})$ is countably compact if and only if every countable family of closed subsets of X with the finite intersection property has a nonempty intersection.

Proof. Suppose $(X, \mathcal{T})$ is countably compact. Let $\mathcal{A} = \{A_\alpha: \alpha \in \Lambda\}$ be a countable family of closed subsets of X with the finite intersection property. Suppose $\bigcap_{\alpha \in \Lambda} A_\alpha = \emptyset$. Let $\mathcal{U} = \{X - A_\alpha: \alpha \in \Lambda\}$. Since $\bigcup_{\alpha \in \Lambda}(X - A_\alpha) = X - \bigcap_{\alpha \in \Lambda} A_\alpha = X - \emptyset = X$, $\mathcal{U}$ is an open cover of X. Since X is countably compact, there exists a finite number $\alpha_1, \alpha_2, \ldots, \alpha_n$ of members of Λ such that $\{X - A_{\alpha_i}: i = 1, 2, \ldots, n\}$

covers X. Thus $X = \bigcup_{i=1}^{n}(X - A_{\alpha_i}) = X - \bigcap_{i=1}^{n} A_{\alpha i}$, and hence $\bigcap_{i=1}^{n} A_{\alpha i} = \emptyset$. This is a contradiction, and hence $\bigcap_{\alpha \varepsilon \Lambda} A_{\alpha} \neq \emptyset$.

Suppose every countable family of closed subsets of X with the finite intersection property has a nonempty intersection. Let $\mathcal{U} = \{U_{\alpha} : \alpha \in \Lambda\}$ be a countable open cover of X. Suppose $\mathcal{U}$ does not have a finite subcover. Let $\mathcal{A} = \{X - U_{\alpha} : \alpha \in \Lambda\}$. Then $\mathcal{A}$ is a countable family of closed subsets of X. Let Γ be a finite subset of Λ. Since $\mathcal{U}$ does not have a finite subcover, $\bigcap_{\alpha \in \Gamma}(X - U_{\alpha}) = X - \bigcup_{\alpha \in \Gamma} U_{\alpha} \neq \emptyset$. Therefore $\mathcal{A}$ has the finite intersection property. Hence $\bigcap_{\alpha \in \Lambda}(X - U_{\alpha}) \neq \emptyset$. This is a contradiction because $\bigcap_{\alpha \in \Lambda}(X - U_{\alpha}) = X - \bigcup_{\alpha \in \Lambda} U_{\alpha} = \emptyset$. Therefore $(X, \mathcal{T})$ is countably compact. ∎

THEOREM 4.6. Every countably compact space has the Bolzano-Weierstrass property.

Proof. Let $(X, \mathcal{T})$ be a countably compact space, and let A be an infinite subset of X. By Theorem F.6, A contains a countably infinite set $B = \{x_i : i \in \mathbb{N}\}$. We may assume that if $i \neq j$ then $x_i \neq x_j$. The proof is by contradiction. Suppose B has no limit point. Then for each $n \in \mathbb{N}$, $C_n = \{x_i \in B : i \geq n\}$ is a closed set. Furthermore, $\{C_n : n \in \mathbb{N}\}$ has the finite intersection property. Therefore, by Theorem 4.5, $\bigcap_{n=1}^{\infty} C_n \neq \emptyset$. But if $x_k \in B$, then $x_k \notin C_{k+1}$, and hence $x_k \notin \bigcap_{n=1}^{\infty} C_n$. Therefore $\bigcap_{n=1}^{\infty} C_n = \emptyset$, and we have a contradiction. Since B has a limit point and $B \subseteq A$, A has a limit point. ∎

THEOREM 4.7 Let $(X, \mathcal{T})$ be a T_1-space. Then X is countably compact if and only if it has the Bolzano-Weierstrass property.

Proof. Suppose $(X, \mathcal{T})$ has the Bolzano-Weierstrass property. The proof that X is countably compact is by contradiction. Suppose $\{U_n : n \in \mathbb{N}\}$ is a countable open cover of X that has no finite subcover. Then for each $n \in \mathbb{N}$, $C_n = X - \bigcup_{i=1}^{n} U_i$ is a nonempty closed set. For each $n \in \mathbb{N}$, let $p_n \in C_n$ and let $A = \{p_n : n \in \mathbb{N}\}$. If A is finite, there exists $p \in A$ such that $p_n = p$ for an infinite number of $n \in \mathbb{N}$. Thus for each $n \in \mathbb{N}$, $p \in C_n$. This is a contradiction since $\{U_n : n \in \mathbb{N}\}$ covers X. Suppose that A is infinite. Then, by hypothesis, there is a point $p \in X$ that is a limit point of A. Since X is a T_1-space, by Theorem 4.3 every neighborhood of p contains an infinite number of members of A. Therefore for each $n \in \mathbb{N}$, p is a limit point of $A_n = \{p_i \in A : i > n\}$. For each $n \in \mathbb{N}$, $A_n \subseteq C_n$. Since for each $n \in \mathbb{N}$, C_n is closed, $p \in C_n$. Once again this is a contradiction, since $\{U_n : n \in \mathbb{N}\}$ covers X. Therefore X is countably compact.

If $(X, \mathcal{T})$ is a countably compact space, then by Theorem 4.6, it has the Bolzano-Weierstrass property. ∎

In Section 4.3, we give an example of a countably compact space that is not compact and an example of a topological space with the Bolzano-Weierstrass property that is not countably compact.

Definition. A topological space $(X, \mathcal{T})$ is **sequentially compact** provided every sequence in X has a subsequence that converges. ∎

It is somewhat surprising that there is a compact space that is not sequentially compact, but we give such an example in Section 4.3.

THEOREM 4.8. Every sequentially compact space is countably compact.

Proof. Suppose $(X, \mathcal{T})$ is a topological space that is not countably compact. Let $\mathcal{U}$ be a countable open cover that does not have a finite subcover. Choose $x_1 \in X$. For each $j > 1$, let $U_j \in \mathcal{U}$ that contains a point x_j that is not in $\bigcup_{i=1}^{j-1} U_i$. We claim the sequence $\langle x_n \rangle$ does not have a subsequence that converges. Let $x \in X$. Then there exists k such that $x \in U_k$. Then $x_j \notin U_k$ for any $j > k$. Thus no subsequence of $\langle x_n \rangle$ converges to x. Since x is an arbitrary point, no subsequence of $\langle x_n \rangle$ converges. Therefore $(X, \mathcal{T})$ is not sequentially compact. ∎

THEOREM 4.9. Every countably compact, first countable space is sequentially compact.

Proof. Let $(X, \mathcal{T})$ be a countably compact, first countable space. By Theorem 1.52, it is sufficient to show that every sequence in X clusters in X. The proof is by contradiction. Suppose there is a sequence $\langle x_n \rangle$ in X that does not cluster in X. For each $n \in \mathbb{N}$, let $U_n = \bigcup \{V \in \mathcal{T} : x_i \notin V \text{ for any } i \geq n\}$. Then $\mathcal{U} = \{U_n : n \in \mathbb{N}\}$ is a countable collection of open sets such that $U_n \subseteq U_{n+1}$ for each $n \in \mathbb{N}$. We show that $\mathcal{U}$ covers X. Let $x \in X$. Since $\langle x_n \rangle$ does not cluster at x, there exists a neighborhood V of x and an $i \in \mathbb{N}$ such that for all $j \in \mathbb{N}$ with $j \geq i$, $x_j \notin V$. Thus $x \in U_i$, and so $\mathcal{U}$ covers X. Since $(X, \mathcal{T})$ is countably compact, there is a finite subcollection $\{U_{i_1}, U_{i_2}, \ldots, U_{i_n}\}$ of $\mathcal{U}$ that covers X. Let $N = \max\{i_1, i_2, \ldots, i_n\}$. Then $U_{i_j} \subseteq U_N$ for each $j = 1, 2, \ldots, n$. Thus $U_N = X$. This is a contradiction because $x_N \notin U_N$. ∎

We are now ready to prove the first theorem that gives some characterizations of compactness in metric spaces.

THEOREM 4.10. Let (X, d) be a metric space. Then the following four statements are equivalent.

(a) (X, d) is compact.

(b) (X, d) is sequentially compact.

(c) (X, d) is countably compact.

(d) (X, d) has the Bolzano-Weierstrass property.

Proof. (a) $\to$ (b). Suppose (X, d) is compact. Then (X, d) is countably compact. Since every metric space is first countable (Theorem 1.14), by Theorem 4.9, (X, d) is sequentially compact.

(b) $\Rightarrow$ (c). Theorem 4.8
(c) $\Rightarrow$ (d). Theorem 4.6
(d) $\Rightarrow$ (a). Suppose (X, d) has the Bolzano-Weierstrass property, and let $\mathcal{O}$ be an open cover of X. By Theorem 4.4, there is a positive number δ such that for each $x \in X$, $B_d(x, \delta)$ is a subset of some member of $\mathcal{O}$. By Theorem 4.2, there is a finite subset $F = \{x_1, x_2, ..., x_n\}$ of X such that $\{B_d(x_i, \delta): i = 1, 2, ..., n\}$ covers X. For each $i = 1, 2, ..., n$, choose a member U_i of $\mathcal{O}$ such that $B_d(x_i, \delta) \subseteq U_i$. Then $\{U_i: i = 1, 2, ..., n\}$ is a finite subcover of $\mathcal{O}$, and hence (X, d) is compact. ∎

Is compactness related to completeness in a metric space? The real line is complete but not compact. We do, however, have the following result.

THEOREM 4.11. A metric space (X, d) is compact if and only if it is complete and totally bounded.

Proof. Let (X, d) be a compact metric space. By Theorem 4.10, (X, d) is sequentially compact. Therefore by Theorem 1.43, (X, d) is complete. We now prove that (X, d) is totally bounded. Let $\varepsilon > 0$, and let $\mathcal{O}$ be a cover of X by open balls of radius ε. Since (X, d) is compact, $\mathcal{O}$ has a finite subcover $\{U_1, U_2, ..., U_n\}$. For each $i = 1, 2, ..., n$, let $p_i \in U_i$. Then $\{p_1, p_2, ..., p_n\}$ is an ε-net for X and hence X is totally bounded.

Suppose that (X, d) is a totally bounded complete metric space. We prove that (X, d) is compact by showing that it is sequentially compact. Let $\langle x_n \rangle$ be a sequence in X. We construct a subsequence of $\langle x_n \rangle$ that is a Cauchy sequence. Since (X, d) is totally bounded, we can cover X with a finite number of open balls of radius 1. There is one of these open balls, call it B_1, such that $x_n \in B_1$ for an infinite number of values of n. Let $A_1 = \{n \in \mathbb{N}: x_n \in B_1\}$. We can also cover X with a finite number of open balls of radius $\frac{1}{2}$, and there is one of these open balls, call it B_2, and an infinite subset C_1 of A_1 such that $x_n \in B_2$ for each $n \in C_1$. Let $A_2 = \{n \in A_1: x_n \in B_2\}$. We proceed inductively. Suppose that for each $i = 2, 3, ..., k$, we have an open ball B_i of radius $\frac{1}{i}$, and an infinite subset C_{i-1} of A_{i-1} such that $x_n \in B_i$ for each $n \in C_{i-1}$ and that $A_i = \{n \in A_{i-1}: x_n \in B_i\}$. There is an open ball B_{k+1} of radius $1/(k+1)$ and an infinite subset C_k of A_k such that $x_n \in B_{k+1}$ for each $n \in C_k$. Let $A_{k+1} = \{n \in A_k: x_n \in B_{k+1}\}$. Therefore by induction, we have an infinite collection of sets $\{A_i: i \in \mathbb{N}\}$ such that $A_{k+1} \subseteq A_k$ for each $k \in \mathbb{N}$. Let $i_1 \in A_1$. Since A_2 is an infinite set, there exists $i_2 \in A_2$ such that $i_2 > i_1$. In general, suppose that for each $j = 2, 3, ..., n$, $i_j \in A_j$ has been chosen so that $i_j > i_{j-1}$. Since A_{n+1} is an infinite set, there exists $i_{n+1} \in A_{n+1}$ such that $i_{n+1} > i_n$. We prove that $\langle x_{i_n} \rangle$ is a Cauchy sequence. Let $\varepsilon > 0$. Then there exists a natural number N such that $\frac{2}{N} < \varepsilon$. Suppose $m, n \geq N$. Since $A_m \subseteq A_N$ and $A_n \subseteq A_N$, both i_m and i_n belong to A_N. Hence both x_{i_m} and x_{i_n} belong to B_N. Therefore $d(x_{i_m}, x_{i_n}) \leq \frac{2}{N} < \varepsilon$, and hence $\langle x_{i_m} \rangle$ is a subsequence of $\langle x_n \rangle$ that is a Cauchy sequence. Since (X, d) is complete, $\langle x_{i_m} \rangle$ converges. We have proved that (X, d) is sequentially compact. By Theorem 4.10, (X, d) is compact. ∎

In calculus, there is a theorem, called the **uniform continuity theorem,** which says: Let $a, b \in \mathbb{R}$ with $a < b$, let $f: [a, b] \to \mathbb{R}$ be continuous, and let ε be a positive number. Then there is a positive number δ such that if $x_1, x_2 \in [a, b]$ and $|x_1 - x_2| < \delta$, then $|f(x_1) - f(x_2)| < \varepsilon$.

This theorem is used, for example, to prove that every continuous function is integrable. We generalize the definition of uniform continuity and prove the uniform continuity theorem in generalized form.

Definition. Let (X, d) and (Y, ρ) be metric spaces and let $f: X \to Y$ be a function. Then f is **uniformly continuous** if for each positive number ε, there is a positive number δ such that if $x, y \in X$ and $d(x, y) < \delta$, then $\rho(f(x), f(y)) < \varepsilon$. ∎

Note the difference between continuity and uniform continuity. If $f: X \to Y$ is a continuous function, then for each $x \in X$ and each positive number ε, there is a positive number $\delta_{x, \varepsilon}$ such that if $y \in X$ and $d(x, y) < \delta_{x, \varepsilon}$, then $\rho(f(x), f(y)) < \varepsilon$. If f is a uniformly continuous function, then for each positive number ε, there is a positive number δ_ε such that if x and y are *any* two members of X such that $d(x, y) < \delta_\varepsilon$, then $\rho(f(x), f(y)) < \varepsilon$.

EXAMPLE 5. Define $f: (0, 1) \to \mathbb{R}$ by $f(x) = \frac{1}{x}$ for each $x \in (0, 1)$. Then f is continuous but not uniformly continuous.

Analysis. The proof that f is continuous is left as Exercise 12.

We show that f if not uniformly continuous. The proof is by contradiction. Let $\varepsilon = 1$ and suppose there is a positive number δ such that if $p, q \in (0, 1)$ and $|p - q| < \delta$ then $\left|\frac{1}{p} - \frac{1}{q}\right| < \varepsilon$. Choose p so that $0 < p < \min\{1, \delta\}$ and let $q = \frac{p}{2}$. Then $|p - q| = \frac{p}{2} < \delta$ but $\left|\frac{1}{q} - \frac{1}{p}\right| = \frac{2}{p} - \frac{1}{p} = \frac{1}{p} > 1 = \varepsilon$. ∎

THEOREM 4.12. Let (X, d) be a compact metric space, let (Y, ρ) be a metric space, and let $f: X \to Y$ be a continuous function. Then f is uniformly continuous.

Proof. Let $\varepsilon > 0$. Then $\mathcal{U} = \{f^{-1}(B(y, \frac{\varepsilon}{2})): y \in Y\}$ is an open cover of X. Since (X, d) is compact, there is a Lebesgue number γ for the open cover $\mathcal{U}$. Suppose $x, x' \in X$ and $d(x, x') < \gamma$. Then $x, x' \in B(x, \gamma) \subseteq U$ for some $U \in \mathcal{U}$. Thus $f(x), f(x') \in f(U) \subseteq B(y, \frac{\varepsilon}{2})$ for some y, and hence $d(f(x), f(x')) < \varepsilon$. Therefore f is uniformly continuous. ∎

We conclude this section with a continuation of the historical remarks in the introduction to this chapter. The theorem that Borel proved, in 1894, was that every countable open cover of $[a, b]$ has a finite subcover; that is, that $[a, b]$ is countably compact. Ernst Lindelöf (1870–1946) proved that every open cover of $[a, b]$ has a countable subcover. Thus it follows from the work of Borel and Lindelöf that $[a, b]$ is compact. Heine's contribution was to prove, in 1872, that every continuous function $f: [a, b] \to \mathbb{R}$ is uniformly continuous. A.M. Schoenflies (1858–1923) gave the Heine-Borel Theorem its present name.

In 1902, W.H. Young (1863–1942) proved that every open cover of a closed and bounded subset of $\mathbb{R}^2$ has a finite subcover. In 1904, Henri Lebesgue (1875–1941) published the same result. Lebesgue was a student of Borel and a professor at the College de France.

In 1904, Fréchet introduced the term *compact* to describe those spaces in which every sequence has a convergent subsequence. As we have seen, such spaces are now called sequentially compact spaces. Fréchet also proved that sequential compactness and compactness are equivalent in a metric space.

Comparison of compactness and the finite intersection property is due to Friedrich Riesz (1876–1959). In 1914, Hausdroff proved the theorem we list as Theorem 4.10.

The definition of compactness that we use today is due to Paul S. Alexandroff and Paul S. Urysohn (1898–1924). They called the property *bicompactness*, and their first paper on the subject was published in 1923. Leopold Vietoris (1891–2002) gave an equivalent definition of compactness in 1921.

EXERCISES 4.1

1. Prove that every subset of a totally bounded metric space is totally bounded.

2. Let $\mathcal{O}$ be a collection of open intervals such that $I \subseteq \bigcup\{O: O \in \mathcal{O}\}$. Prove that there is a finite subset $\{O_1, O_2, ..., O_N\}$ of $\mathcal{O}$ such that $I \subseteq \cup_{n=1}^{N} O_n$.

3. Let (X, d) be a compact metric space, and let ρ be any metric on X such that the topology induced by ρ is the topology induced by d. Prove that (X, ρ) is bounded.

4. Let $(X, \mathcal{T})$ be a metrizable space such that every metric that generates $\mathcal{T}$ is bounded. Prove that X is compact.

5. Let (X, d) be a totally bounded metric space. Prove that X is separable.

6. Give an example of a compact metric space $(X, \mathcal{T})$, a topological space $(Y, \mathcal{U})$ that is not Hausdroff, and a continuous function f that maps X onto Y.

7. Let $\{A_\alpha: \alpha \in \Lambda\}$ be a family of closed subsets of a compact metric space (X, d) such that $\bigcap_{\alpha \in \Lambda} A_\alpha = \emptyset$. Prove that there is a positive number ε such that if B is any subset of X of a diameter less than ε, then there exists $\beta \in \Lambda$ such that $B \cap A_\beta = \emptyset$.

8. Prove that a subspace of $\mathbb{R}$ is compact if and only if it is closed and bounded.

9. Prove that every compact metric space is second countable.

10. Prove that every compact subset of a metric space is closed and bounded.

11. Give an example of a bounded metric space that is not compact.

12. Prove that the function f in Example 5 is continuous.

4.2 Compact Spaces

In this section, we study compactness in a more general setting than in the previous section. Compact Hausdorff space are important in analysis, and sometimes the Hausdorff property is part of the definition of compactness. Because of applications to other fields, it is useful to study compactness without the Hausdorff assumption.

Examples 1(d) and 2(b) in the previous section show that compactness is not hereditary. We do, however, have the following result.

THEOREM 4.13. Every closed subset of a compact space is compact.

Proof. Let $(X, \mathcal{T})$ be a compact space, let A be a closed subset of X, and let $\mathcal{U}$ be an open cover of A. Since A is closed, $X - A$ is open. Therefore $\mathcal{V} = \{X - A\} \cup \mathcal{U}$ is an open cover of X. Since X is compact, $\mathcal{V}$ has a finite subcover $\mathcal{V}'$. Thus $\{V \in \mathcal{V}' : V \in \mathcal{U}\}$ is a finite subcollection of $\mathcal{U}$ that covers A. Therefore A is compact. ∎

EXAMPLE 6. The Cantor set C is the intersection of closed subsets of $\mathbb{R}$. Therefore C is a closed subset of $\mathbb{R}$ and hence of I. Since I is compact, by Theorem 4.13, C is compact.

The following example shows that a compact subset of a topological space is not necessarily closed.

EXAMPLE 7. Let $\mathcal{T}$ be the finite complement topology on $\mathbb{R}$ and let $A = \{x \in \mathbb{R} : x \text{ is rational}\}$. Then A is compact but not closed.

As we shall see, if the space is Hausdorff, then every compact subset is closed.

THEOREM 4.14. Let $(X, \mathcal{T})$ be a Hausdorff space, let A be a compact subset of X, and let $p \in X - A$. Then there exist disjoint open sets U and V such that $A \subseteq U$ and $p \in V$.

Proof. For each $x \in A$, there are disjoint open sets U_x and V_x such that $x \in U_x$ and $p \in V_x$. Then $\mathcal{U} = \{U_x : x \in A\}$ is an open cover A. Since A is compact, there is a finite subcollection $U_{x_1}, U_{x_2}, ..., U_{x_n}$ of $\mathcal{U}$ that cover A. For each $i = 1, 2, ..., n$, U_{x_1} and the corresponding V_{x_1} are disjoint. Therefore $U = \bigcup_{i=1}^{n} U_{x_i}$ and $V = \bigcap_{i=1}^{n} V_{xi}$ are disjoint open sets such that $A \subseteq U$ and $p \in V$. ∎

Corollary 4.15. Every compact subset of a Hausdorff space is closed.

4.2 Compact Spaces 141

Proof. Let A be a compact subset of a Hausdorff space $(X, \mathcal{T})$, and let $p \in X - A$. By Theorem 4.14, there is a neighborhood U of p such that $U \subseteq X - A$. Therefore $X - A$ is open, so A is closed. ■

Corollary 4.16 is an immediate consequence of Theorem 4.13 and Corollary 4.15.

Corollary 4.16. Let $(X, \mathcal{T})$ be a compact Hausdorff space. A subset A of X is compact if and only if it is closed. ■

THEOREM 4.17. Let A and B be disjoint compact subsets of a Hausdorff space $(X, \mathcal{T})$. Then there are disjoint open sets U and V such that $A \subseteq U$ and $B \subseteq V$.

Proof. By Theorem 4.14, for each $x \in B$, there exist disjoint open sets U_x and V_x such that $A \subseteq U_x$ and $x \in V_x$. Then $\mathcal{V} = \{V_x : x \in B\}$ is an open cover of B. Since B is compact, there is a finite subcollection $V_{x_1}, V_{x_2}, ..., V_{x_n}$ of $\mathcal{V}$ that covers B. For each $i = 1, 2, ..., n$, V_{x_i} and the corresponding U_{x_i} are disjoint. Therefore $U = \bigcap_{i=1}^{n} U_{x_i}$ and $V = \bigcup_{i=1}^{n} V_{x_i}$ are disjoint open sets such that $A \subseteq U$ and $B \subseteq V$. ■

Corollary 4.18 is an immediate consequences of Theorems 4.13 and 4.17.

Corollary 4.18. Let A and B be disjoint closed subsets of a compact Hausdorff space $(X, \mathcal{T})$. Then there are disjoint open sets U and V such that $A \subseteq U$ and $B \subseteq V$. ■

A result that is central to the usefulness of compactness is the following theorem.

THEOREM 4.19. The continuous image of a compact space is compact.

Proof. Let $(X, \mathcal{T})$ be a compact space, let $(Y, \mathcal{U})$ be a topological space, let $f : X \to Y$ be a continuous function that maps X onto Y, and let $\mathcal{O}$ be an open cover of Y. Then $\mathcal{O}' = \{f^{-1}(U) : U \in \mathcal{O}\}$ is an open cover of X. Since X is compact, $\mathcal{O}'$ has a finite subcover $\mathcal{O}''$. Thus $\{U \in \mathcal{O} : f^{-1}(U) \in \mathcal{O}''\}$ is a finite subcollection of $\mathcal{O}$ that covers Y. Therefore Y is compact. ■

We now give some consequences of Theorem 4.19.

THEOREM 4.20. Compactness is a topological property. ■

THEOREM 4.21. Let $(X, \mathcal{T})$ be a compact space, let $(Y, \mathcal{U})$ be a Hausdorff space, and let $f : X \to Y$ be a continuous function. Then f is a closed mapping.

Proof. Let C be a closed subset of X. By Theorem 4.13, C is compact. By Theorem 4.19, $f(C)$ is compact. By Corollary 4.15, $f(C)$ is closed. ■

Corollary 4.22. Let $(X, \mathcal{T})$ be a compact space, let $(Y, \mathcal{U})$ be a Hausdorff space, and let $f: X \to Y$ be a continuous surjection. Then f is a quotient map.

Proof. By Theorem 4.21, f is closed. By Theorem 2.55, every closed continuous surjection is a quotient map. ∎

Corollary 4.23. Let $(X, \mathcal{T})$ be a compact space, let $(Y, \mathcal{U})$ be a Hausdorff space, and let f be a one-to-one continuous function that maps X onto Y. Then f is a homeomorphism.

Proof. By Theorem 4.21, f is a closed mapping. Therefore f^{-1} is continuous by Exercise 7 of Section 1.7. ∎

Corollary 4.23 permits us to avoid laborious proofs of continuity of functions that are encountered in calculus.

EXAMPLE 8. Suppose we know that the function $f: \mathbb{R} \to \mathbb{R}$ defined by $f(x) = x^2$ is continuous. Let $g: [0, \infty) \to \mathbb{R}$ be the function defined by $g(x) = \sqrt{x}$. Prove that g is continuous.

Analysis. Let $a \in [0, \infty)$. We restrict f to a compact set that contains a. Let $b \in \mathbb{R}$ such that $b > a$ and consider $f|_{[0, \sqrt{b}]}: [0, \sqrt{b}] \to [0, b]$. Now $f|_{[0, \sqrt{b}]}$ is a one-to-one continuous function from $[0, \sqrt{b}]$ onto $[0, b]$. By Corollary 4.23, $(f|_{[0, \sqrt{b}]})^{-1}: [0, b] \to [0, \sqrt{b}]$ is continuous. In particular, $(f|_{[0, \sqrt{b}]})^{-1}$ is continuous at a. Therefore g is continuous at a. Since a is an arbitrary member of $[0, \infty]$, g is continuous. ∎

The following example shows that one must exercise caution is using Corollary 4.23 to prove that the inverse of a continuous bijection is continuous.

EXAMPLE 9. Let $X = [0, 1)$ and let $S^1 = \{(x, y) \in \mathbb{R}^2: x^2 + y^2 = 1\}$. Define $f: X \to S^1$ by $f(x) = (\cos 2\pi x, \sin 2\pi x)$. Note that $f(0) = (1, 0)$ and that f wraps $[0, 1)$ around S^1 in a counterclockwise direction $[f(\frac{1}{4}) = (0, 1), f(\frac{1}{2}) = (-1, 0)$, and $f(\frac{3}{4}) = (0, -1)]$. The function f is a continuous bijection, but f^{-1} is not continuous.

Analysis. In order to see that f^{-1} is not continuous, we choose a sequence $\langle s_i \rangle = \left(\cos\left(2\pi\left(1 - \frac{1}{i}\right)\right), \sin\left(2\pi\left(1 - \frac{1}{i}\right)\right) \right)$ in S^1 that converges to $(1, 0)$ and observe that the sequence $\langle f^{-1}(s_i) \rangle$ does not converge to 0 in X. ∎

In the previous section we gave a characterization of countable compactness in terms of closed sets (see Theorem 4.5). We now give the analogue for compactness. This result should be anticipated since the topology on a space is as easily described by closed sets as by open sets. Furthermore, we can reformulate the definition of compactness as follows: If $\mathcal{U}$ is a collection of open subsets of a topological space $(X, \mathcal{T})$ such that no finite subcollection of $\mathcal{U}$ covers X, then $\mathcal{U}$ does not cover X. Now a collection $\mathcal{A}$ of subsets of X does not cover X when $X \neq \bigcup_{A \in \mathcal{A}} A$, that is, when $\bigcap_{A \in \mathcal{A}} (X - A) = X - \bigcup_{A \in \mathcal{A}} A \neq \emptyset$. Thus we are led to the following theorem.

THEOREM 4.24. A topological space $(X, \mathcal{T})$ is compact if and only if every family of closed subsets of X with the finite intersection property has a nonempty intersection.

Proof. We give contrapositive arguments. Suppose there is a family $\mathcal{A} = \{A_\alpha : \alpha \in \Lambda\}$ of closed subsets of X with the finite intersection property such that $\bigcap_{\alpha \in \Lambda} A_\alpha = \emptyset$. Let $\mathcal{O} = \{X - A_\alpha : \alpha \in \Lambda\}$. Since $\bigcup_{\alpha \in \Lambda}(X - A_\alpha) = X - \bigcap_{\alpha \in \Lambda} A_\alpha = X - \emptyset = X$, $\mathcal{O}$ is an open cover of X. Let $\{\alpha_1, \alpha_2, ..., \alpha_N\}$ be an arbitrary finite subcollection of Λ. Since $\bigcap_{i=1}^N A_{\alpha_i} \neq \emptyset$, $\bigcup_{i=1}^N (X - A_{\alpha_i}) = X - \bigcap_{i=1}^N A_{\alpha_i} \neq X$. Therefore $\{X - A_{\alpha_i} : i = 1, 2..., N\}$ does not cover X, so X is not compact. ∎

Suppose X is not compact. Then there is an open cover $\mathcal{O} = \{U_\alpha : \alpha \in \Lambda\}$ of X that does not have a finite subcover. Consider the collection $\mathcal{A} = \{X - U_\alpha : \alpha \in \Lambda\}$ of closed subsets of X, and let Γ be a finite subcollection of Λ. Since $\mathcal{O}$ does not have a finite subcover, $\bigcap_{\alpha \in \Gamma}(X - U_\alpha) = X - \bigcup_{\alpha \in \Gamma} U_\alpha \neq \emptyset$. Therefore $\mathcal{A}$ has the finite intersection property. Since $\bigcap_{\alpha = \Lambda}(X - U_\alpha) = X - \bigcup_{\alpha \in \Lambda} U_\alpha = X - X = \emptyset$ the proof is complete.

The following theorem tells us that in order to show that a topological space is compact, it is sufficient to show that every cover of the space by members of a basis for the topology has a finite subcover.

THEOREM 4.25. Let $(X, \mathcal{T})$ be a topological space and let $\mathcal{B}$ be a basis for $\mathcal{T}$. Then $(X, \mathcal{T})$ is compact if and only if every cover of X by members of $\mathcal{B}$ has a finite subcover.

Proof. Let $(X, \mathcal{T})$ be a compact space and let $\mathcal{O}$ be a cover of X by members of $\mathcal{B}$. Then $\mathcal{O}$ is an open cover of X and hence it has a finite subcover.

Suppose every open cover of X by members of $\mathcal{B}$ has a finite subcover, and let $\mathcal{O}$ be an open cover of X. For each $U \in \mathcal{O}$, there is a subcollection $\mathcal{B}_U$ of $\mathcal{B}$ such that $U = \bigcup \{B : B \in \mathcal{B}_U\}$. Now $\{B : B \in \mathcal{B}_U \text{ for some } U \in \mathcal{O}\}$ is an open cover of X by members of $\mathcal{B}$, and hence it has a finite subcover $\mathcal{O}'$. For each $B \in \mathcal{O}'$, choose a member U_B of $\mathcal{O}$ such that $B \subseteq U_B$. Then $\{U_B : B \in \mathcal{O}'\}$ is a finite subcollection of $\mathcal{O}$ that covers X. ∎

A topological space is compact if every cover of the space by members of a subbase has a finite subcover. The proof of this fact, known as the Alexander Subbase Theorem, uses Zorn's Lemma and will be postponed to Chapter 6.

The Alexander Subbase Theorem can be used to show that the product of an arbitrary collection of compact spaces is compact, and this too will be postponed to Chapter 6. In this section we prove that the product of two compact spaces is compact. But first we establish the following result.

THEOREM 4.26. Let $(X, \mathcal{T})$ be a topological space, let $(Y, \mathcal{U})$ be a compact space, let $p \in X$, and let $\mathcal{U}$ be a cover of $\{p\} \times Y$ by sets that are open in $X \times Y$. Then there is a neighborhood V of p such that a finite subcollection of $\mathcal{U}$ covers $V \times Y$.

Proof. Since the function $h: Y \to \{p\} \times Y$ defined by $h(y) = (p, y)$ for each $y \in Y$ is continuous (in fact a homeomorphism), $\{p\} \times Y$ is compact. Thus there is a finite subcollection $\mathcal{F}$ of $\mathcal{U}$ that covers $\{p\} \times Y$. Let $y \in Y$ and let $W_y \in \mathcal{F}$ such that $(p, y) \in W_y$. Then there exist neighborhoods V_y of p and U_y of y such that $V_y \times U_y \subseteq W_y$. The collection $\{U_y: y \in Y\}$ is an open cover of the compact space Y, so there exists a finite subset F of Y such that $\{U_y: y \in F\}$ covers Y. Let $V = \bigcap_{y \in F} V_y$. Then V is a neighborhood of p.

The collection $\{W_y: y \in F\}$ is a finite subcollection of $\mathcal{U}$, and we complete the proof by showing that it covers $V \times Y$. Let $(a, b) \in V \times Y$. Then there exists $y \in F$ such that $b \in U_y$. By the definition of V, $a \in V_y$. Hence $(a, b) \in V_y \times U_y \subseteq W_y$. ∎

THEOREM 4.27. Let $(X, \mathcal{T})$ and $(Y, \mathcal{U})$ be compact spaces. Then $X \times Y$ is compact.

Proof. Let $\mathcal{U}$ be an open cover of $X \times Y$. For each $x \in X$, $\mathcal{U}$ is a cover of $\{x\} \times Y$ by sets that are open in $X \times Y$, so by Theorem 4.26, there is a neighborhood V_x of x and a finite subcollection $\mathcal{F}_x$ of $\mathcal{U}$ such that $V_x \times Y \subseteq \bigcup_{U \in \mathcal{F}_x} U$. Since $\{V_x: x \in X\}$ is an open cover of the compact space X, there is a finite subset F of X such that $\{V_x: x \in F\}$ covers X. Let $\mathcal{F} = \bigcup_{x \in F} \mathcal{F}_x$. Since $\mathcal{F}$ is the finite union of finite subcollections of $\mathcal{U}$, $\mathcal{F}$ is a finite subcollection of $\mathcal{U}$. We complete the proof by showing that $\mathcal{F}$ covers $X \times Y$.

Let $(a, b) \in X \times Y$. Then there exists $x \in F$ such that $a \in V_x$, so $(a, b) \in V_x \times Y \subseteq \bigcup_{U \in \mathcal{F}_x} U$. Hence $(a, b) \in U$ for some $U \in \mathcal{F}$. ∎

THEOREM 4.28. The product of any finite number of compact spaces is compact.

Proof. See Exercise 4. ∎

Corollary 4.29. Let $n \in \mathbb{N}$, let M be a positive real number, and for each $i = 1, 2, \ldots, n$, let $X_i = [-M, M]$. Then $J^n = \prod_{i=1}^{n} X_i$ is compact. ∎

We now characterize the compact subsets of $\mathbb{R}^n$, where $n \in \mathbb{N}$.

THEOREM 4.30. Let $n \in \mathbb{N}$. A subset X of $\mathbb{R}^n$ is compact if and only if it is closed and bounded.

Proof. Suppose X is compact. Since $\mathbb{R}^n$ is Hausdorff, by Corollary 4.15, X is closed. Since a finite subcollection of the collection $\{B((0, 0), n): n \in \mathbb{N}\}$ of open balls centered at the origin covers X, X is bounded.

Suppose X is closed and bounded. Let M be a positive real number such that for each $x \in X$, the distance between x and the origin is less than or equal to M. The set J^n defined in Corollary 4.29 is compact and $X \subseteq J^n$. Since X is closed, by Theorem 4.13, it is compact. ∎

In calculus there is a theorem which says that a continuous real-valued function on a closed interval $[a, b]$ assumes maximum and minimum values. This theorem is a consequence of the following.

THEOREM 4.31. Let $(X, \mathcal{T})$ be a compact space and let $f: X \to \mathbb{R}$ be a continuous function. Then there exist $c, d \in X$ such that for all $x \in X$, $f(c) \leq f(x) \leq f(d)$.

Proof. Since X is compact and f is continuous, by Theorem 4.19, $f(X)$ is compact. Therefore by Theorem 4.30, $f(X)$ is closed and bounded. Thus $f(X)$ contains its least upper bound U and greatest lower bound L. Hence there exist $c, d \in X$ such that $f(c) = L$ and $f(d) = U$. It follows that for any $x \in X$, $f(c) \leq f(x) \leq f(d)$. ∎

Recall that the Cantor set is an uncountable totally disconnected set (Example 1 of Appendix G and Exercises 3 and 5 of Section 3.3). As shown in Example 6 the Cantor set is also compact. Compare this result with the following.

THEOREM 4.32. Let $(X, \mathcal{T})$ be a compact locally connected space. Then $(X, \mathcal{T})$ has a finite number of components.

Proof. Suppose $(X, \mathcal{T})$ has an infinite number of components. By Theorem 3.23, each component of X is an open set. Thus the collection of components of X is an open cover of X that does not have a finite subcover. Since X is compact, this is a contradiction. ∎

We conclude this section with a characterization of total disconnectedness in compact Hausdorff spaces.

THEOREM 4.33. A compact Hausdorff space $(X, \mathcal{T})$ is totally disconnected if and only if whenever $x, y \in X$ and $x \neq y$, there is a set A that is both open and closed such that $x \in A$ and $y \notin A$.

Proof. See Exercise 11. ∎

EXERCISES 4.2

1. Let A be a subset of a topological space $(X, \mathcal{T})$. Prove that A is compact if and only if every cover of A by members of $\mathcal{T}_A$ has a finite subcover.

2. Let $(X, \mathcal{T})$ be a topological space, let $\mathcal{B}$ be a basis for $\mathcal{T}$, and let $\mathcal{A} = \{X - B: B \in \mathcal{B}\}$. Prove that every cover of X by members of $\mathcal{B}$ has a

finite subcover if and only if every subcollection of $\mathscr{A}$ with the finite intersection property has a nonempty intersection.

3. Let $(X, \mathscr{T})$ be a topological space, let $\mathscr{S}$ be a subbasis for $\mathscr{T}$, and let $\mathscr{A} = \{X - S : S \in \mathscr{S}\}$. Prove that every cover of X by members of $\mathscr{S}$ has a finite subcover if and only if every subcollection of $\mathscr{A}$ with the finite intersection property has a nonempty intersection.

4. Prove that the product of a finite number of compact spaces is compact.

5. Let $\{A_\alpha : \alpha \in \Lambda\}$ be a collection of compact subsets of a Hausdorff space $(X, \mathscr{T})$. Prove that $\bigcap_{\alpha \in \Lambda} A_\alpha$ is compact.

6. Let $(X, \mathscr{T})$ be a topological space and let $(Y, \mathscr{U})$ be a compact space. Prove that the projection $\pi_1 : X \times Y \to X$ is a closed map.

7. Let $\mathscr{T}$ and $\mathscr{U}$ be topologies on a set X.

   (a) Suppose $(X, \mathscr{T})$ is compact and $\mathscr{T} \subseteq \mathscr{U}$. Is $(X, \mathscr{U})$ compact? Prove your answer.

   (b) Suppose $(X, \mathscr{U})$ is compact and $\mathscr{T} \subseteq \mathscr{U}$. Is $(X, \mathscr{T})$ compact? Prove your answer.

   (c) Suppose $(X, \mathscr{T})$ and $(X, \mathscr{U})$ are compact Hausdorff spaces. Prove that either $\mathscr{T} = \mathscr{U}$ or $\mathscr{T} \not\subseteq \mathscr{U}$ or $\mathscr{U} \not\subseteq \mathscr{T}$.

8. Let $(X, \leq)$ be a linearly ordered set and let $\mathscr{T}$ denote the order topology on X. Prove that $(X, \mathscr{T})$ is compact if and only if every nonempty subset of X has a greatest lower bound and a least upper bound.

9. If $(X, \mathscr{T})$ and $(Y, \mathscr{U})$ are topological spaces, then a function $f : X \to Y$ is a **perfect map** if it is a closed continuous surjection with the property that for each $y \in Y$, $f^{-1}(y)$ is compact.

   Let $(X, \mathscr{T})$ and $(Y, \mathscr{U})$ be topological spaces, and let $f : X \to Y$ be a perfect map.

   (a) Prove that if X is Hausdorff, then so is Y.

   (b) Prove that if X is second countable, then so is Y.

   (c) Prove that if Y is compact, then so is X.

10. Let $(X, \mathscr{T})$ be a compact Hausdorff space. Prove that the quasi-components of X are the same as the components of X.

11. Prove Theorem 4.33. *Hint:* Use Exercise 10.

12. Let $(X, \mathscr{T})$ and $(Y, \mathscr{U})$ be topological spaces, let $A \subseteq X$ and $B \subseteq Y$ such that $A \times B$ is compact, let W be an open set such that $A \times B \subseteq W$. Prove that there exist $U \in \mathscr{T}$ and $V \in \mathscr{U}$ such that $A \times B \subseteq U \times V \subseteq W$.

13. A **continuum** is a compact, connected Hausdorff space. Let $(X, \mathcal{T})$ be a continuum such that X contains at least two points. Assume that X has a cut point p and let U and V be a separation of $X - \{p\}$. Show by contradiction that each of U and V contains a noncut point.

14. For each nonnegative integer n, let $(X_n, \mathcal{T}_n)$ be a topological space, and for each $n \in \mathbb{N}$, let $f_n: X_n \to X_{n-1}$ be a continuous function. The sequence

$$X_0 \xleftarrow{f_1} X_1 \xleftarrow{f_2} X_2 \xleftarrow{f_3} \ldots ,$$

which we denote by $\langle X_n, f_n \rangle$, is called an **inverse limit sequence**. The **inverse limit**, X_∞, of $\langle X_n, f_n \rangle$ is defined by

$$X_\infty = \left\{ x = \langle x_n \rangle \in \prod_{n=0}^{\infty} X_n : f_n(x_n) = x_{n-1} \text{ for each } n \in \mathbb{N} \right\}.$$

Prove that if, for each $n \in \mathbb{N}$, $X_n \subseteq X_{n-1}$ and $f_n: X_n \to X_{n-1}$ is the injection map (that is, $f_n(x) = x$ for each $x \in X_n$), then X_∞ is homeomorphic to $\bigcap_{n=1}^{\infty} X_n$. (The topology on X_∞ is the subspace topology induced by the product topology on $\prod_{n=0_1}^{\infty} X_n$, and the topology on $\bigcap_{n=1}^{\infty} X_n$ is the subspace topology induced by $\mathcal{T}_0$.)

15. Prove that if $\langle X_n, f_n \rangle$ is an inverse limit sequence of nonempty compact Hausdorff spaces, then the inverse limit space is a nonempty compact Hausdorff space.

16. Let $\langle X_n, f_n \rangle$ and $\langle Y_n, g_n \rangle$ be inverse limit sequences. **A mapping** Φ of $\langle X_n, f_n \rangle$ into $\langle Y_n, g_n \rangle$ is a sequence $\langle \phi_n \rangle$ of functions $\phi_n: X_n \to Y_n$ such that $\phi_{n-1} \circ f_n = g_n \circ \phi_n$ for each $n \in \mathbb{N}$, In other words, the following diagram is commutative.

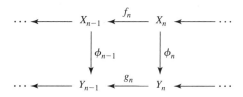

If each ϕ_n is continuous, we say that Φ is **continuous**, and if each ϕ_n maps X_n onto Y_n, we say that Φ maps $\langle X_n, f_n \rangle$ onto $\langle Y_n, g_n \rangle$. The **induced mapping** $\phi: X_\infty \to Y_\infty$ is defined by $\phi(\langle x_n \rangle) = \langle \phi_n(x_n) \rangle$ for each $\langle x_n \rangle \in X_\infty$.

(a) Prove that if $\langle x_n \rangle \in X_\infty$, then $\langle \phi_n(x_n) \rangle \in Y_\infty$.

(b) Prove that the induced mapping ϕ is continuous whenever Φ is continuous.

(c) Prove that if, for each nonnegative n, $(X_n, \mathcal{T}_n)$ and $(Y_n, \mathcal{U}_n)$ are compact Hausdorff spaces and Φ maps $\langle X_n, f_n \rangle$ onto $\langle Y_n, g_n \rangle$, then the induced mapping ϕ maps X_∞ onto Y_∞.

17. For each nonnegative integer n, let $\mathcal{U}_n$ be a partition of a set X, and suppose that, for each $n \in \mathbb{N}$, each $U \in \mathcal{U}_n$ is a subset of some $V \in \mathcal{U}_{n-1}$. Then the **derived sequence** obtained from $\langle \mathcal{U}_n \rangle$ is the inverse limit sequence

$$X_0 \xleftarrow{f_1} X_1 \xleftarrow{f_2} X_2 \xleftarrow{f_3} \ldots,$$

where each X_n is the set whose members are the elements of $\mathcal{U}_n$, the topology on X_n is the discrete topology, and f_n maps each member U of X_n into the member V of X_{n-1} that contains U.

Prove that if $(X, \mathcal{T})$ is a totally disconnected, compact metric space, then:

(a) For each nonnegative integer n, there is a finite open cover $\mathcal{U}_n$ of X such that $\mathcal{U}_n$ is a pairwise disjoint collection of sets of diameter less than $1/2^n$ and each $U \in \mathcal{U}_n$ is a subset of some $V \in \mathcal{U}_{n-1}$.

(b) If $\langle \mathcal{U}_n \rangle$ is a sequence of open covers of X satisfying the condition in (a) and $X_0 \xleftarrow{f_1} X_1 \xleftarrow{f_2} X_2 \xleftarrow{f_3} \ldots$ is the derived sequence obtained from $\langle \mathcal{U}_n \rangle$, then $(X, \mathcal{T})$ is homeomorphic to the inverse limit space X_∞.

18. Prove that if $n \in \mathbb{N}$, $(X, \mathcal{T})$ is a perfect, totally disconnected, compact Hausdorff space, and U is a nonempty open subset of X, then there is a pairwise disjoint collection $\{U_1, U_2, \ldots, U_n\}$ of nonempty open subsets of X such that $U = \bigcup_{i=1}^{n} U_i$.

19. Prove that any two perfect, totally disconnected, compact metric spaces are homeomorphic.

20. Prove that if $(X, \mathcal{T})$ is a perfect, totally disconnected, compact metric space, then $(X, \mathcal{T})$ is homeomorphic to the Cantor set.

21. Prove that every compact metric space is the continuous image of the Cantor set. *Warning*: This is a difficult exercise.

22. Prove that if x and y are distinct points in the Cantor set C, then there is a homeomorphism h of C onto C such that $h(x) = y$.

23. Prove that every totally disconnected, compact metric space is homeomorphic to a subset of the Cantor set.

24. Prove that every perfect set in a complete metric space contains a compact perfect set.

25. Prove that if $(X, \mathcal{T})$ is a compact Hausdorff space that is uncountable, then $(X, \mathcal{T})$ is not perfect.

26. Prove that if (X, d) is a complete metric space that is uncountable, then (X, d) is not perfect.

27. Let $(X, \mathcal{T})$ be a topological space. Prove that if $A \subseteq X$ that has no isolated points, then $\overline{A}$ is a perfect subset of X.

28. Prove that if U is an open subset of the Cantor set C, then there is a sequence (perhaps finite) of pairwise disjoint subsets $\langle A_n \rangle$ of C that are both open and closed such that $C = \bigcup_{n \in \Lambda} A_n$ (Λ is a subset of $\mathbb{N}$).

29. Prove that if U is an open subset of the Cantor set C, then U is homeomorphic to C or to $C - \{0\}$.

4.3 Local Compactness and the Relation between Various Forms of Compactness

In Section 4.1, we defined compactness, countable compactness, sequential compactness, and the Bolzano-Weierstrass property and showed that they are equivalent in metric spaces. We also proved the implications shown in Figure 4.1 in arbitrary topological spaces. In this section we continue the study of the relation between these forms of compactness in non-metric spaces. We also introduce and study a local form of compactness. We begin with an example (promised in Section 4.1) of a compact space that is not sequentially compact.

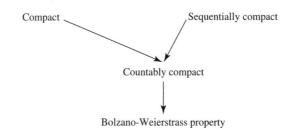

Figure 4.1

EXAMPLE 10. For each $\alpha \in \mathbb{R}$, let $X_\alpha = I$, and let $X = \prod_{\alpha \in \mathbb{R}} X_\alpha$. Then $(X, \mathcal{T})$ is compact but not sequentially compact.

Analysis. By Exercise 2 of Section 4.1 and Theorem 4.25, I is compact. Therefore, since the product of compact spaces is compact (see Section 6.6), $(X, \mathcal{T})$ is compact. Let $\langle a_n \rangle$ be the sequence in $\mathbb{N}$ defined by $a_n = n$, let A be the set of all subsequences of $\langle a_n \rangle$, and let f be a one-to-one function that maps $\mathbb{R}$ onto A. Let $n \in \mathbb{N}$. For each $\alpha \in \mathbb{R}$, denote the subsequence $f(\alpha)$ of $\langle a_n \rangle$ by $i_1, i_2, i_3, \ldots$. If there is an odd integer j such that $n = i_j$, let $x_n(\alpha) = 0$. If there is no such odd integer j, let $x_n(\alpha) = 1$. Then $x_n \in X$, and hence we have a sequence $\langle x_n \rangle$ in X. Suppose $\langle x_{i_n} \rangle$ is a convergent subsequence of $\langle x_n \rangle$, and let $\alpha = f^{-1}(\langle x_{i_n} \rangle)$. Then by Theorems 2.31 and 1.55, the sequence $\langle x_{i_n}(\alpha) \rangle$ converges in I. But this sequence is the sequence whose odd terms are 0 and whose even terms are 1. This is a contradiction, and hence $\langle x_n \rangle$ does not have a subsequence that converges. Therefore X is not sequentially compact. ∎

We now give an example of a countably compact space that is not compact and an example of a topological space with the Bolzano-Weierstrass property that is not countably compact.

EXAMPLE 11. Let $(\Omega, \leq)$ be an uncountable well-ordered set with a maximal element ω_1 having the property that if $x \in \Omega$ and $x \neq \omega_1$, then $\{y \in \Omega: y \leq x\}$ is countable (see Appendix H). Let $\mathcal{T}$ be the order topology on Ω, and let $\Omega_0 = \Omega - \{\omega_1\}$. Then $(\Omega_0, \mathcal{T}_{\Omega_0})$ is countably compact but not compact.

Analysis. By Exercise 17 of Section 1.6, $(\Omega_0, \mathcal{T}_{\Omega_0})$ is a Hausdorff space. Therefore by Theorem 4.7, to show that $(\Omega_0, \mathcal{T}_{\Omega_0})$ is countably compact, it is sufficient to prove that it has the Bolzano-Weierstrass property. Let A be an infinite subset of Ω_0 and let B be a countably infinite subset of A. By Theorem H.11, the least upper bound of B is a member of Ω_0. Let b be the least member of Ω_0 that has infinitely many predecessors in B. The proof that b is a limit point of B is by contradiction. Suppose b is not a limit point of B. Then there exists $a, c \in \Omega_0$ such that $a < b < c$ and $\{x \in \Omega_0: a < x < c\} \cap (B - \{b\}) = \emptyset$. Therefore a has infinitely many predecessors in B, and this is a contradiction since $a < b$. Therefore b is a limit point of B and hence of A, so $(\Omega_0, \mathcal{T}_{\Omega_0})$ is countably compact.

We show that $(\Omega_0, \mathcal{T}_{\Omega_0})$ is not compact by exhibiting an open cover of Ω_0 that does not have a countable subcover. For each $x \in \Omega_0$, let $U_x = [1, x)$. Then $\mathcal{O} = \{U_x: x \in \Omega_0\}$ is an open cover of Ω_0. Suppose $\{U_{x_i}: i \in \mathbb{N}\}$ is a countable subcollection of $\mathcal{O}$. By Theorem H.11, the least upper bound of $\{x_i: i \in \mathbb{N}\}$ is less than ω_1. Therefore $\{U_{x_i}: i \in \mathbb{N}\}$ does not cover Ω_0. Hence $(\Omega_0, \mathcal{T}_{\Omega_0})$ is not compact. ■

EXAMPLE 12. For each, $n \in \mathbb{N}$, let $B_n = \{2n - 1, 2n\}$. Then $\mathcal{B} = \{B_n: n \in \mathbb{N}\}$ is a base for a topology $\mathcal{T}$ on $\mathbb{N}$. The topological space $(\mathbb{N}, \mathcal{T})$ has the Bolzano-Weierstrass property but it is not countably compact.

Analysis. Let A be an infinite subset of $\mathbb{N}$ and let $p \in A$. If p is odd, then every neighborhood of $p + 1$ contains p and hence $p + 1$ is a limit point of A. If p is even, then every neighborhood of $p - 1$ contains p and hence $p - 1$ is a limit point of A. Therefore $(\mathbb{N}, \mathcal{T})$ has the Bolzano-Weierstrass property.

The base $\mathcal{B}$ is an open cover of $\mathbb{N}$ that does not have a finite subcover, so $(\mathbb{N}, \mathcal{T})$ is not countably compact. ■

We now give an example of a sequentially compact space that is not compact, thus showing that there is no direct relationship between compactness and sequential compactness.

EXAMPLE 13. Let $(\Omega_0, \mathcal{T}_{\Omega_0})$ be the topological space in Example 11. Then $(\Omega_0, \mathcal{T}_{\Omega_0})$ is sequentially compact but not compact.

Analysis. We have seen (in Example 11) that $(\Omega_0, \mathcal{T}_{\Omega_0})$ is not compact.

Let $a \in \Omega_0$ and let b be the immediate successor of a. The collection $\{(x, b): x < a\}$ is a countable local basis at a, and hence $(\Omega_0, \mathcal{T}_{\Omega_0})$ is first countable. Therefore by Theorem 4.9 it is sequentially compact. ∎

We saw in Chapter 3 that there are important local forms of connectedness, and the remainder of this section is devoted to the study of a local form of compactness.

Definition. A topological space $(X, \mathcal{T})$ is **locally compact at a point** $p \in X$ provided there is an open set U and a compact subspace K of X such that $p \in U$ and $U \subseteq K$. A topological space is **locally compact** if it is locally compact at each of its points. ∎

Paul Alexandroff and Heinrich Tietze (1890–1964) independently introduced the concept of local compactness.

Every compact space is locally compact, whereas the real line with the usual topology is locally compact but not compact.

THEOREM 4.34. Let $(X, \mathcal{T})$ be a Hausdorff space and let $p \in X$. Then X is locally compact at p if and only if for each neighborhood V of p there is an open set U and a compact subspace K of X such that $p \in U$, $U \subseteq V$, $U \subseteq K$, and $K \subseteq \overline{U}$.

Proof. Suppose X is locally compact at p, and let V be a neighborhood of p. There exist an open set W and a compact set C such that $p \in W$ and $W \subseteq C$. Let $U = V \cap W$ and let $K = \overline{U}$. Then U is open, $p \in U$, $U \subseteq V$, $U \subseteq K$, and $K \subseteq \overline{U}$. We show that K is compact. Note that $K = \overline{U} \subseteq \overline{W}$. Since X is Hausdorff, C is closed. Therefore $\overline{W} \subseteq C$. Thus K is a closed subset of a compact set C and hence K is compact.

The condition clearly implies that X is locally compact. ∎

THEOREM 4.35. Let $(X, \mathcal{T})$ be a Hausdorff space and let $p \in X$. Then X is locally compact at p if and only if there is a neighborhood U of p such that $\overline{U}$ is compact.

Proof. Suppose X is locally compact at p. Then there exists an open set U and a compact set K such that $p \in U$ and $U \subseteq K$. Since K is a compact subset of a Hausdorff space, it is closed. Therefore $\overline{U} \subseteq K$, and hence $\overline{U}$, being a closed subset of a compact space, is compact.

The condition clearly implies that X is locally compact. ∎

For completeness, we state the following theorem. The proof, however, will be given in section 5.2, since it depends upon a concept that will be introduced in that section.

THEOREM 4.36. A Hausdorff space $(X, \mathcal{T})$ is locally compact if and only if for each $p \in X$ and each neighborhood V of p there is a neighborhood U of p such that $\overline{U}$ is compact and $\overline{U} \subseteq V$.

THEOREM 4.37. Every closed subspace of locally compact Hausdorff space is locally compact.

Proof. Let A be a closed subset of a locally compact Hausdorff space $(X, \mathcal{T})$, and let $p \in A$. By Theorem 4.35, there exists $U \in \mathcal{T}$ such that $p \in U$ and $\overline{U}$ is compact. Now $U \cap A \in \mathcal{T}_A$ and $p \in U \cap A$. Since $\overline{U \cap A}$ is a closed subset of $\overline{U}$, it is compact (as a subset of X). But A is closed, so $\overline{U \cap A} \subseteq A$. Therefore $\overline{U \cap A}$ is compact as a subset of A. ∎

The following example shows that the continuous image of a locally compact space need not be locally compact.

EXAMPLE 14. Let $A = \{\frac{1}{n} : n \in \mathbb{N}\}$ and let $\mathcal{B} = \{B \in \mathcal{P}(\mathbb{R}): B$ is an open interval that does not contain 0 or there is a positive number x such that $B = (-x, x) - A\}$. Then $\mathcal{B}$ is a basis for a topology $\mathcal{T}$ on $\mathbb{R}$, and the topological space $(\mathbb{R}, \mathcal{T})$ is a Hausdorff space. Also $A = \{\frac{1}{n} : n \in \mathbb{N}\}$ is a closed subset of $\mathbb{R}$ and $0 \notin A$. In Exercise 1 you are asked to show that there is no neighborhood U of 0 such that $\overline{U} \subseteq \mathbb{R} - A$. Therefore by Theorem 4.36, $(\mathbb{R}, \mathcal{T})$ is not locally compact. Let $\mathcal{U}$ be the discrete topology on $\mathbb{R}$. Then $(\mathbb{R}, \mathcal{U})$ is locally compact. The function $f: (\mathbb{R}, \mathcal{U}) \to (\mathbb{R}, \mathcal{T})$ defined by $f(x) = x$ for each $x \in \mathbb{R}$ is continuous.

THEOREM 4.38. Let $(X, \mathcal{T})$ be a locally compact space, let $(Y, \mathcal{U})$ be a topological space, and let $f: X \to Y$ be an open continuous function from X onto Y. Then $(Y, \mathcal{U})$ is locally compact.

Proof. Let $y \in Y$, and let $x \in f^{-1}(y)$. There is an open set V and a compact set C such that $x \in V$ and $V \subseteq C$. Let $U = f(\text{int}(C))$ and $K = f(C)$. Since $x \in V \subseteq C$ and V is open, $x \in \text{int}(C)$. Since f is open, U is open. Since $x \in \text{int}(C)$, $y \in U$. Since f is continuous and C is compact, K is compact (Theorem 4.19). Since $U \subseteq K$, $(Y, \mathcal{U})$ is locally compact at y. Since y is an arbitrary point of Y, $(Y, \mathcal{U})$ is locally compact. ∎

THEOREM 4.39. Let $\{(X_\alpha, \mathcal{T}_\alpha): \alpha \in \Lambda\}$ be a collection of spaces, and let $\mathcal{T}$ be the product topology on $X = \prod_{\alpha \in \Lambda} X_\alpha$. Then $(X, \mathcal{T})$ is locally compact if and only if for each $\alpha \in \Lambda$, $(X_\alpha, \mathcal{T}_\alpha)$ is locally compact and for all but a finite number of $\alpha \in \Lambda$, $(X_\alpha, \mathcal{T}_\alpha)$ is compact.

Proof. Suppose $(X, \mathcal{T})$ is locally compact. Since projection maps are open and continuous, each $(X_\alpha, \mathcal{T}_\alpha)$ is locally compact (by Theorem 4.38). Let $x \in X$. Then there exist an open set U and a compact set K such that $x \in U$ and $U \subseteq K$. There exist $\beta_1, \beta_2, ..., \beta_n$ in Λ and open sets $V_{\beta_1}, V_{\beta_2}, ..., V_{\beta_n}$ in $X_{\beta_1}, X_{\beta_2}, ..., X_{\beta_n}$ such that $x \in \bigcap_{i=1}^n \pi_{\beta_i}^{-1}(V_{\beta_i}) \subseteq U$. For each $\alpha \in \Lambda$ such that $\alpha \neq \beta_i$ for any $i = 1, 2, ..., n$, $\pi_\alpha(K) = X_\alpha$. Therefore for each $\alpha \in \Lambda$ such that $\alpha \neq \beta_i$ for any $i = 1, 2, ..., n$, $(X_\alpha, \mathcal{T}_\alpha)$ is compact (Theorem 4.19).

Suppose that for each $\alpha \in \Lambda$, $(X_\alpha, \mathcal{T}_\alpha)$ is locally compact and that $\Gamma = \{\beta_1, \beta_2, \ldots \beta_n\}$ is a finite subset of Λ such that for each $\alpha \in \Lambda - \Gamma$, $(X_\alpha, \mathcal{T}_\alpha)$ is compact. Let $x \in X$. For each $i = 1, 2, \ldots n$, there exists a set $U_{\beta_i} \in \mathcal{T}_{\beta_i}$ and a compact subset K_{β_i} of X_{β_i} such that $x_{\beta_i} \in U_{\beta_i}$ and $U_{\beta_i} \subseteq K_{\beta_i}$. Let $V = \bigcap_{i=1}^n \pi_{\beta_i}^{-1}(U_{\beta_i})$ and $C = \bigcap_{i=1}^n \pi_{\beta_i}^{-1}(K_{\beta_i})$. Then V is a neighborhood of x and $V \subseteq C$. Since $C = \prod_{\alpha \in \Lambda} C_\alpha$, where $C_{\beta_i} = K_{\beta_i}$ for each $i = 1, 2, \ldots, n$, $C_\alpha = X_\alpha$ for all $\alpha \notin \Gamma$, and the product of compact space is compact (see Section 6.6), C is compact. Therefore $(X, \mathcal{T})$ is locally compact at x. Since x is an arbitrary point of X, X is locally compact. ∎

EXERCISES 4.3

1. Let $(\mathbb{R}, \mathcal{T})$ be the topological space given in Example 14, and let $A = \{\frac{1}{n} : n \in \mathbb{N}\}$. Show that there is no neighborhood U of 0 such that $U \subseteq \mathbb{R} - A$.

2. Let $(X, \mathcal{T})$ be a countably compact space, let $(Y, \mathcal{U})$ be a topological space, and let $f : X \to Y$ be a continuous function that maps X onto Y. Prove that $(Y, \mathcal{U})$ is countably compact.

3. Let $\mathcal{T}$ be the topology on $\mathbb{N}$ given in Example 12, and let $\mathcal{U}$ be the discrete topology on $\mathbb{N}$. As seen in Example 12, $(\mathbb{N}, \mathcal{T})$ has the Bolzano-Weierstrass property.

    (a) Prove that $(\mathbb{N}, \mathcal{U})$ does not have the Bolzano-Weierstrass property.

    (b) Define $f : (\mathbb{N}, \mathcal{T}) \to (\mathbb{N}, \mathcal{U})$ by $f(n) = (n+1)/2$ if n is odd and $f(n) = \frac{n}{2}$ if n is even. Prove that f is continuos.

4. Prove that the Balzano-Weierstrass property is a topological property.

5. Let $(X, \mathcal{T})$ be a sequentially compact space, let $(Y, \mathcal{U})$ be a topological space, and let $f : X \to Y$ be a continuous function that maps X onto Y. Prove that $(Y, \mathcal{U})$ is sequentially compact.

6. Let $\{(X_i, \mathcal{T}_i) : i \in \mathbb{N}\}$ be a collection of sequentially compact spaces. Prove that if $\mathcal{T}$ is the product topology on $X = \prod_{i \in \mathbb{N}} X_i$, then $(X, \mathcal{T})$ is sequentially compact.

7. Let $(X, \mathcal{T})$ be a compact space and let $(Y, \mathcal{U})$ be a countably compact space. Prove that $X \times Y$ is countably compact.

8. Let $(X, \mathcal{T})$ and $(Y, \mathcal{U})$ be topological spaces, and let $f : X \to Y$ be a perfect map. Prove X is locally compact if and only if Y is locally compact.

9. Let $(X, \mathcal{T})$ be a locally compact space, and, for each $n \in \mathbb{N}$, let U_n be an open dense subset of X. Prove that $\bigcap_{n \in =\mathbb{N}} U_n$ is dense in X.

10. Let $(X, \mathcal{T})$ be a countably compact space, let $(Y, \mathcal{U})$ be a first countable space, and let $f: X \to Y$ be a continuous bijection. Prove that f is a homeomorphism.

11. If $(X, \mathcal{T})$ is a topological space, the **cone over X** is the quotient space $(X \times I)/\sim$, where $\sim$ is the equivalence relation on $X \times I$ defined by $(a, 1) \sim (b, 1)$ for all $a, b \in X$, and if $t \neq 1$, $(a, t) \sim (b, s)$ if and only if $a = b$ and $s = t$. Prove that if the cone over a topological space $(X, \mathcal{T})$ is locally compact, then the cone over X is compact.

12. Let $(X, \mathcal{T})$ be a connected, locally connected, locally compact Hausdorff space and let $a, b \in X$. Prove that there exists a compact connected subset of X that contains both a and b.

13. Let $(X, \mathcal{T})$ be a locally compact space, let $(Y, \mathcal{U})$ be a topological space, let $f: X \to Y$ be an open continuous surjection, and let K be a compact subset of Y. Prove that there exists a compact subset C of X such that $f(C) = K$.

14. Let $(X_1, \mathcal{T}_1)$ and $(X_2, \mathcal{T}_2)$ be topological spaces and suppose $f: X_1 \to X_2$ is a quotient map. Let $(Y, \mathcal{U})$ be a locally compact Hausdorff space. Prove that the function $g: X_1 \times Y \to X_2 \times Y$ defined by $g(x, y) = (f(x), y)$ is a quotient map.

15. Let $(X_1, \mathcal{T}_1)$ and $(Y_2, \mathcal{U}_2)$ be topological spaces, let $(X_2, \mathcal{T}_2)$ and $(Y_1, \mathcal{U}_1)$ be locally compact Hausdorff spaces, and let $f: X_1 \to Y_1$ and $g: X_2 \to Y_2$ be quotient maps. Prove that the function $h: X_1 \times X_2 \to Y_1 \times Y_2$ defined by $h(x_1, x_2) = (f(x_1), g(x_2))$ is a quotient map.

16. Prove that a T_1-space is countably compact if and only if every infinite subset has a cluster point.

4.4 The Weak Topology on a Topological Space

In this section we continue our study of local compactness. We introduce the notion of weak topology since it provides a method of describing the topology in a locally compact Hausdorff space.

Definition. Let X be a set and let $\mathcal{A} = \{A_\alpha: \alpha \in \Lambda\}$ be a family of subsets of X and suppose each A_α has a topology such that:

(a) The topologies of A_α and A_β agree on $A_\alpha \cap A_\beta$.

(b) Either each $A_\alpha \cap A_\beta$ is open in A_α and in A_β or each $A_\alpha \cap A_\beta$ is closed in A_α and in A_β.

The **weak topology** on X induced by $\mathcal{A}$ is $\{U \in \mathcal{P}(X): \text{for each } \alpha \in \Lambda, U \cap A_\alpha \text{ is open in } A_\alpha\}$. ∎

4.4 The Weak Topology on a Topological Space

We first prove that the weak topology is indeed a topology. The two conditions that we have imposed play no role in this proof.

THEOREM 4.40. Let X be a set and let $\mathscr{A} = \{A_\alpha : \alpha \in \Lambda\}$ be a collection of subsets of X and suppose that each A_α has a topology $\mathscr{T}_\alpha$. Then $\mathscr{T} = \{U \in \mathscr{P}(X) :$ for each $\alpha \in \Lambda, U \cap A_\alpha \in \mathscr{T}_\alpha\}$ is a topology on X.

Proof. Since $\emptyset \cap A_\alpha = \emptyset \in \mathscr{T}_\alpha$ for each $\alpha \in \Lambda$, $\emptyset \in \mathscr{T}$. Since $X \cap A_\alpha = A_\alpha \in \mathscr{T}_\alpha$ for each $\alpha \in \Lambda$, $X \in \mathscr{T}$.

Suppose $U, V \in \mathscr{T}$. Then for each $\alpha \in \Lambda$, $(U \cap V) \cap A_\alpha = (U \cap A_\alpha) \cap (V \cap A_\alpha)$. Since $(U \cap A_\alpha) \cap (V \cap V_\alpha)$ is the intersection of two members of $\mathscr{T}_\alpha$, it is a member of $\mathscr{T}_\alpha$. Therefore $U \cap V \in \mathscr{T}$.

Suppose $\{U_\beta : \beta \in \Gamma\}$ is a collection of members of $\mathscr{T}$. Then for each $\alpha \in \Lambda$, $(\bigcup_{\beta \in \Gamma} U_\beta) \cap A_\alpha = \bigcup_{\beta \in \Gamma}(U_\beta \cap A_\alpha)$. Since $\bigcup_{\beta \in \Gamma}(U_\beta \cap A_\alpha)$ is the union of members of $\mathscr{T}_\alpha$, it is a member of $\mathscr{T}_\alpha$. Therefore $\bigcup_{\beta \in \Gamma} U_\beta \in \mathscr{T}$ and hence $\mathscr{T}$ is a topology on X. ■

EXAMPLE 15. Let X be a set and let $\mathscr{A} = \{A_\alpha : \alpha \in \Lambda\}$ be a family of subsets X and suppose each A_α has a topology $\mathscr{T}_\alpha$. Let $B \subseteq X - \bigcup_{\alpha \in \Lambda} A_\alpha$ (see Figure 4.2). Then $B \cap A_\alpha = \emptyset$ for each $\alpha \in \Lambda$. Therefore B is a member of the weak topology $\mathscr{T}$ on X induced by $\mathscr{A}$. Since B is an arbitrary subset of $X - \bigcup_{\alpha \in \Lambda} A_\alpha$, $\mathscr{T}$ is the discrete topology on $X - \bigcup_{\alpha \in \Lambda} A_\alpha$.

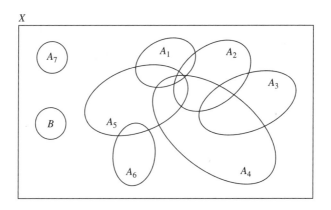

Figure 4.2

Notice that in the definition of the weak topology, we began with a set X and a collection of subset $\mathscr{A}$ of X such that each member A_α of $\mathscr{A}$ has a topology $\mathscr{T}_\alpha$. Then $\mathscr{A}$ induced a topology $\mathscr{T}$ (the weak topology) on X. So for each $\alpha \in \Lambda$, A_α is a subset of the topological space $(X, \mathscr{T})$. Thus $\mathscr{T}$ determines the subspace topology $\mathscr{T}_{A_\alpha}$ on A_α. How are $\mathscr{T}_\alpha$ and $\mathscr{T}_{A_\alpha}$ related? Conditions (a) and (b) in the definition of the weak topology enable us to prove that $\mathcal{T}_{A_\alpha} = \mathcal{T}_\alpha$. Observe in the proof of the following theorem that we do not need either condition (a) or (b) to prove that $\mathcal{T}_{A_\alpha} \subseteq \mathcal{T}_\alpha$.

THEOREM 4.41. Let X be a set and let $\mathcal{A} = \{A_\alpha : \alpha \in \Lambda\}$ be a family of subsets of X, and suppose each A_α has a topology $\mathcal{T}_\alpha$ such that:

(a) The topologies of A_α and A_β agree on $A_\alpha \cap A_\beta$.

(b) Either each $A_\alpha \cap A_\beta$ is open in A_α and in A_β, or each $A_\alpha \cap A_\beta$ is closed in A_α and in A_β.

Let $\mathcal{T}$ be the weak topology on X induced by $\mathcal{A}$. Then for each $\alpha \in \Lambda$, $\mathcal{T}_{A_\alpha} = \mathcal{T}_\alpha$.

Proof. Let $\alpha \in \Lambda$ and let $U \in \mathcal{T}_{A_\alpha}$. Then there exists $V \in \mathcal{T}$ such that $U = A_\alpha \cap V$. Since $V \in \mathcal{T}$, $V \cap A_\alpha \in \mathcal{T}_\alpha$. Therefore $U \in \mathcal{T}_\alpha$ and hence $\mathcal{T}_{A_\alpha} \subseteq \mathcal{T}_\alpha$.

We prove that if each $A_\alpha \cap A_\beta$ is open in A_α and in A_β, then $\mathcal{T}_\alpha \subseteq \mathcal{T}_{A_\alpha}$. The proof that if each $A_\alpha \cap A_\beta$ is closed in A_α and in A_β, then $\mathcal{T}_\alpha \subseteq \mathcal{T}_{A_\alpha}$, is similar and is left as Exercise 3.

Suppose each $A_\alpha \cap A_\beta$ is open in A_α and in A_β. Let $\alpha \in \Lambda$ and let $U \in \mathcal{T}_\alpha$. Let $\beta \in \Lambda$. Since $A_\alpha \cap A_\beta$ has the subspace topology determined by A_α (the topologies of A_α and A_β agree on $A_\alpha \cap A_\beta$), $U \cap (A_\alpha \cap A_\beta)$ is open in $A_\alpha \cap A_\beta$. But $U \subseteq A_\alpha$, so $U \cap (A_\alpha \cap A_\beta) = U \cap A_\beta$. Therefore $U \cap A_\beta$ is open in $A_\alpha \cap A_\beta$. But since $A_\alpha \cap A_\beta$ is open in A_β, $U \cap A_\beta$ is open in A_β. Therefore $U \in \mathcal{T}$ and hence $U \cap A_\alpha = U \in \mathcal{T}_{A_\alpha}$. Thus $\mathcal{T}_\alpha \subseteq \mathcal{T}_{A_\alpha}$ and the proof is complete. ∎

Whenever we speak of the weak topology on a set induced by a collection of subsets, we automatically assume that the subsets have topologies that satisfy the preceding definition.

THEOREM 4.42. Let X be a set, let $\mathcal{A} = \{A_\alpha : \alpha \in \Lambda\}$ be a collection of subsets of X, let $\mathcal{T}$ be the weak topology on X induced by $\mathcal{A}$, and let $(Y, \mathcal{U})$ be a topological space. A function $f : X \to Y$ is continuous if and only if for each $\alpha \in \Lambda$, $f|_{A_\alpha} : A_\alpha \to Y$ is continuous.

Proof. Suppose $f : X \to Y$ is continuous. Then by Theorem 2.13(b), $f|_{A_\alpha} : A_\alpha \to Y$ is continuous for each $\alpha \in \Lambda$.

Suppose that for each $\alpha \in \Lambda$, $f|_{A_\alpha} : A_\alpha \to Y$ is continuous. Let $U \in \mathcal{U}$. Then $(f|_{A_\alpha})^{-1}(U)$ is open in A_α for each $\alpha \in \Lambda$. But for each $\alpha \in \Lambda$, $f^{-1}(U) \cap A_\alpha = (f|_{A_\alpha})^{-1}(U)$ and hence $f^{-1}(U) \cap A_\alpha$ is open in A_α for each $\alpha \in \Lambda$. Therefore by the definition of the weak topology on X, $f^{-1}(U) \in \mathcal{T}$. Therefore f is continuous. ∎

4.4 The Weak Topology on a Topological Space

As Corollary 4.44 indicates, the following theorem permits us to say that a locally compact Hausdorff space has the weak topology induced by the family of compact subsets.

THEOREM 4.43. Let $(X, \mathcal{T})$ be a locally compact Hausdorff space. Then a subset U of X is open if and only if its intersection with each compact subset C of X is open in C.

Proof. If U is open, then its intersection with each subset A of X is open in A.

Suppose U is a subset of X such that for each compact subset C of X, $U \cap C$ is open in C. Let $x \in U$. By Theorem 4.35, there is a neighborhood V of x such that $\overline{V}$ is compact. Now $U \cap \overline{V}$ is open in $\overline{V}$, and hence $U \cap V = (U \cap \overline{V}) \cap V$ is open in V. (Since $U \cap \overline{V}$ is open in $\overline{V}$, there exists an open subset W of X such that $U \cap \overline{V} = \overline{V} \cap W$. Therefore $W \cap V = (W \cap \overline{V}) \cap V = (U \cap \overline{V}) \cap V$ (see Exercise 1)). Since $U \cap V$ is open in V and V is open in X, $U \cap V$ is open in X. Since $x \in U \cap V \subseteq U$ and x is an arbitrary point of U, U is open in X. ∎

Corollary 4.44. A locally compact Hausdorff space has the weak topology induced by the family of compact subsets. ∎

Definition. A Hausdorff space X is a **k-space** if it has the weak topology induced by the family of compact subsets. ∎

The name, k-space, first appears in a paper by David Gale (1921–2008) in 1950. He attributes it to Witold Hurewicz but does not give a reference.

The importance of k-spaces is illustrated by the following theorem, whose proof is left as Exercise 6.

THEOREM 4.45. Let $(X, \mathcal{T})$ be a k-space and let $(Y, \mathcal{U})$ be a topological space. Then $f: X \to Y$ is a continuous if and only if $f|_C$ is continuous for each compact subset C of X. ∎

Notice the every locally compact Hausdorff space is a k-space.

THEOREM 4.46. Every first countable Hausdorff space $(X, \mathcal{T})$ is a k-space.

Proof. Let A be a subset of X such that $A \cap C$ is closed in C for each compact subset C of X. Then $A \cap C$ is closed in X for each compact subset C of X. We show that A is closed by showing that $\overline{A} \subseteq A$. Let $x \in \overline{A}$. By Theorem 1.41, there is a sequence $\langle x_n \rangle$ in A such that $\langle x_n \rangle \to x$. Thus $\langle x_n \rangle \cup \{x\}$ is compact. Thus $A \cap (\langle x_n \rangle \cup \{x\})$ is compact, and hence it is closed. Therefore $x \in A \cap (\langle x_n \rangle \cup \{x\})$, so $x \in A$. Thus $\overline{A} \subseteq A$. ∎

Definition. Let $\{(X_\alpha, \mathcal{T}_\alpha): \alpha \in \Lambda\}$ be a collection of topological spaces. For each $\alpha \in \Lambda$, let $X'_\alpha = \{\alpha\} \times X_\alpha$. Then X'_α is homeomorphic to X_α and the collection $\{X'_\alpha: \alpha \in \Lambda\}$ is pairwise disjoint. The **free union** of $\{X_\alpha: \alpha \in \Lambda\}$ is the set $\bigcup_{\alpha \in \Lambda} X'_\alpha$ with the weak topology determined by the collection $\{X'_\alpha: \alpha \in \Lambda\}$. This space is denoted by $\sum_{\alpha \in \Lambda} X'_\alpha$. ∎

THEOREM 4.47. Let $(X, \mathcal{T})$ be a space with the weak topology induced by the covering $\{A_\alpha : \alpha \in \Lambda\}$. Let $A = \sum_{\alpha \in \Lambda} A'_\alpha$ be the free union of $\{A_\alpha : \alpha \in \Lambda\}$, and for each $\alpha \in \Lambda$, let $h_\alpha : A'_\alpha \to A_\alpha$ be the homeomorphism defined by $h_\alpha((\alpha, \alpha)) = \alpha$. Define $h : \sum_{\alpha \in \Lambda} A'_\alpha \to X$ by $h|_{A'_\alpha} = h_\alpha$. Then h is continuous and $A/\sim_h$ is homeomorphic to X.

Proof. By Theorem 4.42, h is continuous. By Theorem 2.61, it is sufficient to show that h is a quotient map. Let U be a subset of X such that $h^{-1}(U)$ is open in $\sum_{\alpha \in \Lambda} A'_\alpha$. Then $h^{-1}(U) \cap A_\alpha = h_\alpha^{-1}(U \cap A_\alpha)$ is open in A'_α for each α. Since h_α is homeomorphism, $U \cap A_\alpha$ is open in A_α. Thus U is open in X. ∎

THEOREM 4.48 Let $(X, \mathcal{T})$ be a Hausdorff space. Then X is a k-space if and only if it is a quotient space of a locally compact space.

Proof. Suppose X is a k-space. By Theorem 4.47, X is a quotient space of the free union of its compact subspaces. The desired result follows because the free union of compact spaces is locally compact (see Exercise 7).

Let $(Y, \mathcal{U})$ be a locally compact space and suppose $p : Y \to X$ is a quotient map. Let U be a subset of X such that $U \cap C$ is open in C for each compact subset C of X. We need to show that U is open in X. Let V be open subset of Y such that $\overline{V}$ is compact. Then $U \cap p(\overline{V})$ is open in the compact space $p(\overline{V})$. That is, $U \cap p(\overline{V}) = p(\overline{V}) \cap G$, where G is open in X. Since $p^{-1}(U) \cap p^{-1}(p(\overline{V})) = p^{-1}(p(V)) \cap p^{-1}(G)$, $[p^{-1}(U) \cap p^{-1}(p(\overline{V}))] \cap V = [p^{-1}(p(\overline{V})) \cap p^{-1}(G)] \cap V$. Thus $p^{-1}(U) \cap V = V \cap p^{-1}(G)$. Therefore $p^{-1}(U) \cap V$ is open in Y. Since there is a covering $\{V_\alpha : \alpha \in \Lambda\}$ of Y by open sets whose closures are compact and $p^{-1}(U) = \bigcup_{\alpha \in \Lambda}(p^{-1}(U) \cap V_\alpha)$, $p^{-1}(U)$ is open in Y. Therefore U is open in X. ∎

EXERCISES 4.4

1. Let $(X, \mathcal{T})$ be a topological space and let $A \subseteq X$. Prove that the relative topology on A determined by $\mathcal{T}_{\overline{A}}$ is the same as the relative topology on A determined by $\mathcal{T}$.

2. Let X be a set, let $\mathcal{A} = \{A_\alpha : \alpha \in \Lambda\}$ be a family of subsets of X, and let $\mathcal{T}$ be the weak topology on X induced by $\mathcal{A}$. Prove that:

    (a) A subset C of X is closed if and only if its intersection with each A_α is closed in A_α.

    (b) Let $\alpha \in \Lambda$. If, for each $\beta \in \Lambda$, $A_\alpha \cap A_\beta$ is open in A_α and in A_β, then A_α is an open subset of X.

    (c) Let $\alpha \in \Lambda$. If, for each $\beta \in \Lambda$, $A_\alpha \cap A_\beta$ is closed in A_α and in A_β, then A_α is a closed subset of X.

3. This exercise refers to the proof of Theorem 4.41. Prove that if each $A_\alpha \cap A_\beta$ is closed in A_α and in A_β, then $\mathcal{T}_\alpha \subseteq \mathcal{T}_{A_\alpha}$.

4. Let $(X, \mathcal{T})$ and $(Y, \mathcal{U})$ be first countable k-spaces. Prove that $X \times Y$ is a k-space.

5. Let $(X, \mathcal{T})$ and $(Y, \mathcal{U})$ be k-spaces and suppose $(X, \mathcal{T})$ is locally compact. Prove that $X \times Y$ is a k-space.

6. Let $(X, \mathcal{T})$ be a k-space and let $(Y, \mathcal{U})$ be a topological space. Prove that a function $f: X \to Y$ is continuous if and only if $f|_C$ is continuous for each compact subset C of X.

7. Prove that the free union of compact space is locally compact.

4.5 Equicontinuity

Let $(X, \mathcal{T})$ be a compact space. Then $C(X, \mathbb{R}^n)$ denotes the collection of continuous functions that map X into $\mathbb{R}^n$. Since X is compact and the continuous image of a compact space is compact, for each $f \in C(X, \mathbb{R}^n)$, $f(X)$ is a compact subset of $\mathbb{R}^n$. Therefore by Theorem 4.30, $f(X)$ is closed and bounded. Thus we assume in the remainder of this section that the sup metric ρ on $C(X, \mathbb{R}^n)$ is defined by $\rho(f, g) = \max \{d(f(x), g(x)): x \in X\}$, where d is the usual metric on $\mathbb{R}^n$. We might hope that a subset of $(C(X, \mathbb{R}^n), \rho)$ is compact if and only if it is closed and bounded. This result, however, is not true. We heed an additional condition.

In the remainder of this section, we assume that if $(X, \mathcal{T})$ and $(Y, \mathcal{U})$ are topological spaces, then $C(X, Y)$ denotes the collection of continuous functions that map X into Y.

The idea of defining a topology on the set of continuous functions that map one topological space into another plays an important role in topology. There are many possible topologies on the set of functions, and there are many reasons why one studies function spaces. One application of the completeness of the space $C(X, Y)$, with respect to the uniform metric, when Y is a complete metric space (see Section 2.5), is to show the existence of a space-filling curve. A continuous function f mapping $[0, 1]$ onto $[0, 1] \times [0, 1]$ can be constructed as the limit of a sequence of continuous functions. The existence of such a path violates one's intuition.

Fréchet, in 1906, made the first major effort to develop an abstract theory of function spaces.

Definition. Let $(X, \mathcal{T})$ be a topological space, let (Y, d) be a metric space, let $\mathcal{F} \subseteq C(X, Y)$, and let $x_0 \in X$. Then $\mathcal{F}$ is **equicontinuous** at x_0 if for each $\varepsilon > 0$ there exists a neighborhood U of x_0 such that if $f \in \mathcal{F}$ and $x \in U$, then $d(f(x), f(x_0)) < \varepsilon$. If $\mathcal{F}$ is equicontinuous at each point of X, then it is said to be equicontinuous. ■

Note the difference between a collection of continuous functions and a collection of equicontinuous functions. If $\mathscr{F}$ is a collection of continuous functions, then for each $x_0 \in X$, each $\varepsilon > 0$, and each $f \in \mathscr{F}$, there exists a neighborhood U_f of x_0 such that if $x \in U_f$, then $d(f(x), f(x_0)) < \varepsilon$. If $\mathscr{F}$ is a collection of equicontinuous functions, then for each $x_0 \in X$ and each $\varepsilon > 0$, there exists a neighborhood U of x_0 such that if $x \in U$ and f is *any* member of $\mathscr{F}$, then $d(f(x), f(x_0)) < \varepsilon$. (In other words, U works for *every* member of $\mathscr{F}$.)

For compact spaces $(X, \mathscr{T})$ and (Y, d), we have the following characterization of equicontinuity.

THEOREM 4.49. Let $(X, \mathscr{T})$ be a compact space, let (Y, d) be a compact metric space, and let $\mathscr{F} \subseteq C(X, Y)$. Then $\mathscr{F}$ is equicontinuous if and only if $\mathscr{F}$ is totally bounded with respect to ρ.

Proof. Suppose $\mathscr{F}$ is totally bounded with respect to ρ. Let $x_0 \in X$ and let $\varepsilon > 0$. Let $\varepsilon_1 = \frac{\varepsilon}{3}$, and let $\{f_1, f_2, ..., f_n\}$ be an ε_1-net for $\mathscr{F}$. For each $i = 1, 2, ..., n, f_i$ is continuous. Therefore, for each $i = 1, 2, ..., n$, let U_i be a neighborhood of x_0 such that if $x \in U_i$ then $d(f_i(x), f_i(x_0)) < \varepsilon_1$. Let $U = \bigcap_{i=1}^{n} U_i$.

We claim that if $f \in \mathscr{F}$ and $x \in U$, then $d(f_i(x), (x_0)) < \varepsilon$. Let $f \in \mathscr{F}$ and let $x \in U$. Then there exists $i (i = 1, 2, ..., n)$ such that $d(f, f_i) < \varepsilon_1$. Thus $d(f(x), f(x_0)) \leq d(f(x), f_i(x)) + d(f_i(x), f_i(x_0)) + d(f_i(x_0), f(x_0)) < \varepsilon_1 + \varepsilon_1 + \varepsilon_1 = \varepsilon$.

Therefore $\mathscr{F}$ is equicontinuous.

Suppose $\mathscr{F}$ is equicontinuous, and let $\varepsilon > 0$. Let $\varepsilon_1 = \frac{\varepsilon}{3}$. Since $\mathscr{F}$ is equicontinuous, for each $x \in X$ there is a neighborhood U_x of x such that if $z \in U$, then $d(f(z), f(x)) < \varepsilon_1$ for all $f \in \mathscr{F}$. Then $\{U_x : x \in X\}$ is an open cover of X. Since X is compact, there exist $x_1, x_2, ..., x_m$ such that $\{U_{x_1}, U_{x_2}, ..., U_{x_m}\}$ covers X. Now $\{B_d(y, \varepsilon_1) : y \in Y\}$ is an open cover of Y. Since Y is compact, there exist $y_1, y_2, ..., y_n$ such that $\{B_d(y_1, \varepsilon_1), B_d(y_2, \varepsilon_1), ..., B_d(y_n, \varepsilon_1)\}$ covers Y.

Let Λ be the collection of all functions that map $\{1, 2, ..., m\}$ into $\{1, 2, ..., n\}$, and let $\alpha \in \Lambda$. If there exists $f \in \mathscr{F}$ such that for each $i = 1, 2, ..., m, f(x_i) \in B_d(y_{\alpha(i)}, \varepsilon_1)$, choose one such function and label it f_α. Let $\Gamma = \{\alpha \in \Lambda : f_\alpha \text{ exists}\}$. Since Λ is finite and $\Gamma \subseteq \Lambda$, Γ is finite. We claim that $\{B_\rho(f_\alpha, \varepsilon) : \alpha \in \Gamma\}$ covers $\mathscr{F}$.

Let $f \in \mathscr{F}$, and for each $i = 1, 2, ..., m$, let $\alpha(i)$ be an integer such that $f(x_i) \in B_d(y_{\alpha(i)}, \varepsilon_1)$. Then $\alpha \in \Gamma$. We claim that $f \in B_\rho(f_\alpha, \varepsilon)$. Let $x \in X$, and let $i \in \{1, 2, ..., m\}$ such that $x \in U_{x_i}$. Then $d(f(x), f_\alpha(x)) \leq d(f(x), f(x_i)) + d(f(x_i), f_\alpha(x_i)) + d(f_\alpha(x_i), f_\alpha(x)) < \varepsilon_1 + \varepsilon_1 + \varepsilon_1 = \varepsilon$. Since this inequality holds for every $x \in X$, $\rho(f, f_\alpha) = \max\{d(f(x)), f_\alpha(x)) : x \in X\} < \varepsilon$. Hence $f \in B_\rho(f_\alpha, \varepsilon)$. ∎

We prove the following result in order to characterize the compact subsets of $(C(X, \mathbb{R}^n), \rho)$.

THEOREM 4.50. Let $(X, \mathscr{T})$ be a compact space, and let $\mathscr{F}$ be a bounded subset of $(C(X, \mathbb{R}^n), \rho)$. Then there exists a compact subset Y of $\mathbb{R}^n$ such that if $f \in \mathscr{F}$ and $x \in X$, then $f(x) \in Y$.

Proof. Let $f_0 \in \mathscr{F}$. Since $\mathscr{F}$ is bounded, there exists a positive number M such that $\rho(f_0, f) < M$ for all $f \in \mathscr{F}$. Since X is compact and f_0 is continuous, $f_0(X)$ is compact. Hence $f_0(X)$ is a bounded subset of $\mathbb{R}^n$, so there is a positive number N such that $f_0(X) \subseteq B_d((0, 0, ..., 0), N)$. Therefore if $f \in \mathscr{F}$, then $f(X) \subseteq B_d((0, 0, ..., 0), M + N)$. Then $Y = \overline{B_d((0, 0, ..., 0), M + N)}$ is a closed and bounded subset of $\mathbb{R}^n$, and hence it is compact. ∎

THEOREM 4.51 (Ascoli's Theorem). Let $(X, \mathscr{T})$ be a compact space. Then a subset of $(C(X, \mathbb{R}^n), \rho)$ is compact if and only if it is closed, bounded, and equicontinuous.

Proof. Suppose $\mathscr{F}$ is a compact subset of $(C(X, \mathbb{R}^n), \rho)$. Since every compact subset of a Hausdorff space is closed, $\mathscr{F}$ is closed. By Theorem, 4.11, $\mathscr{F}$ is totally bounded, so by Theorem 4.1, $\mathscr{F}$ is bounded. By Theorem 4.50, there exists a compact subset Y of $\mathbb{R}^n$ such that if $f \in \mathscr{F}$ then $f(X) \subseteq Y$. It follows that $\mathscr{F} \subseteq C(X, Y)$. Therefore by Theorem 4.49, $\mathscr{F}$ is equicontinuous.

Suppose $\mathscr{F}$ is a closed, bounded, and equicontinuous subset of $(C(X, \mathbb{R}^n), \rho)$. By Exercise 4 of Section 1.6, $(\mathbb{R}^n, d)$ is complete. Therefore by corollary 2.54, $(C(X, \mathbb{R}^n), \rho)$ is complete. Since $\mathscr{F}$ is closed subset of $(C(X, \mathbb{R}^n), \rho)$, by Theorem 1.47, $\mathscr{F}$ is complete. Since $\mathscr{F}$ is bounded, by Theorem 4.50, there exists a compact subset Y of $\mathbb{R}^n$ such that if $f \in \mathscr{F}$ then $f(X) \subseteq Y$. It follows that $\mathscr{F} \subseteq C(X, Y)$. Therefore by Theorem 4.49, $\mathscr{F}$ is totally bounded. Since $\mathscr{F}$ is complete and totally bounded, by Theorem 4.11, it is compact. ∎

In the exercise that follow, we introduce several new concepts, including the definitions of additional topologies on function spaces. Part (d) of Exercise 3 is often called Arzela's Theorem. Arzela's Theorem was proved by Giulio Ascoli (1843–1896) in 1883 and by Cesare Arzela (1847–1912) in 1895. One version of Ascoli's Theorem was proved by S. B. Myers in 1946, and another version was proved by Gale in 1950.

EXERCISES 4.5

1. Let $(X, \mathscr{T})$ and $(Y, \mathscr{U})$ be topological space and let $\mathscr{F}$ be finite set of continuous functions that map X into Y. Prove that $\mathscr{F}$ is a equicontinuous.

2. For each $n \in \mathbb{N}$, let $f_n: [0, 1] \to \mathbb{R}$ be the function defined by $f_n(x) = x^n$, and let $\mathscr{F} = \{f_n: n \in \mathbb{N}\}$.

    (a) Show that $\mathscr{F}$ is a closed subset of $(C([0, 1], \mathbb{R}), \rho)$.

    (b) Show that $\mathscr{F}$ is a closed subset of $(C([0, 1], \mathbb{R}), \rho)$.

    (c) Show that $\mathscr{F}$ is not equicontinuous.

3. Let $(X, \mathscr{T})$ be a compact space. A subset of $(C(X, \mathbb{R}^n), \rho)$ that is bounded is said to be **uniformly bounded**. A subset $\mathscr{F}$ of $C(X, \mathbb{R}^n)$ is said to be **pointwise bounded** if for each $x \in X$, $\{f(x): f \in \mathscr{F}\}$ is a bounded subset of $\mathbb{R}^n$.

    (a) Prove that if $\mathscr{F}$ is a subset of $(C(X, \mathbb{R}^n), \rho)$ that is pointwise bounded and equicontinuous, then $\mathscr{F}$ is uniformly bounded.

(b) Prove that if $\mathscr{F}$ is an equicontinuous subset of $(C(X, \mathbb{R}^n), \rho)$ then $\overline{\mathscr{F}}$ is equicontinuous.

(c) Let $\mathscr{F}$ be a subset of $(C(X, \mathbb{R}^n), \rho)$. Prove that $\mathscr{F}$ is equicontinuous and pointwise bounded if and only if $\overline{\mathscr{F}}$ is compact.

(d) For each $k \in \mathbb{N}$, let $f_k \in C(X, \mathbb{R}^n)$. Prove that $\{f_k : k \in \mathbb{N}\}$ is an equicontinuous and pointwise bounded subset of $(C(X, \mathbb{R}^n), \rho)$, then the sequence $\langle f_n \rangle$ has a subsequence that converges uniformly.

4. Let X be a set and let $(Y, \mathcal{U})$ be a topological space. For each $x \in X$ and $U \in \mathcal{U}$ let $S(x, U) = \{f \in Y^X : f(x) \in U\}$.

(a) Prove that $\{S(x, U) : x \in X \text{ and } U \in \mathcal{U}\}$ is a subbasis for a topology on Y^X. This topology is called the **topology of pointwise convergence** or the **pointopen topology**.

(b) For each $n \in \mathbb{N}$, let $f_n \in Y^X$, let $f \in Y^X$, and let $\mathcal{T}$ be the topology of pointwise convergence on Y^X. Prove that $\langle f_n \rangle$ converges to f if and only if for each $x \in X$, $\langle f_n(x) \rangle$ converges to $f(x)$.

(c) For each $n \in \mathbb{N}$, let $f_n : [0, 1] \to \mathbb{R}$ be the function defined by $f_n(x) = x^n$; let $f : [0, 1] \to \mathbb{R}$ be the function defined by $f(x) = 0$ if $0 \leq x < 1$ and $f(1) = 1$, and let $\mathcal{T}$ be the topology of pointwise convergence on Y^X. Prove that $\langle f_n \rangle$ converges to f.

5. Let $(X, \mathcal{T})$ be a topological space and let (Y, d) be a metric space. For each $f \in Y^X$, each compact subset C of X, and each $\varepsilon > 0$, let $B_C(f, \varepsilon) = \{g \in Y^X : \text{lub}\{d(f(x), g(x)) : x \in C\} < \varepsilon\}$.

(a) Prove that $\{B_C(f, \varepsilon) : f \in Y^X, C \text{ is a compact subset of } X, \text{ and } \varepsilon > 0\}$ is a basis for a topology on Y^X. This topology is called the **topology of compact convergence**.

(b) For each $n \in \mathbb{N}$, let $f_n \in Y^X$, let $f \in Y^X$, and let $\mathcal{U}$ be the topology of compact convergence on Y^X. Prove that $\langle f_n \rangle$ converges to f if and only if for each compact subset of C of X, $\langle f_n |_C \rangle$ converges to $f |_C$.

6. Let $(X, \mathcal{T})$ be topological space, let (Y, d) be a metric space, let $\mathcal{U}_1$ be the uniform topology on Y^X, let $\mathcal{U}_2$ be the topology of pointwise convergence on Y^X, and let $\mathcal{U}_3$ be the topology of compact convergence on Y^X.

(a) Prove that $\mathcal{U}_2 \subseteq \mathcal{U}_3 \subseteq \mathcal{U}_1$.

(b) Prove that if $(X, \mathcal{T})$ is compact, then $\mathcal{U}_1 = \mathcal{U}_3$.

7. Let $(X, \mathcal{T})$ and $(Y, \mathcal{U})$ be a topological spaces. For each compact subset C of X and each $U \in \mathcal{U}$, let $S(C, U) = \{f \in C(X, Y) : f(C) \subseteq U\}$. Prove that $S(C, U) : C$ is a compact subset of X and $U \in \mathcal{U}\}$ is a basis for a topology on $C(X, Y)$. This topology is called the **compact-open topology**.

Chapter 5
The Separation and Countability Axioms

Whereas connectedness and compactness arise naturally from the study of analysis, the concepts that we study in this chapter arise from a study of topology itself. The importance of first countability (which was introduced in Section 1.3) and Hausdorff was illustrated in the study of convergence (Section 1.6). From an intuitive point of view, second countablity, which was also introduced in Section 1.3, requires that there not be too many open sets. From the same point of view, the properties that are of primary interest in this chapter require that there not be too few open sets. The Hausdorff property, for example, requires that, for each pair of distinct points, there are disjoint open sets that contain them. We have already introduced two properties, T_0- (Section 3.3) and T_1- (Section 4.1), that are weaker than Hausdorff. In this chapter we introduce properties that are stronger than Hausdorff and explore some of their consequences. Some goals are to prove Urysohn's Lemma and the Tietze Extension Theorem and to explore embedding problems such as the embedding of a given topological space in a "cube."

5.1 T_0-, T_1-, and T_2-Spaces

The primary purpose of this section is to present a systematic study of three properties, T_0-, T_1-, and T_2-, that we have already defined. The T_0 property was introduced by A.N. Kolmogorov (1803–1887), and the T_1 property was introduced by Fréchet in 1907. Fréchet used the name *accessible spaces* for T_1-spaces. Hausdorff introduced the T_2 property in 1914. In 1923, Heinrich Tietze introduced the T_i terminology. The separation properties were also known as *Trennungsaxiomen*. As we have seen, the T_0- and T_1-axioms are weaker than the Hausdorff property. For completeness, we recall the definition of T_0- and T_1-spaces. A topological space $(X, \mathcal{T})$ is a **T_0-space** provided that for each pair of distinct points there is an open set that contains one of the points but not the other. If X is a set with at least two members and $\mathcal{T}$ is the trivial topology on X, then $(X, \mathcal{T})$ is not a T_0-space. A topological space $(X, \mathcal{T})$ is a **T_1-space** provided that for each pair x, y of distinct points of X there exist open sets U and V such that $x \in U, y \notin U, y \in V$, and $x \notin V$. It is clear that every T_1-space is a T_0-space. Let $X = \{1, 2\}$ and let $\mathcal{T} = \{\emptyset, \{1\}, X\}$. Then $(X, \mathcal{T})$ is a T_0-space that is not a T_1-space. Let $\mathcal{T}$ be the finite complement topology on $\mathbb{R}$. Then $(\mathbb{R}, \mathcal{T})$ is a T_1-space. However if U and V are any two nonempty open sets, then $U \cap V \neq \emptyset$. Therefore $(\mathbb{R}, \mathcal{T})$ is not a Hausdorff space. Hausdorff spaces are also called **T_2-spaces.** The importance of T_1-spaces, as shown in the following theorem, is that they are spaces in which $\{x\}$ is closed for each x in the space.

THEOREM 5.1. Let $(X, \mathcal{T})$ be a topological space. Then the following statements are equivalent.

(a) $(X, \mathcal{T})$ is a T_1-space.

(b) For each $x \in X$, $\{x\}$ is closed.

(c) If A is any subset of X, then $A = \bigcap \{U \in \mathcal{T} : A \subseteq U\}$.

Proof. (a) $\rightarrow$ (b). Let $x \in X$. For each $y \in X - \{x\}$, there is an open set U_y such that $y \in U_y$ and $x \notin U_y$. Therefore $X - \{x\}$ is open and hence $\{x\}$ is closed.

(b) $\rightarrow$ (c). Let $A \subseteq X$. For each $y \in X - A$, $X - \{y\}$ is an open set containing A. Since $A = \bigcap \{X - \{y\} : y \in X - A\}$, the proof is complete.

(c) $\rightarrow$ (a). Let x and y be distinct members of X. Since $\{x\} = \bigcap \{U \in \mathcal{T} : x \in U\}$, there is an open set V such that $x \in V$ and $y \notin V$. Also since $\{y\} = \bigcap \{U \in \mathcal{T} : y \in U\}$, there is an open set W such that $y \in W$ and $x \notin W$. Therefore $(X, \mathcal{T})$ is a T_1-space. ∎

The Hausdorff case of each of the following two theorems has already been proved (Theorems 2.10 and 2.45). The remaining cases are left as Exercises 3 and 4.

THEOREM 5.2. For each $i = 0, 1, 2$, every subspace of a T_i-space is a T_i-space. ∎

THEOREM 5.3. For each $i = 0, 1, 2$, the product of T_i-spaces is a T_i-space. ∎

The following examples show that the quotient space of a Hausdorff space need not be Hausdorff. Example 2 shows that the closed continuous image of a Hausdorff space need not be Hausdorff, and Examples 1 and 3 show that the open continuous image of a Hausdorff space need not be Hausdorff.

EXAMPLE 1. Let $I = [0, 1]$, and define $x \sim y$ provided $x - y$ is rational. Then $\sim$ is an equivalence relation on I. Let $p : I \rightarrow I/\sim$ be the natural map. In Exercise 5 you are asked to show that $I/\sim$ is not Hausdorff and that p is open.

EXAMPLE 2. Let $A = \{\frac{1}{n} : n \in \mathbb{N}\}$, and let $\mathcal{B} = \{B \in \mathcal{P}(\mathbb{R}) : B$ is an open interval that does not contain 0 or there is a positive number x such that $B = (-x, x) - A\}$. Then $\mathcal{B}$ is a basis for a topology $\mathcal{T}$ on $\mathbb{R}$, and the space $(\mathbb{R}, \mathcal{T})$ is a Hausdorff space. Let $\mathcal{D}$ be the decomposition of $\mathbb{R}$ whose members are A and $\{x\}$ for all $x \notin A$. In

Exercise 6, you are asked to prove that if $\mathcal{U}$ is the quotient topology on $\mathcal{D}$ induced by the natural map $p: \mathbb{R} \to \mathcal{D}$, then p is closed but $(\mathcal{D}, \mathcal{U})$ is not Hausdorff.

EXAMPLE 3. Let $X = \{(x, y) \in \mathbb{R} \times \mathbb{R}: y = 0 \text{ or } y = 1\}$, and let $\mathcal{T}$ be the subspace topology on X. Then $(X, \mathcal{T})$ is a Hausdorff space. For each $a \in \mathbb{R}$ such that $a \neq 0$, let $D_a = \{(a, 0), (a, 1)\}$, and let $\mathcal{D} = \{D_a : a \in \mathbb{R}\} \cup \{(0, 0)\} \cup \{(0, 1)\}$. Let $\mathcal{U}$ be the quotient topology on $\mathcal{D}$ induced by the natural map $p: X \to \mathcal{D}$. (The set $\mathcal{D}$ is shown in Figure 5.1.) In Exercise 7 you are asked to prove that p is open but $(\mathcal{D}, \mathcal{U})$ is not Hausdorff.

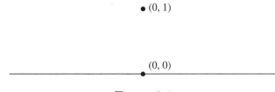

Figure 5.1

Note that every neighborhood of $(0, 1)$ contains a set of the form $\{(x, 0): x \neq 0$ and $-a < x < a$ for some positive number $a\}$. The following sequence of theorems gives various necessary and sufficient conditions for the continuous image of a Hausdorff space to be Hausdorff. (One condition was given in Exercise 9 of Section 4.2.) Recall that if $(X, \mathcal{T})$ is a topological space, $(Y, \mathcal{U})$ is a Hausdorff space, and $f, g: X \to Y$ are continuous, then by Theorem 1.61, $\{x \in X: f(x) = g(x)\}$ is a closed subset of X. As an immediate consequence of this result and Theorem 1.60, we obtain the following theorem.

THEOREM 5.4. Let $(X, \mathcal{T})$ be a topological space, let $(Y, \mathcal{U})$ be a Hausdorff space, and let $f, g: X \to Y$ be continuous functions. If f and g agree on a dense subset of X, then $f = g$. ■

The following result is an immediate corollary of Theorem 5.4.

Corollary 5.5. Let $(X, \mathcal{T})$ be a topological space, let D be a dense subset of X, let $(Y, \mathcal{U})$ be a Hausdorff space, and let $f: (D, \mathcal{T}_D) \to (Y, \mathcal{U})$ be a continuous function. Then f has at most one continuous extension to X. ■

Notice that Corollary 5.5 does not say that a continuous extension exists. In fact if $\mathcal{Q}$ denotes the set of rational numbers, then $\mathcal{Q}$ is a dense subset of $\mathbb{R}$ and the function $f: \mathcal{Q} \to \mathbb{R}$ defined by $f(x) = 1/(x - \pi)$ is continuous, but it does not have a continuous extension to $\mathbb{R}$.

Two important underlying concepts are involved in Theorem 5.4 and Corollary 5.5. One is the question of the existence of a continuous extension of a continuous function, and the other is the question of the uniqueness of a continuous extension of a continuous function. These concepts occurred in the definition of the exponential functions in elementary algebra. There we "defined" the function $f: \mathbb{R} \to \mathbb{R}$ by $f(x) = a^x$ for each positive real number a. We really defined this function for each rational number x and said it had a unique continuous extension to all of $\mathbb{R}$.

Perhaps the most famous result related to the extension problem is the Tietze Extension Theorem (which we discuss in Section 5.5). We prove one extension theorem here, but first we need a definition.

Definition. A subset A of a topological space $(X, \mathcal{T})$ is a **retract** of X provided there is a continuous function $r: X \to A$ such that $r(a) = a$ for each $a \in A$. ∎

The function $r: \mathbb{R} \to [0, 1]$ defined by $r(x) = 0$ if $x \leq 0$, $r(x) = x$ if $x \in [0, 1]$, and $r(x) = 1$ if $x \geq 1$ is continuous and therefore $[0, 1]$ is a retract of $\mathbb{R}$.

THEOREM 5.6. Let $(X, \mathcal{T})$ and $(Y, \mathcal{U})$ be topological spaces, let A be a retract of X, and let $f: A \to Y$ be a continuous function. Then f has a continuous extension $F: X \to Y$.

Proof. Since A is a retract of X, there is a continuous function $r: X \to A$ such that $r(a) = a$ for all $a \in A$. Then $f \circ r: X \to Y$ is continuous and $(f \circ r)(a) = f(r(a)) = f(a)$ for all $a \in A$. Then $f \circ r$ is the desired extension. ∎

THEOREM 5.7. Let $(X, \mathcal{T})$ be a topological space, let $(Y, \mathcal{U})$ be a Hausdorff space, and let $f: X \to Y$ be continuous. Then $\{(x_1, x_2) \in X \times X: f(x_1) = f(x_2)\}$ is a closed subset of $X \times X$.

Proof. Let $A = \{(x_1, x_2) \in X \times X: f(x_1) = f(x_2)\}$, and suppose $(a, b) \notin A$. Then $f(a) \neq f(b)$. Since Y is Hausdorff, there exist disjoint open sets U and V such that $f(a) \in U$ and $f(b) \in V$. Since f is continuous, $f^{-1}(U)$ and $f^{-1}(V)$ are open. Hence $f^{-1}(U) \times f^{-1}(V)$ is a neighborhood of (a,b). Notice that if $(x_1, x_2) \in f^{-1}(U) \times f^{-1}(V)$, then $f(x_1) \in U$ and $f(x_2) \in V$ so $f(x_1) \neq f(x_2)$. Thus $(f^{-1}(U) \times f^{-1}(V)) \cap A = \emptyset$, and so $(X \times X) - A$ is open. Therefore A is closed. ∎

THEOREM 5.8. Let $(X, \mathcal{T})$ and $(Y, \mathcal{U})$ be topological spaces, and let $f: X \to Y$ be an open map of X onto Y. If $\{(x_1, x_2) \in X \times X: f(x_1) = f(x_2)\}$ is a closed subset of $X \times X$, then $(Y, \mathcal{U})$ is a Hausdorff space.

Proof. Suppose $A = \{(x_1, x_2) \in X \times X: f(x_1) = f(x_2)\}$ is a closed subset of $X \times X$. Let c and d be distinct members of Y. Since f maps X onto Y, there exist $a, b \in X$ such that $f(a) = c$ and $f(b) = d$. Since $(a, b) \notin A$, and $(X \times X) - A$ is open, there exist open sets U and V such that $a \in U$, $b \in V$, and $(U \times V) \cap A = \emptyset$. Since f is open, $f(U)$ and $f(V)$ are open subsets of Y. Furthermore $c \in f(U)$, $d \in f(V)$, and $f(U) \cap f(V) = \emptyset$. Therefore $(Y, \mathcal{U})$ is a Hausdorff space. ∎

As an immediate consequence of Theorems 5.7 and 5.8, we obtain the following corollary. Notice the similarity between this result and Exercise 11 of Section 2.2.

Corollary 5.9. Let $(X, \mathcal{T})$ and $(Y, \mathcal{U})$ be topological spaces, and let $f: X \to Y$ be an open continuous map of X onto Y. Then $(Y, \mathcal{U})$ is Hausdorff if and only if $\{(x_1, x_2) \in X \times X : f(x_1) = f(x_2)\}$ is a closed subset of $X \times X$. ∎

The following theorem gives a condition for a quotient space to be a Hausdorff space. In Appendix E, for a relation R on a set X and a subset A of X, we define $R[A]$ to be $\{b \in X : (a, b) \in R \text{ for some } a \in A\}$.

THEOREM 5.10. Let $(X, \mathcal{T})$ be a topological space, and let R be an equivalence relation on X such that R is a closed subset of $X \times X$ and for each open subset U of R, $R[U]$ is open in X. If $\mathcal{U}$ is the quotient topology on X/R induced by the natural map p, then $(X/R, \mathcal{U})$ is a Hausdorff space.

Proof. Let A and B be distinct members of X/R. Then there exist a and b in X such that $p(a) = A$ and $p(b) = B$. By Theorem E.1(c), $(a, b) \notin R$. Since R is a closed subset of $X \times X$, there exist open sets U and V in X such that $a \in U, b \in V$, and $(U \times V) \cap R = \emptyset$. By hypothesis, $R[U]$ and $R[V]$ are open subsets of X. Since $(U \times V) \cap R = \emptyset, R[U] \cap R[V] = \emptyset$ (see Exercise 8). By Exercise 9, $A \subseteq R[U]$ and $B \subseteq R[V]$. Since $R[U]$ is the union of members of X/R, $p(R[U])$ is open in X/R (Theorem 2.62). In the same way, we see that $p(R[V])$ is open in X/R. Furthermore $A \in p(R[U]), B \in p(R[V])$, and $p(R[U]) \cap p(R[V]) = \emptyset$. Therefore $(X/R, \mathcal{U})$ is a Hausdorff space. ∎

EXERCISES 5.1

1. Let $(X, \mathcal{T})$ be a topological space. Prove that $(X, \mathcal{T})$ is a T_0-space if and only if for each pair a and b of distinct members of X, $\overline{\{a\}} \neq \overline{\{b\}}$.

2. Let $(X, \mathcal{T})$ be a topological space, let R be an equivalence relation on X, and let $\mathcal{U}$ be the quotient topology on X/R induced by the natural map. Prove that $(X/R, \mathcal{U})$ is a T_1-space if and only if for each $x \in X, [x]$ is a closed subset of X.

3. For each $i = 0, 1$, prove that every subspace of a T_i-spaces is a T_i-space.

4. For each $i = 0, 1$, prove that the product of T_i-spaces is a T_i-space.

5. Let I, $\sim$, and p be as defined in Example 1.
    (a) Prove that $I/\sim$ is not Hausdorff.
    (b) Prove that p is open.

6. (a) Prove that the space $(\mathscr{D}, \mathscr{U})$ in Example 2 is not Hausdorff.

   (b) Prove that the map p in Example 2 is closed.

7. (a) Prove that the natural map p in Example 3 is open.

   (b) Prove that the space $(\mathscr{D}, \mathscr{U})$ in Example 3 is not Hausdorff.

8. Let R be an equivalence relation on a set X, and let U and V be subsets of X. Prove that if $(U \times V) \cap R = \emptyset$, then $R[U] \cap R[V] = \emptyset$.

9. Let R be an equivalence relation on a set X, let $A \in X/R$, let $a \in A$, and let $U \subseteq X$ such that $a \in U$. Prove that $A \subseteq R[U]$.

10. Let X be a set and let $D \subseteq X$. Define a topology $\mathscr{T}$ on X by saying that a subset C of X is closed whenever $C = C \cup D$ and a subset U of X belongs to $\mathscr{T}$ whenever $X - U$ is closed. For each $i = 0, 1, 2$, under what conditions on D is $(X, \mathscr{T})$ a T_i-space?

11. Let $(X, \mathscr{T})$ be a T_1-space, and let $(Y, \mathscr{U})$ be a topological space, and let f be a closed map of X onto Y. Prove that $(Y, \mathscr{U})$ is a T_1-space.

12. Let $(X, \mathscr{T})$ be a T_1-space and let $(\mathscr{D}, \mathscr{U})$ be a decomposition space of X. Prove that $(\mathscr{D}, \mathscr{U})$ is T_1 if and only if each member of $\mathscr{D}$ is a closed subset of X.

13. Theorem 1.18 says that we can define a topology on a set in terms of closed sets. Define a basis for the closed sets of a topology.

14. Let $n \in \mathbb{N}$, and let $\mathscr{P}$ denote the collection of all polynomials in n variables. For $p \in \mathscr{P}$, let $Z(p) = \{(x_1, x_2, \ldots, x_n) \in \mathbb{R}^n : p((x_1, x_2, \ldots, x_n)) = 0\}$.

    (a) Show that $\{Z(p) : p \in \mathscr{P}\}$ is a basis for the closed sets of some topology (called the Zariski topology) on $\mathbb{R}^n$.

    (b) Show that if $\mathscr{T}$ is the Zariski topology on $\mathbb{R}^n$, then $(\mathbb{R}^n, \mathscr{T})$ is T_1 but not Hausdorff.

    (c) Show that on $\mathbb{R}$, the Zariski topology is the same as the cofinite topology.

    (d) Show that on $\mathbb{R}^2$, the Zariski topology and the cofinite topology are not the same.

5.2 Regular and Completely Regular Spaces

The properties that are studied in the previous section describe the separation of pairs of points by open sets. The properties that are studied in this section describe the separation of a point from a closed set by open sets and, hence, are more restrictive.

Definition. A T_1-space $(X, \mathscr{T})$ is said to be a **regular space** or T_3-**space** provided that whenever C is a closed subset of X and $p \in X - C$, there exist disjoint open sets U and V such that $C \subseteq U$ and $p \in V$. ■

5.2 Regular and Completely Regular Spaces

We require that a regular space be a T_1-space so that every regular space is a Hausdorff space. Notice that if we did not require that singleton sets be closed, then a set with the trivial topology would be regular. Regular spaces were first studied by Vietoris in 1921.

EXAMPLE 4. Let $\mathcal{T}$ be the topology on $\mathbb{R}$ defined in Example 2. As seen in that example, $(\mathbb{R}, \mathcal{T})$ is a Hausdorff space. Also $A = \{\frac{1}{n} : n \in \mathbb{N}\}$ is a closed subset of $\mathbb{R}$ and $0 \notin A$. However if U and V are open sets such that $A \subseteq U$ and $0 \in V$, then $U \cap V \neq \emptyset$ (see Exercise 1). Hence $(\mathbb{R}, \mathcal{T})$ is not regular.

The following two theorems characterize regular spaces.

THEOREM 5.11. A T_1-space $(X, \mathcal{T})$ is regular if and only if for each member p of X and each neighborhood U of p, there is neighborhood V of p such that $\overline{V} \subseteq U$.

Proof. Suppose $(X, \mathcal{T})$ is a regular space. Let $p \in X$ and let U be a neighborhood of p. Then $X - U$ is closed set and $p \notin X - U$. Therefore there exist disjoint open sets V and W such that $p \in V$ and $X - U \subseteq W$. Since $V \subseteq X - W$ and $X - W$ is closed, $\overline{V} \subseteq X - W$. Since $X - W \subseteq U$, V is a neighborhood of p such that $\overline{V} \subseteq U$.

Suppose $(X, \mathcal{T})$ is a T_1-space such that for each $p \in X$ and each neighborhood U of p, there is a neighborhood V of p such that $\overline{V} \subseteq U$. Let C be a closed subset of X and let $p \in X - C$. Then $X - C$ is a neighborhood of p. Hence there is a neighborhood V of p such that $\overline{V} \subseteq X - C$. Thus V and $X - \overline{V}$ are disjoint open sets such that $p \in V$ and $C \subseteq X - \overline{V}$. Therefore $(X, \mathcal{T})$ is regular. ∎

THEOREM 5.12. A T_1-space $(X, \mathcal{T})$ is regular if and only if for each $p \in X$ and each closed set C such that $p \notin C$, there exist open sets U and V such that $C \subseteq U, p \in V$, and $\overline{U} \cap \overline{V} = \emptyset$.

Proof. Suppose $(X, \mathcal{T})$ is a regular space. Let $p \in X$, and let C be a closed set such that $p \notin C$. Then $X - C$ is a neighborhood of p, and hence, by Theorem 5.11, there is a neighborhood W of p such that $\overline{W} \subseteq X - C$. Again by Theorem 5.11, there is a neighborhood V of p such that $\overline{V} \subseteq W$. Let $U = X - \overline{W}$. Since $\overline{W} \subseteq X - C$, $C \subseteq X - \overline{W} \subseteq U$. Furthermore $\overline{V} \cap \overline{U} \subseteq W \cap (X - \overline{W}) = \emptyset$. Therefore U and V are the desired open sets.

It is clear that the condition implies that a T_1-space is regular. ∎

Since, in a Hausdorff space $(X, \mathcal{T})$, $\{p\}$ is closed for each $p \in X$, it follows from Corollary 4.18 that a compact Hausdorff space is regular. We use this fact to prove Theorem 4.36. (Recall that we promised in Section 4.3 that we would prove Theorem 4.36 in this section.)

THEOREM 4.36. A Hausdorff space $(X, \mathcal{T})$ is locally compact if and only if for each $p \in X$ and each neighborhood V of p there is a neighborhood U of p such that $\overline{U}$ is compact and $\overline{U} \subseteq V$.

Proof. Suppose $(X, \mathcal{T})$ is a locally compact Hausdorff space and let $p \in X$. By Theorem 4.35, there is a neighborhood W of p such that $\overline{W}$ is compact. Let V be a neighborhood of p, and let $N = \text{int}(\overline{W} \cap V)$. Then N is a neighborhood of p and $N \subseteq V$. Since $\overline{N} \subseteq \overline{W}$ and $\overline{W}$ is compact, $\overline{N}$ is compact. Since $\overline{N}$ is Hausdorff, it is regular. Since N is a neighborhood of p in the regular space $\overline{N}$, there exists a neighborhood U of p in $\overline{N}$ such that the closure of U in $\overline{N}$ is a subset of N. Since U is open in N and N is open in X, U is open in X. Since the closure of U in $\overline{N}$ is closed in $\overline{N}$, the closure of U in $\overline{N}$ is compact. Thus the closure U in $\overline{N}$, which is the closure of U in X, is a compact subset of V. ∎

The following result is an immediate consequence of Theorem 4.36 and 5.12.

Corollary 5.13. Every locally compact Hausdorff space is regular. ∎

THEOREM 5.14. Every subspace of a regular space is regular.

Proof. Let $(X, \mathcal{T})$ be a regular space, and let $(A, \mathcal{T}_A)$ be a subspace of X. By Theorem 5.2, $(A, \mathcal{T}_A)$ is a T_1-space. Let C be a closed subset of A and let $p \in A - C$. By Theorem 2.5, there is a closed subset D of X such that $C = A \cap D$. Since X is regular, there are disjoint open sets U and V in X such that $D \subseteq U$ and $p \in V$. Thus $U \cap A$ and $V \cap A$ are disjoint open sets in A such that $C \subseteq U \cap A$ and $p \in V \cap A$. ∎

The following example shows that a closed continuous image of a regular space need not be regular and that the quotient space of a regular space need not be regular.

EXAMPLE 5. Let $X = \{(x, y) \in \mathbb{R} \times \mathbb{R} : y \geq 0\}$. Define a basis $\mathcal{B}$ (see Exercise 13 in Section 1.3) for a topology on X as follows: For each $(a, b) \in X$ with $b > 0$, $\{(x, y) \in X : (x - a)^2 + (y - b)^2 < r^2$, where $0 < r < b\} \in \mathcal{B}$. For each real number a and each positive number ε, $\{(a, 0)\} \cup \{(x, y) \in X : (x - a)^2 + (y - \varepsilon)^2 < \varepsilon\} \in \mathcal{B}$. Let $\mathcal{T}$ be the topology on X generated by $\mathcal{B}$. The topological space $(X, \mathcal{T})$ is called the **Moore plane**. The two types of basic open sets are shown in Figure 5.2.

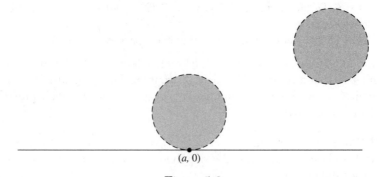

Figure 5.2

The space $(X, \mathcal{T})$ is regular. (The proof will be given later in this section.) Let $A = \{(x, y) \in X : x \text{ is rational and } y = 0\}$ and let $B = \{(x, y) \in X : x \text{ is irrational and } y = 0\}$. Let $\mathcal{D}$ be the decomposition of X that consists of A, B, and the singleton sets $\{x\}$ for all $x \in X - (A \cup B)$. If $\mathcal{U}$ is the quotient topology on $\mathcal{D}$ determined by the natural map $p: X \to \mathcal{D}$, then p is closed (see Exercise 3). The decomposition space $(\mathcal{D}, \mathcal{U})$ is not Hausdorff because if U and V are open sets such that $A \in U$ and $B \in V$, then $U \cap V \neq \emptyset$. Therefore $(\mathcal{D}, \mathcal{U})$ cannot be regular.

The Moore plane is a classical example. It was defined by V. Niemytzki (1900–1967) and is sometimes called the *Niemytzki* plane. It appears in *Topologie* I (1935) by Alexandroff and Hopf.

THEOREM 5.15. Regularity is a topological property.

Proof. Let $(X, \mathcal{T})$ be a regular space, let $(Y, \mathcal{U})$ be a topological space, and let $f: X \to Y$ be a homeomorphism. Let C be a closed subset of Y and let $p \in Y - C$. Then $f^{-1}(C)$ is a closed subset of X and $f^{-1}(p) \in X - f^{-1}(C)$. Thus there exist disjoint open subsets U and V of X such that $f^{-1}(C) \subseteq U$ and $f^{-1}(p) \in V$. So $f(U)$ and $f(V)$ are disjoint open subsets of Y such that $C \subseteq f(U)$ and $p \in f(V)$. Therefore $(Y, \mathcal{U})$ is regular. ∎

THEOREM 5.16. Let $\{(X_\alpha, \mathcal{T}_\alpha) : \alpha \in \Lambda\}$ be a family of topological spaces, and let $X = \prod_{\alpha \in \Lambda} X_\alpha$. Then $(X, \mathcal{T})$ is regular if and only if $(X_\alpha, \mathcal{T}_\alpha)$ is regular for each $\alpha \in \Lambda$.

Proof. Suppose $(X, \mathcal{T})$ is regular and let $\beta \in \Lambda$. By Theorem 5.14, every subspace of $(X, \mathcal{T})$ is regular. By Theorem 2.39, there is a subspace of $(X, \mathcal{T})$ that is homeomorphic to X_β. Since regularity is a topological property (Theorem 5.15), $(X_\beta, \mathcal{T}_\beta)$ is regular.

Suppose $(X_\alpha, \mathcal{T}_\alpha)$ is regular for each $\alpha \in \Lambda$. Since the product of T_1-spaces is T_1 (see Theorem 5.3), $(X, \mathcal{T})$ is T_1. Let $x \in X$ and let U be a neighborhood of x. Then there is a set of the form $\prod_{\alpha \in \Lambda} U_\alpha$, where $U_\alpha \in \mathcal{T}_\alpha$ for each $\alpha \in \Lambda$ and $U_\alpha = X_\alpha$ for all but a finite number of members of Λ, such that $x \in \prod_{\alpha \in \Lambda} U_\alpha$ and $\prod_{\alpha \in \Lambda} U_\alpha \subseteq U$. For each $\alpha \in \Lambda$ such that $U_\alpha = X_\alpha$, let $V_\alpha = X_\alpha$. For each $\alpha \in \Lambda$ such that $U_\alpha \neq X_\alpha$, let V_α be a neighborhood of x_α such that $\overline{V}_\alpha \subseteq U_\alpha$. Let $V = \prod_{\alpha \in \Lambda} V_\alpha$. Then V is a neighborhood of x. By Theorem 2.38, $\overline{V} = \prod_{\alpha \in \Lambda} \overline{V}_\alpha = \prod_{\alpha \in \Lambda} \overline{V}_\alpha \subseteq \prod_{\alpha \in \Lambda} U_\alpha \subseteq U$. Therefore $(X, \mathcal{T})$ is regular. ∎

Definition. A T_1-space $(X, \mathcal{T})$ is said to be **completely regular** or a **Tychonoff space** provided that whenever C is a closed subset of X and $p \in X - C$, there is a continuous function $f: X \to I$ (the closed unit interval $[0, 1]$) such that $f(p) = 1$ and $f(x) = 0$ for all $x \in C$. ∎

Completely regular spaces were introduced by Paul Urysohn in a paper that appeared in 1925. Their importance was established by Tychonoff in 1930, when he proved the result we state as Theorem 5.45. In 1940, the name *Tychonoff space* was suggested by J.W. Tukey (1915–2000).

THEOREM 5.17. Every completely regular space is regular.

Proof. Let $(X, \mathcal{T})$ be a completely regular space, let C be a closed subset of X and let $p \in X - C$. Let $f : X \to I$ be a continuous function such that $f(p) = 1$ and $f(x) = 0$ for all $x \in C$. Then $f^{-1}([0, \frac{1}{3}))$ and $f^{-1}((\frac{2}{3}, 1])$ are disjoint open subsets of X such that $C \subseteq f^{-1}([0, \frac{1}{3}))$ and $p \in f^{-1}((\frac{2}{3}, 1])$. Therefore $(X, \mathcal{T})$ is regular. ∎

An example of a regular space that is not completely regular will be given later.

THEOREM 5.18. Every metric space is completely regular.

Proof. By Theorem 1.36, every metric space is Hausdorff. Let (X, d) be a metric space, let C be a closed subset of X, and let $p \in X - C$. Define $f : X \to I$ by $f(x) = \min\{d(x, C)/d(p, C), 1\}$. Then $f(p) = 1$ and $f(x) = 0$ for all $x \in C$. The proof that f is continuous is Exercise 4. ∎

The Moore plane is a completely regular space that is not metrizable. As shown in Example 5, the subsets A and B of the Moore plane are closed sets with the property that there do not exist disjoint open sets U and V such that $A \subseteq U$ and $B \subseteq V$. Therefore by Theorem 1.38, the Moore plane is not metrizable. It remains to show that the Moore plane is completely regular.

Proof That the Moore Plane Is Completely Regular. Let $(X, \mathcal{T})$ be the Moore plane. Let C be a closed subset of X and let $p \in X - C$. Let B be a member of the basis $\mathcal{B}$ for $\mathcal{T}$ defined in Example 5 such that $p \in B$ and $B \subseteq X - C$. Define $f : X \to I$ by $f(p) = 0$ and $f(x) = 1$ if $x \notin B$. Then for each $x \in B (x \neq p)$, let L denote the line segment in X that begins at p, passes through x, and ends at the boundary of B. Let r be the length of L and define $f(x) = d(p, x)/r$. The proof that f is continuous is left as an exercise. ∎

The proofs that completely regular is a topological property and that every subspace of a completely regular space is completely regular are left as exercises.

THEOREM 5.19. Let $\{(X_\alpha, \mathcal{T}_\alpha) : \alpha \in \Lambda\}$ be a family of topological spaces, and let $X = \prod_{\alpha \in \Lambda} X_\alpha$. Then $(X, \mathcal{T})$ is completely regular if and only if $(X_\alpha, \mathcal{T}_\alpha)$ is completely regular for each $\alpha \in \Lambda$.

Proof. Suppose $(X, \mathcal{T})$ is completely regular and let $\beta \in \Lambda$. By Exercise 7, every subspace of $(X, \mathcal{T})$ is completely regular. By Theorem 2.39, there is a subspace of $(X, \mathcal{T})$ that is homeomorphic to X_β. Therefore by Exercise 6, $(X_\beta, \mathcal{T}_\beta)$ is completely regular.

Suppose $(X_\alpha, \mathcal{T}_\alpha)$ is completely regular for each $\alpha \in \Lambda$. Since the product of T_1-spaces is T_1 (see Theorem 5.3), $(X, \mathcal{T})$ is T_1. Let C be a closed subset of X and let $p \in X - C$. Then there is a finite set $\beta_1, \beta_2, \ldots, \beta_n$ of members of Λ with the property that for each $i = 1, 2, \ldots, n$, there is an open set U_{β_i} of X_{β_i} such that

5.2 Regular and Completely Regular Spaces

$p \in \bigcap_{i=1}^{n} \pi_{\beta_i}^{-1}(U_{\beta_i}) \subseteq X - C$. For each $i = 1, 2, ..., n$, there is a continuous function $f_i: X_{\beta_i} \to I$ such that $f_i(p_{\beta_i}) = 1$ and $f_i(y_{\beta_i}) = 0$ for all $y_{\beta_i} \in X_{\beta_i} - U_{\beta_i}$. Define $f: X \to I$ by $f(x) = \min\{f_i(x_{\beta_i}): i = 1, 2, ..., n\}$. Then $f(p) = 1$ and $f(x) = 0$ for all $x \in X - C$. The proof that f is continuous is Exercise 8. ■

The following example of a regular space that is not completely regular is due to A. Mysior and appeared in the April 1981 issue of the *Proceedings of the American Mathematical Society*. Those who are familiar with the previously known examples of such spaces appreciate this example.

EXAMPLE 6. Let $X = \{(x, y) \in \mathbb{R}^2: y \geq 0\} \cup \{(0, -1)\}$. We define a collection $\mathcal{B}$ of subsets of X as follows: If $(x, y) \in X$ and $y > 0$, then $\{(x, y)\} \in \mathcal{B}$. For each $x \in \mathbb{R}$, let $M_x = \{(x, y): 0 \leq y < 2\}$, let $N_x = \{(x + y, y): 0 \leq y < 2\}$, and let P_x be a finite subset of $(M_x \cup N_x) - \{(x, 0)\}$ (see Figure 5.3(a)). Then for each $x \in \mathbb{R}$, $((M_x \cup N_x) - P_x) \in \mathcal{B}$. Finally for each $n \in \mathbb{N}$, $(\{(x, y) \in X: x > n\} \cup \{(0, -1)\}) \in \mathcal{B}$ (see Figure 5.3(b)). Then $\mathcal{B}$ is a basis for a topology $\mathcal{T}$ on X. The space $(X, \mathcal{T})$ is regular but not completely regular.

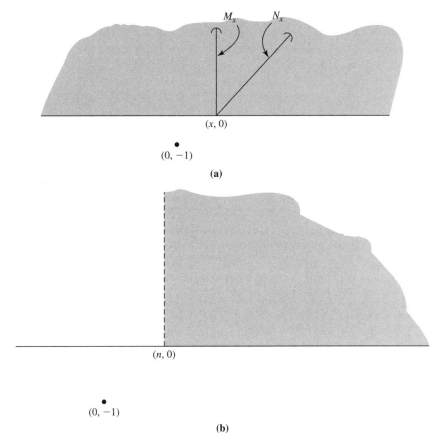

Figure 5.3

Analysis. In Exercise 9, you are asked to show that $\mathcal{B}$ is a basis for a topology $\mathcal{T}$ on X. Note that $\mathcal{B}$ has three types of members: (1) If $y > 0$, then $\{(x, y)\} \in \mathcal{B}$. Thus $\mathcal{T}$ is the discrete topology on the subset $\{(x, y) \in X : y > 0\}$. (2) A member of $\mathcal{B}$ that contains $(x, 0)$ is the union of two line segments (shown in Figure 5.3 (a)) with a finite number of points removed. (3) A member of $\mathcal{B}$ that contains $(0, -1)$ is the union of $\{(0, -1)\}$ and the set consisting of all members of X that lie to the right of the vertical line $x = n$, where n is a natural number (see Figure 5.3(b)).

We show that $(X, \mathcal{T})$ is regular. Let $(x, y) \in X$. We consider three cases.

Case 1. Suppose $y > 0$ and let U be a neighborhood of (x, y). Then $\{(x, y)\}$ is a neighborhood of (x, y) and $\overline{\{(x, y)\}} = \{(x, y)\} \subseteq U$.

Case 2. Suppose $y = 0$ and let U be a neighborhood of $(x, 0)$. Then there is a finite subset P_x of $M_x \cup N_x$ such that $(M_x \cup N_x) - P_x \subseteq U$. Then $(M_x \cup N_x) - P_x$ is a neighborhood of $(x, 0)$ and $\overline{(M_x \cup N_x) - P_x} \subseteq (M_x \cup N_x) - P_x \subseteq U$.

Case 3. Suppose $y = -1$ and let U be a neighborhood of $(0, -1)$. Then there is a natural number N such that $\{(x, y) \in X : x > N\} \subseteq U$. Then $V = \{(x, y) \in X : x > N + 2\} \cup \{(0, -1)\}$ is a neighborhood of $(0, -1)$ such that $\overline{V} \subseteq U$. Therefore $(X, \mathcal{T})$ is regular.

We prove that $(X, \mathcal{T})$ is not completely regular. Let $C = \{(x, 0) : x \leq 1\}$. Then C is a closed subset of X and $(0, -1) \notin C$. Let $f : X \to I$ be a continuous function such that $f(x) = 0$ for all $x \in C$. For each $n \in \mathbb{N}$, let $A_n = f^{-1}(\{0\}) \cap \{(x, 0) : n - 1 \leq x \leq n\}$, and let $S = \{n \in \mathbb{N} : A_n \text{ is an infinite set}\}$. Since $C \subseteq f^{-1}(\{0\})$, $A_1 = I$. Therefore $1 \in S$. Suppose $n \in S$, let D be a countably infinite subset of A_n such that $(n - 1, 0) \notin D$, and let $(d, 0) \in D$. Since $D \subseteq A_n$ and $A_n \subseteq f^{-1}(\{0\})$, $(d, 0) \in f^{-1}(\{0\})$. Since $\{0\} = \bigcap_{n=1}^{\infty}(-\frac{1}{n}, \frac{1}{n})$, $f^{-1}(\{0\}) = f^{-1}(\bigcap_{n=1}^{\infty}(-\frac{1}{n}, \frac{1}{n})) = \bigcap_{n=1}^{\infty} f^{-1}((-\frac{1}{n}, \frac{1}{n}))$. Therefore, since f is continuous, $f^{-1}(\{0\})$ is a G_δ-set. Notice that N_d is a closed set. Therefore $N_d - f^{-1}(\{0\})$ is an F_σ-set. Let $\{E_i : i \in \mathbb{N}\}$ be a collection of closed sets such that $N_d - f^{-1}(\{0\}) = \bigcup_{i=1}^{\infty} E_i$. For each $i \in \mathbb{N}$, E_i is a finite set because if it is infinite, then $(d, 0) \in \overline{E_i} = E_i$. Therefore $N_d - f^{-1}(\{0\})$ is a countable set. Thus $P = \pi_1(N_d - f^{-1}(\{0\}))$ is countable. Let $F = \{(x, 0) : n \leq x \leq n + 1\} - P$. Then F is an infinite set. Let $(x, 0) \in F$ and let $(d, 0) \in D$. Then $n - 1 \leq d \leq n$, $n \leq x \leq n + 1$, and $(x, 0) \notin P$. Since $n - 1 \leq d \leq n$ and $n \leq x \leq n + 1$, there is a real number y such that $(x, y) \in N_d$. Since $(x, 0) \notin P$, $(x, y) \in f^{-1}(\{0\})$. Since $(x, y) \in M_x$, $(x, y) \in M_x \cap (N_d \cap f^{-1}(\{0\}))$. Therefore for each $(x, 0) \in F$ and each $(d, 0) \in D$, $M_x \cap (N_d \cap f^{-1}(\{0\})) \neq \emptyset$. Hence $(x, 0) \in \overline{f^{-1}(\{0\})}$. But $f^{-1}(\{0\})$ is closed, so $(x, 0) \in f^{-1}(\{0\})$. Thus $F \subseteq f^{-1}(\{0\})$. Since F is an infinite subset of both $f^{-1}(\{0\})$ and $\{(x, 0) : n \leq x \leq n + 1\}$, A_{n+1} is an infinite set. Therefore $n + 1 \in S$ and hence $S = \mathbb{N}$. Let U be an open set such that $(0, -1) \in U$. Then there exists $N \in \mathbb{N}$ such that $\{(x, y) \in X : x > N\} \subseteq U$. Therefore $U \cap A_{N+1}$ is an infinite set, so $(0, -1) \in \overline{f^{-1}(\{0\})}$. Since $f^{-1}(\{0\})$ is closed, $(0, -1) \in f^{-1}(\{0\})$. Hence $f((0, -1)) = 0$. Therefore $(X, \mathcal{T})$ is not completely regular. ∎

THEOREM 5.20. Completely regular is a topological property.

Proof. See Exercise 6. ∎

THEOREM 5.21. *Every subspace of a completely regular space is completely regular.*

Proof. See Exercise 7. ∎

EXERCISES 5.2

1. Let $\mathcal{T}$ be the topology on $\mathbb{R}$ defined in Example 2, and let $A = \{\frac{1}{n} : n \in \mathbb{N}\}$. Show that if U and V are open sets such that $A \subseteq U$ and $0 \in V$, then $U \cap V \neq \emptyset$.

2. Let A be a compact subset of a regular space and let U be an open set such that $A \subseteq U$. Prove that there is an open set V such that $A \subseteq V \subseteq \overline{V} \subseteq U$.

3. Prove that the function p in Example 5 is closed.

4. Prove that the function f in the proof of Theorem 5.18 is continuous.

5. Prove that the function f in the proof that the Moore plane is completely regular is continuous.

6. Prove that completely regular is a topological property.

7. Prove that every subspace of a completely regular space is completely regular.

8. Prove that the function f in the proof of Theorem 5.19 is continuous.

9. Show that the collection $\mathcal{B}$ defined in Example 6 is a basis for a topology on X.

10. Let $X = \{1, 2, 3\}$. Find all topologies $\mathcal{T}$ on X such that $(X, \mathcal{T})$ is regular.

11. Let $X = \{(0, 0)\} \cup ((0, 1) \times (0, 1)) \cup \{(1, 0)\}$. Define a collection $\mathcal{B}$ of subsets of X as follows: For each $(a, b) \in (0, 1) \times (0, 1)$ and each positive number r that is less than the distance between (a, b) and the boundary of $(0, 1) \times (0, 1)$, $\{(x, y) \in (0, 1) \times (0, 1) : (x - a)^2 + (y - b)^2 < r^2\} \in \mathcal{B}$. For each r such that $0 < r < 1$, $\{(0, 0)\} \cup ((0, \frac{1}{2}) \times (0, r)) \in \mathcal{B}$ and $\{(1, 0)\} \cup ((\frac{1}{2}, 1) \times (0, r)) \in \mathcal{B}$.

    (a) Show that $\mathcal{B}$ is a basis for a topology $\mathcal{T}$ on X.

    (b) Is $(X, \mathcal{T})$ Hausdorff? Prove your answer.

    (c) Is $(X, \mathcal{T})$ regular? Prove your answer.

12. Define a collection $\mathcal{B}$ of subsets of $\mathbb{R}$ as follows: For each $a, b \in \mathbb{R}$ with $a < b$, $(a, b) \in \mathcal{B}$ and $(a, b) \cap \mathbb{Q} \in \mathcal{B}$ ($\mathbb{Q}$ is the set of rational numbers).

    (a) Show that $\mathcal{B}$ is a basis for a topology $\mathcal{T}$ on $\mathbb{R}$.

(b) Prove that $(\mathbb{R}, \mathcal{T})$ is Hausdorff.

(c) Show that $\mathbb{R} - \mathbb{Q}$ is closed.

(d) Let $\mathcal{U}$ be the usual topology on $\mathbb{R}$ and let $f: (\mathbb{R}, \mathcal{T}) \to (\mathbb{R}, \mathcal{U})$ be a continuous function such that $f(x) = 0$ for all $x \in \mathbb{R} - \mathbb{Q}$. Prove that $f(x) = 0$ for all $x \in \mathbb{R}$.

(e) Prove that $(\mathbb{R}, \mathcal{T})$ is not regular.

13. Let $X = \mathbb{R}$ and let $\mathcal{T}$ be the usual topology on X. Let $Y = \mathbb{R}$ and let $\mathcal{U}$ be the topology on Y consisting of all sets of the form (a, ∞), where $a \in \mathbb{R}$. Let $\mathcal{V}$ be the product topology on $X \times Y$. Is $(X \times Y, \mathcal{V})$ regular? Prove your answer.

14. Let $X_1 = \mathbb{R}$ and let $\mathcal{T}_1$ be the usual topology on X_1. Let $X_2 = \{0, 1\}$ and let $\mathcal{T}_2$ be the trivial topology on X_2. Let $X_3 = \{1, 2, 3\}$ and let $\mathcal{T}_3$ be the discrete topology on X_3. Let $\mathcal{U}$ be the product topology on $X_1 \times X_2 \times X_3$. Is $(X_1 \times X_2 \times X_3, \mathcal{U})$ regular? Prove your answer.

15. Let $X = \{(x, y) \in \mathbb{R}^2 : y \geq 0\}$. Define a basis $\mathcal{B}$ for a topology on X as follows: For each $(a, b) \in X$ with $b > 0$, $\{(x, y) \in X : (x - a)^2 + (y - b)^2 < r^2$, where $0 < r < b\} \in \mathcal{B}$. For each number a and each positive number ε, $\{(a, 0)\} \cup (x, y) \in X : y > 0$ and $(x - a)^2 + y^2 < \varepsilon\} \in \mathcal{B}$. Let $\mathcal{T}$ be the topology on X generated by $\mathcal{B}$. Is $(X, \mathcal{T})$ Hausdorff? Is $(X, \mathcal{T})$ regular? Prove your answers.

16. Let $(X, \leq)$ be a linearly ordered set, and let $\mathcal{T}$ be the order topology on X. Prove that $(X, \mathcal{T})$ is regular.

17. Let X be a set and let $D \subseteq X$. Define a collection $\mathcal{T}$ of subsets of X by saying that a subset U of X belongs to $\mathcal{T}$ provided $U = \emptyset$ or $D \subseteq U$.

(a) Show that $\mathcal{T}$ is a topology on X.

(b) Under what conditions is $(X, \mathcal{T})$ regular?

(c) Under what conditions is $(X, \mathcal{T})$ completely regular?

18. For each $(a, b) \in \mathbb{R}^2$, an **open disk** with center at (a, b) is a set of the form $\{(x, y) \in \mathbb{R}^2 : (x - a)^2 + (y - b^2) < \varepsilon^2\}$, where ε is a positive number. A **slotted open disk** is an open disk with a finite number of straight lines through the center removed. Let $\mathcal{B}$ be the collection of all subsets of $\mathbb{R}^2$ of the form $\{(a, b)\} \cup D$, where D is a slotted open disk with center at (a, b).

(a) Show that $\mathcal{B}$ is a basis for a topology $\mathcal{T}$ on $\mathbb{R}^2$.

(b) Is $(\mathbb{R}^2, \mathcal{T})$ regular? Prove your answer.

19. Let $X = \{(m, n) \in \mathbb{Z} \times \mathbb{Z} : m \text{ and } n \text{ are nonnegative integers}\}$. Define a collection $\mathcal{T}$ of subsets of X as follows: If $(m, n) \in X$ and either $m \neq 0$ or $n \neq 0$, then $\{(m, n)\} \in \mathcal{T}$. A subset U of X such that $(0, 0) \in U$ belongs to $\mathcal{T}$

provided that for all but a finite number of integers m, the sets $S_m = \{n: (m, n) \in U\}$ are finite.

   (a) Show that $\mathcal{T}$ is a topology on X.

   (b) Is $(X, \mathcal{T})$ a T_1-space? Prove your answer.

   (c) Is $(X, \mathcal{T})$ regular? Prove your answer.

20. Let $(X, \mathcal{T})$ and $(Y, \mathcal{U})$ be topological spaces and let $f: X \to Y$ be a perfect map (see Exercise 9 in Section 4.2 for the definition of perfect map). Prove that if $(X, \mathcal{T})$ is regular, then so is $(Y, \mathcal{U})$.

5.3 Normal and Completely Normal Spaces

We studied the separation of a closed set and a point in the previous section, and we discovered that the two separation properties in that section were well behaved with respect to subspaces and products. The two separation properties studied in this section are concerned with the separation of two closed sets, and as we shall see these properties are not well behaved with respect to subspaces and products.

Definition. A T_1-space $(X, \mathcal{T})$ is said to be a **normal space** or a **T_4-space** provided that whenever C and D are disjoint closed subsets of X, there are disjoint open sets U and V such that $C \subseteq U$ and $D \subseteq V$. ∎

Normal spaces were introduced by Vietoris in 1921 and by Tietze in 1923. They were also studied independently by Alexandroff and Urysohn in 1925.

We require that a normal space be a T_1-space so that every normal space is a regular space. The following is an example of a topological space that is not regular but does have the property that any two disjoint closed sets can be separated by disjoint open sets.

EXAMPLE 7. For each $a \in \mathbb{R}$, let $U_a = \{x \in \mathbb{R}: x > a\}$ and let $\mathcal{T} = \{U_a: a \in \mathbb{R}\}$. Then $\mathcal{T}$ is a topology on $\mathbb{R}$. Since no two nonempty closed subsets of $(\mathbb{R}, \mathcal{T})$ are disjoint, any two disjoint closed sets can be separated by disjoint open sets. However $(\mathbb{R}, \mathcal{T})$ is not regular because $C = \{x \in \mathbb{R}: x \leq 0\}$ is a closed set, $1 \notin C$, and if U is any open set that contains C, then $U = \mathbb{R}$ and hence $1 \in U$.

The following is an immediate consequence of Corollary 4.18.

Corollary 5.22. Every compact Hausdorff space is normal. ∎

As we have seen, the Moore plane is completely regular. In order to prove that the Moore plane is not normal, we prove Theorem 5.23. The following notation, which is introduced in Appendices F and G, is needed. For sets X and Y, $X \sim Y$ means there is a one-to-one function mapping X onto Y, and $X \leq Y$ means there is a subset Y' of Y such that $X \sim Y'$.

THEOREM 5.23. Let $(X, \mathcal{T})$ be a topological space with a dense subset D and a closed, relatively discrete subset C such that $\mathcal{P}(D) \lesssim C$. Then $(X, \mathcal{T})$ is not normal.

Proof. Suppose $(X, \mathcal{T})$ is a normal space, D is a dense subset of X, and C is a closed, relatively discrete subset of X. Let A be a nonempty proper subset of C. We assert that A is closed in C: Let $x \in C - A$. There exists $U \in \mathcal{T}$ such that $U \cap C = \{x\}$. Now $U \cap C \in \mathcal{T}_C$ and $(U \cap C) \cap A = \emptyset$. Therefore $C - A$ is open in C and hence A is closed in C. Therefore, since C is closed in X, A is closed in X. Of course $C - A$ is also a nonempty closed subset of X. Since $(X, \mathcal{T})$ is normal, there are disjoint open sets $U(A)$ and $V(A)$ such that $A \subseteq U(A)$ and $C - A \subseteq V(A)$.

Now suppose A_1 and A_2 are nonempty proper subsets of C such that $A_1 - A_2 \neq \emptyset$. Let $U(A_1)$ and $V(A_1)$ be disjoint open sets such that $A_1 \subseteq U(A_1)$ and $C - A_1 \subseteq V(A_1)$, and let $U(A_2)$ and $V(A_2)$ be disjoint open sets such that $A_2 \subseteq U(A_2)$ and $C - A_2 \subseteq V(A_2)$. Then $U(A_1) \cap V(A_2) \neq \emptyset$. Since D is dense in X, $U(A_1) \cap V(A_2) \cap D \neq \emptyset$. But $U(A_1) \cap V(A_2) \cap D \subseteq U(A_1) \cap D$ and $U(A_1) \cap V(A_2) \cap D \not\subseteq U(A_2) \cap D$. Thus if A_1 and A_2 are distinct nonempty proper subsets of C, then $U(A_1) \cap D$ and $U(A_2) \cap D$ are distinct subsets of D. Therefore $\mathcal{P}(C) \lesssim \mathcal{P}(D)$. Since $C \not\sim \mathcal{P}(C)$ and $C \lesssim \mathcal{P}(C)$, it is not the case that $\mathcal{P}(D) \lesssim C$. ∎

EXAMPLE 8. The Moore plane is not normal.

Analysis. Let $(X, \mathcal{T})$ be the Moore plane, let $C = \{(x, y) \in X: y = 0\}$, and let $D = \{(x, y) \in X: x \text{ and } y \text{ are rational}\}$. Then C is a closed, relatively discrete subset of $(X, \mathcal{T})$ and D is a dense subset of $(X, \mathcal{T})$. Furthermore $2^{|D|} \leq |C|$. Therefore by Theorem 5.23, $(X, \mathcal{T})$ is not normal. ∎

By Theorem 1.38, every metric space is normal.
The following two theorems characterize normal spaces in the same way that Theorems 5.11 and 5.12 characterized regular spaces. Their proofs are also similar to the proofs of those two theorems and consequently they are left as Exercises 1 and 2.

THEOREM 5.24. A T_1-space $(X, \mathcal{T})$ is normal if and only if for each closed subset C of X and each open set U such that $C \subseteq U$, there is an open set V such that $C \subseteq V$ and $\overline{V} \subseteq U$. ∎

THEOREM 5.25. A T_1-space $(X, \mathcal{T})$ is normal if and only if for each pair of disjoint closed subsets C and D of X, there are open sets U and V such that $C \subseteq U$, $D \subseteq V$, and $\overline{U} \cap \overline{V} = \emptyset$. ∎

In the next section we give an example to show that the product of two normal spaces need not be normal, and in Section 6.7 we give an example to show that a subspace of a normal space need not be normal. Example 3 shows that the quotient space of a normal space need not be normal and that the open continuous image of a normal space need not be normal. However we have the following two results. The proof of the first result is left as Exercise 3.

THEOREM 5.26. Let C be a closed subset of a normal space $(X, \mathcal{T})$. Then $(C, \mathcal{T}_C)$ is normal. ∎

THEOREM 5.27. Let $(X, \mathcal{T})$ be a normal space, let $(Y, \mathcal{U})$ be a topological space, and let $f: X \to Y$ be a closed continuous function that maps X onto Y. Then $(Y, \mathcal{U})$ is normal.

Proof. Let C and D be disjoint closed subsets of $(Y, \mathcal{U})$. Then $f^{-1}(C)$ and $f^{-1}(D)$ are disjoint closed subsets of $(X, \mathcal{T})$. Therefore there exist disjoint open sets U and V in $(X, \mathcal{T})$ such that $f^{-1}(C) \subseteq U$ and $f^{-1}(D) \subseteq V$. Since f is closed, $f(X - U)$ and $f(X - V)$ are closed subsets of $(Y, \mathcal{U})$. Therefore $M = Y - f(X - U)$ and $N = Y - f(X - V)$ are open. The proof that $C \subseteq M$, $D \subseteq N$, and $M \cap N = \emptyset$ is left as Exercise 4. ∎

The following theorem is useful in dealing with examples.

THEOREM 5.28. Let $(X, \leq)$ be a well-ordered set, and let $\mathcal{T}$ denote the order topology on X. Then $(X, \mathcal{T})$ is a normal space.

Proof. By Exercise 17 of Section 1.6, $(X, \mathcal{T})$ is T_1. We assert that every interval in X of the form $(a, b]$ is open in X: If $(X, \leq)$ has a largest member and b is that member, then $(a, b]$ is a basic open set. If b is not the largest member of $(X, \leq)$, let $A = \{x \in X: x > b\}$. Since $(X, \leq)$ is well-ordered, A has a least member c. Thus $(a, b] = (a, c)$ and hence $(a, b]$ is open.

Since $(X, \leq)$ is well-ordered, X has a least member a_0. Suppose C and D are disjoint closed subsets of X.

Case 1. Suppose $a_0 \notin C \cup D$. Since D is closed, $C \cap D = \emptyset$ and $a_0 \notin C$, for each $a \in C$, there is an interval $(x_a, a]$ such that $(x_a, a] \cap D = \emptyset$. Also for each $b \in D$, there is an interval $(y_b, b]$ such that $(y_b, b] \cap C = \emptyset$. Let $U = \bigcup_{a \in C}(x_a, a]$ and $V = \bigcup_{b \in D}(y_b, b]$. Then $C \subseteq U$ and $D \subseteq V$. We assert that $U \cap V = \emptyset$: Suppose $z \in U \cap V$. Then there exists $a \in C$ and $b \in D$ such that $z \in (x_a, a]$ and $z \in (y_b, b]$. We may assume without loss of generality that $a < b$. We consider two cases: If $a \leq y_b$, then $(x_a, a] \cap (y_b, b] = \emptyset$. If $a > y_b$, then $a \in (y_b, b]$. This is impossible because $(y_b, b] \cap C = \emptyset$. Therefore we have a contradiction and hence $U \cap V = \emptyset$.

Case 2. Suppose $a_0 \in C$. We assert that $\{a_0\}$ is open: Let $A = \{x \in X: x > a_0\}$ and let c be the least member of A. Then $\{a_0\} = \{x \in X: x < c\}$, and hence $\{a_0\}$ is open. Therefore $C - \{a_0\}$ is closed. By Case 1 there are disjoint open sets U and V such that $C - \{a_0\} \subseteq U$ and $D \subseteq V$. Therefore $U \cup \{a_0\}$ and V are disjoint open sets such that $C \subseteq U \cup \{a_0\}$ and $D \subseteq V$. ∎

Definition. A T_1-space $(X, \mathcal{T})$ is **completely normal** provided that whenever A and B are subsets of X such that $A \cap \overline{B} = \overline{A} \cap B = \emptyset$, there are disjoint open sets U and V such that $A \subseteq U$ and $B \subseteq V$. ∎

Completely normal spaces were introduced in 1923 by Tietze.

It is clear that every completely normal space is normal. The following characterization of completely normal shows that whenever we give an example of a normal space with a subspace that is not normal, we will also have an example of a normal space that is not completely normal.

THEOREM 5.29. A T_1-space $(X, \mathcal{T})$ is completely normal if and only if every subspace of X is normal.

Proof. Suppose $(X, \mathcal{T})$ is a topological space with the property that every subspace is normal. Let A and B be subsets of X such that $A \cap \overline{B} = \overline{A} \cap B = \emptyset$. Let $E = X - (\overline{A} \cap \overline{B})$. Then A and B are subsets of E. Let C be the closure of A in $(E, \mathcal{T}_E)$ and let D be the closure of B in $(E, \mathcal{T}_E)$. Then C and D are disjoint closed subsets of $(E, \mathcal{T}_E)$. Since $(E, \mathcal{T}_E)$ is normal, there exist $U, V \in \mathcal{T}_E$ such that $C \subseteq U, D \subseteq V$, and $U \cap V = \emptyset$. Since E is an open subset of X, $U, V \in \mathcal{T}$. Therefore $(X, \mathcal{T})$ is completely normal.

Suppose $(X, \mathcal{T})$ is completely normal. Let $(A, \mathcal{T}_A)$ be a subspace of X, and let C and D be disjoint closed subsets of A. Then $\overline{C} \cap D = C \cap \overline{D} = \emptyset$. (If $x \in \overline{C} - C$, then $x \notin A$ and hence $x \notin D$). Therefore there are disjoint open subsets U and V of X such that $C \subseteq U$ and $D \subseteq V$. Hence $U \cap A$ and $V \cap A$ are disjoint open subsets of $(A, \mathcal{T}_A)$ such that $C \subseteq U \cap A$ and $D \subseteq V \cap A$. Therefore $(A, \mathcal{T}_A)$ is normal. ∎

In Exercise 13 you are asked to prove that every metric space is completely normal. In Section 5.5 we show that every normal space is completely regular.

Thus we have the following hierarchy:

metrizable → completely normal → normal
  → completely regular → regular → Hausdorff → T_1 → T_0.

Definition. A T_1-space $(X, \mathcal{T})$ is **perfectly normal** provided that for each pair of disjoint closed subsets C and D in X there is a continuous function $f: X \to I$ such that $C = f^{-1}(0)$ and $D = f^{-1}(1)$. ∎

Perfectly normal spaces were introduced in 1924 by Urysohn.

THEOREM 5.30. A topological space $(X, \mathcal{T})$ is perfectly normal if and only if it is normal and each closed subset of X is a G_δ-set.

Proof. See Exercise 11. ∎

By Exercise 13 of Section 1.5, every closed subset of a metric space is a G_δ-set. Therefore every metric space is perfectly normal.

EXERCISES 5.3

1. Prove that a T_1-space $(X, \mathcal{T})$ is normal if and only if for each closed subset C of X and each open set U such that $C \subseteq U$, there is an open set V such that $C \subseteq V$ and $\overline{V} \subseteq U$.

2. Prove that a T_1-space $(X, \mathcal{T})$ is normal if and only if for each pair of disjoint closed subsets C and D of X, there are open sets U and V such that $C \subseteq U, D \subseteq V$, and $\overline{U} \cap \overline{V} = \emptyset$.

3. Prove that every closed subset of a normal space is normal.

4. Complete the proof of Theorem 5.27 by showing that $C \subseteq M$, $D \subseteq N$, and $M \cap N = \emptyset$.

5. Consider the proof that every subspace of a regular space is regular (Theorem 5.14). Why can't the same method of proof be used to show that every subspace of a normal space is normal?

6. Let $X = \{1, 2, 3\}$. Find all topologies $\mathcal{T}$ on X such that $(X, \mathcal{T})$ is normal.

7. Let $X = \{1, 2, 3, 4\}$. Determine whether each of the following topological spaces is normal.

   (a) $(X, \mathcal{T})$, where $\mathcal{T} = \{\emptyset, \{1, 2\}, \{3, 4\}, X\}$.

   (b) $(X, \mathcal{T})$, where $\mathcal{T} = \{\emptyset, \{1\}, \{1, 2\}, \{1, 2, 3\}, X\}$.

   (c) $(X, \mathcal{T})$, where $\mathcal{T} = \{\emptyset, \{1\}, \{2\}, \{1, 2\}, \{1, 3\}, \{1, 2, 3\}, \{1, 2, 4\}, X\}$.

8. Determine whether those topological spaces in Exercise 7 that are not normal are regular.

9. Let $\mathcal{T}$ be the usual topology on $\mathbb{R}$. Let $\mathcal{U} = \{U \in \mathcal{P}(\mathbb{R}): U = V \cup A$, where $V \in \mathcal{T}$ and A is a subset of the irrationals$\}$.

   (a) Prove that $\mathcal{U}$ is a topology on $\mathbb{R}$.

   (b) Is $(\mathbb{R}, \mathcal{U})$ normal? Prove your answer.

10. Let $(X, \leq)$ be a well-ordered set and let $\mathcal{T}$ be the order topology on X. Prove that $(X, \mathcal{T})$ is completely normal.

11. Prove Theorem 5.30.

12. Let $\mathcal{T} = \{U \in \mathcal{P}(\mathbb{R}): 0 \notin U$ or $R - U$ is finite$\}$.

    (a) Prove that $(\mathbb{R}, \mathcal{T})$ is completely normal.

    (b) Prove that $(\mathbb{R}, \mathcal{T})$ is not perfectly normal.

13. Prove that every metric space is completely normal.

14. For each $n \in \mathbb{N}$, let $(X_n, \mathcal{T}_n)$ be a normal space. Furthermore, suppose that for each $n \in \mathbb{N}$, X_n is a closed subset of $(X_{n+1}, \mathcal{T}_{n+1})$. Let $X = \bigcup_{n \in \mathbb{N}} X_n$, and let $\mathcal{T}$ be the weak topology on X induced by $\{X_n : n \in \mathbb{N}\}$. Prove that $(X, \mathcal{T})$ is normal. *Note:* Section 4.4 is required for this exercise.

15. Let X be the open interval $(0, 1)$, and let $\mathcal{T} = \{\emptyset\} \cup \{X\} \cup \{(\frac{1}{n}, 1) : n \in \mathbb{N} \text{ and } n > 1\}$.

    (a) Show that $\mathcal{T}$ is a topology on X.

    (b) Is $(X, \mathcal{T})$ Hausdorff?

    (c) Is $(X, \mathcal{T})$ T_1?

    (d) Is it true that if A is a closed subset of X and $p \in X - A$, then there are disjoint open sets U and V such that $A \subseteq U$ and $p \in V$?

    (e) Is it true that if A and B are disjoint closed subsets of X, then there are disjoint open sets U and V such that $A \subseteq U$ and $B \subseteq V$?

5.4 The Countability Axioms

The second axiom of countability, the first axiom of countability, and separability are countability axioms. We know that a second countable space is first countable, a metric space is first countable, a second countable space is separable, and a separable metric space is second countable. In this section we introduce another countability axiom and examine relationships between the various countability and separation axioms.

Definition. A space $(X, \mathcal{T})$ is a **Lindelöf space** or has the **Lindelöf property** provided every open cover has a countable subcover. ∎

THEOREM 5.31. Every second countable space is Lindelöf.

Proof. Let $(X, \mathcal{T})$ be a second countable space, let $\mathcal{B}$ be a countable basis for $\mathcal{T}$, and let $\mathcal{O}$ be an open cover of X. For each $x \in X$, let $O_x \in \mathcal{O}$ such that $x \in O_x$ and let $B_x \in \mathcal{B}$ such that $x \in B_x$ and $B_x \subseteq O_x$. Since $\mathcal{B}$ is countable, $\mathcal{B}' = \{B_x : x \in X\}$ is a countable open cover of X. For each $B_x \in \mathcal{B}'$, let $O'_x \in \mathcal{O}$ such that $B_x \subseteq O'_x$. Let $\mathcal{O}'$ be the collection of all such O'_x. Then $\mathcal{O}'$ is a countable subcover of X. Therefore $(X, \mathcal{T})$ is Lindelöf. ∎

Lindelöf spaces were first studied by Ernst Lindelöf. He proved Theorem 5.31 in 1903. The term *Lindelöf space* is due to K. Kuratowski and W. Sierpeński (1882–1969). They studied these spaces in 1921.

EXAMPLE 9. Let $\mathcal{T}$ be the lower-limit topology on $\mathbb{R}$. Then by Example 34 of Chapter 1, $(\mathbb{R}, \mathcal{T})$ is not second countable. However $(\mathbb{R}, \mathcal{T})$ is Lindelöf.

Analysis. If $\mathcal{O}$ is an open cover of $\mathbb{R}$, then for each $x \in \mathbb{R}$, there is an $O_x \in \mathcal{O}$ such that $x \in O_x$, and there is an interval $[a_x, b_x)$ such that $x \in [a_x, b_x)$ and $[a_x, b_x) \subseteq O_x$. Therefore it is sufficient to show that if $\mathcal{O}$ is an open cover of $\mathbb{R}$ by sets of the form $[a, b)$, then $\mathcal{O}$ has a countable subcover. Let $\mathcal{O} = \{[a_\alpha, b_\alpha): \alpha \in \Lambda\}$ be an open cover of $\mathbb{R}$. Let $A = \bigcup_{\alpha \in \Lambda}(a_\alpha, b_\alpha)$, and let $\mathcal{U}$ be the usual topology on $\mathbb{R}$. By Example 34 of Chapter 1 and Theorem 2.10, $(A, \mathcal{U}_A)$ is second countable. Thus by Theorem 5.13, $(A, \mathcal{U}_A)$ is Lindelöf. Therefore the collection $\{(a_\alpha, b_\alpha): \alpha \in \Lambda\}$ has a countable subcover; that is, there exist $\alpha_1, \alpha_2, \ldots$ in Λ such that the collection $\{(a_{\alpha_i}, b_{\alpha_i}): i \in \mathbb{N}\}$ covers A. Hence the collection $\{[a_{\alpha_i}, b_{\alpha_i}): i \in \mathbb{N}\}$ covers A. Let $x \in \mathbb{R} - A$. Then there exists $\alpha \in \Lambda$ such that $x = a_\alpha$. Let r_x be a rational number in the interval (a_α, b_α). Since $(a_\alpha, b_\alpha) \subseteq A$, $(a_\alpha, r_x) \subseteq A$. Define a function $f : (\mathbb{R} - A) \to \mathbb{Q}$ (the set of rationals) by $f(x) = r_x$. We assert that f is one-to-one: Suppose $x, y \in \mathbb{R} - A$, $x < y$ and $r_y \leq r_x$. Then $x < y = a_\beta < r_y \leq r_x$. Therefore $y \in (x, r_x)$. But $y \in \mathbb{R} - A$ and $(x, r_x) \subseteq A$, so we have a contradiction. Therefore $r_x < r_y$ and hence f is one-to-one. By Exercise 13 of Appendix F and Theorem F.5, $\mathbb{R} - A$ is countable. For each $x \in \mathbb{R} - A$, choose $\beta_x \in \Lambda$ such that $x \in [a_{\beta_x}, b_{\beta_x})$. Then $\{[a_{\beta_x}, b_{\beta_x}): x \in \mathbb{R} - A\} \cup \{[a_{\alpha_i}, b_{\alpha_i}): i \in \mathbb{N}\}$ is a countable subcollection of $\mathcal{O}$ that covers $\mathbb{R}$. ∎

EXAMPLE 10. Let $\mathcal{T}$ be the finite complement topology on $\mathbb{R}$. By Example 35 of Chapter 1, $(\mathbb{R}, \mathcal{T})$ is not first countable. However, it is easy to see that $(\mathbb{R}, \mathcal{T})$ is Lindelöf.

Example 11 is an example of a Lindelöf space with a subspace that is not Lindelöf. Since the subspace is first countable, it also provides an example of a first countable space that is not Lindelöf.

EXAMPLE 11. Let $(\Omega, \leq)$ be an uncountable well-ordered set with a maximal element ω_1 having the property that if $x \in \Omega$ and $x \neq \omega_1$, then $\{y \in \Omega: y \leq x\}$ is countable (see Appendix H). Let $\mathcal{T}$ be the order topology on Ω, and let $\Omega_0 = \Omega - \{\omega_1\}$. Then $(\Omega, \mathcal{T})$ is Lindelöf while $(\Omega_0, \mathcal{T}_{\Omega_0})$ is first countable but not Lindelöf.

Analysis. Recall that 1 denotes the least element of Ω. Let $\mathcal{O}$ be an open cover of Ω, and let $U \in \mathcal{O}$ such that $\omega_1 \in U$. Then there is an $x \in \Omega_0$ such that $(x, \omega_1] \subseteq U$. Since $[1, x]$ is countable, there is a countable subcollection $\mathcal{O}'$ of $\mathcal{O}$ that covers $[1, x]$. Hence $\mathcal{O}' \cup \{U\}$ is a countable subcover of Ω, and hence $(\Omega, \mathcal{T})$ is Lindelöf.

In Example 11 of Chapter 4, we exhibit an open cover of Ω_0 that does not have a countable subcover. Therefore $(\Omega_0, \mathcal{T}_{\Omega_0})$ is not Lindelöf. In Example 13 of Chapter 4, we show that $(\Omega_0, \mathcal{T}_{\Omega_0})$ is first countable. ∎

Example 12 is an example of a Lindelöf space that is not separable.

EXAMPLE 12. Let $A = \{(x, y) \in \mathbb{R} \times \mathbb{R}: y = 0\}$, let $X = A \cup \{(0, 1)\}$, and let $\mathcal{T} = \{U \in \mathcal{P}(X): U \subseteq A, \text{ or } (0, 1) \in U \text{ and } A - U \text{ is finite}\}$. Then $\mathcal{T}$ is a topology on X, and $(X, \mathcal{T})$ is Lindelöf but not separable.

Analysis. Let $\mathcal{O}$ be an open cover of X and let $U \in \mathcal{O}$ such that $(0, 1) \in U$. Since $A - U$ is finite, there is a finite subcollection $\mathcal{O}'$ of $\mathcal{O}$ that covers $A - U$. Hence $\mathcal{O}' \cup \{U\}$ is a finite (and hence countable) subcover of X. Therefore $(X, \mathcal{T})$ is Lindelöf.

Let B be a countable subset of X. Since A is uncountable, there exists $z \in A - B$. Then $\{z\} \in \mathcal{T}$ and $\{z\} \cap B = \emptyset$. Therefore $\overline{B} \neq X$ and hence $(X, \mathcal{T})$ is not separable. ∎

We use the following theorem to give an example (Example 13) of a separable space that is not Lindelöf. Example 13 also shows that the product of two Lindelöf spaces need not be Lindelöf and that the product of two normal spaces need not be normal.

THEOREM 5.32. Every regular Lindelöf space is normal.

Proof. Let $(X, \mathcal{T})$ be a regular Lindelöf space, and let A and B be disjoint closed subsets of X. By Theorem 5.12, for each $a \in A$, there is a neighborhood U_a of a such that $\overline{U}_a \cap B = \emptyset$, and for each $b \in B$, there is a neighborhood V_b of b such that $\overline{V}_b \cap A = \emptyset$. Let $\mathcal{O} = \{U_a : a \in A\} \cup \{V_b : b \in B\} \cup \{X - (A \cup B)\}$. Then $\mathcal{O}$ is an open cover of X. Since $(X, \mathcal{T})$ is Lindelöf, $\mathcal{O}$ has a countable subcover. Therefore there is a sequence $a_1, a_2, \ldots$ of members of A and a sequence $b_1, b_2, \ldots$ of members of B such that $A \subseteq \bigcup_{i=1}^{\infty} U_{a_i}$ and $B \subseteq \bigcup_{i=1}^{\infty} V_{b_i}$. For each $n \in \mathbb{N}$, let $U'_n = U_{a_n} - \bigcup_{i=1}^{n} \overline{V}_{b_i}$ and $V'_n = V_{b_n} - \bigcup_{i=1}^{n} \overline{U}_{a_i}$. Then for each $n \in \mathbb{N}$, U'_n and V'_n are open sets. Since $\overline{V}_{b_n} \cap A = \emptyset$ and $\overline{U}_{a_i} \cap B = \emptyset$ for each $i \in \mathbb{N}$, $A \subseteq \bigcup_{i=1}^{\infty} U'_n$ and $B \subseteq \bigcup_{n=1}^{\infty} V'_n$. We complete the proof by showing that for all $m, n \in \mathbb{N}, U'_m \cap V'_n = \emptyset$: If $m \leq n$, then $V'_n \cap U'_m \subseteq V'_n \cap \overline{U}_{a_m}$ and $V'_n \cap \overline{U}_{a_m} = \emptyset$. Therefore $V'_n \cap U'_m = \emptyset$. If $m > n$, then $U'_m \cap V'_n \subseteq U'_m \cap \overline{V}_{b_n}$ and $U'_m \cap \overline{V}_{b_n} = \emptyset$. Therefore $U'_m \cap V'_n = \emptyset$. ∎

The following corollary is an immediate consequence of Theorems 5.31 and 5.32.

Corollary 5.33. Every second countable regular space is normal. ∎

EXAMPLE 13. Let $\mathcal{T}$ be the lower-limit topology on $\mathbb{R}$, and let $\mathcal{U}$ be the product topology (with respect to $\mathcal{T}$) on $\mathbb{R}^2$. Then $(\mathbb{R}, \mathcal{T})$ is a normal Lindelöf space, and $(\mathbb{R}^2, \mathcal{U})$ is a separable space that is neither Lindelöf nor normal.

Analysis. By Example 9, $(\mathbb{R}, \mathcal{T})$ is Lindelöf. We assert that $(\mathbb{R}, \mathcal{T})$ is regular: Let C be a closed subset of $\mathbb{R}$ and let $p \in \mathbb{R} - C$. Since C is closed there is a $b \in \mathbb{R}$ such that $[p, b) \subseteq \mathbb{R} - C$. Since $\mathbb{R} - [p, b) = (-\infty, p) \cup [b, \infty)$, $[p, b)$ is closed. Therefore $[p, b)$ and $\mathbb{R} - [p, b)$ are disjoint open sets such that $p \in [p, b)$ and $C \subseteq \mathbb{R} - [p, b)$. Therefore $(\mathbb{R}, \mathcal{T})$ is regular. By Theorem 5.32, $(\mathbb{R}, \mathcal{T})$ is normal.

The topology $\mathcal{U}$ has as a basis all sets of the form $[a, b) \times [c, d)$. We assert that the set $C = \{(x, y) \in \mathbb{R}^2 : y = -x\}$ is a closed, relatively discrete subset of $\mathbb{R}^2$ (see Figure 5.4): Let $(x, -x) \in C$. Then if b and d are positive real numbers, $U = [x, x + b) \times [-x, -x + d)$ is an open set such that $U \cap C = \{(x, -x)\}$.

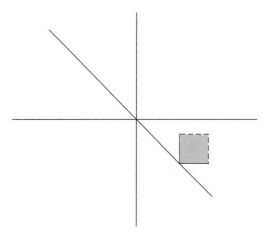

Figure 5.4

Therefore C is a relatively discrete subset of $\mathbb{R}^2$. Let $(x, y) \in \mathbb{R}^2 - C$. If $y > -x$ and b and d are positive real numbers, then $[x, x + b) \times [y, y + d) \subseteq \mathbb{R}^2 - C$. If $y < -x$, then $[x, x + \frac{1}{2}|x + y|) \times [y, y + \frac{1}{2}|x + y|) \subseteq \mathbb{R}^2 - C$. Therefore C is closed. Let $D = \{(x, y) \in \mathbb{R}^2 : x \text{ and } y \text{ are rational}\}$. Then D is a countable dense subset of $\mathbb{R}^2$. Therefore by Exercise 7 of Appendix G and Theorem 5.23, $(\mathbb{R}^2, \mathcal{U})$ is not normal. By Theorem 5.16, $(\mathbb{R}^2, \mathcal{U})$ is regular. Hence by Theorem 5.32, $(\mathbb{R}^2, \mathcal{U})$ is not Lindelöf. ∎

EXERCISES 5.4

1. Prove that every closed subspace of a Lindelöf space is Lindelöf.

2. Prove that every Lindelöf metric space is second countable.

3. Prove that every uncountable subset of a Lindelöf space has a limit point.

4. Prove that a regular space $(X, \mathcal{T})$ is Lindelöf if and only if each open cover $\mathcal{O}$ has a countable subcollection $\mathcal{O}' = \{U_n : n \in \mathbb{N}\}$ such that $\{\overline{U_n} : U_n \in \mathcal{O}'\}$ covers X.

5. Let $(X, \mathcal{T})$ be a second countable space. Prove that every subspace of X is Lindelöf.

6. Let $\mathcal{T}$ be the topology on $\mathbb{R}^2$ defined in Exercise 18 of Section 5.2. Prove that $(\mathbb{R}^2, \mathcal{T})$ is separable, but neither first countable nor Lindelöf.

7. Is the Moore plane Lindelöf? Prove your answer.

8. Is the open continuous image of a Lindelöf space Lindelöf? Prove your answer.

9. Let X be the closed interval $[-1, 1]$, and let $\mathcal{T} = \{U \in \mathcal{P}(X): 0 \notin U$ or $(-1, 1) \subseteq U\}$.

   (a) Prove that $\mathcal{T}$ is a topology on X.

   (b) Prove that $(X, \mathcal{T})$ is Lindelöf.

   (c) Let $A = X - \{0\}$. Prove that $(A, \mathcal{T}_A)$ is not Lindelöf.

   (d) Prove that $(X, \mathcal{T})$ is not separable.

10. Let $\{(X_\alpha, \mathcal{T}_\alpha): \alpha \in \Lambda\}$ be a collection of T_1-spaces. Prove that if the product space $\prod_{\alpha \in \Lambda} X_\alpha$ is Lindelöf, then, for each $\alpha \in \Lambda$, $(X_\alpha, \mathcal{T}_\alpha)$ is Lindelöf.

11. A point p in a topological space $(X, \mathcal{T})$ is a **condensation point** of a subset A of X provided every neighborhood of p contains an uncountable number of points of A. Prove that if A is an uncountable subset of a Lindelöf space $(X, \mathcal{T})$, then there exists a condensation point of A.

5.5 Urysohn's Lemma and the Tietze Extension Theorem

As mentioned in the introduction to this chapter, the separation properties guarantee a sufficient supply of open sets for whatever purpose we have in mind. One of the purposes served is to guarantee a plentiful supply of continuous functions. The fundamental theorem in this direction is Urysohn's Lemma. This lemma is named in honor of Paul Urysohn, a Russian mathematician who accidentally drowned at the age of 25. The proof is an inductive proof and involves an ingenious idea. We begin by considering the dyadic numbers in I.

Definition. A **dyadic number** is a number that can be expressed as the quotient of two integers, where the denominator is a power of 2. ∎

Every dyadic number in I can be written in the form $m/2^n$, where $n \in \mathbb{N} \cup \{0\}$ and $m = 0, 1, 2, ..., 2^n$. By Exercise 13 in Appendix F, the set of all rational numbers is countable. Thus the dyadic numbers in $[0, 1]$, denoted by P, are certainly countable. Since the proof of Urysohn's Lemma is an inductive proof, we describe a specific ordering of $P - \{0\}$: $\frac{1}{2}, \frac{1}{4}, \frac{3}{4}, \frac{1}{8}, \frac{3}{8}, \frac{5}{8}, \frac{7}{8}, ...$ The ordering we have in mind is the one that, for each $n \in \mathbb{N}$, groups the members of P with 2^n in the denominator and within that group arranges the numbers in order according to the numerators $1, 3, 5, ..., 2^n - 1$. Then the immediate successor of $(2^n - 1)/2^n$ is $1/(2^n + 1)$. We denote the members of P by $p_1, p_2, p_3, ...$ Thus $p_1 = \frac{1}{2}, p_2 = \frac{1}{4}, p_3 = \frac{3}{4}, p_4 = \frac{1}{8}, ...$ The proof of Urysohn's Lemma involves choosing an open set U_{p_i} for each $i \in \mathbb{N}$. In the inductive proof we assume that $U_{p_1}, U_{p_2}, ..., U_{p_i}$ have been chosen and choose U_{p_i+1}. We use two theorems. The proof of the first is left as Exercise 1.

THEOREM 5.34. The set of dyadic numbers in I is dense in I. ∎

5.5 Urysohn's Lemma and the Tietze Extension Theorem

THEOREM 5.35. Let $(X, \mathcal{T})$ be a topological space and let D be a dense subset of I. Suppose that for each $t \in D$, there is an open set U_t in X such that: **(1)** if $t_1 < t_2$ then $\overline{U}_{t_1} \subseteq U_{t_2}$ and **(2)** $X = \bigcup_{t \in D} U_t$. Define $f: X \to I$ by $f(x) = \text{glb}\{t \in D: x \in U_t\}$ for each $x \in X$. Then f is continuous.

Proof. Let $\mathcal{S}$ be the subbasis for the topology on I that consists of all sets of the form $[0, a)$ and $(a, 1]$, where $a \in (0, 1)$. By Theorem 1.54, it is sufficient to show that $f^{-1}(S) \in \mathcal{T}$ for each $S \in \mathcal{S}$. Let $a \in (0, 1)$. Then $f^{-1}([0, a)) = \{x \in X: f(x) < a\}$. Since $f(x) < a$ if and only if there is a $t \in D$ such that $t < a$ and $x \in U_t$, $f^{-1}([0, a)) = \bigcup \{U_t: t \in D \text{ and } t < a\}$. Therefore $f^{-1}([0, a))$ is the union of open sets and hence it is open. To show that $f^{-1}((a, 1])$ is open, we show that $X - f^{-1}((a, 1])$ is closed. Since $X - f^{-1}((a, 1]) = \{x \in X: f(x) \le a\}$, it is sufficient to show that $\{x \in X: f(x) \le a\} = \bigcap \{\overline{U}_t: t \in D \text{ and } a < t\}$. Let $y \in \{x \in X: f(x) \le a\}$, and let $t \in D$ such that $a < t$. Then there is an $s < t$ such that $y \in U_s$. Since $U_s \subseteq \overline{U}_t$, $y \in \overline{U}_t$ and hence $\{x \in X: f(x) \le a\} \subseteq \bigcap \{\overline{U}_t: t \in D \text{ and } a < t\}$. Let $y \in \bigcap \{\overline{U}_t: t \in D \text{ and } a < t\}$, and let $\varepsilon > 0$. Since D is dense in I, there exists $s_1 \in D$ such that $a < s_1 < a + \varepsilon$. Now $y \in \overline{U}_{s_1}$. Again since D is dense in I, there exists $s_2 \in D$ such that $s_1 < s_2 < a + \varepsilon$. Since $\overline{U}_{s_1} \subseteq U_{s_2}$, $y \in U_{s_2}$. Therefore $f(y) \le s_2$. Since $s_2 < a + \varepsilon$, $f(y) < a + \varepsilon$. Then since ε is an arbitrary positive number, $f(y) \le a$. Therefore $\bigcap \{\overline{U}_t: t \in D \text{ and } a < t\} \subseteq \{x \in X: f(x) \le a\}$. This completes the proof. ∎

THEOREM 5.36 (Urysohn's Lemma). A T_1-space $(X, \mathcal{T})$ is normal if and only if for each pair A, B of disjoint closed subsets of X there is a continuous function $f: X \to I$ such that $f(x) = 0$ for all $x \in A$ and $f(x) = 1$ for all $x \in B$. ∎

The illustration in Figure 5.5 is designed to make the proof of Uysohn's Lemma easier to follow. In this figure, $U_{3/8}$ is the last set that has been chosen. (This is the fifth step).

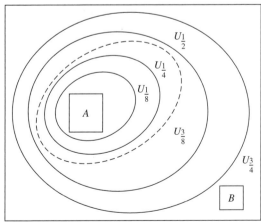

Figure 5.5

Proof. Suppose $(X, \mathcal{T})$ is normal and let A and B be disjoint closed subsets of X. Let P denote the dyadic numbers in $I - \{0\}$, ordered according to the discussion preceding Theorem 5.34. Since $X - B$ is open and $A \subseteq X - B$, by Theorem 5.24, there is an open set $U_{1/2}$ such that $A \subseteq U_{1/2}$, and $\overline{U_{1/2}} \subseteq X - B$. We continue to use Theorem 5.24. Since $A \subseteq U_{1/2}$, there is an open set $U_{1/4}$ such that $A \subseteq U_{1/4}$ and $\overline{U_{1/4}} \subseteq U_{1/2}$. Since $\overline{U_{1/2}} \subseteq X - B$, there is an open set $U_{3/4}$ such that $\overline{U_{1/2}} \subseteq U_{3/4}$ and $\overline{U_{3/4}} \subseteq X - B$. Thus we have the following:

$$A \subseteq U_{1/4} \subseteq \overline{U_{1/4}} U_{1/2} \subseteq \overline{U_{1/2}} \subseteq U_{3/4} \subseteq \overline{U_{3/4}} \subseteq X - B.$$

Then we continue. The next four sets, listed in the order in which they are chosen, are: $U_{1/8}, U_{3/8}, U_{5/8}$, and $U_{7/8}$, and they are chosen so that

$$A \subseteq U_{1/8} \subseteq \overline{U_{1/8}} \subseteq U_{1/4} \subseteq \overline{U_{1/4}} \subseteq U_{3/8} \subseteq \overline{U_{3/8}} \subseteq U_{1/2} \subseteq \overline{U_{1/2}} \subseteq U_{5/8}$$
$$\subseteq \overline{U_{5/8}} \subseteq U_{3/4} \subseteq \overline{U_{3/4}} \subseteq U_{7/8} \subseteq \overline{U_{7/8}} \subseteq X - B.$$

We can make this idea precise by using the second principle of mathematical induction to select recursively the open set U_t for each $t \in P$. Suppose $U_{p_1}, U_{p_2}, ..., U_{p_i}$ have been chosen. There exist an odd natural number m and a natural number n such that $p_i = m/(2^n)$. We consider three cases in order to choose $U_{p_{i+1}}$.

Case 1. If $m = 2^n - 1$, then $p_{i+1} = 1/(2^{n+1})$. Let $q = 1/(2^n)$ and let $U_{p_{i+1}}$ be an open set such that $A \subseteq U_{p_{i+1}}$ and $\overline{U_{p_{i+1}}} \subseteq U_q$.

Case 2. If $m = 2^n - 3$, then $p_{i+1} = (2^n - 1)/2^n$. Let $q = (2^{n-1} - 1)/2^{n-1}$ and let $U_{p_{i+1}}$ be an open set such that $\overline{U_q} \subseteq U_{p_{i+1}}$ and $\overline{U_{p_{i+1}}} \subseteq X - B$.

Case 3. If $m < 2^n - 3$, then $p_{i+1} = (m + 2)/2^n$. Let $q = (m + 1)/2^n$ and $r = (m + 3)/2^n$ and let $U_{p_{i+1}}$ be an open set such that $\overline{U_q} \subseteq U_{p_{i+1}} \overline{U_{p_{i+1}}} \subseteq U_r$.

After choosing U_{p_i} for each $p_i \in P$, we let $U_1 = X$ so that each member of X will be in some U_j. Then we define $f : X \to I$ by $f(x) = \text{glb}\{t \in P \cup \{1\}: x \in U_t\}$. By Theorem 5.35, f is continuous. If $x \in A$, then $x \in U_t$ for all $t \in P \cup \{1\}$ and hence $f(x) = 0$. If $x \in B$, then $x \in U_1$ but $x \notin U_t$ for any $t \in P$, so $f(x) = 1$.

Suppose $(X, \mathcal{T})$ is a T_1-space and the condition holds. Let A and B be disjoint closed subsets of X. Then there is a continuous function $f : X \to I$ such that $f(x) = 0$ for all $x \in A$ and $f(x) = 1$ for all $x \in B$. Hence $f^{-1}([0, \frac{1}{2}))$ and $f^{-1}((\frac{1}{2}, 1])$ are disjoint open sets such that $A \subseteq f^{-1}([0, \frac{1}{2}))$ and $B \subseteq f^{-1}((\frac{1}{2}, 1])$. ∎

Corollary 5.37. Every normal space is completely regular. ∎

Another characterization of normal spaces is given by the Tietze Extension Theorem. This theorem is probably the most well-known of all results related to finding a continuous extension of a continuous function. The Tietze Extension Theorem is named in honor of Heinrich Tietze. It was proved for closed subsets of $\mathbb{R}^2$ by Henri Lebesgue in 1907 and extended to metric spaces by Tietze in 1915. Then Urysohn proved it for normal spaces in 1924. In order to prove the Tietze Extension Theorem, we observe that if $a, b \in \mathbb{R}$ and $a < b$, then $[a, b]$ is homeomorphic to I. Therefore we can replace I by $[a, b]$ in Urysohn's Lemma. This fact is used in the proof of Theorem 5.38. We give two versions of the Tietze Extension Theorem.

5.5 Urysohn's Lemma and the Tietze Extension Theorem

THEOREM 5.38 (Tietze Extension Theorem). A T_1-space $(X, \mathcal{T})$ is normal if and only if whenever A is a closed subset of X and $f: A \to [-1, 1]$ is a continuous function, then there is a continuous function $F: X \to [-1, 1]$ such that $F|_A = f$.

Proof. Suppose the condition holds, and let C and D be disjoint closed subsets of X. Then $C \cup D$ is a closed subset of X, and the function $f: C \cup D \to [-1, 1]$ defined by $f(x) = -1$ for all $x \in C$ and $f(x) = 1$ for all $x \in D$ is continuous. Thus by the condition, there is a continuous extension $F: X \to [-1, 1]$ of f. Therefore by Urysohn's Lemma, $(X, \mathcal{T})$ is normal.

Suppose $(X, \mathcal{T})$ is a normal space, A is a closed subset of X, and $f: A \to [-1, 1]$ is continuous. Let $C_1 = \{x \in A: f(x) \leq -\frac{1}{3}\}$ and let $D_1 = \{x \in A: f(x) \geq \frac{1}{3}\}$. Then C_1 and D_1 are closed subsets of X. By Urysohn's Lemma, there is a continuous function $f_1: X \to [-\frac{1}{3}, \frac{1}{3}]$ such that $f_1(x) = -\frac{1}{3}$ for all $x \in C_1$ and $f_1(x) = \frac{1}{3}$ for all $x \in D_1$. Notice that for each $x \in A$, $|f(x) - f_1(x)| \leq \frac{2}{3}$. Define $g_1: A \to [-\frac{2}{3}, \frac{2}{3}]$ by $g_1(x) = f(x) - f_1(x)$ for each $x \in A$. Let $C_2 = \{x \in A: g_1(x) \leq -\frac{2}{9}\}$ and let $D_2 = \{x \in A: g_1(x) \geq \frac{2}{9}\}$. By Urysohn's Lemma, there is a continuous function $f_2: X \to [-\frac{2}{9}, \frac{2}{9}]$ such that $f_2(x) = -\frac{2}{9}$ for all $x \in C_2$ and $f_2(x) = \frac{2}{9}$ for all $x \in D_2$. Notice that for each $x \in A$, $|f(x) - (f_1(x) + f_2(x))| = |g_1(x) - f_2(x)| \leq \frac{4}{9} = (\frac{2}{3})^2$. Define $g_2: A \to [-\frac{4}{9}, \frac{4}{9}]$ by $g_2(x) = g_1(x) - f_2(x)$ for all $x \in A$. Let $C_3 = \{x \in A: g_2(x) \leq -\frac{4}{27}\}$ and let $D_3 = \{x \in A: g_2(x) \geq \frac{4}{27}\}$. By Urysohn's Lemma, there is a continuous function $f_3: X \to [-\frac{4}{27}, \frac{4}{27}]$ such that $f_3(x) = -\frac{4}{27}$ for all $x \in C_3$ and $f_3(x) = \frac{4}{27}$ for all $x \in D_3$. Proceeding inductively, we obtain a sequence $f_1, f_2, \ldots$ of continuous functions such that: (1) for each $n \in \mathbb{N}$, $f_n: X \to [-2^{n-1}/3^n, 2^{n-1}/3^n]$ and (2) $|f(x) - \sum_{i=1}^{n} f_i(x)| \leq (\frac{2}{3})^n$ for each $x \in A$. Since the series $\sum_{n=1}^{\infty} 2^{n-1}/3^n$ converges and $|f_n(x)| \leq 2^{n-1}/3^n$ for all $x \in X$, the series $\sum_{n=1}^{\infty} f_n(x)$ converges to a number in the interval $[-\sum_{n=1}^{\infty} 2^{n-1}/3^n, \sum_{n=1}^{\infty} 2^{n-1}/3^n] = [-1, 1]$. Define $F: X \to [-1, 1]$ by $F(x) = \sum_{n=1}^{\infty} f_n(x)$. Since $|f(x) - \sum_{i=1}^{n} f_i(x)| \leq (\frac{2}{3})^n$ for each $x \in A$, $F(x) = f(x)$ for each $x \in A$. It remains to show that F is continuous: Let $x \in X$ and let $\varepsilon > 0$. There is a natural number N such that $\sum_{n=N+1}^{\infty} (\frac{2}{3})^n < \frac{\varepsilon}{2}$. For each $i = 1, 2, \ldots, N$, f_i is continuous, so there is a neighborhood U_i of x such that for each $y \in U_i$, $|f_i(x) - f_i(y)| < \frac{\varepsilon}{2N}$. Let $U = \bigcap_{i=1}^{N} U_i$. Then U is a neighborhood of x and for each $y \in U$,

$$|F(x) - F(y)| = \left|\sum_{i=1}^{\infty} f_i(x) - \sum_{i=1}^{\infty} f_i(y)\right| \leq \sum_{i=1}^{N} |f_i(x) - f_i(y)| + \sum_{i=N+1}^{\infty} |f_i(x) - f_i(y)|$$

$$< N(\tfrac{\varepsilon}{2N}) + \sum_{i=N+1}^{\infty} (\tfrac{2}{3})^i < \tfrac{\varepsilon}{2} + \tfrac{\varepsilon}{2} = \varepsilon.$$

Therefore F is continuous. ∎

THEOREM 5.39 (Tietze Extension Theorem). A T_1-space $(X, \mathcal{T})$ is normal if and only if whenever A is a closed subset of X and $f: A \to \mathbb{R}$ is a continuous function, then there is a continuous function $F: X \to \mathbb{R}$ such that $F|_A = f$.

Proof. Suppose the condition holds and let C and D be disjoint closed subsets of X. Define $f: C \cup D \to \mathbb{R}$ by $f(x) = 0$ for all $x \in C$ and $f(x) = 1$ for all $x \in D$. Then f is continuous. By the condition, there is a continuous extension $F: X \to \mathbb{R}$ of f. Define $h: \mathbb{R} \to I$ by $h(x) = 0$ for all $x < 0$, $h(x) = 1$ for all $x > 1$ and $h(x) = x$ for all $x \in I$. Then h is continuous and $h \circ F: X \to I$ is a continuous function such that $(h \circ F)(x) = 0$ for all $x \in C$ and $(h \circ F)(x) = 1$ for all $x \in D$. Hence by Urysohn's Lemma, $(X, \mathcal{T})$ is normal.

Suppose $(X, \mathcal{T})$ is normal, A is a closed subset of X, and $f: A \to \mathbb{R}$ is a continuous function. There is a homeomorphism $h: \mathbb{R} \to (-1, 1)$ (see Exercise 2). Thus $h \circ f: A \to [-1, 1]$. By Theorem 5.38, $h \circ f$ as a continuous extension. $G: X \to [-1, 1]$. Let $B = \{x \in X: G(x) = 1\}$. If $a \in A$, then $G(a) = (h \circ f)(a) \neq 1$. Therefore A and B are disjoint closed subsets of X. By Urysohn's Lemma, there is a continuous function $g: X \to I$ such that $g(x) = 0$ for all $x \in B$ and $g(x) = 1$ for all $x \in A$. Then for each $x \in X$, $g(x)G(x) \in (-1, 1)$ because if $G(x) = 1$ then $g(x) = 0$ and if $g(x) = 1$ then $G(x) = h \circ f(x) \in (-1, 1)$. [The point is that $G(x)$ can be -1 or 1 for some x, but $g(x)G(x)$ cannot be either.] Define $F: X \to \mathbb{R}$ by $F(x) = h^{-1}(g(x)G(x))$ for each $x \in X$. Then F is continuous and for each $a \in A$,

$$F(a) = h^{-1}(g(a)G(a)) = h^{-1}(1 \cdot G(a)) = h^{-1}((h \circ f)(a)) = f(a).$$

Hence F is the desired extension of f. ∎

EXERCISES 5.5

1. Prove that the set of dyadic numbers in I is dense in I.

2. Prove that there is a homeomorphism $h: \mathbb{R} \to (-1, 1)$.

3. Assume the Tietze Extension Theorem (one or both versions) and prove Urysohn's Lemma.

4. Let $(X, \mathcal{T})$ be a regular second countable space, and let $U \in \mathcal{T}$.

    (a) Prove that U is an F_σ-set; that is, prove that there is a countable collection $\{C_i : i \in \mathbb{N}\}$ of closed sets such that $U = \bigcup_{i=1}^\infty C_i$.

    (b) Prove that there is a continuous function $f: X \to I$ such that $f(x) = 0$ for all $x \notin U$ and $f(x) > 0$ for all $x \in U$.

5. Let A and B be disjoint closed subsets of a normal space $(X, \mathcal{T})$. Prove that A is a G_δ-set in X if and only if there is a continuous function $f: X \to I$ such that $f(x) = 1$ for all $x \in B$ and $f^{-1}(\{0\}) = A$.

6. A normal space $(Y, \mathcal{U})$ is an **absolute retract** if, for each normal space $(X, \mathcal{T})$ and each closed subset A of X, each continuous function $f: A \to Y$ has a continuous extension $F: X \to Y$.

(a) Prove that any finite set with the discrete topology is an absolute retract.

(b) Prove that if $n \in \mathbb{N}$ and, for each $i = 1, 2, ..., n$, $(X_i, \mathcal{T}_i)$ is an absolute retract such that the product space $\prod_{i=1}^{n} X_i$ is normal, then the product space is an absolute retract.

(c) Let $n \in \mathbb{N}$ and, for each $i = 1, 2, ..., n$, let X_i be the closed unit interval. Prove that the product space $\prod_{i=1}^{n} X_i$ is an absolute retract.

(d) Show that the property of being an absolute retract is a topological invariant.

(e) Prove that a normal space $(X, \mathcal{T})$ is an absolute retract if and only if, whenever X is homeomorphic to a closed subset A of a normal space $(Y, \mathcal{U})$, then A is a retract of Y.

5.6 Embeddings

One purpose of this section is to explore further the concepts of a collection of functions separating points and separating points from closed sets. Another purpose is to give two characterizations of completely regular spaces and to define cardinal functions. But first we prove theorems about the natural map and the open and closed continuous image of a regular space.

It is remarkable that with an extra condition on the collection of functions, a space with the weak topology induced by this collection can be embedded in a product space. The following definition is needed.

Definition. Let $\{(X_\alpha, \mathcal{T}_\alpha): \alpha \in \Lambda\}$ be a collection of topological spaces, let X be a set, and for each $\alpha \in \Lambda$ let $f_\alpha: X \to X_\alpha$ be a function. The **evaluation map** $e: X \to \prod_{\alpha \in \Lambda} X_\alpha$ **induced by** $\{f_\alpha: \alpha \in \Lambda\}$ is defined as follows: for each $x \in X$, $e(x)$ is the member of $\prod_{\alpha \in \Lambda} X_\alpha$ whose αth coordinate is $f_\alpha(x)$ for each $\alpha \in \Lambda$. ∎

We use the notation $[e(x)]_\alpha = f_\alpha(x)$ to indicate that $f_\alpha(x)$ is the αth coordinate of $e(x)$.

THEOREM 5.40. Let $\{(X_\alpha, \mathcal{T}_\alpha): \alpha \in \Lambda\}$ be a collection of topological spaces, let X be a set, and for each $\alpha \in \Lambda$ let $f_\alpha: X \to X_\alpha$ be a function. Then the evaluation map $e: X \to \prod_{\alpha \in \Lambda} X_\alpha$ induced by $\{f_\alpha: \alpha \in \Lambda\}$ is an embedding if and only if X has the weak topology induced by $\{f_\alpha: \alpha \in \Lambda\}$ and $\{f_\alpha: \alpha \in \Lambda\}$ separates points in X.

Proof. Suppose $e: X \to \prod_{\alpha \in \Lambda} X_\alpha$ is an embedding. By Theorems 2.33 and 2.41, $e(X)$ has the weak topology induced by $\{\pi_\alpha|_{e(X)}: \alpha \in \Lambda\}$. Since $e: X \to e(X)$ is a homeomorphism, X has the weak topology induced by $\{\pi_\alpha \circ e: \alpha \in \Lambda\}$. But for each $\alpha \in \Lambda$, $\pi_\alpha \circ e = f_\alpha$. Thus X has the weak topology induced by $\{f_\alpha: \alpha \in \Lambda\}$. Suppose $x, y \in X$ and $x \neq y$. Then $e(x) \neq e(y)$ and hence there exists $\beta \in \Lambda$ such that

$[e(y)]_\beta \neq [e(y)]_\beta$. But $f_\beta(x) = (\pi_\beta \circ e)(x) = [e(x)]_\beta$ and $f_\beta(y) = (\pi_\beta \circ e)(x) = [e(y)]_\beta$. Therefore $f_\beta(x) \neq f_\beta(y)$ and hence $\{f_\alpha : \alpha \in \Lambda\}$ separates points in X.

Supppose X has the weak topology induced by $\{f_\alpha : \alpha \in \Lambda\}$ and $\{f_\alpha : \alpha \in \Lambda\}$ separates points in X. Then for each $\alpha \in \Lambda$, f_α is continuous. But $f_\alpha = \pi_\alpha \circ e$ and hence by Theorem 2.37, e is continuous. Suppose $x, y \in X$ and $x \neq y$. Then there exists $\beta \in \Lambda$ such that $f_\beta(x) \neq f_\beta(y)$. Thus $[e(x)]_\beta \neq [e(y)]_\beta$. Therefore $e(x) \neq e(y)$ and hence e is one-to-one. It remains to show that e is an open map. Since e is one-to-one, it is sufficient to show that if U is a member of a subbasis for the weak topology on X, then $e(U)$ is open. So suppose there exists $\beta \in \Lambda$ and $V \in \mathcal{T}_\beta$ such that $U = f_\beta^{-1}(V)$. Then $U = [\pi_\beta|_{e(X)} \circ e]^{-1}(V) = e^{-1}[(\pi_\beta|_{e(X)})^{-1}(V)]$, so $e(U) = [\pi_\beta|_{e(X)}]^{-1}(V) = \pi_\beta^{-1}(V) \cap e(X)$ is an open set in $e(X)$. Therefore $e: X \to \prod_{\alpha \in \Lambda} X_\alpha$ is an embedding. ∎

Theorem 5.40 is important in the study of questions of the following type: If $(X, \mathcal{T})$ is a topological space and $\mathcal{P}$ is a topological property, can X be embedded in a topological space that has property $\mathcal{P}$?

It is of interest to know when the natural map from a topological space $(X, \mathcal{T})$ into a decomposition space of X is closed. In order to give a condition on the decomposition that will make the natural map closed, we introduce the following concepts.

Definition. Let $(X, \mathcal{T})$ be a topological space and let $(\mathcal{D}, \mathcal{U})$ be a decomposition space of X. An open subset V of X is **saturated relative to** $\mathcal{D}$ provided V is the union of members of $\mathcal{D}$ (that is, $V = p^{-1}(U)$ for some open set U in $\mathcal{D}$). The decomposition $\mathcal{D}$ is **upper semicontinuous** provided that for each $D \in \mathcal{D}$ and each open set V in X containing D there is a saturated set W in X such that $D \subseteq W \subseteq V$. ∎

THEOREM 5.41. Let $(X, \mathcal{T})$ be a topological space and let $(\mathcal{D}, \mathcal{U})$ be a decomposition space of X. The natural map $p: X \to \mathcal{D}$ is closed if and only if $\mathcal{D}$ is upper semicontinuous.

Proof. Suppose p is closed. Let $D \in \mathcal{D}$ and let V be an open set in X containing D. Then $p(X - V)$ is a closed set in $\mathcal{D}$. Hence $p^{-1}(p(X - V))$ is a closed set in X, and it is the union of members of $\mathcal{D}$. So $W = X - p^{-1}(p(X - V))$ is a saturated open set in X.

It remains to show that $D \subseteq W \subseteq V$. Let $x \in D$. Since $D \in \mathcal{D}$, $p(x) = D$. Furthermore if $y \in X - V$, then $p(y) \neq D$, so $D \notin p(X - V)$. It follows that $x \notin p^{-1}(p(X - V))$, so $x \in W$. Therefore $D \subseteq W$. Let $x \in W$. Then $x \notin p^{-1}(p(X - V))$, so $x \notin X - V$. Hence $x \in V$ and $W \subseteq V$.

Suppose $\mathcal{D}$ is upper semicontinuous, and let C be a closed subset of X. Let $D \in \mathcal{D} - p(C)$. Then $D \subseteq X - C$. Since $\mathcal{D}$ is upper semicontinuous, there is a saturated set W in X such that $D \subseteq W \subseteq X - C$. Now $p(W)$ is an open set, $D \in p(W)$, and $p(W) \subseteq \mathcal{D} - p(C)$. Therefore $p(C)$ is closed. ∎

Theorem 5.40 gives a characterization of when the evaluation map is an embedding. The following theorem gives another condition that ensures that the evaluation map is an embedding.

THEOREM 5.42. Let $(X, \mathcal{T})$ be a T_1-space, let $\{(X_\alpha, \mathcal{T}_\alpha): \alpha \in \Lambda\}$ be a collection of topological spaces, and for each $\alpha \in \Lambda$ let $f_\alpha: X \to X_\alpha$ be a function. If $\{f_\alpha: \alpha \in \Lambda\}$ separates points from closed sets, then the evaluation map $e: X \to \prod_{\alpha \in \Lambda} X_\alpha$ induced by $\{f_\alpha: \alpha \in \Lambda\}$ is an embedding.

Proof. Suppose $\{f_\alpha: \alpha \in \Lambda\}$ separates points from closed sets. By Corollary 2.43, $\mathcal{T}$ is the weak topology on X induced by $\{f_\alpha: \alpha \in \Lambda\}$. Since X is a T_1-space and $\{f_\alpha: \alpha \in \Lambda\}$ separates points from closed sets, $\{f_\alpha: \alpha \in \Lambda\}$ separates points in X. Therefore by Theorem 5.40, the evaluation map $e: X \to \prod_{\alpha \in \Lambda} X_\alpha$ induced by $\{f_\alpha: \alpha \in \Lambda\}$ is an embedding. ■

The space $(X, \mathcal{T})$ in Example 3 is a regular space, and hence Example 3 shows that an open continuous image of a regular space need not be regular. The following theorem provides a partial result concerning the image of regular spaces.

THEOREM 5.43. Let $(X, \mathcal{T})$ be a regular space, let $(Y, \mathcal{U})$ be a topological space, and let $f: X \to Y$ be a continuous map of X onto Y that is closed and open. Then $(Y, \mathcal{U})$ is Hausdorff.

Proof. By Corollary 5.9 it is sufficient to show that $A = \{(x, y) \in X \times X: f(x) = f(y)\}$ is a closed subset of $X \times X$. Let $(a, b) \in (X \times X) - A$. Since X is T_1, $\{a\}$ is a closed set. Thus since f is a closed mapping, $f(a)$ is closed. Therefore since f is continuous, $f^{-1}(f(a))$ is closed. Since $b \notin f^{-1}(f(a))$ and X is regular, there exist disjoint open sets U and V such that $b \in U$ and $f^{-1}(f(a)) \subseteq V$. Since f is an open continuous surjection, a subset U of Y is open if and only if $f^{-1}(U)$ is open. Therefore $\mathcal{U}$ is the quotient topology on Y induced by f. Since f is closed, by Theorem 5.41, there is a saturated open set W such that $f^{-1}(f(a)) \subseteq W \subseteq V$. Then $W \times U$ is a neighborhood of (a, b) such that $W \cap U = \emptyset$. Hence $(W \times U) \cap A = \emptyset$. Therefore $(X \times X) - A$ is open and hence A is closed. ■

The following two theorems illustrate the importance of completely regular spaces. If $(X, \mathcal{T})$ is a topological space, we let $C^*(X) = \{f: X \to \mathbb{R}: f \text{ is continuous and bounded}\}$.

THEOREM 5.44. A T_1-space $(X, \mathcal{T})$ is completely regular if and only if $\mathcal{T}$ is the weak topology induced by $C^*(X)$.

Proof. Suppose $(X, \mathcal{T})$ is completely regular. Then $C^*(X)$ separates points from closed sets. Therefore by Corollary 2.43, $\mathcal{T}$ is the weak topology induced by $C^*(X)$.

Suppose $\mathcal{T}$ is the weak topology on X induced by $C^*(X)$. Let C be a closed subset of X and let $p \in X - C$. Let $\mathcal{S}$ be the subbasis for the usual topology on $\mathbb{R}$ consisting of all sets of the form $(-\infty, a)$ and (b, ∞). Since $X - C \in \mathcal{T}$ and $\mathcal{T}$ is the weak topology on X induced by $C^*(X)$, there are functions $f_1, f_2, ..., f_n$ in $C^*(X)$ and members $V_1, V_2, ..., V_n$ of $\mathcal{S}$ such that $p \in \bigcap_{i=1}^{n} f_i^{-1}(V_i) \subseteq X - C$. For each $i \in \{1, 2, ..., n\}$

such that V_i is of the form $(-\infty, a)$, replace V_i by $(-a, \infty)$ and f_i by $-f_i$. Then $f_i^{-1}(V_i) = (-f_i)^{-1}((-a, \infty))$, so we may assume that each V_i is of the form (b_i, ∞). For each $i = 1, 2, \ldots, n$ define $g_i: X \to \mathbb{R}$ by $g_i(x) = \max\{0, f_i(x) - b_i\}$ for each $x \in \mathbb{R}$. Then $g_i(x) \geq 0$ for all $x \in X$ and $g_i^{-1}((0, \infty)) = f_i^{-1}((b_i, \infty))$. Therefore $p \in \bigcap_{i=1}^n g_i^{-1}((0, \infty)) \subseteq X - C$. Define $g: X \to \mathbb{R}$ by $g(x) = g_1(x)g_2(x)\ldots g_n(x)$ for each $x \in X$. Then $g(p) > 0$, so $p \in g^{-1}((0, \infty))$. Suppose $y \in X$ and $g(y) > 0$. Then for each $i = 1, 2, \ldots, n$, $g_i(y) > 0$, and hence $y \in \bigcap_{i=1}^n g_i^{-1}((0, \infty)) \subseteq X - C$. Thus if $y \in C$, then $g(y) = 0$. Therefore there is an interval $[0, d]$ such that $g: X \to [0, d]$, $g(p) > 0$ and $g(y) = 0$ for all $y \in C$. Since there is a continuous function $h: [0, d] \to [0, 1]$ such that $h(0) = 0$ and $h(g(p)) = 1$, $(X, \mathcal{T})$ is completely regular. ∎

Definition. A **cube** is a product of closed and bounded intervals (in $\mathbb{R}$). ∎

THEOREM 5.45. A topological space $(X, \mathcal{T})$ is completely regular if and only if it is homeomorphic to a subspace of a cube.

Proof. Suppose $(X, \mathcal{T})$ is homeomorphic to a subspace of a cube. Each subset of $\mathbb{R}$ is completely regular, so by Theorem 5.19, a cube is completely regular. By Theorem 5.21, every subspace of a cube is completely regular. By Theorem 5.20, completely regular is a topological property. Therefore $(X, \mathcal{T})$ is completely regular.

Suppose $(X, \mathcal{T})$ is completely regular. Then by Theorem 5.44, $\mathcal{T}$ has the weak topology induced by $C^*(X)$. If $f \in C^*(X)$, then there exist real numbers a_f and b_f such that $f(X) \subseteq [a_f, b_f]$. Now $\prod_{f \in C^*(X)} [a_f, b_f]$ is a cube. Since $(X, \mathcal{T})$ is completely regular, $C^*(X)$ separates points from closed sets. Therefore by Theorem 5.42, the evaluation map $e: X \to \prod_{f \in C^*(X)} [a_f, b_f]$ is an embedding. Hence $(X, \mathcal{T})$ is homeomorphic to a subspace of a cube. ∎

Cardinal functions extend properties such as second countable, first countable, and separable to higher cardinals.

Definition. A **cardinal function** is a function ϕ from the collection of all topological spaces (or some subcollection) into the collection of all infinite cardinals such that $\phi(X) = \phi(Y)$ whenever X and Y are homeomorphic. ∎

It is customary to require that $\phi(X)$ be an infinite cardinal to simplify statements of theorems and place the emphasis on infinite cardinal arithmetic. One example of a cardinal function is **cardinality**. If X is a topological space we define the cardinality of X, denoted by $|X|$, to be the number of members of X plus ω_0. Another example of a cardinal function is **weight**, defined by $w(X) = \min\{|\mathcal{B}|: \mathcal{B}$ is a basis for $X\} + \omega_0$. Yet another example of a cardinal function is the function o defined by taking $o(X)$ to be the number of open sets in X plus ω_0. We state, without proof, the following theorem.

THEOREM 5.46. Let $(X, \mathcal{T})$ be a topological space.

(a) Then $w(X) \leq o(X) \leq 2^{|X|}$.

(b) If $(X, \mathcal{T})$ is a T_0-space, then $|X| \leq 2^{w(X)}$ and $|X| \leq o(X)$. ■

We refer those interested in studying cardinal functions to a paper by R. Hodel called "Cardinal Functions I." This paper can be found in *Handbook of Set Theoretic Topology* edited by K. Kunen and J.E. Vaughan, North-Holland, 1984, 1–62.

Chapter 6
Special Topics

The purpose of this chapter is to study some additional topics that are not covered elsewhere in this text. The first section is concerned with contractions in metric spaces. The second section, which is independent of the first, deals with a special type of metric space, namely a normed linear space. The third section, which depends on the first two, provides an introduction to the Fréchet derivative. Then the fourth and fifth sections, which are independent of each other and of the first three, provide an introduction to manifolds and fractals. The sixth section, which is independent of the first five, discusses the embedding of a non-compact space into a compact space. The last section, which is independent of the first six, gives a proof that the product of compact spaces in compact.

6.1 Contraction Mappings in Metric Spaces

If $(X, \mathcal{T})$ is a topological space and T is a function from X into X, then a point $x \in X$ is a **fixed point** of T if $T(x) = x$. The purpose of this section is to show that certain maps have fixed points. This result is used in Section 6.3 to obtain an Implicit Function Theorem for Banach spaces.

Definition. Let (X, d) be a metric space. A function $T: X \to X$ is a **contraction** if there exists $\alpha \in \mathbb{R}$ such that $0 < \alpha < 1$ and $d(T(x_1), T(x_2)) \leq \alpha d(x_1, x_2)$ for all $x_1, x_2 \in X$. ∎

We first prove that a contraction cannot have more than one fixed point.

THEOREM 6.1. Let (X, d) be a metric space, and let $T: X \to X$ be a contraction. Then T has at most one fixed point.

Proof. Suppose x_1 and x_2 are fixed points of T. Since T is a contraction, there exists $\alpha \in \mathbb{R}$ such that $0 < \alpha < 1$ and $d(T(x_1), T(x_2)) \leq \alpha d(x_1, x_2)$. Since $T(x_1) = x_1$ and $T(x_2) = x_2$, we have $d(x_1, x_2) \leq \alpha d(x_1, x_2)$, so $(1 - \alpha)d(x_1, x_2) \leq 0$. Thus $d(x_1, x_2) = 0$ and hence $x_1 = x_2$. ∎

We now show that a contraction on a complete metric space has a fixed point. The Contraction Mapping Theorem was proved by Stefan Banach (1892–1945).

THEOREM 6.2 (Contraction Mapping Theorem). Let (X, d) be a complete metric space, and let $T: X \to X$ be a contraction. Then T has exactly one fixed point.

Proof. We have already proved that T can have at most one fixed point, so it remains to show the existence of a fixed point. We begin by defining inductively a sequence of functions from X into X. Define $T^0: X \to X$ by $T^0(x) = x$. Suppose $n \in \mathbb{N}$ and $T^0, T^1, \ldots, T^n$ have been defined. Define $T^{n+1}: X \to X$ by $T^{n+1}(x) = T^n(T(x))$. Note that $T^1 = T$.

Let $\alpha \in \mathbb{R}$ such that $0 < \alpha < 1$ and $d(T(x_1), T(x_2)) \le \alpha d(x_1, x_2)$ for all $x_1, x_2 \in X$. Let $x \in X$. We prove by induction that for each $p \in \mathbb{N}$, $d(T^p(x), T^{p+1}(x)) \le \alpha^p d(x, T(x))$. We first note that $d(T^0(x), T^1(x)) = d(x, T(x)) = \alpha^0 d(x, T(x))$ and $d(T^1(x), T^2(x)) \le \alpha d(T^0(x), T^1(x)) = \alpha d(x, T(x))$. Suppose $n \in \mathbb{N}$ and $d(T^n(x), T^{n+1}(x)) \le \alpha^n d(x, T(x))$. Then

$$d(T^{n+1}(x), T^{n+2}(x)) = d(T^n(T(x)), T^{n+1}(T(x)))$$
$$\le \alpha^n(T(x), (T(T(x))))$$
$$= \alpha^n(T^1(x), T^2(x))$$
$$\le \alpha^{n-1}(x, T(x)).$$

This completes the induction.

It follows from the above and the triangle inequality that if $0 \le n \le n + k$, then

$$d(T^n(x), T^{n+k}(x)) \le \sum_{i=0}^{k-1} d(T^{n+i}(x), T^{n+i+1}(x))$$
$$\le \sum_{i=0}^{k-1} \alpha^{n+i} d(x, T(x)) = d(x, T(x)) \sum_{i=0}^{k-1} \alpha^{n+i}.$$

Since $\sum_{i=0}^{k-1} \alpha^i = (1 - \alpha^k)/(1 - \alpha) < 1/(1 - \alpha)$, $\sum_{i=0}^{k-1} \alpha^{n+i} < \alpha^n/(1 - \alpha)$. Therefore $d(T^n(x), T^{n+k}(x)) < (\alpha^n/(1 - \alpha)) d(x, T(x))$. Hence for each $x \in X$, $\langle T^n(x) \rangle$ is a Cauchy sequence in X. Since (X, d) is complete, $\langle T^n(x) \rangle$ converges to a member x' of X. Recall that $d(T(T^n(x)), T(x')) \le \alpha d(T^n(x), x')$. Therefore

$$d(x', T(x')) = \lim_{n \to \infty} d(T^{n+1}(x), T(x'))$$
$$= \lim_{n \to \infty} d(T(T^n(x)), T(x'))$$
$$\le \lim_{n \to \infty} \alpha d(T^n(x), T(x'))$$
$$= \alpha d(x', x') = 0.$$

Thus $T(x') = x'$, so x' is a fixed point of T. ∎

Corollary 6.3. Let (X, d) be a complete metric space and let $T: X \to X$ be a contraction. Let $\alpha \in \mathbb{R}$ such that $0 < \alpha < 1$ and $d(T(x_1), T(x_2)) \le \alpha d(x_1, x_2)$ for all

$x_1, x_2 \in X$, and let x' be the fixed point of T. Then for all $x \in X$ and $n \geq 0$, $d(T^n(x), x') \leq (\alpha^n/(1-x))d(x, T(x))$.

Proof. In the proof of Theorem 6.2, we obtained the inequality $d(T^n(x), T^{n+k}(x)) < (\alpha^n/(1-\alpha))d(x, T(x))$. Therefore $\lim_{k\to\infty} d(T^n(x), T^{n+k}(x)) \leq (\alpha^n/1-\alpha))d(x, T(x))$, and the proof is complete. ∎

THEOREM 6.4. Let (X, d) be a complete metric space, let (Y, ρ) be a metric space, and let $\alpha \in \mathbb{R}$ such that $0 < \alpha < 1$. Suppose $T: Y \times X \to X$ is a continuous function such that $d(T(y, x_1), T(y, x_2)) \leq \alpha d(x_1, x_2)$ for all $y \in Y$ and $x_1, x_2 \in X$. Then

(a) for each $y \in Y$, there exists a unique member x'_y of X such that $T(y, x'_y) = x'_y$, and

(b) if $y_1, y_2 \in Y$ and x'_{y_1} and x'_{y_2} are the unique members of X given by (a), then $d(x'_{y_1}, x'_{y_2}) \leq (1/(1-\alpha))d(x'_{y_2}, T(y_1, x'_{y_2}))$.

Proof. Since, for each $y \in Y$, the function $T|_{\{y\} \times X}$ is a contraction mapping of X into itself, (a) is an immediate consequence of the Contraction Mapping Theorem (Theorem 6.2).

Let $y_1, y_2 \in Y$. In Corollary 6.3, take $n = 0$, $T = T|_{\{y_1\} \times X}$, $x' = x'_{y_1}$ and $x = x'_{y_2}$. Then, substituting in the formula given by Corollary 6.3, we obtain the formula in (b). ∎

EXERCISES 6.1

1. Let $A = (a_{ij})$ be an $n \times n$ matrix, let $b \in \mathbb{R}^n$, and define $T: \mathbb{R}^n \to \mathbb{R}^n$ by $T(x) = Ax + b$ for each $x \in \mathbb{R}^n$. Prove that if the sum of the squares of the entries of A, $\sum_{i,j} a_{ij}^2$, is less than 1, then T is a contraction of $\mathbb{R}^n$.

2. An **affine function** is a function from a linear space into itself, which is the sum of a linear transformation and a fixed vector. Suppose that M is an $n \times n$ matrix such that $A = (I - \alpha M) = (a_{ij})$ has the property that $\sum_{i,j} a_{ij}^2 < 1$ for some scalar α. Prove that a solution of the linear system $Mx + m = 0$, $m \in \mathbb{R}^n$, in $\mathbb{R}^n$ is the fixed point of some affine function that is a contraction.

3. Let (X, d) be a nonempty compact metric space, and let $T: X \to X$ be a function such that $d(T(x_1), T(x_2)) < d(x_1, x_2)$ for all $x_1, x_2 \in X$ with $x_1 \neq x_2$. Prove that T has a unique fixed point.

4. Let (X, d) be a complete metric space, and let $T: X \to X$ be a function such that T^p is a contraction for some $p \geq 1$. Show that T has a unique fixed point.

6.2 Normed Linear Spaces

Many useful metric spaces are subsets of vector spaces endowed with a metric that arises from a "norm." In Section 1.1, we defined a norm on $\mathbb{R}^n$ and on a subset of $\mathbb{R}^\omega$. The purpose of this section is to define a norm on a vector space and study its properties.

Definition. A **norm** on a linear space V is a nonnegative real-valued function $\|\cdot\|$ on V that satisfies the conditions:

(a) $\|x\| = 0$ if and only if $x = 0$.
(b) $\|\alpha x\| = |\alpha| \|x\|$ for each scalar α and vector x.
(c) $\|x + y\| \leq \|x\| + \|y\|$ for any vectors x and y.

A **normed linear space** is a vector space with a norm. ∎

Condition (c) in the preceding definition is a version of the triangle inequality. Note that $\mathbb{R}^n$ and ℓ^2 with the norms defined in Section 1.1 are normed linear spaces. The notion of a normed linear space is due to many, but notably Banach, Hahn, Eduard Helly (1884–1943), and Norbert Wiener (1894–1964).

EXAMPLE 1. Let X be a nonempty set, and let $B(X) = \{f : X \to \mathbb{R} : f \text{ is bounded}\}$. Then $B(X)$ is a vector space and the function defined by $\|f\| = \sup_{x \in X} |f(x)|$ is a norm on $B(X)$. This norm is called the **sup norm** over X or the **norm of uniform convergence** on X. The space $B(\mathbb{N})$ is of special interest, and it is denoted by ℓ^∞.

The proof of the following theorem is left as Exercise 2.

THEOREM 6.5. If V is a normed linear space, then the function d defined by $d(x, y) = \|x - y\|$ is a metric on V. ∎

Definition. If V is a normed linear space the topology on V induced by the metric d in Theorem 6.5 is called the **norm topology** on V. ∎

Definition. A sequence $\langle x_x \rangle$ in a normed linear space **converges in norm** to x if it converges to x with the norm topology. ∎

The proof of the following theorem is left as Exercise 3.

THEOREM 6.6. The operations of addition and scalar multiplication on a normed linear space are continuous. ∎

Recall that a **linear transformation**, or **linear operator** from a linear space V into a linear space W is a function $T: V \to W$ such that:

(a) $T(\alpha x) = \alpha T(x)$ for each scalar α and each vector x.
(b) $T(x + y) = T(x) + T(y)$ for any vectors x and y.

If W is either $\mathbb{R}$ or the set of complex numbers, then the linear transformation $T: V \to W$ is called a **linear functional.**

Definition. The **closed unit ball** of a normed linear space V is $\{x \in V: \|x\| \leq 1\}$. A **bounded linear operator** T from a normed linear space V into a normed linear space W is a linear operator from V into W such that $\sup_{x \in V}\{\|T(x)\|: \|x\| \leq 1\}$ is a real number. ∎

THEOREM 6.7. If a bounded linear operator $T: V \to W$ is bounded on all of V, then $T(x) = 0$ for all $x \in V$.

Proof. Suppose there exists $x \in V$ such that $T(x) \neq 0$. Since $\|T(\alpha x)\| = \|\alpha T(x)\| = |\alpha| \|T(x)\|$, for each real number a, we can choose α large enough so that $|\alpha| \|T(x)\| > a$. Therefore T is not bounded on all of V. ∎

Definition. If $T: V \to W$ is a bounded linear operator we define $\|T\|$ to be $\sup_{x \in V} \{\|T(x)\|: \|x\| \leq 1\}$. Then $\|\cdot\|$ is called the **norm** of the set of bounded linear operators. ∎

The proof of the following theorem is left as Exercise 4.

THEOREM 6.8. If V and W are normed linear spaces, $T: V \to W$ is a bounded linear operator, and $x \in V$, then $\|T(x)\| \leq \|T\| \|x\|$. ∎

THEOREM 6.9. Let V and W be normed linear spaces and let $\mathcal{B}(V, W) = \{T: V \to W: T \text{ is a bounded linear operator}\}$. Then, with respect to the norm defined above, $\mathcal{B}(V, W)$ is a normed linear space.

Proof. Let $T, T' \in \mathcal{B}(V, W)$, and let α and β be scalars.

Since $(\alpha T + \alpha T')(x) = \alpha T(x) + \alpha T'(x) = \alpha(T(x) + T'(x))$, $\alpha T + \alpha T' = \alpha(T + T')$.

Since $(\alpha T + \beta T)(x) = \alpha T(x) + \beta T(x) = (\alpha + \beta)T(x)$, $\alpha T + \beta T = (\alpha + \beta)T$.

Since $(\alpha\beta)T(x) = T((\alpha\beta)x) = T(\alpha(\beta x)) = \alpha T(\beta x) = \alpha(\beta T(x))$, $(\alpha\beta)T = \alpha(\beta T)$.

It is clear that $1 \cdot T = T$, so $\mathcal{B}(V, W)$ is a linear space.

For $T \in \mathcal{B}(V, W)$ and $x \in V$, $\|T(x)\| \geq 0$. Therefore $\|T\| \geq 0$. By Theorem 6.8, $\|T\| = 0$ if and only if $T = 0$.

Let $T \in \mathcal{B}(V, W)$ and let α be a scalar. Then $\|(\alpha T)(x)\| = \|\alpha T(x)\| = |\alpha| \|T(x)\|$. By taking the supremum over x in the unit ball of V, we see that αT is bounded and $\|\alpha T\| = |\alpha| \|T\|$.

Let $T, T' \in \mathcal{B}(V, W)$. Then $\|(T + T')(x)\| = \|T(x) + T'(x)\| \leq \|T(x)\| + \|T'(x)\|$. So by Theorem 6.8, $\|(T + T')(x)\| \leq \|T\| \|x\| + \|T'\| \|x\|$.

Therefore $\mathcal{B}(V, W)$ is a normed linear space. ∎

Definition. The norm topology on $\mathcal{B}(V, W)$ is called the **uniform operator topology.** We denote $\mathcal{B}(V, V)$ by $\mathcal{B}(V)$. If W is either $\mathbb{R}$ or the set of complex numbers, then the set of bounded linear functionals on V is called the **conjugate space** of V and is denoted by V^*. ∎

We now show that boundedness and continuity are the same for linear operators.

THEOREM 6.10. Let V and W be normed linear spaces and let $T: V \to W$ be a linear operator. Then the following are equivalent:

(a) T is continuous.

(b) T is continuous at 0.

(c) T is bounded.

Proof. (a) → (b). By definition, if T is continuous, then it is continuous at 0.

(b) → (c). Suppose T is continuous at 0. Then there is a positive number δ such that if $\|x\| \leq \delta$, then $\|T(x)\| \leq 1$. Suppose x is a member of the closed unit ball of V. Then $\|x\| \leq 1$, so $\|\delta x\| \leq \delta$. Therefore $\|T(\delta x)\| \leq 1$, so $\|T(x)\| \leq 1/\delta$. Thus T is bounded.

(c) → (a). Suppose T is bounded. Let $x_0 \in V$, and let $\varepsilon > 0$. Since T is bounded there is a real number M such that $\|T\| = M$. Let $x \in V$ such that $\|x - x_0\| < \varepsilon/M$. By Theorem 6.8, $\|T(x - x_0)\| \leq \|T\| \|x - x_0\| < M \cdot \varepsilon/M = \varepsilon$. Since $T(x) - T(x_0) = T(x - x_0)$, T is continuous at x_0. But x_0 is an arbitrary member of X, so T is continuous. ∎

Definition. A **Banach space** is a normed linear space that is complete with respect to the metric defined by the norm. ∎

Banach spaces were introduced in 1923 by Banach.

By Exercise 4 of Section 1.6, $\mathbb{R}^n$ is a Banach space. In Exercise 5 you are asked to prove that ℓ^2 (see Example 6 in Section 1.1) is a Banach space. In Exercise 6 you are asked to show that the set of bounded real-valued functions on a nonempty set X with sup norm is a Banach space. In Exercise 7 you are asked to show that $C([0, 1], \mathbb{R})$, the set of real-valued, continuous functions on I, with the norm defined by $\|f\| = (\int_0^1 |f(x)|^2 \, dx)^{1/2}$ is not a Banach space.

Definition. Let $\langle a_n \rangle$ be a bounded sequence of real numbers, and let $A = \{x \in \mathbb{R}: x$ is a cluster point of $\langle a_n \rangle\}$. Then **limit superior** and **limit inferior** are defined by $\limsup \langle a_n \rangle = \sup A$ and $\liminf \langle a_n \rangle = \inf A$. ∎

Note that $A = \{x \in \mathbb{R}: x$ is the limit point of a subsequence of $\langle a_n \rangle\}$, $\limsup \langle a_n \rangle$ is the maximum cluster point of $\langle a_n \rangle$, and $\liminf \langle a_n \rangle$ is the minimum cluster point of $\langle a_n \rangle$.

THEOREM 6.11. Let $\langle T_n \rangle$ be a sequence of bounded linear operators from a normed linear space V into a Banach space W such that:

(1) $\sup_{n \in \mathbb{N}} \|T_n\| < \infty$, and

(2) there is a dense subset D of V such that if $x \in D$ then $\lim_{n \to \infty} T_n(x)$ exists.

Then:

(a) there exists a bounded linear operator $T: V \to W$ such that for each $x \in V$, $T(x) = \lim_{n \to \infty} T_n(x)$, and

(b) $\|T\| \leq \liminf \|T_n\|$.

Proof. **(a)** For each $x \in V$ such that $\lim_{n \to \infty} T_n(x)$ exists, let $T(x) = \lim_{n \to \infty} T_n(x)$. We show that $\lim_{n \to \infty} T_n(x)$ exists for all $x \in V$.

Let $z \in V$, let $M = \sup \|T_n\|$, and let $\varepsilon > 0$. Since $\lim_{n \to \infty} T_n(x)$ exists for each x in the dense set D, there is an $x \in D$ such that $\|x - z\| < \varepsilon$ and $\lim_{n \to \infty} T_n(x)$ exists. Since $\lim_{n \to \infty} T_n(x)$ exists, there exists $N \in \mathbb{N}$ such that $\|T_n(x) - T_m(x)\| < \varepsilon$ for all $m, n \geq N$. Thus for $m, n \geq N$, $\|T_m(z) - T_n(z)\| \leq \|T_m(z - x)\| + \|(T_m - T_n)(x)\| + \|T_n(x - z)\| \leq M\varepsilon + \varepsilon + M\varepsilon$. Therefore $\langle T_n(z) \rangle$ is a Cauchy sequence in W. Since W is complete, $\lim_{n \to \infty} T_n(z)$ exists and therefore T is defined for all $x \in V$.

Since the operations of addition and scalar multiplication are continuous on V, T is linear.

Suppose x is in the closed unit ball of V. Then $\|x\| \leq 1$, so $\|T(x)\| = \lim_{n \to \infty} \|T_n(x)\| \leq \limsup \|T_n\|$. Therefore T is bounded, and $\|T\| \leq \limsup \|T_n\|$.

The proof of **(b)** is left as Exercise 10. ∎

THEOREM 6.12. *Let V be a normed linear space and let W be a Banach space. The set $\mathcal{B}(V, W)$ is a Banach space.*

Proof. By Theorem 6.9, $\mathcal{B}(V, W)$ is a normed linear space. Thus in order to prove that $\mathcal{B}(V, W)$ is a Banach space, it is sufficient to show that every Cauchy sequence converges. Let $\langle T_n \rangle$ be a Cauchy sequence in $\mathcal{B}(V, W)$. We first show that for each $x \in V$, $\langle T_n(x) \rangle$ is a Cauchy sequence in W. Let $x \in V$ and let $\varepsilon > 0$. Since $\langle T_n \rangle$ is a Cauchy sequence, there exists $N \in \mathbb{N}$ such that if $m, n \geq N$, $\|T_m - T_n\| < \varepsilon/\|x\|$. Thus, by Theorem 6.8,

$$\|T_m(x) - T_n(x)\| = \|(T_m - T_n)(x)\| \leq \|T_m - T_n\| \|x\| \leq (\varepsilon/\|x\|) \cdot \|x\| = \varepsilon.$$

Hence $\langle T_n(x) \rangle$ is a Cauchy sequence. Since W is a Banach space, $\langle T_n(x) \rangle$ converges. Moreover, $\sup \|T_n\| < \infty$, so by Theorem 6.11, there exists $T \in \mathcal{B}(V, W)$ such that for each $x \in V$, $T(x) = \lim_{n \to \infty} T_n(x)$. Also, for each $n \in \mathbb{N}$, $\langle (T_m - T_n)(x) \rangle$ converges to $(T - T_n)(x)$. So, by Theorem 6.11, for each $n \in \mathbb{N}$, $\|T - T_n\| \leq \liminf \|T_m - T_n\|$. Therefore, since $\langle T_n \rangle$ is a Cauchy sequence, $\lim_{n \to \infty} \|T - T_n\| = 0$. Hence $\langle T_n \rangle$ converges to T. ∎

The Principle of Uniform Boundedness (Theorem 6.13) is one of the fundamental theorems of functional analysis.

THEOREM 6.13. *Let V and W be Banach spaces, and let $\{T_\alpha : V \to W : \alpha \in \Lambda\}$ be a collection of bounded linear operators. If, for each $x \in V$, $\sup \|T_\alpha(x)\| < \infty$, then $\sup \|T_\alpha\| < \infty$.*

Proof. For each $n \in \mathbb{N}$, let $C_n = \{x \in V:$ for each $\alpha \in \Lambda$, $\sup \|T_\alpha(x)\| \leq n\}$. Then, for each $n \in \mathbb{N}$, C_n is a closed set and $C_n \subseteq C_{n+1}$ (see Exercise 11(a)). Moreover, $V = \bigcup_{n \in \mathbb{N}} C_n$ (see Exercise 11(b)). In Exercise 12, you are asked to show that there exists $p \in \mathbb{N}$ such that C_p has an interior point x_0. Let $\varepsilon > 0$ such that the open ball with center at x_0 and radius ε is a subset of C_p. Let $M = \sup\{\|T_\alpha(x_0)\|: \alpha \in \Lambda\}$, and let x be a member of the closed unit ball of V. Then $\|x\| \leq 1$, so $\|(x_0 + \varepsilon x) - x_0\| < \varepsilon$. Therefore $x_0 + \varepsilon x \in C_p$. Thus, for each $\alpha \in \Lambda$, $\varepsilon \|T_\alpha(x)\| = \|T_\alpha(x_0 + \varepsilon x) - T_\alpha(x_0)\| \leq p + M$. Hence for each $\alpha \in \Lambda$, $\|T_\alpha\| \leq (p + M)/\varepsilon$, and the proof is complete. ∎

THEOREM 6.14. Let $U, V,$ and W be normed linear spaces, let $T \in \mathscr{B}(U, V)$ and $T' \in \mathscr{B}(V, W)$. Then $T' \circ T$ is bounded and $\|T' \circ T\| \leq \|T'\| \|T\|$.

Proof. See Exercise 13. ∎

EXERCISES 6.2

1. Recall that $C(I, \mathbb{R})$ is the set of real-valued, continuous functions on I. (This set of functions is bounded). Show that each of the following defines a norm on $C(I, \mathbb{R})$.

    (a) $\|f\| = \sup_{x \in I} |f(x)|$
    (b) $\|f\| = \int_0^1 |f(x)| dx$
    (c) $\|f\| = (\int_0^1 |f(x)|^2 dx)^{1/2}$

2. Prove Theorem 6.5.

3. Prove Theorem 6.6.

4. Prove Theorem 6.8.

5. Prove that ℓ^2 is a Banach space.

6. Prove that if X is a nonempty set, then the set of bounded, real-valued functions on X with sub norm is a Banach space.

7. Prove that $C(I, \mathbb{R})$ is not complete with respect to the norm defined in Exercise 1(c).

8. Prove that if V is a finite dimensional normed linear space and T is a linear functional on V, then T is bounded.

9. Prove that if V is a finite dimensional normed linear space, W is a normed linear space, and T is a linear operator form V into W, then T is bounded. *Hint:* Use Exercise 8.

10. Prove part (b) of Theorem 6.11.

11. Let $\{C_n: n \in \mathbb{N}\}$ be the collection of sets defined in the proof of Theorem 6.13.
    (a) Show that, for each $n \in \mathbb{N}$, C_n is a closed set and $C_n \subseteq C_{n+1}$.
    (b) Show that $V = \bigcup_{n \in \mathbb{N}} C_n$.
12. Let $\{C_n: n \in \mathbb{N}\}$ be the collection of sets defined in the proof of Theorem 6.13. Show that there exists $p \in \mathbb{N}$ such that C_p has an interior point. *Hint:* Use the Baire Category Theorem.
13. Prove Theorem 6.14.

6.3 The Fréchet Derivative

In this section we define and study some of the properties of the Fréchet derivative, culminating with the implicit and inverse functions theorems. The results are the basis for the study of differential topology, Morse theory, differential equations, partial differential equations, and analysis. The fundamental concepts of abstract differentials were introduced by Fréchet around 1909. The ideas of Fréchet were applied to concrete problems by R. Gâteaux, whose brilliant work was published in 1919 and 1922.

Recall that if V and W are normed linear spaces, then $\mathscr{B}(V, W)$ has sup norm: $\|T\| = \sup_{\|x\| \leq 1} \|T(x)\|$ for each $T \in \mathscr{B}(V, W)$.

Definition. Let V and W be Banach spaces, let M be an open subset of V, and let $F: M \to W$ be a function. Then F is **differentiable** at $x_0 \in M$ if there is a bounded linear operator $T: V \to W$ such that

$$\lim_{x \to x_0} \frac{F(x) - F(x_0) - T(x - x_0)}{\|x - x_0\|} = 0. \tag{6.1}$$

The linear operator T is the **Fréchet derivative** of F at x_0, and it is denoted by $DF(x_0)$. The function F is **differentiable** if it is differentiable at each point of M. The function F is **continuously differentiable** if it is differentiable and the derivative $DF: M \to \mathscr{B}(V, W)$ is continuous; that is, for each $x_0 \in M$ and $\varepsilon > 0$, there exists a $\delta > 0$ such that if $\|x - x_0\| < \delta$, then $\|DF(x)(t) - DF(x_0)(t)\| < \varepsilon$ for all $t \in V$ with $\|t\| \leq 1$. ∎

We first list some basic properties of the Fréchet derivative. Formula (6.1) can be stated in the form $F(x) = F(x_0) + DF(x_0)(x - x_0) + R(x)$, where $\lim_{x \to x_0} R(x)/\|x - x_0\| = 0$.

The derivative of F is unique because if T and T' are bounded linear operators that satisfy the formula (6.1), the $\lim_{x \to x_0}(T - T')(x - x_0)/\|x - x_0\| = 0$. If we let $tz = x - x_0$, then $z = (x - x_0)/t$, so $\lim_{t \to 0}(T - T')(tz)/\|tz\| = 0$. Therefore $(T - T')(z) = 0$, and hence $T = T'$.

The following three theorems are immediate consequences of the definition of the Fréchet derivative, and we leave their proofs as exercises.

THEOREM 6.15. Let V and W be Banach spaces, let M be an open subset of V, and let $F: M \to W$ be a function. If F is differentiable at $x_0 \in M$, then F is continuous at x_0. ∎

THEOREM 6.16. Let V and W be Banach spaces, let M be an open subset of V, let $F, G: M \to W$ be functions, and let α be a scalar. If F and G are differentiable at x_0, then $F + G$ and αF are differentiable at x_0, $D(F + G)(x_0) = DF(x_0) + DG(x_0)$, and $D(\alpha F)(x_0) = \alpha DF(x_0)$. ∎

THEOREM 6.17. Let V and W be Banach spaces, let $w_0 \in W$, and let $T \in \mathcal{B}(V, W)$. If $F: V \to W$ is defined by $F(x) = T(x) + w_0$ for each $x \in V$, then, for each $x_0 \in V$, F is differentiable at x_0 and $DF(x_0) = T$. ∎

Definition. A bounded linear operator T from a Banach space V into a Banach space W is **invertible** if there is a bounded linear operator T' from W into V such that $T' \circ T$ is the identity of V and $T \circ T'$ is the identity on W. ∎

We leave as Exercise 4 the proof that if $T: V \to W$ is invertible and T' and T'_1 are bounded linear operators from W into V such that $T' \circ T$ and $T'_1 \circ T$ are the identity on V and $T \circ T'$ and $T \circ T'_1$ are the identity of W, then $T' = T'_1$. Therefore T' is unique, and we denote it by T^{-1}.

THEOREM 6.18. A bounded linear operator T from a Banach space V into a Banach space W is invertible if and only if it is one-to-one, maps V onto W, and has the property that there is a positive number α such that $\|T(x)\| \geq \alpha \|x\|$ for each $x \in V$.

Proof. See Exercise 5. ∎

THEOREM 6.19. Let T be an invertible bounded linear operator from a Banach space V into a Banach space W and let $\alpha > 0$ such that $\|T(x)\| \geq \alpha \|x\|$ for each $x \in V$. Then $\|T^{-1}\| \leq 1/\alpha$.

Proof. See Exercise 6. ∎

THEOREM 6.20. Let V and W be Banach spaces, let $M = \{T \in \mathcal{B}(V, W): T$ is invertible$\}$, and define $F: M \to \mathcal{B}(W, V)$ by $F(T) = T^{-1}$ for each $T \in M$. Then M is an open subset of $\mathcal{B}(V, W)$, F is differentiable on M, and the derivative of F at T is given by $DF(T)(S) = -T^{-1}ST^{-1}$ for each $S \in \mathcal{B}(V, W)$.

Proof. We first show that M is an open subset of $\mathcal{B}(V, W)$. Suppose $0 < \alpha < 1$, and let $T \in M$. Let $S \in \mathcal{B}(V, W)$ such that $\|S\| \leq \alpha/\|T^{-1}\|$. Then, by Theorem 6.14, $\|ST^{-1}\| \leq \alpha$, so, for each $n \geq 0$, $\|(ST^{-1})^n\| \leq \alpha^n$. Thus the partial sums of the series $A = \sum_{n=0}^{\infty}(-1)^n(ST^{-1})^n$ form a Cauchy sequence in $\mathcal{B}(W)$. By Theorem

6.12, $\mathcal{B}(W)$ is a Banach space, so the series converges to a member of $\mathcal{B}(W)$; that is, $A \in \mathcal{B}(W)$. Thus $T^{-1}A \in \mathcal{B}(W, V)$.

We now prove that $T^{-1}A$ is an inverse for $T + S$. By multiplying the partial sums for $T^{-1}A$ by $T + S$, we obtain $(T + S)T^{-1}\sum_{n=0}^{N}(-1)^n(ST^{-1})^n = I + (-1)^N(ST^{-1})^{N+1}$, where I is the identity on W. By taking the limit as N approaches ∞, we conclude that $(T + S)T^{-1}A$ is the identity on W. By a similar proof, we can show that $T^{-1}A(T + S)$ is the identity on W. Therefore $T + S$ is invertible and $(T + S)^{-1} = F^{-1}\sum_{n=0}^{\infty}(-1)^n(ST^{-1})^n$. In Exercise 7, you are asked to complete the proof that there is an open ball with center at T that is contained in M.

Notice that F is differentiable at T_0 if $\lim_{T \to T_0}(F(T) - F(T_0) - DF(T_0)(T - T_0))/\|T - T_0\| = 0$. Think of S as representing $T - T_0$. Then F is differentiable at T if $\lim_{S \to 0}(G(T + S) - G(T) - DF(T)(S))/\|S\| = 0$ or $\lim_{S \to 0}((T + S)^{-1} - T^{-1} + T^{-1}ST^{-1})/\|S\| = 0$. Let $G(S) = (T + S)^{-1} - T^{-1} + T^{-1}ST^{-1}$. Then F is differentiable at T_0 if $\lim_{S \to 0} G(S)/\|S\| = 0$. Now

$$G(S) = (T + S)^{-1} - T^{-1} + T^{-1}ST^{-1} = T^{-1} + \sum_{n=0}^{\infty}(-1)^n(ST^{-1})^n - T^{-1} + T^{-1}ST^{-1}$$

$$= T^{-1}(I - ST^{-1} + (ST^{-1})^2 - \cdots) - T^{-1} + T^{-1}ST^{-1}$$

$$= T^{-1}((ST^{-1})^2 - (ST^{-1})^3 + \cdots)$$

$$= T^{-1}((ST^{-1})^2(I - ST^{-1} + (ST^{-1})^2 - \cdots)).$$

$$= T^{-1}(ST^{-1})^2 \sum_{n=0}^{\infty}(-1)^n(ST^{-1})^n.$$

Therefore $\|G(S)\| \le \|S\|^2 \|T^{-1}\|^3 \sum_{n=0}^{\infty}\alpha^n$, and hence $\lim_{S \to 0} G(S)/\|S\| = 0$.

In order to prove the Implicit Function Theorem, we first establish the following two theorems. A definition is needed. ∎

Definition. Let $h: [0, 2) \to \mathbb{R}$ be a continuous function, let $x \in [0, 1)$, and define $f_x: (0, 1] \to \mathbb{R}$ by $f_x(\delta) = (h(x + \delta) - h(x))/\delta$. Then $\lim \sup_{\delta \to 0^+} f_x(\delta) = \inf \{\sup \{f_x(\delta): 0 < \delta \le 1/n\}: n \in \mathbb{N}\}$. ∎

THEOREM 6.21. Suppose $0 < \alpha < 1$. Let $h: [0, 2) \to \mathbb{R}$ be a continuous function such that $h(0) = 0$ and $\lim \sup_{\delta \to 0^+}(h(x + \delta) - h(x))/\delta \le \alpha$ for all $x \in [0, 1)$. Then $h(x) \le \alpha x$ for all $x \in [0, 1]$.

Proof. For each $x \in [0, 1)$, define $f_x: (0, 1] \to \mathbb{R}$ by $f_x(\delta) = (h(x + \delta) - h(x))/\delta$. Let $\varepsilon > 0$. Since $\lim \sup_{\delta \to 0^+} f_x(\delta) \le \alpha$, there exists $n \in \mathbb{N}$ such that $\sup \{f_x(\delta): 0 < \delta \le 1/n\} < \alpha + \varepsilon/2$. So $(h(x + \delta) - h(x))/\delta < \alpha + \varepsilon - \varepsilon/2$ for all δ such that $0 < \delta \le 1/n$. Therefore $h(x + \delta) - h(x) < (\alpha + \varepsilon)\delta - \varepsilon\delta/2$ for all δ such that $0 < \delta \leq 1/n$. Therefore $h(x + \delta) - (\alpha + \varepsilon)\delta < h(x) - \varepsilon\delta/2$ for all δ such that $0 < \delta \leq 1/n$. So $h(x + \delta) - (\alpha + \varepsilon)(x + \delta) < h(x) - (\alpha + \varepsilon)x - \varepsilon\delta/2$ for all δ such that $0 < \delta \leq 1/n$. Define $g: [0, 1] \to \mathbb{R}$ by $g(x) = h(x) - (\alpha + \varepsilon)x$. Then $g(x + \delta) < g(x) - \varepsilon\delta/2$ for all δ such that $0 < \delta \leq 1/n$. So $(g(x + \delta) - g(x))/\delta < -\varepsilon/2$ for all δ such that $0 < \delta \leq 1/n$. Thus $\sup\{(g(x + \delta) - g(x))/\delta : 0 < \delta \leq 1/n\} \leq -\varepsilon/2 < 0$. Therefore

$$\limsup_{\delta \to 0^+} \frac{g(x + \delta) - g(x)}{\delta} < 0. \tag{6.2}$$

Suppose $s \in [0, 1]$, and let $x \in [0, s]$, such that g attains its minimum value over the interval $[0, s]$ at x. If $x < s$, then, by formula (6.2), there exists $\delta > 0$ such that $x + \delta < s$ and $g(x + \delta) - g(x) < 0$. This contradicts the choice of x, and thus the minimum value of g over the interval $[0, s]$ is attained at s. In particular, $g(s) \leq g(0) = 0$. Therefore $h(s) \leq (\alpha + \varepsilon)s$, and hence $h(s) \leq \alpha s$ for all $s \in [0, 1]$. ∎

THEOREM 6.22. Let V and W be Banach spaces, let M be an open ball in V, and let $F: M \to W$ be a function that is differentiable at each point of M. Let $P > 0$ and let $T \in \mathcal{B}(V, W)$ such that $\|(DF)(y) - T\| \leq P$ for all y in M. Then $\|F(y_2) - F(y_1) - T(y_2 - y_1)\| \leq P\|y_2 - y_1\|$ for all $y_1, y_2 \in M$.

Proof. Let $G = F - T$. Since T is a linear operator, for each $y_0 \in M$,

$$\lim_{y \to y_0} \frac{G(y) - G(y_0) - ((DF)(y_0) - T)(y - y_0)}{\|y - y_0\|}$$
$$= \lim_{y \to y_0} \frac{F(y) - T(y) - F(y_0) + T(y_0) - (DF)(y_0)(y - y_0) + T(y - y_0)}{\|y - y_0\|}$$
$$= \lim_{y \to y_0} \frac{F(y) - F(y_0) - (DF)(y_0)(y - y_0) - T(y - y_0) - T(y - y_0)}{\|y - y_0\|}$$
$$= 0.$$

Therefore $(DG)(y_0) = (DF)(y_0) - T$. Since y_0 is an arbitrary point of M, $(DG)(y) = (DF)(y) - T$ for all $y \in M$. Thus $\|(DG)(y)\| \leq P$ for all $y \in M$.

We complete the proof by showing that $\|G(y_2) - G(y_1)\| \leq P\|y_2 - y_1\|$ for all $y_1, y_2 \in M$. Let $y_1, y_2 \in M$. We may assume, without loss of generality, that $y_1 = 0$. Define $h: [0, 1] \to \mathbb{R}$ by $h(x) = \|G(xy_2) - G(0)\|$ for each $x \in [0, 1]$. Then for each $x \in [0, 1)$ and positive number δ such that $x + \delta \in [0, 1]$,

$$h(x + \delta) - h(x) \leq \|G((x + \delta)y_2) - G(xy_2)\|$$
$$\leq \|G((x + \delta)y_2) - G(xy_2) - \delta(DG)(xy_2)(y_2)\|$$
$$+ \delta\|(DG)(xy_2)\|\|y_2\|.$$

So

$$\frac{h(x+\delta) - h(x)}{\delta} \leq \frac{\|G((x+\delta)y_2) - G(xy_2) - \delta(DG)(xy_2)(y_2)\|}{\delta}$$
$$+ \|(DG)(xy_2)\|\|y_2\|.$$

Thus,
$$\lim_{\delta \to 0^+} \sup \frac{h(x+\delta) - h(x)}{\delta} \leq \|(DG)(xy_2)\|\|y_2\| \leq P\|y_2\|.$$

Therefore, by Theorem 6.21, $h(x) \leq P\|y_2\|x$ for all $x \in I$. In particular, for $x = 1$, we have $\|G(y_2) - G(0)\| \leq P\|y_2\|$, and so the proof is complete. ∎

THEOREM 6.23 (The Implicit Function Theorem). Let V and W be Banach spaces, let (X, d) be a metric space, and let $(x_0, y_0) \in X \times V$. Suppose $M \times N$ is a neighborhood of (x_0, y_0) such that:

(1) There is a continuous function $F: M \times N \to W$ such that $F(x_0, y_0) = 0$.

(2) For each $x \in M$, the function $f_x: N \to W$ defined by $f_x(y) = F(x, y)$ is differentiable with derivative $G: M \times N \to \mathcal{B}(V, W)$ that is continuous in x and y.

(3) $G(x_0, y_0)$ is an invertible operator from V to W.

Then:

(a) There exist positive numbers r and ρ such that $B(x_0, r) \subseteq M$ and for each $x \in B(x_0, r)$, there exists a unique $y_x \in B(y_0, \rho)$ such that $F(x, y_x) = 0$.

(b) The function $f: B(x_0, r) \to B(y_0, \rho)$ defined by $f(x) = y_x$ is continuous.

(c) Now suppose X is a Banach space (that is, suppose there is a norm $\|\cdot\|$ on X such that $d(x_1, x_2) = \|x_1 - x_2\|$ for all $x_1, x_2 \in X$ and (X, d) is complete) and the function $g: B(x_0, r) \to W$ defined by $g(x) = F(x, y_0)$ is differentiable at x_0 with derivative $H(x_0, y_0)$. Then f is differentiable at x_0 and $(DF)(x_0) = -G(x_0, y_0)^{-1}H(x_0, y_0)$.

Proof. Let $T = G(x_0, y_0)$. Let r and ρ be positive numbers such that $B(x_0, r) \subseteq M$, $B(y_0, \rho) \subseteq N$, and if $(x, y) \in B(x_0, r) \times B(y_0, \rho)$, then:

$$\|G(x, y) - T\| < 1/(2\|T^{-1}\|) \tag{6.3}$$
$$\text{and } \|F(x, y_0)\| < \rho/(2\|T^{-1}\|). \tag{6.4}$$

For each $x \in B(x_0, r)$, define $S_x: B(y_0, \rho) \to V$ by $S_x(y) = y - T^{-1}F(x, y)$. Then $F(x, y) = 0$ if and only if $S_x(y) = y$ (that is, y is a fixed point of S_x).

We now prove that for each $x \in B(x_0, r)$, S_x is a contraction. Let $x \in B(x_0, r)$ and let $y_1, y_2 \in B(y_0, \rho)$. Then, by Formula (6.3) and Theorem 6.22, $\|F(x, y_2) - F(x, y_1) - T(y_2 - y_1)\| \leq \|y_2 - y_1\|/(2\|T^{-1}\|)$. Thus $S_x(y_2) - S_x(y_1) = -T^{-1}(F(x, y_2) - F(x, y_1) - T(y_2 - y_1))$, so $\|S_x(y_2) - S_x(y_1)\| \leq \frac{1}{2}\|y_2 - y_1\|$. Hence S_x is a contraction.

It remains to show that $S_x: B(y_0, \rho) \to B(y_0, \rho)$. Let $y \in B(y_0, \rho)$. Then

$$\|S_x(y) - y_0\| \le \|S_x(y) - S_x(y_0)\| + \|S_x(y_0) - y_0\|$$
$$\le \|y - y_0\|/2 + \|T^{-1}F(x, y_0)\|$$
$$\le \rho/2 + \rho/2 = \rho.$$

(Note: By Formula (6.4), $\|T^{-1}F(x, y_0)\| \le \rho/2$). Therefore $S_x(y) \in B(y_0, \rho)$ and $S_x: B(y_0, \rho) \to B(y_0, \rho)$.

By the Contraction Mapping Theorem (Theorem 6.2), $S_x: B(y_0, \rho) \to B(y_0, \rho)$ has a unique fixed point, y_x. So $F(x, y_x) = T(y_x - S_x(y_x)) = T(0) = 0$, and y_x is the only point with this property. This establishes part (**a**).

Part (**b**) follows immediately from part (**b**) of Theorem 6.4. Note that from part (**b**) of Theorem 6.4 with $\alpha = \frac{1}{2}$, for each $x \in B(x_0, r)$,

$$\|f_x - y_0\| \le 2\|T^{-1}\|\|F(x, y_0)\|. \tag{6.5}$$

Now suppose X is a Banach space and the function $g: B(x_0, r) \to W$ defined by $g(x) = F(x, y_0)$ is differentiable at x_0 with derivative $H(x_0, y_0)$. Let $\varepsilon > 0$. Since $G: M \times N \to \mathscr{B}(V, W)$ is continuous, there exists $\delta > 0$ such that $\delta < r$ and if $d((x, y), (x_0, y_0)) < \delta$ then $\|G(x, y) - G(x_0, y_0)\| < \varepsilon$. Then, by Formula (6.3) and Theorem 6.22, we have $\|F(x, y) - F(x, y_0) - T(y - y_0)\| \le \varepsilon\|y - y_0\|$. Let $x \in B(x_0, r)$ such that $f(x) = y$. Then $\|F(x, f(x)) - F(x, y_0) - T(f(x) - y_0)\| \le \varepsilon\|f(x) - y_0\|$, so $\|F(x, y_0) + T(f(x) - y_0)\| \le \varepsilon\|f(x) - y_0\|$. Using Formula (6.5), we obtain

$$\|f(x) - y_0 + T^{-1}F(x, y_0)\| \le \|T^{-1}\|\|F(x, y_0) + T(f(x) - y_0)\|$$
$$\le 2\varepsilon\|T^{-1}\|^2\|F(x, y_0)\|.$$

Now $F(x, y_0) = H(x_0, y_0)(x - x_0) + R(x)$, where $\lim_{x \to x_0} \|R(x)\|/\|x - x_0\| = 0$, so

$$\|f(x) - y_0 + T^{-1}H(x_0, y_0)(x - x_0)\| \le 2\varepsilon\|T^{-1}\|^2\|H(x_0, y_0)\|\|x - x_0\|$$
$$+ 2\varepsilon\|T^{-1}\|^2\|R(x)\| + \|T^{-1}\|\|R(x)\|.$$

Therefore

$$\limsup_{x \to x_0} \frac{\|f(x) - y_0 + T^{-1}H(x_0, y_0)(x - x_0)\|}{\|x - x_0\|} \le 2\varepsilon\|T^{-1}\|^2\|H(x_0, y_0)\|.$$

Hence

$$\limsup_{x \to x_0} \frac{\|f(x) - y_0 + T^{-1}H(x_0, y_0)(x - x_0)\|}{\|x - x_0\|} = 0.$$

Therefore f is differentiable at x_0 and $(DF)(x_0) = -T^{-1}H(x_0, y_0)$. This establishes part (c). ∎

THEOREM 6.24 (The Inverse Function Theorem). Let V and W be Banach spaces, and let $x_0 \in V$. Suppose there is a neighborhood M of x_0 and a continuously differentiable function $F: M \to W$ such that $(DF)(x_0)$ is invertible. Then:

(a) There exists a neighborhood N of x_0 such that $F(N)$ is open in W and $F|_N$ is a one-to-one function from N onto $F(N)$.

(b) The inverse function of $F|_N$ is continuously differentiable.

Proof. Let $y_0 = F(x_0)$. By Theorem 6.20, $\{T \in \mathcal{B}(V, W): T \text{ is invertible}\}$ is open in $\mathcal{B}(V, W)$, so we may assume that $(DF)(x)$ is invertible for each $x \in M$. Define $G: M \times W \to W$ by $G(x, y) = F(x) - y$ for each $(x, y) \in M \times W$. The partial derivatives, G_1 and G_2, of G are given by $G_1(x, y) = (DF)(x)$ and $G_2(x, y) = -I$, where $I: W \to W$ is the identity. By part (**a**) of the Implicit Function Theorem, there exist positive numbers r and ρ such that for each $y \in B(y_0, r)$, there exists a unique $x_y \in B(x_0, \rho)$ such that $G(x_y, y) = 0$. Thus $F(x_y) = y$. Define $f: B(y_0, r) \to B(x_0, \rho)$ by $f(y) = x_y$. Let $N = F^{-1}(B(y_0, r)) \cap B(x_0, \rho)$. Then $F|_N$ is a one-to-one function from N onto $F(N) = B(y_0, r)$,, and part (**a**) is established.

Since f is the inverse of $F|_N$, part (**b**) follows from part (**c**) of Theorem 6.23 (see Exercise 8). ∎

EXERCISES 6.3

1. Prove Theorem 6.15.

2. Prove Theorem 6.16.

3. Prove Theorem 6.17.

4. Let T be an invertible, bounded linear operator from a Banach space V into a Banach space W, and let T' and T_1' be bounded linear operators from W into V such that $T' \circ T$ and $T_1' \circ T$ are the identity on V and $T \circ T'$ and $T \circ T_1'$ are the identity on W. Prove that $T' = T_1'$.

5. Prove Theorem 6.18.

6. Prove Theorem 6.19.

7. In the proof of Theorem 6.20, complete the proof that there is an open ball with center at T that is contained in M.

8. Complete the proof of part (**b**) of Theorem 6.24.

9. Let V and W be Banach spaces, and let $M = \{T \in \mathcal{B}(V, W): T \text{ is invertible}\}$. Define $F: M \to \mathcal{B}(W, V)$ by $F(T) = T^{-1}$. Prove that DF is continuous.

10. Let $n \in \mathbb{N}$, and for each $i = 1, 2, ..., n$ let $F_i: \mathbb{R} \to \mathbb{R}$ be a continuous function. Define $F: \mathbb{R}^n \to \mathbb{R}^n$ by $F((x_1, x_2, ..., x_n)) = (F(x_1), F(x_2), ..., F(x_n))$. Suppose that for each $i, j = 1, 2, ..., n$, the partial derivatives $\partial G_i/\partial x_j$ exist (everywhere) and are continuous. Prove that G is continuously differentiable.

11. Define $F: \mathbb{R}^2 \to \mathbb{R}$ by $F((0, 0)) = 0$ and $F((x, y)) = xy/(x^2 + y^2)^{1/2}$ if $(x, y) \neq (0, 0)$. Prove that F is not differentiable (in the sense of Fréchet).

12. Let V and W be Banach spaces and suppose $T \in \mathcal{B}(V, W)$ is invertible. Is $\|T^{-1}\| = 1/\|T\|$? Prove your answer.

6.4 Manifolds

Some of the most important topological spaces are locally like $\mathbb{R}^n$. Such spaces, called manifolds, occurs frequently and have applications in many branches of mathematics. The systematic study of manifolds began in the mid-nineteenth century with the work of Enrico Betti, A.F. Möbius, Bernard Riemann, and others. In this section we give a brief introduction. For those who wish to continue their study of these spaces, references are given at the end of the section. The references in the text refer to this list.

For convenience if $x \in \mathbb{R}^n$, we let $|x|$ denote the distance between x and the origin.

Definition. A **topological n-dimensional manifold** or **n-manifold** is a second countable Hausdorff space in which each point has a neighborhood that is homeomorphic to the open disc $U^n = \{x \in \mathbb{R}^n : |x| < 1\}$. A 1-manifold is called a **curve** and a 2-manifold is called a **surface**. ∎

The 2-sphere, $S^2 = \{x \in \mathbb{R}^3 : |x| = 1\}$, shown in Figure 6.1, is an example of a surface.

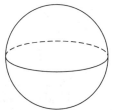

Figure 6.1

Using methods of algebraic topology, one can prove that U^m is not homeomorphic to U^n unless $m = n$. This fact, called the **Invariance of Domain Theorem**, establishes that the dimension of an n-manifold is well-defined. Proofs can be found in Croom [1] and Munkrees [7].

For each $n \in \mathbb{N}$, $\mathbb{R}^n$ is an n-manifold.

EXAMPLE 2. For each $n \in \mathbb{N}$, the n-sphere $S^n = \{x \in \mathbb{R}^{n+1} : |x| = 1\}$ is an n-manifold.

Analysis. The orthogonal projection of $V = \{(x_1, x_2, ..., x_{n+1}) \in \mathbb{R}^{n+1} : x_1 > 0\}$ onto the hyperplane in $\mathbb{R}^{n+1}$ defined by $x_1 = 0$ is a homeomorphism. Therefore the point $(1, 0, 0, ..., 0) \in S^n$ has a neighborhood that is homeomorphic to U^n. If $x \in S^n$ and $x \neq (1, 0, 0, ..., 0)$, the rotation of S^n onto itself that maps x into $(1, 0, 0, ..., 0)$ is a homeomorphism of S^n onto itself (In Exercise 1 you are asked to specify the rotation). The set that is mapped onto V by this homeomorphism is the required neighborhood of x. ∎

In Exercise 2 you are asked to show that every open subset of an n-manifold is an n-manifold, and in Exercise 3 you are asked to show that every in n-manifold is locally compact. Note that an n-manifold may be either compact or noncompact, connected or disconnected.

EXAMPLE 3. Let $f: \mathbb{R} \to \mathbb{R}$ be a continuous function. Then $X = \{(x, f(x)): x \in \mathbb{R}\}$ is a curve.

Analysis. See Exercise 4. ∎

EXAMPLE 4. The torus, projective plane, and Klein bottle are surfaces.

Analysis. Each of these can be constructed as the quotient space of the unit square (see Figure 6.2). We construct the torus by identifying pairs of points of the form $(a, 0)$ and $(a, 1)$, where $0 \leq a \leq 1$, and pairs of points of the form $(0, a)$ and $(1, a)$, where $0 \leq a \leq 1$. We construct the projective plane by identifying pairs of points of the form $(a, 0)$ and $(1 - a, 1)$, where $0 \leq a \leq 1$, and pairs of points of the form $(0, a)$ and $(1, 1 - a)$, where $0 \leq a \leq 1$. We construct the Klein bottle by identifying pairs of points of the form $(a, 0)$ and $(a, 1)$, where $0 \leq a \leq 1$, and pairs of points of the form $(0, a)$ and $(1, 1 - a)$, where $0 \leq a \leq 1$. ∎

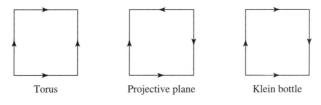

Torus   Projective plane   Klein bottle

Figure 6.2

We can construct additional examples of compact surfaces by taking connected sums of tori and projective planes. Intuitively the connected sum of two surfaces is formed by cutting a circular hole in each surface and pasting the surfaces together along the boundaries of these holes.

Definition. Let X_1 and X_2 be surfaces, and for each $i = 1, 2$, let D_i be a closed disk (that is, D_i is homeomorphic to $\{x \in \mathbb{R}^2: |x| \leq 1\}$) in X_i. For each $i = 1, 2$, let $Y_i = X_i - \text{int}(D_i)$, and let $f: \text{bd}(D_1) \to \text{bd}(D_2)$ be a homeomorphism. The **connected sum**, $X_1 \# X_2$ of X_1 and X_2, is the quotient space obtained from $Y_1 \cup Y_2$ by identifying, for each $x \in \text{bd}(D_1)$, x and $f(x)$. ∎

We leave as Exercise 6 the proof that the connected sum of two surfaces is a surface, and we give two additional examples of "cutting and pasting" techniques that can be used to construct surfaces.

EXAMPLE 5. Let D_1 and D_2 be disjoint closed disks in the 2-sphere, S^2, and let $X = S^2 - (\text{int}(D_1) \cup \text{int}(D_2))$. Let $Y = S^1 \times I$ (S^1 is a circle), and let $f: (S^1 \times$

$\{0\}) \cup (S^1 \times \{1\}) \to \mathrm{bd}(D_1) \cup \mathrm{bd}(D_2)$ be a homeomorphism. The quotient space obtained from $X \cup Y$ by identifying, for each $x \in (S^1 \times \{0\}) \cup (S^1 \times \{1\})$, x and $f(x)$ is called a **sphere with one handle** (See Figure 6.3). In Exercise 9, you are asked to show that the sphere with one handle is homeomorphic to the torus.

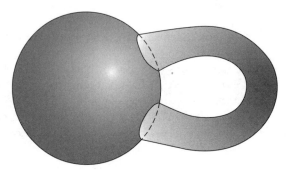

Sphere with one handle

Figure 6.3

If n is any natural number, we can obtain a sphere with n-handles by choosing a collection of $2n$ pairwise disjoint closed disks in S^2 and proceeding in the same manner as in Example 5.

EXAMPLE 6. Let D be a closed disk in S^2, and let $X = S^2 - \mathrm{int}(D)$. Let Y denote the Möbius strip, and note that $\mathrm{bd}(Y)$ is homeomorphic to a circle. Let $f: \mathrm{bd}(D) \to \mathrm{bd}(Y)$ be a homeomorphism. In Exercise 10, you are asked to show that the quotient space, obtained from $X \cup Y$ by identifying, for each $x \in \mathrm{bd}(D)$, x and $f(x)$, is homeomorphic to the projective plane.

If n is any natural number, we can obtain a surface by choosing a collection of n pairwise disjoint closed disks in S^2, removing the interiors of these n disks, and taking the quotient space obtained from the complement of the interiors of these n disks in S^2 and the union n Möbius strips in the same manner as in Example 6.

The remarkable thing is that we can classify the compact surfaces. A proof of this can be found in Massey [4]. We state the result and discuss it.

THEOREM 6.25. If X is a compact surface, then X is homeomorphic to exactly one of the following.

(a) the 2-sphere

(b) a connected sum of tori

(c) a connected sum of projective planes. ∎

We begin our discussion with a description of the connected sum of two tori. As we have seen, each torus can be represented by a rectangle with opposite sides identified (see Figure 6.4).

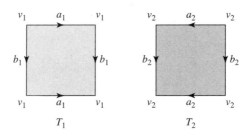

Figure 6.4

We remove the interior of a closed disk in each torus (see Figure 6.5).

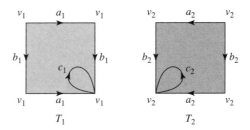

Figure 6.5

Note that the four vertices of each rectangle are identified and become a single point in the torus. The two tori with open disks removed can be pictured as shown in Figure 6.6:

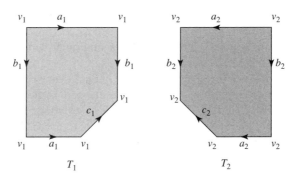

Figure 6.6

To complete the connected sum of the two tori, we paste them together along the circles c_1 and c_2, as in Figure 6.7.

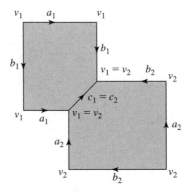

Figure 6.7

Thus the connected sum of two tori can be pictured as the octagon shown in Figure 6.8 with sides identified in pairs as shown.

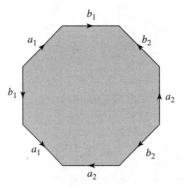

Figure 6.8

In Exercise 11, you are asked to draw corresponding pictures that illustrate the construction of the connected sum of three tori.

The topological structure (that it be locally Euclidean) of a surface does not give us much leverage in proving a geometric result such as the classification theorem for surfaces. As we shall see in Chapter 8, we can define algebraic invariants, such as the fundamental group, for arbitrary topological spaces. But such invariants are useful only if we can calculate them for a large collection of spaces. These problems can be dealt with effectively if one works with the collection of spaces that can be written as the union of "nice pieces" that fit together "nicely." Such a space is called a **triangulable space**.

Definition. A **triangulation** of a compact surface X consists of a finite collection $\{T_1, T_2, ..., T_n\}$ of closed subsets of X that cover X and a collection $\{\phi_i: T'_i \to T_i : \neq i = 1, 2, ..., n\}$ of homeomorphisms, where each T'_i is a triangle in the plane. Each T_i is also called a **triangle.** The members of T_i that are images of the vertices of T'_i are called **vertices,** and the subsets of T_i that are images of the edges of T'_i are called **edges.** In addition, for each pair T_i and T_j of distinct triangles, $T_i \cap T_j = \emptyset$, $T_i \cap T_j$ is a vertex, or $T_i \cap T_j$ is an edge. ∎

In Figure 6.9, we exhibit triangulations of the torus and the projective plane.

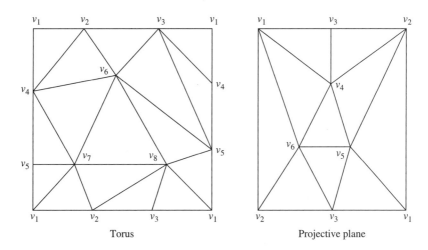

Figure 6.9

In the early days of topology, it was apparently assumed that all manifolds could be triangulated. The first rigorous proof that compact surfaces could be triangulated was given by Tibor Radó in 1925. Radó also gave an example of a surface that could not be triangulated. In 1952, Edwin Moise proved that 3-manifolds can be triangulated. It follows from work of Michael Freedman (1982) and Andrew Casson (1985) that some 4-manifolds cannot be triangulated.

A tool that is useful in showing that two compact surfaces are not homeomorphic is a numerical invariant called the **Euler characteristic.**

Definition. Let X be a compact surface with triangulation $\{T_1, T_2, ..., T_n\}$. If t is the number of distinct triangles, e is the number of distinct edges, and v is the number of distinct vertices, then $\chi(X) = t - e + v$ is called the **Euler characteristic** of X. ∎

It can be shown that the Euler characteristic is a topological invariant. From Figure 6.9, one can see that the Euler characteristic of the torus is 0, and the Euler characteristic of the projective plane is 1. Under the assumption that compact surfaces can be triangulated, we prove the following theorem.

THEOREM 6.26. If X and Y are compact surfaces, then $\chi(X \# Y) = \chi(X) + \chi(Y) - 2$.

Proof. Assume X and Y are triangulated. Form $X \# Y$ by removing the interior of a triangle from each and identifying the edges and vertices of the boundaries of the triangles that have been removed. (For example, if the boundaries of the triangles that have been removed are as shown in Figure 6.10, then, for each $i = 1, 2, 3$, u_i is identified with v_i and e_i is identified with f_i).

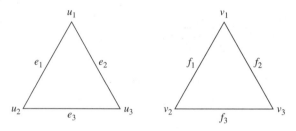

Figure 6.10

We establish the formula by counting the triangles, edges, and vertices before and after taking the connected sum. ∎

We now return to the study of n-manifolds, where n is an arbitrary natural number.

THEOREM 6.27. The product of an m-manifold and an n-manifold is an $(m + n)$-manifold.

Proof. Let X be an m-manifold and let Y be an n-manifold. Since the product of two second countable Hausdorff spaces is a second countable Hausdorff space (Theorems 2.23 and 2.26), $X \times Y$ is a second countable Hausdorff space. Let $(x, y) \in X \times Y$. Then there exist neighborhoods V of x and W of y such that V is homeomorphic to U^m and W is homeomorphic to U^n. By Exercise 5, U^m and U^n are homeomorphic to $\mathbb{R}^m$ and $\mathbb{R}^n$, respectively. By Exercise 16, $U^m \times U^n$ is homeomorphic to $\mathbb{R}^m \times \mathbb{R}^n$. By Exercise 17, $\mathbb{R}^m \times \mathbb{R}^n$ is homeomorphic to $\mathbb{R}^{m+n}$. By Exercise 5, $\mathbb{R}^{m+n}$ is homeomorphic to U^{m+n}. Therefore $V \times W$ is a neighborhood of (x, y) that is homeomorphic to U^{m+n}, and hence $X \times Y$ is an $(m + n)$-manifold. ∎

An n-manifold is also called an **n-manifold without boundary.**

Definition. A **topological n-dimensional manifold with boundary** or **n-manifold with boundary** is a second countable Hausdorff space X such that if $x \in X$ then: (1) x has a neighborhood that is homeomorphic to U^n, or (2) x has a neighborhood that is homeomorphic to $U^n_+ = \{x = (x_1, x_2, ..., x_n) \in \mathbb{R}^n : |x| < 1 \text{ and } x_n \geq 0\}$. If $x \in X$ and x has a neighborhood that is homeomorphic to U^n, then x is called an **interior**

point of X, and the set of all interior points of X is called the **interior** of X. If $x \in X$ and x has a neighborhood that is homeomorphic to U_+^n, then x is called a **boundary point** of X, and the set of all boundary points of X is called the **boundary** of X. If X is an n-manifold with boundary, we denote the boundary of X by ∂X. ∎

EXAMPLE 7. The closed interval $[0, 1]$ is a 1-manifold with boundary. The interior is $(0, 1)$, and the boundary is $\{0, 1\}$.

EXAMPLE 8. The square, $(\{(x, y) \in \mathbb{R}^2: -1 \leq x \leq 1 \text{ and } -1 \leq y \leq 1\})$, the annulus, $(\{(x, y) \in \mathbb{R}^2: 1 \leq x^2 + y^2 \leq 4\})$, and the cylinder, $(S^1 \times [0, 1])$, are 2-manifolds with boundary (see Figure 6.11). The boundary of the square is $\{(x, y) \in \mathbb{R}^2: x = \pm 1 \text{ and } -1 \leq y \leq 1\} \cup \{(x, y) \in \mathbb{R}^2: -1 \leq x \leq 1 \text{ and } y = \pm 1\}$, and the interior of the square is $\{(x, y) \in \mathbb{R}^2: -1 < x < 1 \text{ and } -1 < y < 1\}$. The boundary of the annulus is $\{(x, y) \in \mathbb{R}^2: x^2 + y^2 = 1 \text{ or } x^2 + y^2 = 4\}$, and the interior of the annulus is $\{(x, y) \in \mathbb{R}^2: 1 < x^2 + y^2 < 4\}$. The boundary of the cylinder is $S^1 \times \{0, 1\}$, and the interior of the cylinders is $S^1 \times (0, 1)$.

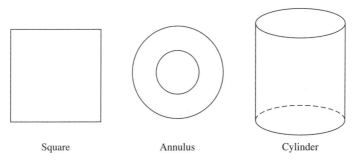

Square Annulus Cylinder

Figure 6.11

The definition of an n-manifold may be considered to be a special case of the definition of an n-manifold with boundary. That is, an n-manifold with empty boundary is an n-manifold. It follows from the Invariance of Domain Theorem and the definition of a manifold, that if X is an n-manifold with nonempty boundary, then the boundary of X is an $(n - 1)$-manifold without boundary.

Definition. A **closed n-manifold** is a compact connected n-manifold. ∎

EXAMPLE 9. For each $n \in \mathbb{N}$, $S^n = \{x \in \mathbb{R}^{n+1}: |x| = 1\}$ is a closed n-manifold.

THEOREM 6.28. Let X be an m-manifold and let Y be an n-manifold with boundary. Then $X \times Y$ is an $(m + n)$-manifold with boundary, and the boundary of $X \times Y$ is $X \times \partial Y$.

Proof. As in the proof of Theorem 6.27, $X \times Y$ is a second countable Hausdorff space. Since $Y = \text{int}(Y) \cup \partial Y$, $X \times Y = (X \times \text{int}(Y)) \cup (X \times \partial Y)$. By Theorem 6.27, each point of $X \times \text{int}(Y)$ has a neighborhood that is homeomorphic to U^{m+n}. Let $(x, y) \in X \times \partial Y$. Then x has a neighborhood that is homeomorphic to U^m and y has a neighborhood that is homeomorphic to U^n_+. Therefore, by Exercise 18, (x, y) has a neighborhood that is homeomorphic to U^{m+n}_+, and hence $X \times \partial Y$ is the boundary of $X \times Y$. ∎

There is a theorem, whose proof is beyond the scope of this text, which shows that every n-manifold can be considered as a subspace of $\mathbb{R}^m$, for some $m \in \mathbb{N}$.

Definition. Let $m, n \in \mathbb{N}$ and let X be a closed n-manifold. **A topological immersion** of X in $\mathbb{R}^m$ is a continuous function $f: X \to \mathbb{R}^m$ such that, for each $x \in X$, there is a neighborhood U of x such that $f|_U: U \to f(U)$ is a homeomorphism. ∎

The usual manner of constructing the Klein bottle (see Figure 2.10 or Figure 6.2) represents a topological immersion of the Klein bottle in $\mathbb{R}^3$. The Klein bottle is, however, a closed 2-manifold that cannot be embedded in $\mathbb{R}^3$. The picture of the Klein bottle shown in Figure 6.12 is the best 3-dimensional representation available:

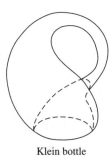

Klein bottle

Figure 6.12

The study of manifolds and related topological spaces is called **geometric topology**. Some manifolds satisfy a smoothness condition similar to that satisfied by differentiable functions. Such manifolds are called **smooth manifolds,** and **differentiable topology** is the study of smooth manifolds.

Definition. Let $m, n \in \mathbb{N}$, and let U be an open subset of $\mathbb{R}^n$. A function $f: U \to \mathbb{R}^m$ is **smooth** if the partial derivatives of all orders exist and are continuous. More generally, if $A \subseteq \mathbb{R}^n$, then a function $f: A \to \mathbb{R}^m$ is **smooth** provided that for each $x \in A$ there exists a neighborhood U of x and a smooth function $g: U \to \mathbb{R}^m$ such that $g(x) = F(x)$ for all $x \in U \cap A$. ∎

In Exercise 21, you are asked to show that if $A \subseteq \mathbb{R}^n \subseteq \mathbb{R}^m$, then the function $i: A \to \mathbb{R}^m$ defined by $i(x) = x$ for each $x \in A$ is smooth. In Exercise 22, you are asked to show that if $A \subseteq \mathbb{R}^n$, $f: A \to \mathbb{R}^m$ is smooth, $B \subseteq \mathbb{R}^m$ such that $f(A) \subseteq B$, and $g: B \to \mathbb{R}^p$ is smooth, then $g \circ f: A \to \mathbb{R}^p$ is smooth.

Definition. Let $m, n \in \mathbb{N}$, let $A \subseteq \mathbb{R}^n$, and let $B \subseteq \mathbb{R}^m$. A **diffeomorphism** from A to B is a one-to-one function f mapping A onto B such that f and f^{-1} are smooth. If there is a diffeomorphism from A onto B, then A and B are **diffeomorphic spaces**. ∎

Definition. Let $m, n \in \mathbb{N}$ and let $X \subseteq \mathbb{R}^m$. Then X is an ***n*-dimensional smooth manifold** or a **smooth *n*-manifold** provided that, for each $x \in X$, there is a neighborhood U of x that is diffeomorphic to U^n. ∎

EXAMPLE 10. S^1 is a smooth 1-manifold.

Analysis. Let $(a, b) \in S^1$ such that $b > 0$. Then $U = \{(x, y) \in S^1 : y > 0\}$ is a neighborhood of x. Define $f: U \to (-1, 1)$ by $f((x, y)) = x$ for each $(x, y) \in U$. Then f is a smooth one-to-one function from U onto $(-1, 1)$. The inverse of f is the smooth function $f^{-1}: (-1, 1) \to U$ defined by $f^{-1}(x) = (x, \sqrt{1 - x^2})$ for each $x \in (-1, 1)$. In Exercise 23, you are asked to complete the analysis in order to show that S^1 is a smooth 1-manifold. ∎

EXAMPLE 11. Let $X = \{(x, y) \in \mathbb{R}^2 : y = |x|\}$. Then X is a 1-manifold that is not smooth because of the "sharp" point at the origin.

The analogue of Theorem 6.27 carries over to smooth manifolds.

THEOREM 6.29. The product of a smooth *m*-manifold and a smooth *n*-manifold is a smooth $(m + n)$-manifold.

Proof. See Exercise 28. ∎

Definition. Let $m, n \in \mathbb{N}$. A subset X of $\mathbb{R}^m$ is an ***n*-dimensional smooth manifold with boundary** or **smooth *n*-manifold with boundary** provided that if $x \in X$ then: (1) x has a neighborhood that is diffeomorphic to U^n, or (2) x has a neighborhood that is diffeomorphic to U^n_+ under a diffeomorphism that maps x into a point of U^n_+ whose *n*th coordinate is 0. ∎

The definition of interior points, interior, boundary points, and boundary of *n*-manifolds carry over (with the obvious changes) to *n*-dimensional smooth manifolds with boundary. The analogue of Theorem 6.28 also carries over to smooth manifolds.

THEOREM 6.30. Let X be a smooth *n*-manifold and let Y be a smooth *n*-manifold with boundary. Then $X \times Y$ is a smooth $(m + n)$-manifold with boundary, and the boundary of $X \times Y$ is $X \times \partial Y$.

Proof. See Exercise 29. ∎

EXERCISES 6.4

1. Specify the rotation in the analysis of Example 2.
2. Show that every open subset of an n-manifold is an n-manifold.
3. Show that every n-manifold is locally compact. *Note:* Section 4.3 is required for this exercise.
4. Prove that if $f: \mathbb{R} \to \mathbb{R}$ is a continuous function, then $X = \{(x, f(x)): x \in \mathbb{R}\}$ is a curve.
5. Prove that if $n \in \mathbb{N}$ then U^n is homeomorphic to $\mathbb{R}^n$.
6. Prove that the connected sum of two surfaces is a surface.
7. Let X be a surface and let S^2 denote the 2-sphere. Prove that $X \# S^2$ is homeomorphic to S^2.
8. Prove that the connected sum of two projective planes is homeomorphic to the Klein bottle.
9. Show that the sphere with one handle is homeomorphic to the torus.
10. Show that the quotient space constructed in Example 6 is homeomorphic to the projective plane.
11. Start with three tori and follow the pattern of the construction shown in Figures 6.4–6.8 to illustrate the construction of the connected sum of three tori.
12. Start with two projective planes (see Figures 6.2) and follow the pattern of the construction shown in Figures 6.4–6.8 to illustrate the construction of the connected sum of two projective planes.
13. Exhibit triangulations of the:
    (a) 2-sphere
    (b) Klein bottle
    (c) Möbius strip
14. What is the Euler characteristic of the:
    (a) 2-sphere
    (b) Klein bottle
    (c) Möbius strip
15. Let X denote the following equilateral triangle and its interior, and let $\mathcal{T}$ denote the subspace topology on X induced by the usual topology on the plane.

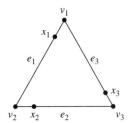

Form a quotient space $(Y, \mathcal{U})$ of $(X, \mathcal{T})$ as follows:

For any three points x_1, x_2, and x_3 such that x_i is a member of e_i and $d(v_1, x_1) = d(v_2, x_2) = d(v_3, x_3)$, identify the points x_1, x_2, and x_3. (Intuitively we are placing the edges e_1 and e_2 on top of e_3). Is $(Y, \mathcal{U})$ a manifold? Prove your answer.

16. Let $m, n \in \mathbb{N}$. Prove that $U^m \times U^n$ is homeomorphic to $\mathbb{R}^m \times \mathbb{R}^n$.

17. Let $m, n \in \mathbb{N}$. Show that $\mathbb{R}^m \times \mathbb{R}^n$ is homeomorphic to $\mathbb{R}^{m+n}$.

18. Let X be an m-manifold, let Y be an n-manifold with boundary, and let $(x, y) \in X \times \partial Y$. Prove that (x, y) has a neighborhood that is homeomorphic to U_+^{m+n}.

19. Prove that every n-manifold is regular. *Note:* Section 5.2 is required for this exercise.

20. Let X be an m-manifold with boundary and let Y be an n-manifold with boundary. Is $X \times Y$ an $(m + n)$-manifold with boundary?

21. Let $m, n \in \mathbb{N}$, and suppose $A \subseteq \mathbb{R}^n \subseteq \mathbb{R}^m$. Prove that the function $i: A \to \mathbb{R}^m$ defined by $i(x) = x$ for each $x \in A$ is smooth.

22. Let $m, n, p \in \mathbb{N}$, and suppose $A \subseteq \mathbb{R}^n$, $f: A \to \mathbb{R}^m$ is smooth, $B \subseteq \mathbb{R}^m$ such that $f(A) \subseteq B$, and $g: B \to \mathbb{R}^p$ is smooth. Prove that $g \circ f : A \to \mathbb{R}^p$ is smooth.

23. In Example 10, complete the proof that S^1 is a smooth 1-manifold.

24. Show that S^2 is a smooth 2-manifold.

25. Show that $\{(x, y) \in \mathbb{R}^2 : x \neq 0 \text{ and } y = \sin(1/x)\}$ is a smooth 1-manifold.

26. Prove that the free union of disjoint n-manifolds is an n-manifold. *Note:* Section 4.4 is required for this exercise.

27. Let $n \in \mathbb{N}$ and define a relation $\sim$ on $X = \mathbb{R}^{n+1} - \{0\}$ as follows: Let $x = (x_1, x_2, ..., x_{n+1})$ and $y = (y_1, y_2, ..., y_{n+1})$ be members of X. Then $x \sim y$ if and only if there exists a nonzero real number λ such that $y_i = \lambda x_i$ for each $i = 1, 2, ..., n + 1$.

(a) Show that $\sim$ is an equivalence relation of X.

(b) Show that the quotient space $P^n(\mathbb{R}) = X/\sim$ is an n-manifold.

The quotient space $P^n(\mathbb{R})$ is called the **n-dimensional real projective space.**

28. Prove Theorem 6.29.

29. Prove Theorem 6.30.

30. Give an example to show that the product of two smooth 1-manifolds with boundary may fail to be smooth.

References

[1] Croom, F.H., *Basic Concepts of Algebraic Topology*, Springer-Verlag, New York, 1978
[2] Guillemin, V., and Pollack, A., *Differential Topology*, Prentice-Hall, Englewood Cliffs, New Jersey, 1974
[3] Hu, S.T., *Differential Manifolds*, Holt, Rinehart, and Winston, New York, 1969
[4] Massey, W.S., *A Basic Course in Algebraic Topology*, Springer-Verlag, New York, 1991
[5] Milnor, J., *Topology from a Differential Viewpoint*, University of Virginia Press, Charlottesville, Virginia, 1965
[6] Moise, E.E., *Geometric Topology in Dimensions Two and Three*, Springer-Verlag, New York, 1977
[7] Munkrees, J.R., *Topology: A First Course*, Prentice-Hall, Englewood Cliffs, New Jersey 1975

6.5 Fractals

Many natural forms, despite their irregular appearance, share a remarkable feature in that they repeat themselves within the same object. A fragment of rock, for example, looks like the mountain in which it is located. A tree's twigs may have the same branching patterns as the tree's trunk. Benoit B. Mandelbrot was the first person to recognize how widespread this type of structure is in nature. He introduced the term self-similar to describe such features, and in 1975, he coined the word **fractal** as a label for irregular and fragmented self-similar shapes. Properties such as those of the Cantor set are characteristic of fractals.

The main tool of fractal geometry is dimension. In classical geometry and topology, objects have a dimension expressed as an integer. We know that a curve is a 1-dimensional object and a surface is 2-dimensional. Fractal curves are so irregular that they fall in the gap between two dimensions. If an object is rather smooth and closely resembles a line, then its fractal dimension is close to 1. It is not so clear, but for many purposes the Cantor set should be regarded as having dimension $\log 2/\log 3 = 0.631$ (see Exercise 6). If an object zigzags terribly and comes close to filling a plane, it has fractal dimension close to 2. Other objects have dimension between 2 and 3. A mountainous fractal scene has fractal dimension close to 2 if it is large with small bumps. Such a scene that is particularly rough would have fractal dimension close to 3.

The notion of fractal dimension is an extension of the concept of dimension that is normally used. A dimension is a measure of the prominence of the irregularities of a set when viewed in very small scales. The idea is to determine the number of small objects needed to cover a large object. If an object of fractal size P is composed of smaller objects of size p, the number, N, of small objects that fits into the object is the ratio of the sizes raised to a power, and the power, d, is called the **Hausdorff dimension.** This dimension is named in honor of Felix Hausdorff, who introduced the idea of fractional dimension in 1919, and it can be written $N = (P/p)^d$ or $d = \log N/\log(P/p)$. Originally, Mandelbrot defined a fractal to be a set whose Hausdorff dimension is strictly greater than its topological dimension. This definition excluded some sets that ought to be regarded as fractals. In some sense, the definition of a "fractal" should be regarded in the same way that a biologist regards the definition of "life"; that is, it is a list of properties characteristic of a living thing.

For completeness, we give formal definitions of Hausdorff dimension and topological dimension.

Definition. Let $n \in \mathbb{N}$, let $A \subseteq \mathbb{R}^n$, and let d be the usual metric on $\mathbb{R}^n$. We denote the **diameter** of A by $|A|$; that is, $|A| = \sup\{d(x, y)\ x, y \in A\}$. If $X \subseteq \mathbb{R}^n, \delta > 0$, and $\{A_i : i \in \mathbb{N}\}$ is a countable collection of sets that $X \subseteq \bigcup_{i \in \mathbb{N}} A_i$ and $0 < |A_i| \leq \delta$ for each $i \in \mathbb{N}$, then we say that $\{A_i : i \in \mathbb{N}\}$ is a δ-cover of X. ∎

Definition. If $n \in \mathbb{N}, X \subseteq \mathbb{R}^n, s \geq 0$, and $\delta > 0$, we define

$$\mathcal{D}^s_\delta(X) = \inf\left\{\sum_{i \in N} |A_i|^s : \{A_i\} \text{ is a } \delta\text{-cover of } X\right\},$$

and we let $\mathcal{D}^s(X) = \lim_{\delta \to 0} \mathcal{D}^s_\delta(X)$. (We consider all covers of X by sets of diameter at most δ and minimize the sum of the sth powers of the diameters. For any subset X of $\mathbb{R}^n$, $\lim_{\delta \to 0} \mathcal{D}^s_\delta(X)$ exists but it may be 0 or ∞). Then $\mathcal{D}^s(X)$ is called the **s-dimensional Hausdorff measure of X**. The **Hausdorff dimension of X**, $\dim_H X$, is defined to be $\inf\{s: \mathcal{D}^s(X) = 0\} = \sup\{s: \mathcal{D}^s(X) = \infty\}$. ∎

We leave as exercises the proofs of the following properties of the s-dimensional Hausdorff measure of a subset of $\mathbb{R}^n$.

THEOREM 6.31. Let $n \in \mathbb{N}$, let $X \subseteq \mathbb{R}^n$, let $s \geq 0$, let $\lambda > 0$, and let $\lambda X = \{\lambda x : x \in X\}$. Then $\mathcal{D}^s(\lambda X) = \lambda^s \mathcal{D}^s(X)$. ∎

THEOREM 6.32. Let $m, n \in \mathbb{N}$, let $X \subseteq \mathbb{R}^n$, let $\alpha, \beta > 0$, and let $f : X \to \mathbb{R}^m$ be a function such that $d(f(x), f(y)) \leq \alpha(d(x, y))^\beta$ for all $x, y \in X$. Then for each $s \geq 0$, $\mathcal{D}^{s/\beta}(f(X)) \leq \alpha^{s/\beta} \mathcal{D}^s(X)$. ∎

THEOREM 6.33. Let $n \in \mathbb{N}$, let $X \subseteq \mathbb{R}^n$, let $\delta > 0$, let $s \geq 0$, and let $t > s$. If $\{A_i : i \in \mathbb{N}\}$ is a δ-cover of X, then $\sum_{i \in N} |A_i|^t \leq \delta^{t-s} \sum_{i \in N} |A_i|^s$. ∎

THEOREM 6.34 Let $n \in \mathbb{N}$, let $X \subseteq \mathbb{R}^n$, and let $s \geq 0$.

(a) If $s < \dim_H X$, then $\mathscr{D}^s(X) = \infty$.

(b) If $s < \dim_H X$, then $\mathscr{D}^s(X) = 0$. ∎

We also note that if $s = \dim_H X$, then $\mathscr{D}^s(X)$ may be infinite or a nonnegative real number.

The topological definition of dimension is an inductive definition. The concept of 0-dimensional is defined in Section 3.3. Suppose $n \in \mathbb{N}$. A topological space $(X, \mathscr{T})$ has **dimension less than or equal to n at a point $p \in X$** provided that if U is any neighborhood of p then there is a neighborhood V of p such that $V \subseteq U$ and the dimension of the boundary of V is less than or equal to $n - 1$. Then $(X, \mathscr{T})$ has **dimension n at p** if the dimension of X at p is less than or equal to n and the dimension of X at p is not less than or equal to $n - 1$. The topological space $(X, \mathscr{T})$ has **dimension less than or equal to n** if the dimension of X at each point p in X is less than or equal to n, and $(X, \mathscr{T})$ has **dimension n** if the dimension of X is less than or equal to n and the dimension of X is not less than or equal to $n - 1$.

As we have indicated, the Cantor set is an example of a fractal. We now give some other examples.

EXAMPLE 12. One fractal, known as the **Sierpinski gasket,** starts as a triangle. Then successively smaller triangles are removed as shown in Figure 6.13, leaving a figure with zero area.

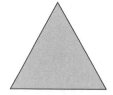 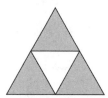 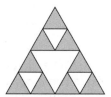

Figure 6.13

EXAMPLE 13. Another fractal, known as the **Sierpinski carpet,** starts as a square. Then successively smaller squares are removed as shown in Figure 6.14.

 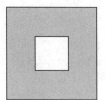 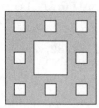

Figure 6.14

6.5 Fractals

EXAMPLE 14. Another fractal, known as the **Menger sponge,** is the 3-dimensional analog of the Sierpinski carpet.

In 1904, the Swedish mathematician Helge von Koch created a set that is now known to be a fractal and is usually called the Koch snowflake. We describe this set in Example 15.

EXAMPLE 15. In this example, triangle means solid triangle; that is, a triangle plus its interior. We start with an equilateral triangle. Then we add an equilateral triangle, one-third the size of the original, to the middle of each side of the original triangle (see Figure 6.15). The boundary of the resulting figure has twelve segments, and the length of the boundary is four-thirds times the perimeter of the original triangle. At the next stage, we add an equilateral triangle, one-third the size of the triangles added at the second stage, to the middle of each of the twelve segments that constitute the boundary of the second set shown in Figure 6.15. The result is the third set shown 6.15. We continue this process in order to obtain the Koch snowflake. The first four stages in the construction of the Koch snowflake are shown in Figure 6.15.

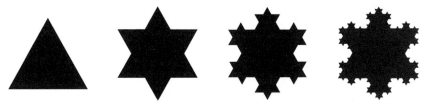

First four stages in the construction of the Koch snowflake

Figure 6.15

The boundary of the Koch snowflake is continuous, but there are an infinite number of zigzags between any two points on the boundary. Thus the length between any two points on the boundary is infinite. Yet the boundary bounds a set whose area is eight-fifths that of the original triangle.

Mandelbrot recognized fractals as a useful tool for analyzing a variety of physical phenomena. The first application was to solve the problem of noise during data transmission. Mandelbrot considered the pattern of errors that occur when data is transmitted as a group of on/off switches. The errors showed up in bursts, and he found that each burst was itself intermittent. The best mathematical model for these bursts was the Cantor set.

The notion of self-similarity does not cover all possible fractals. In 1980, Mandelbrot described a fractal that John Hubbard later named the **Mandelbrot set.** This set arose by taking an expression such as $x^2 - 3x$, starting with an initial value for x, substituting the resulting answer back into the original expression, and continuing. The resulting Mandelbrot set is complicated because the small Mandelbrot sets within the larger one have more "hair" than the larger one. The Mandelbrot set is shown in Figure 6.16.

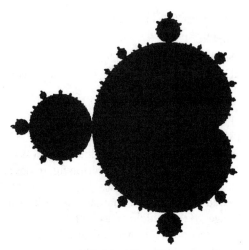

The Mandelbrot set

Figure 6.16

Computer graphs and fractal geometry are linked because computer graphics provides a way of picturing fractals, and fractal geometry is useful for creating computer images. Of course, the fractal image on a computer screen is only an approximate picture of the fractal.

If we start with a real number, the set generated by the expression $x^2 - 1$ isn't particularly exciting. It is a different matter, however, if we start with a complex number and consider the set generated by the expression $z^2 - 1$. For some choices of the complex number, the set that is generated grows without bound. For other choices of the complex number, the set that is generated stays within a well-defined boundary, and the boundary is a fractal (see Figure 6.17).

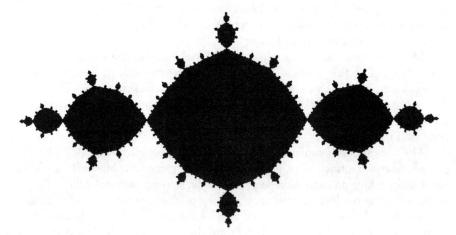

Figure 6.17

The kind of boundary shown in Figure 6.17 is known as a **Julia set**. It is named in honor of a French mathematician, Gaston Julia.

In general, suppose c is a complex number and consider the expression $z^2 + c$. If $c = 0$ and we start with a point whose distance from the origin is less than 1, the successive squarings produce orbits that approach 0.

For those who are interested in pursuing the study of fractals, we provide a list of references at the end of this section. Since metric spaces provide the framework in which fractal geometry is studied, in the remainder of this section, we outline the necessary background in order to describe a topological space in which fractals are found. The details are left as exercises. Throughout this section we assume that (X, d) is a complete metric space, and we let $\mathcal{H}(X)$ denote the family of nonempty compact subsets of X.

EXAMPLE 16. Let d be the usual metric on $\mathbb{R}$. Then the Cantor set and every closed interval of the form $[a, b]$, where $a < b$, belongs to $\mathcal{H}(\mathbb{R})$.

Let A be a subset of X and let $x \in A$. In Section 1.5, we defined $d(x, A) = \text{glb}\{d(x, a)\, a \in A\}$ Let $A, B \in \mathcal{H}(X)$ and define $d(A, B) = \text{lub}\{d(a, B): a \in A\}$. The **Hausdorff distance** between $A, B \in \mathcal{H}(X)$ is defined by $h(A, B) = \max\{d(A, B), d(B, A)\}$. In Exercise 16 you are asked to show that h is a metric on $\mathcal{H}(X)$. The metric space $(\mathcal{H}(X), h)$ is called the **space of fractals**.

We turn our attention to showing that $(\mathcal{H}(X), h)$ is complete.

Definition. Let $S \subseteq X$ and let $r \geq 0$. Then $S + r = \{y \in X: d(x, y) \leq r$ for some $x \in S\}$ is called the **dilation of S by a ball of radius r**. ∎

Note that $S + r$ is just the set of points in X that are within r units of S.

THEOREM 6.35. Let $A, B \in \mathcal{H}(X)$ and let $r > 0$. Then $h(A, B) \leq r$ if and only if $A \subseteq B + r$ and $B \subseteq A + r$.

Proof. See Exercise 20. ∎

Let $\{A_n: n \in \mathbb{N}\}$ be a Cauchy sequence in $(\mathcal{H}(X), h)$. Then for each positive number ε there exists $N \in \mathbb{N}$ such that if $m, n \geq N$, then $h(A_m, A_n) \leq \varepsilon$. That is, $\{A_n: n \in \mathbb{N}\}$ is a Cauchy sequence in $(\mathcal{H}(X), h)$ if and only if for each positive number ε there exists $N \in \mathbb{N}$ such that if $m, n \geq N$, then $A_m \subseteq A_n + \varepsilon$ and and $A_n \subseteq A_m + \varepsilon$.

The following theorem is called an *extension lemma* because it permits the extension of a Cauchy subsequence.

THEOREM 6.36. (The Extension Lemma). Let $\{A_n: n \in \mathbb{N}\}$ be a Cauchy sequence in $(\mathcal{H}(X), h)$, and let $\{n_i: i \in \mathbb{N}\}$ be a sequence of natural numbers such that $n_i < n_{i+1}$ for each $i \in \mathbb{N}$. Furthermore, suppose there is a Cauchy sequence $\{x_{n_i}: i \in \mathbb{N}\}$ in X such that $x_{n_i} \in A_{n_i}$ for each $i \in \mathbb{N}$. Then there exists a Cauchy sequence $\{\tilde{x}_n: i \in \mathbb{N}\}$ in X such that $\tilde{x}_{n_i} = x_{n_i}$ for all $i \in \mathbb{N}$.

Proof. See Exercise 21. ∎

The following theorem shows that $(\mathcal{H}(X), h)$ is complete.

THEOREM 6.37. Let $\{A_n: n \in \mathbb{N}\}$ be a Cauchy sequence in $\mathcal{H}(X)$ and let $A = \{x \in X: \text{there is a Cauchy sequence } \{x_n: n \in \mathbb{N}\} \text{ in } X, \text{ with } x_n \in A_n \text{ for each } n \in \mathbb{N} \text{ that converges to } x\}$. Then $A \in \mathcal{H}(X)$ and $\{A_n: n \in \mathbb{N}\}$ coverges to A.

Proof. By Exercise 22(a), $A \neq \emptyset$. By Exercise 22(b), A is closed. Therefore by Theorem 1.47, A is complete. By Exercise 22(d), A is totally bounded. Therefore by Theorem 4.11, A is compact. Hence $A \in \mathcal{H}(X)$. By Exercise 22(e), $\{A_n: n \in \mathbb{N}\}$ converges to A. ∎

EXERCISES 6.5

1. Prove Theorem 6.31.

2. Prove Theorem 6.32.

3. Prove Theorem 6.33.

4. Prove Theorem 6.34.

5. Let $X = \{(x, y, z) \in \mathbb{R}^3: x^2 + y^2 \leq 1 \text{ and } z = 0\}$.

    (a) Show that $\mathcal{D}^1(X) = \infty$, $0 < \mathcal{D}^2(X) < \infty$, and $\mathcal{D}^3(X) = 0$.

    (b) What is $\dim_H(X)$?

6. Let X be the Cantor set and let $s = \log 2/\log 3 = 0.6309\ldots$. Show that $\dim_H X = s$ and $\frac{1}{2} \leq \mathcal{D}^s(X) \leq 1$.

7. Let $A \in \mathcal{H}(X)$ and let $x \in X$. Prove that $d(x, A) = \min\{d(x, a): a \in A\}$.

8. Let $A \in \mathcal{H}(X)$ and let $x \in X$. Define $f: A \to \mathbb{R}$ by $f(a) = d(x, a)$ for all $a \in A$. Prove that there exists $\tilde{a} \in A$ such that $f(\tilde{a}) = \text{glb}\{f(a): a \in A\}$.

9. Let $A, B \in \mathcal{H}(X)$. Show that there exists $\tilde{a} \in A$ and $\tilde{b} \in B$ such that $d(A, B) = d(\tilde{a}, b)$.

10. Suppose $A, B \in \mathcal{H}(X)$, $A \subseteq B$, and $x \in X$. Show that $d(x, B) \leq d(x, A)$.

11. Give an example of a complete metric space (X, d) and compact subsets A and B of X such that $d(A, B) \neq d(B, A)$.

6.5 Fractals   231

12. Let $A, B \in \mathcal{H}(X)$ such that $A \neq B$.

    (a) Show that either $d(A, B) \neq 0$ or $d(B, A) \neq 0$.

    (b) Show that if $A \subseteq B$, then $d(A, B) = 0$.

13. Suppose $A, B, C \in \mathcal{H}(X)$ and $B \subseteq C$. Show that $d(A, C) \leq d(A, B)$. *Hint:* Use Exercise 10.

14. Show that if $A, B, C \in \mathcal{H}(X)$ then $d(A \cup B, C) = \max\{d(A, C), d(B, C)\}$.

15. Show that if $A, B, C \in \mathcal{H}(X)$, then $d(A, B) \leq d(A, C) + d(C, B)$.

16. Show that h is a metric on $\mathcal{H}(X)$. (See the discussion following Example 16).

17. Let $A, B, C, D \in \mathcal{H}(X)$. Show that $h(A \cup B, C \cup D) \leq \max\{h(A, C), h(B, D)\}$.

18. Show that if (X, d) is compact, then so is $(\mathcal{H}(X), h)$.

19. Show that if $A, B \in \mathcal{H}(X)$, then there exist $a \in A$ and $b \in B$ such that $h(A, B) = d(a, b)$.

20. Let $A, B \in \mathcal{H}(X)$ and let $r > 0$. Prove that $h(A, B) \leq r$ if and only if $A \subseteq B + r$ and $B \subseteq A + r$.

21. Prove Theorem 6.36.

22. Let $\{A_n : n \in \mathbb{N}\}$ be a Cauchy sequence in $\mathcal{H}(X)$ and let $A = \{x \in X:$ there exists a Cauchy sequence $\{x_n : n \in \mathbb{N}\}$ in X, with $x_n \in A_n$ for each $n \in \mathbb{N}$, that converges to $x\}$.

    (a) Prove that $A \neq \emptyset$. *Hint:* There exists a sequence $\{N_i : i \in \mathbb{N}\}$ of natural numbers such that $N_i < N_{i+1}$ for each $i \in \mathbb{N}$ and $h(A_m, A_n) < 1/2^i$ for all $m, n > N_i$. Find a Cauchy sequence $\{x_{N_i}\}$ such that $x_{N_i} \in A_{N_i}$ for each $i \in \mathbb{N}$. Then use the Extension Lemma.

    (b) Show that A is closed. *Hint:* Suppose $\{a_i : i \in \mathbb{N}\}$ is a sequence for members of A that converges to $a \in X$. Show that for each $i \in \mathbb{N}$ there exists a sequence $\{X_{i, n} : n \in \mathbb{N}\}$ with $X_{i, n} \in A_n$, that converges to a_i. Find an increasing sequence $\{N_i : i \in \mathbb{N}\}$ of natural numbers such that $d(a_{N_i}, a) < \frac{1}{i}$. Find a subsequence $\{m_i : i \in \mathbb{N}\}$ of natural numbers such that $d(x_{N_i, m_i}, a_{N_i}) \leq \frac{1}{i}$. Then use the Extension Lemma.

    (c) Show that for each $\varepsilon > 0$, there exists $N \in \mathbb{N}$ such that if $n \geq N$, then $A \subseteq A_n + \varepsilon$.

(d) Show that A is totally bounded. *Hint:* Suppose A is not totally bounded. Then use **(c)**.

(e) Show that $\{A_n: n \in \mathbb{N}\}$ converges to A. *Hint:* Use (c) and Theorem 6.35.

23. Suppose (X, d) is compact, and let $A \in \mathcal{H}(X)$. Show that the dilation of A is a member of $\mathcal{H}(X)$.

References

[1] Barnsley, M., *Fractals Everywhere*, Academic Press, San Diego, 1988.
[2] Devaney, R.L., and Keen, L., *Chaos and Fractals: The Mathematics Behind the Computer Graphics*, American Mathematical Society, Providence, Rhode Island, 1989.
[3] Falconer, K., *Fractal Geometry: Mathematical Foundations and Applications*, John Wiley & Sons, New York, 1990.
[4] Mandelbrot, B., *The Fractal Geometry of Nature*, W.H. Freeman & Co., New York. 1983.

6.6 Compactifications

Compact spaces have important properties. At times it is possible to embed a non-compact topological space $(X, \mathcal{T})$ into a compact space $(Y, \mathcal{U})$ and then use the properties of Y to gain information about X. Such a space Y is called a **compactification** of X. The simplest such compactification is made by adjoining a single point to X, and we begin our study by considering these one-point compactifications. The one-point compactification was introduced by Alexandroff and Urysohn in 1924. Alexandroff proved the result that we state as Theorem 6.38. Work on compactification continued with Tychonoff, who proved the result we state as Theorem 6.41.

Definition. Let $(X, \mathcal{T})$ be a topological space, let p be an object that does not belong to X, and let $Y = X \cup \{p\}$. Let $\mathcal{U} = \{U \in \mathcal{P}(Y): U \in \mathcal{T} \text{ or } Y - U \text{ is a closed compact subspace of } X\}$. Then $\mathcal{U}$ is a topology on Y and $(Y, \mathcal{U})$ is called the **one-point compactification** of X. ∎

The proof that $\mathcal{U}$ is a topology on Y is left as Exercise 1.

EXAMPLE 17. Let X be the open unit interval $(0, 1)$ with the usual topology $\mathcal{T}$, let p be an object that does not belong to X, let $Y = X \cup \{p\}$, and let $\mathcal{U}$ be the topology on Y given by the preceding definition. Members of $\mathcal{U}$ that do not contain p are members of $\mathcal{T}$, and members of $\mathcal{U}$ that contain p are sets whose complements in Y are closed compact subsets of Y. An example of a member of U is given in Figure 6.18.

6.6 Compactifications   233

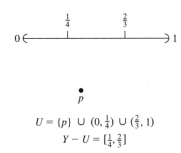

Figure 6.18

The following theorem says that the one-point compactification of a topological space accomplishes the stated purpose and has additional desired properties.

THEOREM 6.38. Let $(X, \mathcal{T})$ be a topological space and let $(Y, \mathcal{U})$ be its one-point compactification. Then the following conditions hold:

(a) $(Y, \mathcal{U})$ is compact.
(b) $(X, \mathcal{T})$ is a subspace of $(Y, \mathcal{U})$.
(c) $(Y, \mathcal{U})$ is Hausdorff if and only if $(X, \mathcal{T})$ is Hausdorff and locally compact.
(d) X is a dense subset of Y if and only if X is not compact.

Proof. Let p denote the member of Y such that $Y = X \cup \{p\}$.

(a) Let $\mathcal{O}$ be an open cover of Y. Then there is a member U of $\mathcal{O}$ such that $p \in U$. Since $p \in U$. and U is open, $X - U$ is a compact subspace of X. Therefore a finite collection $\{U_1, U_2, ..., U_n\}$ of members of $\mathcal{O}$ covers $X - U$, and hence $\{U, U_1, U_2, ..., U_n\}$ is a finite subcollection of $\mathcal{O}$ that covers Y. Thus $(Y, \mathcal{U})$ is compact.

(b) Let $U \in \mathcal{T}$. Then $U \subseteq X$ and $U \in \mathcal{U}$. Thus $U = X \cap U \in \mathcal{U}_x$, h and hence $\mathcal{T} \subseteq \mathcal{U}_x$. Let $V \in \mathcal{U}_x$. Then there exists $U \in \mathcal{U}$ such that $V = U \cap X$. Either $U \subseteq X$ or $Y - U$ is a closed compact subset of X. If $U \subseteq X$, then $V = U$ and hence $V \in \mathcal{T}$. Suppose $Y - U$ is a closed compact subset of X. Then $X \cap U = X - (Y - U)$ is open in X. Therefore $V = X \cap U \in \mathcal{T}$, so $\mathcal{U}_x \subseteq \mathcal{T}$.

(c) Suppose $(Y, \mathcal{U})$ is Hausdorff. Since every subspace of a Hausdorff space is Hausdorff, $(X, \mathcal{T})$ is Hausdorff. Let $x \in X$. Then there exist disjoint open sets U and V such that $p \in U$ and $x \in V$. Thus $V \subseteq X - U$ and $X - U$ is compact. Therefore $(X, \mathcal{T})$ is locally compact.

Suppose $(X, \mathcal{T})$ is locally compact and Hausdorff. It is sufficient to show that if $x \in X$ then there are disjoint open sets U and V such that $x \in U$ and $p \in V$. Let $x \in X$. By Theorem 4.35, there is a neighborhood U of x such that $\overline{U}$ is compact. Thus $X - \overline{U}$ is a neighborhood of p. Since $X - \overline{U}$ and U are disjoint, $(Y, \mathcal{U})$ is Hausdorff.

(d) Suppose X is not dense in Y. Then $\{p\} \in \mathcal{U}$ and hence X is compact. Suppose X is compact. Since X is a closed subset of X, $\{p\}$ is open. Therefore X is not dense in Y. ∎

We return to Example 17. Let $(X, \mathcal{T})$ and $(Y, \mathcal{U})$ be the spaces given in that example. We claim that $(Y, \mathcal{U})$ is homeomorphic to S^1. Define $f : Y \to S^1$ by $f(p) = (1, 0)$ and $f(t) = (\cos 2\pi t, \sin 2\pi t)$ for all $t \in (0, 1)$. Then, by Exercise 2, f is a one-to-one continuous function that maps Y onto S^1. By Corollary 4.23, f is a homeomorphism. Thus S^1 is the one-point compactification of the open unit interval.

THEOREM 6.39. The one-point compactifications of two homeomorphic topological spaces are homeomorphic.

Proof. Let $(X_1, \mathcal{T}_1)$ and $(X_2, \mathcal{T}_2)$ be topological spaces, let $f : X_1 \to X_2$ be a homeomorphism, and for each $i = 1, 2$, let $(Y_i, \mathcal{U}_i)$ be the one-point compactifications of $(X_i, \mathcal{T}_i)$, where $Y_i = X_i \cup \{p_i\}$. Define $g: Y_1 \to Y_2$ by $g(p_1) = p_2$ and $g(x) = f(x)$ for all $x \in X_1$. Then g is a one-to-one function from Y_1 onto Y_2, and by Exercise 3, g is continuous. Therefore by Corollary 4.23, g is a homeomorphism. ∎

EXAMPLE 18. The one-point compactification of $\mathbb{R}$ is homeomorphic to S^1.

Analysis. Since $\mathbb{R}$ and $(0, 1)$ are homeomorphic, the desired result follows immediately from theorem 6.39 and Example 17. We can also construct a homeomorphism of the one-point compactification of $\mathbb{R}$ onto a circle. Let $C = \{(x, y) \in \mathbb{R}^2 : x^2 + (y - 1)^2 = 1\}$. If we think of $\mathbb{R}$ as $\{(x, y) \in \mathbb{R}^2 : y = 0\}$, then C is tangent to $\mathbb{R}$ at $(0, 0)$. (See Figure 6.19.)

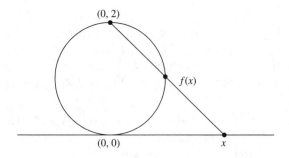

Figure 6.19

6.6 Compactifications

Let $(Y, \mathcal{U})$ be the one-point compactification of $\mathbb{R}$, where $Y = \mathbb{R} \cup \{p\}$. Define $f : Y \to C$ by $f(p) = (0, 2)$ and for each $x(= (x, 0)) \in \mathbb{R}$, let $f(x)$ be the point of intersection of the line segment joining $(x, 0)$ and $(0, 2)$ with $C - \{(0, 2)\}$. Then f is a continuous (see Exercise 4) bijection and hence by Corollary 4.23, f is a homeomorphism. ∎

EXAMPLE 19. Let $\mathcal{T}$ be the lower limit topology on $\mathbb{R}$. If $a, b \in \mathbb{R}$ and $a < b$, then

$$\left\{ \left[\frac{a + (2^{n-1} - 1)b}{2^{n-1}}, \frac{a + (2^n - 1)b}{2^n} \right) : n \in \mathbb{N} \right\}$$

is an open cover of $[a, b)$ that does not have a finite subcover. Therefore $[a, b)$ is not compact, and so $(\mathbb{R}, \mathcal{T})$ is not locally compact. By Theorem 6.38, the one-point compactification $(Y, \mathcal{U})$ of $(\mathbb{R}, \mathcal{T})$ is not Hausdorff. Notice that if $U \in \mathcal{U}$ and the point $p \in Y - \mathbb{R}$ belongs to U, then the complement of U does not contain an interval.

In order for the one-point compactification of a topological space to be Hausdorff, it is necessary for the space to be a locally compact Hausdorff space (Theorem 6.38(c)). Thus many mathematicians define the one-point compactification only for locally compact Hausdorff spaces.

We generalize the notion of compactification as follows. We assume that the base space is locally compact and Hausdorff. Then, by Theorem 6.38, the one-point compactification is Hausdorff.

Definition. A **compactification** of a locally compact Hausdorff space $(X, \mathcal{T})$ is an ordered pair $((Y, \mathcal{U}), h)$ or (Y, h), where $(Y, \mathcal{U})$ is a compact Hausdorff space and h is an embedding of X as a dense subset of Y. ∎

Notice that if $(Y, \mathcal{U})$ is the one-point compactification of a topological space $(X, \mathcal{T})$, then the function $i: X \to Y$ defined by $i(x) = x$ each $x \in X$ is a one-to-one continuous function that maps X onto the subspace X of Y and has the property that $i^{-1}: X \to X$ is continuous.

Quite often, the embedding h is an inclusion map so $X \subseteq Y$. In any case X and $h(X)$ are homeomorphic, so we can write X rather than $h(X)$ and think of X as being a subset of Y.

By Corollary 5.13, every locally compact Hausdorff space is regular. We can now prove a stronger result, and this result illustrates what was stated in the introduction.

THEOREM 6.46. Every locally compact Hausdorff space is completely regular.

Proof. Let $(X, \mathcal{T})$ be a locally compact Hausdorff space. By Theorem 6.38, its one-point compactification $(Y, \mathcal{U})$ is a compact Hausdorff space. Thus by Corollary 5.22, $(Y, \mathcal{U})$ is normal and therefore completely regular (Corollary 5.37). Therefore by Exercise 7 of Section 5.2, $(X, \mathcal{T})$ is completely regular. ∎

The proof of the following theorem is a modification of the proof of Theorem 5.45.

THEOREM 6.41. Every completely regular space has a compactification.

Proof. Let $(X, \mathcal{T})$ be a completely regular space. By Theorem 5.44, $\mathcal{T}$ has the weak topology induced by $C^*(X)$. If $f \in C^*(X)$, then there exist real numbers a_f and b_f such that $f(X) \subseteq [a_f, b_f]$. Now $\prod_{f \in C^*(X)} [a_f, b_f]$ is a cube and hence it is a compact Hausdorff space. Since $(X, \mathcal{T})$ is completely regular, $C^*(X)$ separates points from closed sets. Therefore by Theorem 5.42, the evaluation map $e\colon X \to \prod_{f \in C^*(X)} [a_f, b_f]$ is an embedding. Let $\beta(X) = \overline{e(X)}$. Then $\beta(X)$ is a closed subset of a compact Hausdorff space, and hence it is a compact Hausdorff space. The ordered pair $(\beta(X), e)$ is a compactification of X. ∎

The compactification (given in the proof of Theorem 6.41) that was constructed by Tychonoff is name in honor of Eduard Čech (1893–1960) and M.H. Stone (1903–2008) who, in 1937, independently developed its properties by proving the results we state as Theorems 6.42, 6.47, and 6.48.

Definition. Let $(X, \mathcal{T})$ be a completely regular space. The compactification $(\beta(X), e)$ is the **Stone-Čech compactification** of X. ∎

The following theorem tells us that a continuous function from a completely regular space $(X, \mathcal{T})$ into a compact Hausdorff space can be extended to $\beta(X)$.

THEOREM 6.42. Let $(X, \mathcal{T})$ be a completely regular space, let $(Y, \mathcal{U})$ be a compact Hausdorff space, and let $h\colon X \to Y$ be a continuous function. Then there is a continuous function $H\colon \beta(X) \to Y$ such that $H \circ e = h$.

Proof. Since X and Y are completely regular, the evaluation maps $e\colon X \to \prod_{f \in C^*(X)} [a_f, b_f]$ and $e'\colon Y \to \prod_{g \in C^*(Y)} [a_g, b_g]$ are embeddings. The diagram in Figure 6.20 illustrates the situation, and we first define the function K illustrated in that diagram.

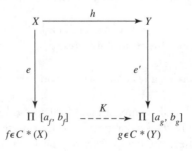

Figure 6.20

Let $t \in \prod_{f \in C^*(X)}[a_f, b_f]$. Then $t: C^*(X) \to \bigcup_{f \in C^*(X)}[a_f, b_f]$. Let $g \in C^*(Y)$. Then $g: Y \to \mathbb{R}$. We define $K(t)$ to be the function whose value at g is $t(g \circ h)$; that is $[K(t)]_g = t(g \circ h)$. Notice that $g \circ h: X \to \mathbb{R}$ and so $K(t: C^*(Y) \to \bigcup_{g \in C^*(Y)}[a_g, b_g]$. Let $g \in C^*(Y)$ and observe that for each $t \in \prod_{f \in C^*(X)}[a_f, b_f]$, $(\pi_g \circ K)(t) = [K(t)]_g = t(g \circ h) = \pi_{g \circ h}(t)$. Therefore for each $g \in C^*(Y)$, $\pi_g \circ K$ is continuous. Hence by Theorem 2.37, K is continuous. Let $g \in C^*(Y)$ and $y \in e(X)$. Then there is an $x \in X$ such that $e(x) = y$ and $K[e(x)]_g = [e(x)]_{g \circ h} = (g \circ h)(x) = [e'(h(x))]_g$. Thus $K[e(x)] = [e'(h(x)) \in e'(Y)$ and hence $K[e(x)] \subseteq e'(Y)$. Since $e(X)$ is dense in $\beta(X)$, $K[e(x)]$ is dense in $K[\beta(x)]$. By Exercise 5, $e'(Y)$ is closed; that is $\beta(Y) = e'(Y)$. Therefore $K[\beta(X)] \subseteq e'(Y)$. Let $H = (e')^{-1} \circ (K|_{\beta(X)})$. Then $H: \beta(X) \to Y$ is continuous, and for each $x \in X$, $(H \circ e)(x) = (e^1)^{-1}[K(e(x))] = (e')^{-1}[e'(h(x))] = h(x)$. ∎

We introduce some terminology to prove that Theorem 6.42 characterizes the Stone-Čech compactification.

Definition. Let $(X, \mathcal{T})$ be a completely regular space, and let (Y_1, h_1) and (Y_2, h_2) be compactifications of X. We write $(Y_1, h_1) \leq (Y_2, h_2)$ to mean that there is a continuous function $H: Y_2 \to Y_1$ such that $H \circ h_2 = h_1$. If $(Y_1, h_1) \leq (Y_2, h_2)$ and $(Y_2, h_2) \leq (Y_1, h_1)$ we say that the pairs (Y_1, h_1) and (Y_2, h_2) are **topologically equivalent compactifications**. ∎

Notice that if $(Y_1, h_1) \leq (Y_2, h_2)$ then there is a continuous functions H such that the following diagram is commutative.

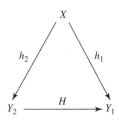

Also observe that H is an extension of the homeomorphism $h_1 \circ (h_2)^{-1}$. Thus if h_1 and h_2 are inclusion maps, then $(Y_1, h_1) \leq (Y_2, h_2)$ if and only if there is a continuous function $F: Y_2 \to Y_1$ such that $F|_x$ is the identity. We write $H: (Y_1, h_1) \leq (Y_2, h_2)$ to denote that $(Y_1, h_1) \leq (Y_2, h_2)$ and $H: Y_2 \to Y_1$, is a continuous function such that $H \circ h_2 = h_1$.

The following theorem tells us that if the compactifications (Y_1, h_1) and (Y_2, h_2) of a completely regular space are topologically equivalent, then Y_1 and Y_2 are homeomorphic. Thus the terminology, topologically equivalent compactifications, is justified.

THEOREM 6.43. Let $(X, \mathcal{T})$ be a completely regular space, and let (Y_1, h_1) and (Y_2, h_2) be compactifications of X. Then (Y_1, h_1) and (Y_2, h_2) are topologically equivalent if and only if there is a homeomorphism $H: Y_2 \to Y_1$ such that $H \circ h_2 = h_1$.

Proof. See Exercise 6. ∎

THEOREM 6.42. Let $(X, \mathcal{T})$ be a Hausdorff space, let $(Y, \mathcal{U})$ be a topological space, let A be a dense subset of X, and let $f : X \to Y$ be a continuous function such that $f|_A$ is a homeomorphism. Then $f(X - A) \subseteq Y - f(A)$.

Proof. The proof is by contradiction. Suppose $f(X - A) \not\subseteq Y - f(A)$. Then there is an $x_1 \in X - A$ such that $f(x_1) \in f(A)$. Thus there exists $x_2 \in A$ such that $f(x_2) = f(x_1)$. Let U and V be disjoint open sets such that $x_1 \in U$ and $x_2 \in V$. Now $V \cap A$ is a neighborhood of x_2 in A. Since $f|_A$ is a homeomorphism, $f(V \cap A)$ is a neighborhood of $f(x_2)$ in $f(A)$. Thus there is a neighborhood W of $f(x_2)$ in Y such that $f(V \cap A) = f(A) \cap W$. Let N be a neighborhood of x_1. Then $N \cap U$ is a neighborhood of x_1 and since A is dense in X, there exists $z \in N \cap U \cap A$. Thus $f(z) \in f(A)$ but $f(z) \notin f(V \cap A)$. Hence $f(z) \notin W$. Therefore $f(N) \not\subseteq W$ and hence f is not continuous at x_1. This is a contradiction. ∎

THEOREM 6.45. Suppose $H: (Y_1, h_1) \leq (Y_2, h_2)$. Then the following conditions hold:

(a) $H|_{h_2(X)}$ is a homeomorphism of $h_2(X)$ onto $h_1(X)$.

(b) H maps $Y_2 - h_2(X)$ onto $Y_1 - h_1(X)$.

Proof. See Exercise 7. ∎

THEOREM 6.46. Let $(X, \mathcal{T})$ be a completely regular space, let (K_1, h_1) and (K_2, h_2) be compactification of X, and suppose each continuous function f that maps X into a compact Hausdorff space $(Y, \mathcal{U})$ can be extended to continuous function $F: K_2 \to Y$ such that $F \circ h_2 = f$. Then $(K_1, h_1) \leq (K_2, h_2)$.

Proof. See Exercise 8. ∎

THEOREM 6.47. Let $(X, \mathcal{T})$ be a completely regular space, let (K, h) be a compactification of X with the property that if (Y, k) is any compactification of X such that each continuous function $f: X \to Y$ can be extended to a continuous function $F: K \to Y$ where $F \circ h = f$. Then K is homeomorphic to $\beta(X)$.

Proof. See Exercise 9. ∎

Thus $\beta(X)$ is the only compactification of X with the extension property of Theorem 6.42. The last theorem of the section provides additional information about the Stone-Čech compactification and uses the following terminology.

Definition. A subset A of a topological space $(X, \mathcal{T})$ is **C*-embedded** in X if every member of $C^*(A)$ can be extended to a member of $C^*(X)$. ∎

6.6 Compactifications

By Theorem 6.42, $e(X)$ is C^*-embedded in $\beta(X)$. The following theorem establishes that this property also characterizes $\beta(X)$.

THEOREM 6.48. Let $(X, \mathcal{T})$ be a completely regular space and let (Y, h) be a compactification of X such that $h(X)$ is C^*-embedded in Y. Then (Y, h) is the Stone-Čech compactification of X.

Proof. It is sufficient to show that the extension property of Theorem 6.42 holds for (Y, h). Let Y' be a compact Hausdorff space, and let $f: X \to Y'$ be a continuous function. Since Y' is completely regular, $C^*(Y')$ separates points from closed sets. Therefore by Theorem 5.42, the evaluation map $e: Y' \to \prod_{g \in C^*(Y')}[a_g, b_g]$ is an embedding. Thus we have the following situations.

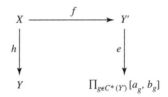

Since $h(X)$ is C^*-embedded in Y, for each $g \in C^*(Y')$, the function $g \circ f \circ h^{-1}: h(X) \to [a_g, b_g]$ has a continuous extension $k_g: Y \to [a_g, b_g]$. Define $G: Y \to \prod_{g \in C^*(Y')}[a_g, b_g]$ by $[G(y)]_g = k_g(y)$. Then for each $g \in C^*(Y')$ and each $y \in Y$, $(\pi_g \circ G)(y) = k_g(y)$ so $\pi_g \circ G$ is continuous. Therefore by Theorem 2.37, G is continuous. Let $y \in h(X)$. Then there exists $x \in X$ such that $h(x) = y$ and hence, for each $g \in C^*(Y')$,

$$[G(y)]_g = [G(h(x))]_g = k_g(h(x)) = (g \circ f \circ h^{-1})(h(x)) = g(f(X)) = e[f(x)]_g. \quad (*)$$

Therefore G maps $h(X)$ into $e(Y')$. Since $G(h(X))$ is dense in $G(Y)$ and $e(Y')$ is compact, $G(Y) \subseteq e(Y')$. Therefore $e^{-1} \circ G$ maps Y into Y'. By $(*)$, $(e^{-1} \circ G) \circ h = f$, and the proof is complete. ∎

Theorem 6.48 is useful in showing that a given compact Hausdorff space is or is not the Stone-Čech compactification of a given completely regular space. For example, I is not the Stone-Čech compactification of the open interval $(0, 1)$ because the function $f: (0, 1) \to \mathbb{R}$ defined by $f(x) = \sin \frac{1}{x}$ is a member of $C^*((0, 1))$ that does not have a continuous extension on I. (See Exercise 10.)

EXERCISES 6.6

1. Let $(X, \mathcal{T})$ be a topological space, let p be an object that does not belong to X, and let $Y = X \cup \{p\}$. Prove that $\mathcal{U} = \{U \in \mathcal{P}(Y): U \in \mathcal{T}$ or $Y - U$ is a closed compact subspace of $X\}$ is a topology on Y.

2. Let X be the open unit interval with the usual topology, and let $(Y, \mathcal{U})$ be the one-point compactification of X, where $Y = X \cup \{p\}$. Prove that the function $f: Y \to S^1$ defined by $f(p) = (1, 0)$ and $f(t) = (\cos 2\pi t, \sin 2\pi t)$ for all $t \in I$ is a one-to-one continuous function that maps Y onto S^1.

3. Prove that the function g defined in the proof of Theorem 6.39 is continuous.

4. Prove that the function f defined in the analysis of Example 18 is continuous.

5. Let e' be the function defined in the proof of Theorem 6.42. Prove that $e'(Y)$ is closed.

6. Prove Theorem 6.43.

7. Prove Theorem 6.45.

8. Prove Theorem 6.46.

9. Prove Theorem 6.47.

10. Prove that the function $f: (0, 1) \to \mathbb{R}$ defined by $f(x) = \sin\left(\frac{1}{x}\right)$ cannot be extended to a continuous function with domain I.

6.7 The Alexander Subbase and the Tychonoff Theorems

The last characterization of compactness is given in terms of subbasic open sets and is known as the Alexander Subbase Theorem. It was proved by J.W. Alexander (1888–1971) in 1939. The proof is difficult and uses Zorn's Lemma, and we introduce two new concepts that we use in the proof.

Definition. A collection of subsets of a topological space $(X, \mathcal{T})$ is **inadequate** provided the collection does not cover X. A collection of subsets of X is **finitely inadequate** provided no finite subcollection covers X. ∎

THEOREM 6.49. (Alexander Subbase Theorem). Let $(X, \mathcal{T})$ be a topological space and let $\mathcal{S}$ be a subbasis for $\mathcal{T}$. Then $(X, \mathcal{T})$ is compact if and only if every cover of X by members of $\mathcal{S}$ has a finite subcover.

Proof. Let $(X, \mathcal{T})$ be compact space and let $\mathcal{O}$ be a cover of X by members of $\mathcal{S}$. Then $\mathcal{O}$ is an open cover of X and hence it has a finite subcover.

Suppose every cover of X by members of $\mathcal{S}$ has a finite subcover. We prove that $(X, \mathcal{T})$ is compact by proving that every finitely inadequate collection of open sets is inadequate. Let $\mathcal{B}$ be a finitely inadequate collection of open sets, and let $\mathcal{P}$ be the

6.7 The Alexander Subbase and the Tychonoff Theorems 241

set of all finitely inadequate collections $\mathscr{G}$ of open sets such that $\mathscr{B} \subseteq \mathscr{G}$. Then $(\mathscr{P}, \subseteq)$ is a partially ordered set. Let C be a chain in $\mathscr{P}$, and let $\mathscr{G}^*$ be the union of all the members of C. Then $\mathscr{G}^*$ is a collection of open sets and $\mathscr{B} \subseteq \mathscr{G}^*$ because each member of C contains $\mathscr{B}$. Furthermore for each $\mathscr{G} \in C, \mathscr{G} \subseteq \mathscr{G}^*$. Suppose there is a finite subcollection $U_1, U_2, ..., U_N$ of $\mathscr{G}^*$ that covers X. For each $i = 1, 2, ..., N$, there exists $\mathscr{G}_i \in C$ such that $U_i \in \mathscr{G}_i$. Since C is a chain, there exists k with $1 \leq k \leq N$ such that $\mathscr{G}_k$ contains all the other members of C. Therefore for each $i = 1, 2, ..., N$, $U_i \in \mathscr{G}_k$. Since $\mathscr{G}_k$ is finitely inadequate, we have a contradiction. Therefore $\mathscr{G}^*$ is finitely inadequate. Hence $\mathscr{G}^* \in \mathscr{P}$, and $\mathscr{G}^*$ is an upper bound of C. By Zorn's Lemma, $\mathscr{P}$ has a maximal element $\mathscr{A}$.

We want to show that $\mathscr{B}$ is inadequate. Since $\mathscr{B} \subseteq \mathscr{A}$, it is sufficient to show that $\mathscr{A}$ is inadequate. We establish that $\mathscr{A}$ has the following properties:

(a) If $U \in \mathscr{T}$ and $U \notin \mathscr{A}$, then there exists a finite subcollection $U_1, U_2, ..., U_n$ of $\mathscr{A}$ such that $X = U \cup (\bigcup_{i=1}^{n} U_i)$.

(b) If $U_1, U_2, ..., U_n$ is a finite collection of open sets none of which are members of $\mathscr{A}$, then $\bigcap_{i=1}^{n} U_i \notin \mathscr{A}$.

(c) If $U_1, U_2, ..., U_n$ is a finite collection of open sets and V is a member of $\mathscr{A}$ such that $\bigcap_{i=1}^{n} U_i \subseteq V$, then there exists j with $1 \leq j \leq n$ such that $U_j \in \mathscr{A}$.

Since $\mathscr{A}$ is a maximal finitely inadequate collection of open sets, if $U \in \mathscr{T}$ and $U \notin \mathscr{A}$, then $\mathscr{A} \cup \{U\}$ is not finitely inadequate. Therefore *(a)* is true.

In order to establish *(b)* it is sufficient to show that if U_1 and U_2 are open sets that do not belong to $\mathscr{A}$, then $U_1 \cap U_2 \notin \mathscr{A}$. Let U_1 and U_2 be open sets that do not belong to $\mathscr{A}$. By *(a)*, there are finite subcollections $V_1, V_2, ..., V_m$ and $W_1, W_2, ..., W_n$ of $\mathscr{A}$ such that $X = U_1 \cup \bigcup_{i=1}^{m} V_i = U_2 \cup \bigcup_{j=1}^{n} W_j$. Hence $X = (U_1 \cap U_2) \cup \bigcup_{i=1}^{m} V_i \cup \bigcup_{j=1}^{n} W_j$. Since $\mathscr{A}$ is a maximal finitely inadequate collection of open sets, $U_1 \cap U_2 \notin \mathscr{A}$. Therefore *(b)* is true.

The proof of *(c)* is by contradiction. Suppose $U_1, U_2, ..., U_n$ is a finite collection of open sets none of which are members of $\mathscr{A}$ and V is a member of $\mathscr{A}$ such that $\bigcap_{i=1}^{n} U_i \subseteq V$. By *(b)*, $\bigcap_{i=1}^{n} U_i \notin \mathscr{A}$. By *(a)*, there is a finite subcollection $V_1, V_2, ..., V_m$ of $\mathscr{A}$ such that $X = (\bigcap_{i=1}^{n} U_i) \cup (\bigcup_{j=1}^{m} V_j)$. Since $\bigcap_{i=1}^{n} U_i \subseteq V$, $X = V \cup (\bigcup_{j=1}^{m} V_j)$. This is a contradiction since $\mathscr{A}$ is finitely inadequate. Therefore *(c)* is true.

In Exercise 1, you are asked to explain why $\mathscr{S} \cap \mathscr{A} \neq \emptyset$. The collection $\mathscr{S} \cap \mathscr{A}$ is finitely inadequate because it is a subset of $\mathscr{A}$ and $\mathscr{A}$ is finitely inadequate. Therefore $\mathscr{S} \cap \mathscr{A}$ is inadequate because it is a subset of $\mathscr{S}$, and we are assuming that every cover of X by members of $\mathscr{S}$ has a finite subcover. Thus, in order to show that $\mathscr{A}$ is inadequate, it is sufficient to show that $\bigcup_{A \in \mathscr{A}} A \subseteq \bigcup_{S \in \mathscr{S} \cap \mathscr{A}} S$. Let $p \in \bigcup_{A \in \mathscr{A}} A$. Then there is a member V of $\mathscr{A}$ such that $p \in V$. Since $\mathscr{S}$ is a subbasis for $\mathscr{T}$ and $V \in \mathscr{T}$, there exists a finite subcollection $V_1, V_2, ..., V_n$ of $\mathscr{S}$ such that $p \in \bigcup_{i=1}^{n} V_i \subseteq V$. By *(c)*, there exists k with $1 \leq k \leq n$ such that $V_k \in \mathscr{A}$. Therefore $p \in V_k \in \mathscr{S} \cap \mathscr{A}$. Hence $\bigcup_{A \in \mathscr{A}} A \subseteq \bigcup_{S \in \mathscr{S} \cap \mathscr{A}} S$. Thus $\mathscr{A}$ is inadequate, and the proof is complete. ∎

The definition of product spaces was formulated by A.N. Tychonoff in 1930, and in 1935 he proved that the product of compact spaces is compact. This result is one of the most important theorems in topology. We use the Alexander Subbase Theorem to prove this result.

THEOREM 6.50. (Tychonoff Theorem). The product of compact spaces is compact.

Proof. Let $\{(X_\alpha, \mathcal{T}_\alpha): \alpha \in \Lambda\}$ be a collection of compact spaces, and let $X = \prod_{\alpha \in \Lambda} X_\alpha$. Let $\mathcal{S}$ be the subbase for the product topology on X that consists of all sets of the form $\pi_\alpha^{-1}(U_\alpha)$, where $U_\alpha \in \mathcal{T}_\alpha$. By the Alexander Subbase Theorem, it is sufficient to show that every cover of X by members of $\mathcal{S}$ has a finite subcover.

We prove the contrapositive. Suppose $\mathcal{O}$ is a collection of members of $\mathcal{S}$ that has no finite subcover. For each $\alpha \in \Lambda$, let $\mathcal{O}_\alpha = \{U_\alpha \in \mathcal{T}_\alpha: \pi_\alpha^{-1}(U_\alpha) \in \mathcal{O}\}$. Notice that if there is a finite subcollection $\mathcal{O}_\alpha'$ of $\mathcal{O}_\alpha$ that covers X_α, then $\{\pi_\alpha^{-1}(U_\alpha): U_\alpha \in \mathcal{O}_\alpha'\}$ is a finite subcollection of $\mathcal{O}$ that covers X. Therefore no finite subcollection of $\mathcal{O}_\alpha$ covers X_α. Since X_α is compact, $\mathcal{O}_\alpha$ does not cover X_α. Therefore for each $\alpha \in \Lambda$, there is member x_α of X_α such that x_α does not belong to any member of $\mathcal{O}_\alpha$. Thus $\langle x_\alpha \rangle_{\alpha \in \Lambda}$ does not belong to any member of $\mathcal{O}$. Thus $\mathcal{O}$ does not cover X. ∎

We conclude this chapter with an example of a normal space that is not completely normal.

EXAMPLE 20. Let $(\Omega, \leq)$ be an uncountable well-ordered set with a maximal element ω_1 having the property that if $x \in \Omega$ and $x \neq \omega_1$, then $\{y \in \Omega: y \leq x\}$ is countable (see Appendix H). Let ω_0 be the first infinite ordinal, and let $[1, \omega_0]$ denote $\{y \in \Omega: y \leq \omega_0\}$. Let $\mathcal{T}$ and $\mathcal{U}$ denote the order topologies on Ω and $[1, \omega_0]$ respectively. Then the product space $T = \Omega \times [1, \omega_0]$ is called the **Tychonoff plank**, and it is a normal space with a subspace $T_\infty = T - \{(\omega_1, \omega_0)\}$, called the **deleted Tychonoff plank,** that is not normal.

Analysis. Since every nonempty subset of Ω, and of $[1, \omega_0]$, has a greatest lower bound and a least upper bound, by Exercise 8 of Section 4.2, $(\Omega, \mathcal{T})$ and $([1, \omega_0], \mathcal{U})$ are compact. Therefore by Theorem 4.27, T is compact. By Exercise 17 of Section 1.6, $(\Omega, \mathcal{T})$ and $([1, \omega_0], \mathcal{U})$ are Hausdorff. Therefore by Theorem 2.45, T is Hausdorff. Hence by Corollary 5.22, T is normal.

We now prove that T_∞ is not normal. Let $A = \{(\omega_1, y): y \in [1, \omega_0] \text{ and } y \neq \omega_0\}$ and $B = \{(x, \omega_0): x \in \Omega \text{ and } x \neq \omega_1\}$. Then A and B are disjoint closed subsets of T_∞. Let U be an open set in T_∞ such that $A \subseteq U$. For each $(\omega_1, y) \in A$, there exists $\alpha_y \in \Omega - \{\omega_1\}$ such that $\{(x, y) \in T_\infty: \alpha_y < x \leq \omega_1\} \subseteq U$. Let $\bar{\alpha} = \text{lub}\{\alpha_y: (\omega_1, y) \in A\}$. Since $\bar{\alpha}$ has only countably many predecessors while ω_1 has uncountably many, $\bar{\alpha} \in \omega_1$. (Why?) Therefore $\{(x, y): \bar{\alpha} < x \leq \omega_1 \text{ and } y \in [1, \omega_0] - \{\omega_0\}\} \subseteq U$. Let $\bar{\alpha} + 1$ denote the immediate successor of $\bar{\alpha}$. Then $(\bar{\alpha} + 1, \omega_0) \in B$, and every neighborhood of $(\bar{\alpha} + 1, \omega_0)$ in T_∞ has a nonempty intersection with U. Therefore if V is an open set in T_∞ such that $B \subseteq V$, then $U \cap V \neq \emptyset$. Therefore T_∞ is not normal.

By Theorem 5.29, T is not completely normal. ∎

EXERCISES 6.7

1. In the proof of the Alexander Subbase Theorem, explain why $\mathcal{S} \cap \mathcal{A} \neq \emptyset$.

2. A topological space $(X, \mathcal{T})$ is **realcompact** if there exists a nonempty set Λ such that X can be imbedded in $\prod_{\alpha \in \Lambda} R_\alpha$, Where $R_\alpha = \mathbb{R}$ for each $\alpha \in \Lambda$.

   (a) Prove that every compact Hausdorff space is realcompact.

   (b) Let $(X, \mathcal{T})$ be a topological space and for each α in a nonempty set Λ, let A_α, be a realcompact subset of X. Prove that $\bigcap_{\alpha \in \Lambda} A_\alpha$ is realcompact.

   (c) Prove that the product of realcompact spaces is realcompact.

   (d) Let $(X, \mathcal{T})$ be a completely regular space. Construct a realcompact space that contains a dense subset that is homeomorphic to X.

3. For a topological space $(X, \mathcal{T})$, let 2^X denote the set whose members are the closed subsets of X. For each finite collection, $U_1, U_2, ..., U_n$, of open subsets of X, let $<U_1, U_2, ..., U_n>$ denote $\{A \in 2^X : A \cap U_i \neq \emptyset$ for each $i = 1, 2, ..., n$ and $A \subseteq \bigcup_{i=1}^n U_i\}$.

   (a) Prove that the collection $\{<U_1, U_2, ..., U_n> : U_1, U_2, ..., U_n$ are open subsets of $X\}$ forms a basis for a topology $\mathcal{U}$ on 2^X.

   (b) Prove that if $(X, \mathcal{T})$ is compact, then so is $(2^X, \mathcal{U})$. Hint: Use the Alexander Subbase Theorem.

4. Is the Tychonoff plank perfectly normal? Prove your answer.

Chapter 7
Metrizability and Paracompactness

We have studied metric spaces, and we have seen that a metric on a set generates a topology on that set. A natural question that arises is: which topologies are generated by metrics? More precisely, can we find conditions, stated in terms of the topology (that is, open sets) which guarantee that there is a metric that generates the given topology? There are many nice results along this line. Early results, due to Urysohn, were proved in 1923 and 1924. The final question has not been answered, but significant progress has been made. In particular, in the early 1950s, J. Nagata, Y. Smirnov, and R.H. Bing independently proved the general metrization theorem.

7.1 Urysohn's Metrization Theorem

The proof of Urysohn's metrization Theorem involves imbedding a space that satisfies certain conditions in a metric space. Then, since every subspace of a metric space is a metric space (Theorem 2.9) and metrizability is a topological property (Theorem 1.57), the desired result follows. We know that the product of two metric spaces is metrizable (Theorem 2.16). Thus it follows that the product of any finite number of metric spaces is metrizable. We also know that $\mathbb{R}^\omega$ is metrizable (Theorem 2.50). We begin by showing that the product of a countable number of metric spaces is metrizable.

THEOREM 7.1. For each $n \in \mathbb{N}$, let (X_n, d_n) be a metric space, let $X = \prod_{n \in \mathbb{N}} X_n$, and let $\mathcal{T}$ be the product topology on X. Then $(X, \mathcal{T})$ is metrizable.

Proof. By Theorem 1.30, we may assume that for each $n \in \mathbb{N}$, $d_n(x_n, y_n) \leq 1$ for all $x_n, y_n \in X_n$. Define $d(x, y) = \sum_{n=1}^{\infty} d_n(x_n, y_n)/2^n$ for all $x = (x_1, x_2, \ldots), y = (y_1, y_2, \ldots) \in X$. In Exercise 1 you are asked to prove that d is a metric on X. Let $\mathcal{U}$ be the topology generated by d. We claim that $\mathcal{U} = \mathcal{T}$.

Let $U \in \mathcal{T}$ and let $p \in U$. Then there exist integers $i_1, i_2, \ldots, i_m$ and open sets $U_{i_1}, U_{i_2}, \ldots, U_{i_m}$ in $X_{i_1}, X_{i_2}, \ldots, X_{i_m}$ such that $p \in \bigcap_{j=1}^{m} \pi_{i_j}^{-1}(U_{i_j}) \subseteq U$. For each $j = 1, 2, \ldots, m$, there exists $\varepsilon_{i_j} > 0$ such that $B_{d_{i_j}}(p_{i_j}, \varepsilon_{i_j}) \subseteq U_{i_j}$. Let $\varepsilon = \min\{\varepsilon_{i_j}/2^{i_j} : j = 1, 2, \ldots, m\}$. Then if $x \in X$ and $d(p, x) < \varepsilon$, $d_{i_j}(p_{i_j}, x_{i_j}) < \varepsilon_{i_j}$ for each $j = 1, 2, \ldots, m$. Therefore $B_d(p, \varepsilon) \subseteq U$ and so $\bigcup_{p \in U} B_d(p, \varepsilon) = U \in \mathcal{U}$.

Now let $U \in \mathcal{U}$ and the $p \in U$. Then there exists $\varepsilon > 0$ such that $B_d(p, \varepsilon) \subseteq U$, and there exists $N \in \mathbb{N}$ such that $\sum_{i=N+1}^{\infty} 1/2^i < \varepsilon/2$. Then $\bigcap_{i=1}^{N} \pi_i^{-1}(B_{d_i}(p_i, \varepsilon/2N))$ $\in \mathcal{T}$ such that $p \in \bigcap_{i=1}^{N} \pi_i^{-1}(B_{d_i}(p_i, \varepsilon/2N))$. In Exercise 2, you are asked to prove that $\bigcap_{i=1}^{N} \pi_i^{-1}(B_{d_i}(p_i, \varepsilon/2N)) \subseteq U$. It follows that $U \in \mathcal{T}$.

EXAMPLE 1. It follows from Theorem 7.1 that the space I^ω, obtained by taking the countably infinite product of I with itself, is metrizable. This space is called the **Hilbert cube**. Some mathematicians would rather think of the Hilbert cube as being the product $[0, 1] \times [0, \frac{1}{2}] \times [0, \frac{1}{3}] \times \cdots$. To us it makes no difference.

Urysohn's Metrization Theorem tells us that every regular second countable space is metrizable. The proof involves showing that every such space can be imbedded in the Hilbert cube.

THEOREM 7.2 (Urysohn's Metrization Theorem). Let $(X, \mathcal{T})$ be a T_1-space. Then the following are equivalent:

(a) $(X, \mathcal{T})$ is regular and second countable.

(b) X can be imbedded as a subspace of I^ω.

(c) $(X, \mathcal{T})$ is a separable metric space.

Proof. (a) $\to$ (b). Suppose $(X, \mathcal{T})$ is a regular second countable space. Then, by Corollary 5.33, $(X, \mathcal{T})$ is normal. Let $\mathcal{B}$ be a countable basis for $\mathcal{T}$, and let $\mathcal{A} = \{(U, V) : U, V \in \mathcal{B} \text{ and } \overline{U} \subseteq V\}$. Then $\mathcal{A}$ is countable, and, by Urysohn's Lemma (Theorem 5.36), for each $(U, V) \in \mathcal{A}$ there exists a continuous function $f_{UV} : X \to I$ such that $f_{UV}(x) = 0$ for all $x \in \overline{U}$ and $f_{UV}(x) = 1$ for all $x \in X - V$. Let $\mathcal{F} = \{f_{UV} : (U, V) \in \mathcal{A}\}$. Then $\mathcal{F}$ is a countable collection of continuous functions that separates points from closed sets. Therefore, by Theorem 5.42, the evaluation map $e : X \to I^\mathcal{F}$ defined by $[e(x)]_f = f(x)$ is an imbedding. Since $\mathcal{F}$ is countable, $I^\mathcal{F}$ is homeomorphic to I^ω and the proof is complete.

(b) $\to$ (c) Suppose X can be imbedded as a subspace of I^ω. Since I^ω is the product of a countable number of second countable spaces, it is second countable (Theorem 2.48). Since every subspace of a second countable space is second countable (Theorem 2.10) and second countability is a topological property (Exercise 5 of Section 1.7), $(X, \mathcal{T})$ is second countable. Since every second countable space is separable (Theorem 1.23), $(X, \mathcal{T})$ is separable. By Example 1, I^ω is a metric space. Since every subspace of a metric space is a metric space (Theorem 2.9), $(X, \mathcal{T})$ is a metric space.

(c) → (a) Suppose $(X, \mathcal{T})$ is a separable metric space. Then by Theorem 1.31, $(X, \mathcal{T})$ is second countable. Since every metric space is regular (Theorems 5.17 and 5.18), the proof is complete. ∎

Theorem 7.2. completely characterizes the second countable regular spaces; that is, it shows that the collection of second countable regular spaces is the same as the collection of separable metric spaces. The **(a)** implies **(b)** part of the proof of Theorem 7.2 uses a result about the evaluation map that was established in an optional section. In Exercise 3 we outline a proof of **(a)** implies **(b)** that does not depend on this result.

The following theorem is an important consequence of the Urysohn Metrization Theorem.

THEOREM 7.3. Let (X, d) be a compact metric space, let $(Y, \mathcal{U})$ be a Hausdorff space, and let $f: X \to Y$ be a continuous function that maps X onto Y. Then $(Y, \mathcal{U})$ is metrizable.

Proof. Since the continuous image of a compact space is compact (Theorem 4.19), $(Y, \mathcal{U})$ is compact. Since every compact Hausdorff space is normal (Corollary 5.22), $(Y, \mathcal{U})$ is regular. By Urysohn's Metrization Theorem, it is sufficient to show that $(Y, \mathcal{U})$ is second countable.

By Exercise 9 of Section 4.1, the topology $\mathcal{T}$ generated by d has a countable basis $\mathcal{B}$. Let $\mathcal{A}$ be the collection of all finite unions of members of $\mathcal{B}$, and let $\mathcal{C} = \{Y - f(X - A): A \in \mathcal{A}\}$. By Theorem 4.21, f is a closed mapping. Therefore $\mathcal{C}$ is a countable collection of members of $\mathcal{U}$. We claim that $\mathcal{C}$ is a basis for $\mathcal{U}$. Let $U \in \mathcal{U}$ and let $y \in U$. Then $f^{-1}(y) \subseteq f^{-1}(U)$. Since $f^{-1}(y)$ is a closed subset of a compact space, it is compact (Corollary 4.16). Since $\mathcal{B}$ is a basis for $\mathcal{T}$, for each $x \in f^{-1}(y)$, there exists $B_x \in \mathcal{B}$ such that $x \in B_x \subseteq f^{-1}(U)$. Then $\{B_x \in \mathcal{B}: x \in B_x \subseteq f^{-1}(U)\}$ is an open cover of $f^{-1}(y)$, and hence there exist $x_1, x_2, ..., x_n \in f^{-1}(y)$ such that $f^{-1}(y) \subseteq \bigcup_{i=1}^n B_{x_i} \subseteq f^{-1}(U)$. Now $A = \bigcup_{i=1}^n B_{x_i} \in \mathcal{A}$, so $Y - f(X - A) \in \mathcal{C}$. We complete the proof by showing that $y \in Y - f(X - A)$ and $Y - f(X - A) \subseteq U$. Since $f^{-1}(y) \subseteq A$, there does not exist $x \in X - A$ such that $f(x) = y$. Thus $f^{-1}(y) \cap (X - A) = \emptyset$, and hence $y \in Y - f(X - A)$. Let $z \in Y - f(X - A)$. Then $z \notin f(X - A)$. Since f maps X onto Y, there exists $x \in X$ such that $f(x) = z$. Since $f(x) \notin f(X - A)$, $x \in A$. Since $A \subseteq f^{-1}(U), f(x) = z \in U$. Therefore $Y - f(X - A) \subseteq U$. ∎

EXERCISES 7.1

1. Show that the function d defined in the proof of Theorem 7.1 is a metric.

2. Complete the proof of Theorem 7.1 by showing that $\bigcap_{i=1}^N \pi_i^{-1}(B_{d_i}(p_i, \varepsilon/2N)) \subseteq U$.

3. Let $(X, \mathcal{T})$ be a regular second countable space and let $\mathcal{B} = \{B_i : i \in \mathbb{N}\}$ be a countable basis for $\mathcal{T}$. By Corollary 5.33, $(X, \mathcal{T})$ is normal. Let $\Lambda = \{(i, j) \in \mathbb{N} \times \mathbb{N} : \overline{B_i} \subseteq B_j\}$. Then Λ is countable, so we may assume that $\Lambda = \mathbb{N}$. Let $k \in \mathbb{N}$. Then there exist $i, j \in \mathbb{N}$ such that $k = (i, j)$. By Urysohn's Lemma, there exists a continuous function $f_k : X \to [0, 1]$ such that $f_k(x) = 0$ for all $x \in \overline{B_i}$ and $f_k(x) = 1$ for all $x \in X - B_j$. Define $f : X \to I^\omega$ by

$$f(x) = \left(\frac{f_1(x)}{2}, \frac{f_2(x)}{2^2}, \ldots, \frac{f_k(x)}{2^k}, \ldots \right)$$

for each $x \in X$.

(a) Show that f is one-to-one.

(b) Show that f is continuous.

(c) Show that f is an open mapping from X onto $f(X)$.

4. Prove that every locally compact, second countable Hausdorff space is metrizable.

5. Prove that every n-manifold is metrizable.

6. Give an example of a second countable Hausdorff space that is not metrizable.

7. Let Λ be a nonempty set and for each $\alpha \in \Lambda$, let $(X_\alpha, \mathcal{T}_\alpha)$ be a topological space. For each $\alpha \in \Lambda$, let $X_\alpha^* = X_\alpha \times \{\alpha\}$. Let $\mathcal{T}_\alpha^*$ be the product topology on X_α^*. (Note that $X_\alpha^* \cap X_\beta^* = \emptyset$ for all $\alpha, \beta \in \Lambda$). Let $X = \bigcup_{\alpha \in \Lambda} X_\alpha^*$, and and let $\mathcal{T} = \{U \in \mathcal{P}(X) : U \cap X_\alpha^* \in \mathcal{T}_\alpha^* \text{ for each } \alpha \in \Lambda\}$.

(a) Prove that $(X_\alpha^*, \mathcal{T}_\alpha^*)$ is homeomorphic to $(X_\alpha, \mathcal{T}_\alpha)$.

(b) Prove that $\mathcal{T}$ is a topology on X.

The space $(X, \mathcal{T})$ is called the **disjoint union** of $\{(X_\alpha, \mathcal{T}_\alpha) : \alpha \in \Lambda\}$.

8. Let Λ be an infinite set and for each $\alpha \in \Lambda$, let $I_\alpha = [0, 1]$ and let $\mathcal{T}_\alpha$ be the usual topology on I_α. Let $(X, \mathcal{T})$ be the disjoint union of $\{(I_\alpha, \mathcal{T}_\alpha) : \alpha \in \Lambda\}$. We define an equivalence relation $\sim$ on X by $x \sim y$ if and only if: (1) there exists $\alpha \in \Lambda$ such that $x, y \in I_\alpha$ and $x = y$, or (2) $x = y = 0$. Let $(Y, \mathcal{U})$ denote the resulting quotient space. (Intuitively we have a disjoint collection of closed unit intervals and we are identifying the left endpoints.) Define $d : Y \times Y \to \mathbb{R}$ by $d(x, y) = |x - y|$ if there exists $\alpha \in \Lambda$ such that $x, y \in I_\alpha$ and $d(x, y) = |x| + |y|$ if there does not exist $\alpha \in \Lambda$ such that $x, y \in I_\alpha$.

(a) Prove that d is a metric on Y.

(b) Is $\mathcal{U}$ the topology generated by d? Prove your answer.

7.2 Paracompactness

Dieudonné introduced paracompact spaces in 1944 as a natural generalization of compact spaces. In 1948, A.H. Stone proved that every metric space is paracompact, and, with this result and its subsequent use in the solutions of the general metrization problem by Bing, Nagata, and Smirnov, paracompactness gained in stature. Since every compact Hausdorff space is paracompact, two important classes of spaces, the compact Hausdorff spaces and the metric spaces, are paracompact. We begin our study by introducing some terminology and obtaining some preliminary results.

Definition. Let X be a set and let $\mathcal{U}$ and $\mathcal{V}$ be covers of X. We say that $\mathcal{U}$ is a **refinement** of $\mathcal{V}$, or $\mathcal{U}$ **refines** $\mathcal{V}$, and write $\mathcal{U} < \mathcal{V}$, provided that for each $U \in \mathcal{U}$ there exists $V \in \mathcal{V}$ such that $U \subseteq V$. ∎

EXAMPLE 2. Let $\mathcal{U}$ be the collection of squares shown in Figure 7.1, and let $\mathcal{V}$ be the collection of intersections of X with circles as shown in Figure 7.1. Then $\mathcal{U}$ and $\mathcal{V}$ cover **X** and $\mathcal{U}$ is a refinement of $\mathcal{V}$.

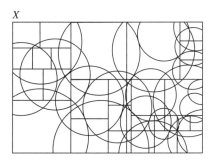

Figure 7.1

EXAMPLE 3. Let $\mathcal{U} = \{(a, b): a, b \in \mathbb{R} \text{ and } a < b\}$ and let $\mathcal{V} = \{[a, b]: a, b, \in \mathbb{R} \text{ and } a < b\}$. Then $\mathcal{U}$ and $\mathcal{V}$ cover $\mathbb{R}$ and $\mathcal{U}$ refines $\mathcal{V}$.

Definition. Let X be a set, let A be a subset of X, and let $\mathcal{U}$ be a cover of X. The **star of A with respect to** $\mathcal{U}$, denoted by $\text{St}(A, \mathcal{U})$, is $\bigcup \{U \in \mathcal{U}: A \cap U \neq \emptyset\}$. If $x \in X$, we write $\text{St}(x, \mathcal{U})$ to mean $\text{St}(\{x\}, \mathcal{U})$. ∎

EXAMPLE 4. If A is the rectangle shown in Figure 7.2, X is the union of all the circles shown in Figure 7.2, and $\mathcal{U}$ is the collection of circles shown in Figure 7.2, then $\text{St}(A, \mathcal{U})$ is the shaded area.

Definition. Let $\mathcal{U}$ and $\mathcal{V}$ be covers of a set X. Then $\mathcal{U}$ is a **star-refinement** of $\mathcal{V}$, or $\mathcal{U}$ **star-refines** $\mathcal{V}$, denoted by $\mathcal{U}^* < \mathcal{V}$, provided that for each $U \in \mathcal{U}$ there exists $V \in \mathcal{V}$ such that $\text{St}(U, \mathcal{U}) \subseteq V$. ∎

250   Chapter 7   ■   Metrizability and Paracompactness

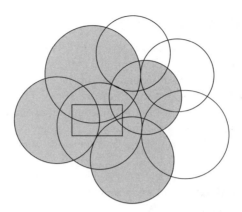

Figure 7.2

Definition. Let $\mathcal{U}$ and $\mathcal{V}$ be covers of a set X. Then $\mathcal{U}$ is a **barycentric refinement** of $\mathcal{V}$, denoted by $\mathcal{U} \Delta \mathcal{V}$, provided $\{St(x, \mathcal{U}): x \in X\}$ is a refinement of $\mathcal{V}$. ■

The relation between star-refinement and barycentric refinement is given by the following two theorems.

THEOREM 7.4. Let $\mathcal{U}$ and $\mathcal{V}$ be covers of a set X such that $\mathcal{U}^* < \mathcal{V}$. Then $\mathcal{U} \Delta \mathcal{V}$.

Proof. Let $x \in X$. Since $\mathcal{U}$ covers X, there exists $U \in \mathcal{U}$ such that $x \in U$. Since $\mathcal{U}^* < \mathcal{V}$, there exists $V \in \mathcal{V}$ such that $St(U, \mathcal{U}) \subseteq V$. Since $x \in U$, $St(x, \mathcal{U}) \subseteq St(U, \mathcal{U})$. Hence $St(x, \mathcal{U}) \subseteq V$, and so $\mathcal{U} \Delta \mathcal{V}$. ■

THEOREM 7.5. Let $\mathcal{U}, \mathcal{V}$, and $\mathcal{W}$ be covers of a set X such that $\mathcal{U} \Delta \mathcal{V}$ and $\mathcal{V} \Delta \mathcal{W}$. Then $\mathcal{U}^* < \mathcal{W}$.

Proof. See Exercise 1. ■

Definition. A collection $\mathcal{U}$ of subsets of a set X is **point finite** provided each $x \in X$ belongs to only a finite number of members of $\mathcal{U}$. ■

Definition. A collection $\mathcal{U}$ of subsets of a topological space $(X, \mathcal{T})$ is **locally finite** provided each $x \in X$ has a neighborhood that intersects only a finite number of members of $\mathcal{U}$. ■

Notice that every locally finite collection of subsets of a topological space is point finite.

EXAMPLE 5. Let $\mathcal{T}$ be the usual topology on $\mathbb{R}$. Then $\{\{x\}: x \in \mathbb{R}\}$ is a point finite collection of subsets of $\mathbb{R}$ that is not locally finite.

7.2 Paracompactness

Definition. A collection $\mathcal{U}$ of subsets of a topological space $(X, \mathcal{T})$ is **discrete** provided each $x \in X$ has a neighborhood that intersects at most one member of $\mathcal{U}$. ∎

Notice that every discrete collection of subsets of a topological space is locally finite.

EXAMPLE 6. Let $\mathcal{T}$ be the usual topology on $\mathbb{R}$. Then $\{[n, n+1): n \in \mathbb{Z}\}$ is a locally finite collection of subsets that is not discrete.

Definition. A collection $\mathcal{U}$ of subsets of a topological space $(X, \mathcal{T})$ is σ-**locally finite** provided $\mathcal{U} = \bigcup_{n \in \mathbb{N}} \mathcal{U}_n$, where each $\mathcal{U}_n$ is a locally finite collection of subsets of X. ∎

EXAMPLE 7. Let $\mathcal{T}$ be the usual topology on $\mathbb{R}$. For each $n \in \mathbb{N}$, let $\mathcal{U}_n = \left\{ \left[p + \frac{1}{n}, p + 1 + \frac{1}{n} \right) : p \in \mathbb{Z} \right\}$. Since, for each $n \in \mathbb{N}$, $\mathcal{U}_n$ is locally finite, $\mathcal{U} = \bigcup_{n \in \mathbb{N}} \mathcal{U}_n$ is σ-locally finite.

Definition. A collection $\mathcal{U}$ of subsets of a topological space $(X, \mathcal{T})$ is σ-**discrete** provided $\mathcal{U} = \bigcup_{n \in \mathbb{N}} \mathcal{U}_n$, where each $\mathcal{U}_n$ is a discrete collection of subsets of X. ∎

EXAMPLE 8. Let $\mathcal{T}$ be the lower-limit topology on $\mathbb{R}$. For each $n \in \mathbb{N}$, let $\mathcal{U}_n = \left\{ \left[p + \frac{1}{n}, p + 1 + \frac{1}{n} \right) : p \in \mathbb{Z} \right\}$. Since, for each $n \in \mathbb{N}$, $\mathcal{U}_n$ is descrete, $\mathcal{U} = \bigcup_{n \in \mathbb{N}} \mathcal{U}_n$ is σ-discrete.

You should be aware that if $\mathcal{U}$ is a σ-locally finite (σ-discrete) cover of a topological space $(X, \mathcal{T})$, then the locally finite (discrete) subcollection $\mathcal{U}_n$ of $\mathcal{U}$ need not cover X.

THEOREM 7.6. Let $\mathcal{U}$ be a locally finite collection of subsets of a topological space $(X, \mathcal{T})$. Then $\{\overline{U} : U \in \mathcal{U}\}$ is locally finite.

Proof. Let $x \in X$. Then there exists a neighborhood V of x that intersects only a finite number of members of $\mathcal{U}$. Suppose $U \in \mathcal{U}$ and $V \cap U = \emptyset$. Then no member of V belongs to $\overline{U}$, so $V \cap \overline{U} = \emptyset$. Therefore V intersects only a finite number of members of $\mathcal{U}$. ∎

If $\{A_\alpha : \alpha \in \Lambda\}$ is a collection of closed sets, then $\bigcup_{\alpha \in \Lambda} A_\alpha$ is not necessarily closed. However, the following result establishes that if $\{A_\alpha : \alpha \in \Lambda\}$ is locally finite, then $\bigcup_{\alpha \in \Lambda} A_\alpha$ is closed.

THEOREM 7.7. Let $\{A_\alpha : \alpha \in \Lambda\}$ be a locally finite collection of subsets of a topological space. Then $\bigcup_{\alpha \in \Lambda} \overline{A_\alpha} = \overline{\bigcup_{\alpha \in \Lambda} A_\alpha}$.

Proof. We do not need local finiteness in order to prove that $\bigcup_{\alpha \in \Lambda} \overline{A_\alpha} \subseteq \overline{\bigcup_{\alpha \in \Lambda} A_\alpha}$. Let $x \in \bigcup_{\alpha \in \Lambda} \overline{A_\alpha}$. Then there exists $\beta \in \Lambda$ such that $x \in \overline{A_\beta}$. Thus every neighborhood of x intersects A_β, and hence every neighborhood of x intersects $\bigcup_{\alpha \in \Lambda} A_\alpha$. Therefore $x \in \overline{\bigcup_{\alpha \in \Lambda} A_\alpha}$ and so $\bigcup_{\alpha \in \Lambda} \overline{A_\alpha} \subseteq \overline{\bigcup_{\alpha \in \Lambda} A_\alpha}$.

Now let $x \in \overline{\bigcup_{\alpha \in \Lambda} A_\alpha}$. Since $\{A_\alpha : \alpha \in \Lambda\}$ is locally finite, there exists a neighborhood U of x that intersects only a finite number of members $A_{\alpha_1}, A_{\alpha_2}, \ldots, A_{\alpha_n}$ of $\{A_\alpha : \alpha \in \Lambda\}$. Suppose there exists a neighborhood V of x that does not intersect $\bigcup_{i=1}^n A_{\alpha_i}$. Then $U \cap V$ is a neighborhood of x that does not intersect $\bigcup_{\alpha \in \Lambda} A_\alpha$. This is a contradiction, since $x \in \overline{\bigcup_{\alpha \in \Lambda} A_\alpha}$. Therefore every neighbourhood of x intersects $\bigcup_{i=1}^n A_{\alpha_i}$. Hence $x \in \overline{\bigcup_{i=1}^n A_{\alpha_i}}$. But $\overline{\bigcup_{i=1}^n A_{\alpha_i}} = \bigcup_{i=1}^n \overline{A_{\alpha_i}}$, so $x \in \overline{A_{\alpha_i}}$ for some $i = 1, 2, \ldots, n$. Thus $x \in \bigcup_{\alpha \in \Lambda} \overline{A_\alpha}$, and so $\overline{\bigcup_{\alpha \in \Lambda} A_\alpha} \subseteq \bigcup_{\alpha \in \Lambda} \overline{A_\alpha}$. ∎

Definition. A **paracompact space** $(X, \mathcal{T})$ is a Hausdorff space with the property that every open cover of X has an open locally finite refinement. ∎

In the 1950s, E. Michael did a lot of work on paracompactness, and the following theorem, due to him, gives several characterizations of paracompactness in the class of regular spaces.

THEOREM 7.8. Let $(X, \mathcal{T})$ be a regular space. Then the following are equivalent:

(a) $(X, \mathcal{T})$ is paracompact.

(b) Every open cover of X has an open σ-locally finite refinement.

(c) Every open cover of X has a locally finite refinement.

(d) Every open cover of X has a closed locally finite refinement.

Proof. (a) → (b). Let $\mathcal{U}$ be an open cover of X. Then there exists an open locally finite refinement $\mathcal{V}$ of $\mathcal{U}$. Now $\mathcal{V}$ is an open σ-locally finite refinement, and the proof is complete.

(b) → (c). Let $\mathcal{U}$ be an open cover of X. Then there exists an open σ-locally finite refinement $\mathcal{V}$ of $\mathcal{U}$. So $\mathcal{V} = \bigcup_{n \in \mathbb{N}} \mathcal{V}_n$, where each $\mathcal{V}_n$ is locally finite. For each $n \in \mathbb{N}$, let $W_n = \bigcup\{V : V \in \mathcal{V}_n\}$. Then $\{W_n : n \in \mathbb{N}\}$ is an open cover of X. For each $n \in \mathbb{N}$, let $A_n = W_n - \bigcup_{i=1}^{n-1} W_i$. It is clear that $\{A_n : n \in \mathbb{N}\}$ is a refinement of $\{W_n : n \in \mathbb{N}\}$. Let $x \in X$, and let n_x be the smallest member of $\{n \in \mathbb{N} : x \in W_n\}$. Then $x \in A_{n_x}$, and hence $\{A_n : n \in \mathbb{N}\}$ covers X. Also W_{n_x} is a neighborhood of x that does not intersect A_n for any $n > n_x$, and so $\{A_n : n \in \mathbb{N}\}$ is locally finite. Let $\mathcal{A} = \{A_n \cap V : n \in \mathbb{N} \text{ and } V \in \mathcal{V}_n\}$. Since $\mathcal{V}$ is a refinement of $\mathcal{U}$, $\mathcal{A}$ refines $\mathcal{U}$. Let $x \in X$. Since $\{A_n : n \in \mathbb{N}\}$ is locally finite, there exists a neighborhood M of x that intersects only a finite number of members $A_{n_1}, A_{n_2}, \ldots, A_{n_k}$ of $\{A_n : n \in \mathbb{N}\}$. For each $i = 1, 2, \ldots, k$, there exists a neighborhood N_{n_i} of x that intersects only a finite number of members of $\mathcal{V}_{n_i}$. Then $M \cap \bigcap_{i=1}^k N_{n_i}$ is a neighborhood of x that intersects only a finite number of members of $\mathcal{A}$. Therefore $\mathcal{A}$ is locally finite, and so $\mathcal{A}$ is the desired locally finite refinement of $\mathcal{U}$.

(c) → (d). Let $\mathcal{U}$ be an open cover of X. For each $x \in X$, let $U_x \in \mathcal{U}$ such that $x \in U_x$. Since $(X, \mathcal{T})$ is regular, for each $x \in X$, there exists a neighborhood V_x of x such that $\overline{V_x} \subseteq U_x$. Then $\{V_x : x \in X\}$ is an open cover of X, and so, by (c), it has a locally finite refinement $\{A_\alpha : \alpha \in \Lambda\}$. By Theorem 7.6, $\{\overline{A_\alpha} : \alpha \in \Lambda\}$ is locally finite. For each $\alpha \in \Lambda$, there exists $x \in X$ such that $A_\alpha \subseteq V_x$. Therefore since $\overline{V_x} \subseteq U_x$ for each $x \in X$, $\overline{A_\alpha} \subseteq U_x$. Thus $\{\overline{A_\alpha} : \alpha \in \Lambda\}$ is a closed locally finite refinement of $\mathcal{U}$.

(d) → (a). Let $\mathcal{U}$ be an open cover of X. Then there exists a closed locally finite refinement $\mathcal{A}$ of $\mathcal{U}$. For each $x \in X$, let V_x be a neighborhood of x that intersects only a finite number of members of $\mathcal{A}$. Then $\{V_x : x \in X\}$ is an open cover of X, so there exists a closed locally finite refinement $\mathcal{C}$ of $\{V_x : x \in X\}$. For each $A \in \mathcal{A}$, let $A^* = X - \bigcup\{C \in \mathcal{C} : A \cap C = \varnothing\}$. Since $\mathcal{C}$ is locally finite, by Theorem 7.7 $\overline{\bigcup\{C \in \mathcal{C} : A \cap C = \varnothing\}} = \bigcup\{\overline{C} \in \mathcal{C} : A \cap C = \varnothing\}$. But each member of $\mathcal{C}$ is closed, so $\bigcup\{\overline{C} \in \mathcal{C} : A \cap C = \varnothing\} = \bigcup\{C \in \mathcal{C} : A \cap C = \varnothing\}$. Therefore $\bigcup\{C \in \mathcal{C} : A \cap C = \varnothing\}$ is closed, and so A^* is open. For each $A \in \mathcal{A}$, $A \subseteq A^*$. Therefore $\{A^* : A \in \mathcal{A}\}$ is a cover of X. We claim that $\{A^* : A \in \mathcal{A}\}$ is locally finite.

Let $x \in X$. There exists a neighborhood W of x that intersects only a finite number of members $C_1, C_2, ..., C_n$ of $\mathcal{C}$. Since $\mathcal{C}$ covers X, $W \subseteq \bigcup_{i=1}^n C_i$. Therefore if $W \cap A^* \neq \varnothing$, then there exists $k (1 \leq k \leq n)$ such that $C_k \cap A^* \neq \varnothing$. But $C_k \cap A^* \neq \varnothing$ implies $C_k \cap A \neq \varnothing$. Since each C_i intersects only a finite number of members of $\mathcal{A}$, $W \cap A^* = \varnothing$ for all but a finite number of members of $\{A^* : A \in \mathcal{A}\}$. Therefore $\{A^* : A \in \mathcal{A}\}$ is locally finite.

Now for each $A \in \mathcal{A}$, choose $U_A \in \mathcal{U}$ such that $A \subseteq U_A$. Then $\{A * \cap U_A : A \in \mathcal{A}\}$ is an open locally finite refinement of $\mathcal{U}\}$. ∎

Corollary 7.9. Every regular Lindelöf space is paracompact.

Proof. Let $(X, \mathcal{T})$ be a regular Lindelöf space, and let $\mathcal{U}$ be an open cover of X. Since $(X, \mathcal{T})$ is Lindelöf, there exists a countable subcollection $\mathcal{V}$ of $\mathcal{U}$ that covers X. Then $\mathcal{V}$ is an open σ-locally finite refinement of $\mathcal{U}$. Since $(X, \mathcal{T})$ is regular, by Theorem 7.8, it is paracompact. ∎

Corollary 7.10. Every compact Hausdorff space is paracompact. ∎

Now we prove that every metric space is paracompact and every paracompact space is normal. Thus it follows that we have the hierarchy:

regular Lindelöf
↘
metric → paracompact → normal
↗
compact Hausdorff

You should also recall the hierarchy at the end of Section 4.3.

THEOREM 7.11. Every metric space is paracompact.

Proof. Let (X, d) be a metric space and let $\mathcal{U}$ be an open cover of X. For each $U \in \mathcal{U}$ and $n \in \mathbb{N}$, let $U_n = \{x \in U : d(x, X - U) \geq 1/2^n\}$.

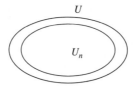

Then by Exercise 2, $d(U_n, X - U_{n+1}) \geq 1/2^{n+1}$. Let $<$ be a well-ordering of $\mathcal{U}$. For each $U \in \mathcal{U}$ and $n \in \mathbb{N}$, let $U_n^* = U_n - \bigcup \{V_{n+1} : V \in \mathcal{U}$ and $V < U\}$. For each $U, V \in \mathcal{U}$ and $n \in \mathbb{N}$, we have $U_n^* \subseteq X - V_{n+1}$ or $V_n^* \subseteq X - U_{n+1}$, depending on which comes first in the well-ordering. In either case, $d(U_n^*, V_n^*) \geq 1/2^{n+1}$. For each $U \in \mathcal{U}$ and $n \in \mathbb{N}$, let $U_n' = \{x \in X : d(x, U_n^*) < 1/2^{n+3}\}$. Then U_n' is an open set that is contained in U. For each $n \in \mathbb{N}$, let $\mathcal{V}_n = \{U_n' : U \in \mathcal{U}\}$, and let $\mathcal{V} = \bigcup_{n \in \mathbb{N}} \mathcal{V}_n$. By Exercise 3, for each $n \in \mathbb{N}$ and each $U, V \in \mathcal{U}$, $d(U_n', V_n') \geq 1/2^{n+2}$. Therefore for each $n \in \mathbb{N}$, $\mathcal{V}_n$ is discrete. Therefore $\mathcal{V}$ is σ-discrete. We complete the proof that $\mathcal{V}$ is an open σ-locally finite refinement of $\mathcal{U}$ by showing that $\mathcal{V}$ covers X. Let $x \in X$, and let U be the first member of $\mathcal{U}$ that contains x. Then there exists $n \in \mathbb{N}$ such that $x \in U_n'$. ∎

In order to prove that every paracompact space is normal, we first prove that every paracompact space is regular.

THEOREM 7.12. Every paracompact space is regular.

Proof. Let $(X, \mathcal{T})$ be a paracompact space, let A be a closed subset of X, and let $p \in X - A$. Since $(X, \mathcal{T})$ is Hausdorff, for each $x \in A$, there exists an open set U_x such that $x \in U_x$ and $p \notin \overline{U_x}$. Then $\mathcal{U} = \{X - A\} \cup \{U_x : x \in A\}$ is an open cover of X, and so there exists an open locally finite refinement $\mathcal{V}$ of $\mathcal{U}$. Let $W = \bigcup \{V \in \mathcal{V} : V \cap A \neq \emptyset\}$. Then W is open, $A \subseteq W$, and, since $\mathcal{V}$ is locally finite, $\overline{W} = \bigcup \{\overline{V} \in \mathcal{V} : V \cap A \neq \emptyset\}$. Let $V \in \mathcal{V}$ such that $V \cap A \neq \emptyset$. Then there exists $x \in A$ such that $V \subseteq U_x$, so $\overline{V} \subseteq \overline{U_x}$. Therefore $p \notin \overline{V}$ and $p \notin \overline{W}$. Therefore W and $X - \overline{W}$ are disjoint open sets such that $A \subseteq W$, and $p \in X - \overline{W}$. Therefore $(X, \mathcal{T})$ is regular. ∎

THEOREM 7.13. Every paracompact space is normal.

Proof. Let $(X, \mathcal{T})$ be a paracompact space, and let A and B be disjoint closed subsets of X. Since $(X, \mathcal{T})$ is regular, for each $x \in A$, there exists a neighborhood U_x of x such that $\overline{U_x} \cap B = \emptyset$. The remainder of the proof is essentially the same as the proof of Theorem 7.12, and so we leave it as Exercise 4. ∎

Now we give an example of a space that is paracompact but not metrizable and an example of a space that is normal but not paracompact. We introduce a new concept in order to prove that the space in Example 10 is not paracompact.

Definition. A **metacompact space** is a Hausdorff space with the property that every open cover has an open point finite refinement. ∎

It is clear that every paracompact space is metacompact.

EXAMPLE 9. Let $(\Omega, \leq)$ be an uncountable well-ordered set with a maximal element ω_1 having the property that if $x \in \Omega$ and $x \neq \omega_1$, then $\{y \in \Omega: y \leq x\}$ is countable. Let $\mathcal{T}$ be the order topology on Ω. Then $(\Omega, \mathcal{T})$ is a compact Hausdorff space (and hence paracompact).

Analysis. Let $\mathcal{U} = \{U_\alpha: \alpha \in \Lambda\}$ be an open cover of Ω. Recall that $\{(a, b]: a, b \in \Omega \text{ and } a < b\} \cup \{[1, a): a \in \Omega \text{ and } 1 < a\}$ is a basis for $\mathcal{T}$ (see Section 1.3). Let $f: \Omega \to \Omega$ be a function chosen so that $f(1) = 1$, and, for each $a \in \Omega$ such that $a \neq 1$, $f(a) \in \Omega$ such that $f(a) < a$ and $(f(a), a]$ is a subset of some member of $\mathcal{U}$. Define a sequence $\langle a_i \rangle$ by induction as follows: $a_1 = \omega_1$, and, for each $i \in \mathbb{N}$ such that $i > 1$, let $a_i = f(a_{i-1})$. Then $a_{i+1} < a_i$ for each $i \in \mathbb{N}$, so by Theorem H.1 part **(10)**, there exists $n \in \mathbb{N}$ such that $a_i = a_n$ for all $i \geq n$. Observe that a_n must be 1 (otherwise we could choose $a_{n+1} = f(a_n) < a_n$ such that $(a_{n+1}, a_n]$ is a subset of some member of $\mathcal{U}$). Therefore $(1, \omega_1] \subseteq \bigcup_{i=1}^{n-1}(a_{i+1}, a_i]$. For each $i = 1, 2, \ldots, n-1$, choose $U_{\alpha_i} \in \mathcal{U}$ such that $(a_{i+1}, a_i) \subseteq U_{\alpha_i}$, and choose $U_{\alpha_0} \in \mathcal{U}$ such that $1 \in U_{\alpha_0}$. Then $\{U_{\alpha_i}: i = 0, 1, \ldots, n-1\}$ is a finite subcollection of $\mathcal{U}$ that covers Ω. Therefore $(\Omega, \mathcal{T})$ is compact. By Exercise 17 of Section 1.6, it is Hausdorff. So by Corollary 7.10, $(\Omega, \mathcal{T})$ is paracompact. ∎

We use Example 10 to show that $(\Omega, \mathcal{T})$ is not metrizable, but first we prove the following theorem.

THEOREM 7.14. Let $\Omega_0 = \Omega - \{\omega_1\}$, and let $f: \Omega_0 - \{1\} \to \Omega_0$ be a function such that $f(a) < a$ for all $a \in \Omega_0 - \{1\}$. Then there exists $a_0 \in \Omega_0$ such that if $b \in \Omega_0$ then there exists $c \in \Omega_0$ such that $c \geq b$ and $f(c) < a_0$.

Proof. The proof is by contradiction. Suppose there is no such c. Then for each $a \in \Omega_0$, there exists $b(a) \in \Omega_0$ such that for all $c \in \Omega_0$ with $c \geq b(a), a \leq f(c)$. Define a sequence $\langle a_i \rangle$ inductively by $a_1 = 1$ and $a_{n+1} = b(a_n)$ for each $n \in \mathbb{N}$, and let $a^* = \sup\{a_n: n \in \mathbb{N}\}$. By Theorem H.11, $a^* \in \Omega_0$. Now for each $n \in \mathbb{N}, a^* \geq a_{n+1} = b(a_n)$, and so $a_n \leq f(a^*)$. Therefore $a^* \leq f(a^*) < a^*$, and we have a contradiction. ∎

EXAMPLE 10. The space $(\Omega_0, \mathcal{T}_{\Omega_0})$ is normal but not paracompact.

Analysis. By Theorem 5.28, $(\Omega_0, \mathcal{T}_{\Omega_0})$ is normal. We show that it is not paracompact by showing that it is not metacompact. For each $a \in \Omega_0 - \{1\}$, let

$U_a = \{b \in \Omega_0 : b < a\}$. Then $\mathcal{U} = \{U_a : a \in \Omega_0 - \{1\}\}$ is an open cover of Ω_0. Let $\mathcal{V}$ be any open refinement of $\mathcal{U}$. Let $f : \Omega_0 \to \Omega_0$ be a function such that $f(1) = 1$ and, for each $a \in \Omega_0 - \{1\}$, $f(a) < a$ and $(f(a), a]$ is a subset of some member V_a of $\mathcal{V}$. By Theorem 7.14, there exists $a_0 \in \Omega_0$ such that if $b \in \Omega_0$ then there exists $c \in \Omega_0$ such that $c \geq b$ and $f(c) < a_0$. Therefore we can find a sequence $\langle c_i \rangle$ such that $c_1 > a_0$, for each $i \in \mathbb{N}$, $c_{i+1} > c_i$ and $f(c_i) < a_0$, for each $i \in \mathbb{N}$. Then $a_0 \in \bigcap_{i \in \mathbb{N}} V_{c_i}$, and hence $\mathcal{V}$ is not point finite. ∎

We conclude this section with an investigation of subspaces and products. Recall that an F_σ-subset of a topological space is a subset that can be expressed as the union of a countable number of closed sets.

THEOREM 7.15. Every F_σ-subset of a paracompact space is paracompact.

Proof. Let $(X, \mathcal{T})$ be a paracompact space, and let F be and F_σ-subset of X. Then $F = \bigcup_{n \in \mathbb{N}} F_n$, where each F_n is closed. Let $\mathcal{U} = \{U_\alpha : \alpha \in \Lambda\}$ be an open cover of F, and, for each $\alpha \in \Lambda$, let V_α be an open subset of X such that $U_\alpha = F \cap V_\alpha$. Then for each $n \in \mathbb{N}$, $\{X - F_n\} \cup \{V_\alpha : \alpha \in \Lambda\}$ is an open cover of X, and so it has an open locally finite refinement $\mathcal{W}_n$. Let $\mathcal{A} = \{F \cap W : W \in \mathcal{W}_n\}$. Then $\mathcal{A}_n$ is a locally finite collection of open subsets of F, so $\mathcal{A} = \bigcup_{n \in \mathbb{N}} \mathcal{A}_n$ is an open σ-locally finite refinement of $\mathcal{U}$. Since F is a subset of a regular space, F is regular. Therefore by Theorem 7.8, $(F, \mathcal{T}_F)$ is paracompact. ∎

The following result is an immediate consequence of Theorem 7.15.

Corollary 7.16. Every closed subset of a paracompact space is paracompact. ∎

As we have seen, $(\Omega_0, \mathcal{T}_{\Omega_0})$ is an open subset of the paracompact space $(\Omega, \mathcal{T})$ that is not paracompact. In Exercise 5, you are asked to prove that if every open subset of a paracompact space is paracompact, then every subspace is paracompact.

THEOREM 7.17. The product of a paracompact space and a compact Hausdorff space is paracompact.

Proof. Let $(X, \mathcal{T})$ be a paracompact space and let $(Y, \mathcal{U})$ be a compact Hausdorff space. By Theorem 2.23, $X \times Y$ is Hausdorff. Let $\mathcal{U}$ be an open cover of $X \times Y$. For each $x \in X$, $\{x\} \times Y$ is compact, so there exist a finite number of numbers $U^x_{\alpha_1}, U^x_{\alpha_2}, \ldots, U^x_{\alpha_{n(x)}}$ of $\mathcal{U}$ that covers $\{x\} \times Y$. By Theorem 4.26, there exists a neighborhood V_x of x such that $V_x \times Y \subseteq \bigcup_{i=1}^{n(x)} U^x_{\alpha_i}$. Then $\mathcal{V} = \{V_x : x \in X\}$ is an open cover of X. Let $\mathcal{W}$ be an open locally finite refinement of $\mathcal{V}$. For each $W \in \mathcal{W}$, choose V_x such that $W \subseteq V_x$ and let $\mathcal{A}_W = \{(W \times Y) \cap U^x_{\alpha_i} : i = 1, 2, \ldots, n(x)\}$. Then $\mathcal{A} = \{A_W : W \in \mathcal{W}\}$ is an open refinement of $\mathcal{U}$. Also for each $(x, y) \in X \times Y$, there exists a neighborhood U of x that intersects only a finite number of members of $\mathcal{W}$. Hence the neighborhood $U \times Y$ of (x, y) intersects only a finite number of members of $\mathcal{A}$. Then $\mathcal{A}$ is locally finite, and so $X \times Y$ is paracompact. ∎

The following example shows that the product of two paracompact spaces need not be paracompact.

EXAMPLE 11. Let $\mathcal{T}$ be the lower-limit topology on $\mathbb{R}$. By Example 13 of Chapter 5, $(\mathbb{R}, \mathcal{T})$ is a normal Lindelöf space. Since every normal space is regular, by Corollary 7.9, $(\mathbb{R}, \mathcal{T})$ is a paracompact space. By Example 13 of Chapter 5, $\mathbb{R} \times \mathbb{R}$ is not normal. Therefore by Theorem 7.13 it is not paracompact.

EXERCISES 7.2

1. Let $\mathcal{U}$, $\mathcal{V}$, and $\mathcal{W}$ be covers of a set X such that $\mathcal{U} \Delta \mathcal{V}$ and $\mathcal{V} \Delta \mathcal{W}$. Prove that $\mathcal{U}^* < \mathcal{W}$.

2. Let U_n and U_{n+1} be the subsets of a metric space (X, d) defined in the proof of Theorem 7.11. Prove that $d(U_n, X - U_{n+1}) \geq 1/2^{n+1}$.

3. Let U'_n and V'_n be the subsets of a metric space (X, d) defined in the proof of Theorem 7.11. Prove that $d(U'_n, V'_n) \geq 1/2^{n+1}$.

4. Complete the proof of Theorem 7.13.

5. Let $(X, \mathcal{T})$ be a paracompact space with the property that every open subset of X is paracompact. Prove that every subset of X is paracompact.

6. Let $\mathcal{U}$ be a point-finite cover of a topological space $(X, \mathcal{T})$. Prove that there is a subcover $\mathcal{V}$ of $\mathcal{U}$ with the property that no proper subcollection of $\mathcal{V}$ covers X.

7. Prove that every countably compact metacompact space is compact.

8. Let $(X, \mathcal{T})$ be a paracompact space that has a dense Lindelöf subspace. Prove that $(X, \mathcal{T})$ is Lindelöf.

9. Let $\mathcal{T} = \{U \in \mathcal{P}(\mathbb{R}): U = V \cup W$, where V is open with respect to the usual topology on $\mathbb{R}$, and W is a subset of the irrationals$\}$.

    (a) Prove that $\mathcal{T}$ is a topology on $\mathbb{R}$.

    (b) Prove that $(\mathbb{R}, \mathcal{T})$ is paracompact.

    (c) Prove that if $X = \{x \in \mathbb{R}: x$ is irrational$\}$ and $\mathcal{U}$ is the usual topology on X, then $\mathbb{R} \times X$ is not normal.

7.3 The Nagata-Smirnov Metrization Theorem

We have established one metrization theorem—Urysohn's Theorem. In this section we prove four more such theorems, and the section culminates with the Nagata-Smirnov Theorem, also called the "general metrization theorem." We begin with some terminology.

Definition. A **normal sequence** in a topological space $(X, \mathcal{T})$ is a sequence $\langle \mathcal{U}_n \rangle$ of open covers of X such that for each $n \in \mathbb{N}$, $\mathcal{U}_{n+1}$ is a star-refinement of $\mathcal{U}_n$. A normal sequence $\langle \mathcal{U}_n \rangle$ in X is said to be **compatible** if for each $x \in X$, $\{St(x, \mathcal{U}_n): n \in \mathbb{N}\}$ is a local basis at x. An open cover $\mathcal{U}$ of $(X, \mathcal{T})$ is a **normal cover** if there is a normal sequence $\langle \mathcal{U}_n \rangle$ in X such that $\mathcal{U} = \mathcal{U}_1$. ∎

Notice that every member of a normal sequence in a topological space is a normal cover.

THEOREM 7.18. *A $\mathcal{T}_0$-space is metrizable if and only if it has a compatible sequence.*

Proof. Let (X, d) be a metric space. For each $n \in \mathbb{N}$, let $\mathcal{U}_n = \{B_d(x, 1/3^n): x \in X\}$. Then $\langle \mathcal{U}_n \rangle$ is a sequence of open covers of X. Let $n \in \mathbb{N}$ and let $U \in \mathcal{U}_{n+1}$. Then there exists $x \in X$ such that $U = B_d(x, 1/3^{n+1})$. Now $V = B_d(x, 1/3^n) \in \mathcal{U}_n$. We show that $\langle \mathcal{U}_n \rangle$ is a normal sequence in X by showing that $St(U, \mathcal{U}_{n+1}) \subseteq V$. Let $y \in St(U, \mathcal{U}_{n+1})$. Then there exists $W \in \mathcal{U}_{n+1}$ such that $y \in W$ and $U \cap W \neq \emptyset$. Let $z \in U \cap W$. Then $d(x, y) \leq d(x, z) + d(z, y) < 1/3^{n+1} + 2/3^{n+1} = 1/3^n$, and hence $y \in V$.

Now we show that $\langle \mathcal{U}_n \rangle$ is compatible. Let $x \in X$ and let V be a neighborhood of x. Then there exists $n \in \mathbb{N}$ such that $B_d(x, 1/3^n) \subseteq V$. Let $W = St(x, \mathcal{U}_{n+1})$. We complete the proof by showing that $W \subseteq B_d(x, 1/3^n)$. Let $y \in W$. Then there exists $U \in \mathcal{U}_{n+1}$ such that $x, y \in U$. So $d(x, y) < 2/3^{n+1} < 1/3^n$, and hence $y \in B_d(x, 1/3^n)$.

Now suppose the $\mathcal{T}_0$-space $(X, \mathcal{T})$ has a compatible sequence $\langle \mathcal{U}_n \rangle$. Define $\rho: X \times X \to [0, 1]$ by

$$\rho(x, y) = 0 \quad \text{if } y \in \bigcap_{n \in \mathbb{N}} St(x, \mathcal{U}_n)$$
$$\rho(x, y) = 1 \quad \text{if } y \notin St(x, \mathcal{U}_1)$$
$$\rho(x, y) = \frac{1}{2^n} \quad \text{if } y \in St(x, \mathcal{U}_n) - St(x, \mathcal{U}_{n+1}).$$

For each pair x, y of members of X, let $S(x, y)$ denote the set of all finite sequences $\langle x_1, x_2, \ldots, x_n \rangle$ of members of X such that $x_1 = x$ and $x_n = y$, or $x_1 = y$ and $x_n = x$. Define $d: X \times X \to \mathbb{R}$ by $d(x, y) = \text{glb}\{\sum_{i=1}^{n-1} \rho(x_i, x_{i+1}): \langle x_1, x_2, \ldots, x_n \rangle \in S(x, y)\}$. In Exercise 1, you are asked to prove that d is a metric on X.

We must show that the topology generated by d is $\mathcal{T}$. For each $n \in \mathbb{N}$, let $\mathcal{V}_n = \{B_d(x, 1/2^n): x \in X\}$. Then each $\mathcal{V}_n$ is an open cover of (X, d) and, by the first part of the proof of this theorem, for each $x \in X$, $\{St(x, \mathcal{U}_n): n \in \mathbb{N}\}$ is a local basis at x (with respect to the topology generated by d). In order to show that the topology generated by d is $\mathcal{T}$, it is sufficient to show that for each $n \in \mathbb{N}$, $\mathcal{U}_{n+1} < \mathcal{V}_n$ and $\mathcal{V}_{n+1} < \mathcal{U}_n$.

Let $n \in \mathbb{N}$, let $U \in \mathcal{U}_{n+1}$, and let $x \in U$. If $y \in U$ then $y \in St(x, \mathcal{U}_{n+1})$, so $\rho(x, y) \leq 1/2^{n+1}$. Therefore $d(x, y) \leq 1/2^{n+1} < 1/2^n$, and hence $y \in B_d(x, 1/2^n)$. Thus $U \subseteq B_d(x, 1/2^n)$, and hence $\mathcal{U}_{n+1} < \mathcal{V}_n$.

7.3 The Nagata-Smirnov Metrization Theorem

Now let $n \in \mathbb{N}$ and let $V \in \mathcal{V}_{n+1}$. Then there exists $x \in X$ such that $V = B_d(x, 1/2^{n+1})$. Let $y \in V$. Then $d(x, y) < 1/2^{n+1}$, so $\text{glb}\{\sum_{i=1}^{n-1} \rho(x_i, x_{i+1}) : \langle x_1, x_2, \ldots, x_n \rangle \in S(x, y)\} < 1/2^{n+1}$. Hence there exists a member $\langle x_1, x_2, \ldots, x_k \rangle \in S(x, y)$ such that $\sum_{i=1}^{k-1} \rho(x_i, x_{i+1}) < 1/2^{n+1}$. It is sufficient to show that there exists $W \in \mathcal{U}_{n+1}$ such that $x, y \in W$, because then there exists $N \in \mathcal{U}_n$ such that $B_d(x, 1/2^{n+1}) \subseteq \text{St}(x, \mathcal{U}_{n+1}) \subseteq N$. The proof that there exists $W \in \mathcal{U}_{n+1}$ such that $x, y \in W$ is by induction on the length of the sequence $\langle x_1, x_2, \ldots, x_k \rangle$ such that $\sum_{i=1}^{k-1} \rho(x_i, x_{i+1}) < 1/2^{n+1}$. Suppose $k = 2$. Then $\rho(x, y) < 1/2^{n+1}$, so there exists $m > n + 1$ such that $y \in \text{St}(x, \mathcal{U}_m) - \text{St}(x, \mathcal{U}_{m+1})$. In particular, $y \in \text{St}(x, \mathcal{U}_{n+2})$, so there exists $U \in \mathcal{U}_{n+2}$ such that $x, y \in U$. Thus it is certainly true that there exists $W \in \mathcal{U}_{n+1}$ such that $x, y \in W$.

Suppose $k \geq 2$ and for all sequences $\langle x_1, x_2, \ldots, x_j \rangle \in S(x, y)$ of length $j (2 \leq j \leq k)$ such that $\sum_{i=1}^{j-1} \rho(x_i, x_{i+1}) < 1/2^{n+1}$, there exists $W \in \mathcal{U}_{n+1}$ such that $x, y \in W$. Suppose $\sum_{i=1}^{k+1} \rho(x_i, x_{i+1}) < 1/2^{n+1}$. Let j be the largest number $(2 \leq j \leq k)$ such that $\sum_{i=1}^{j} \rho(x_i, x_{i+1}) < 1/2^{n+2}$. Then $\sum_{i=1}^{j+1} \rho(x_i, x_{i+1}) \geq 1/2^{n+2}$, so $\sum_{i=j+2}^{k+1} \rho(x_i, x_{i+1}) < 1/2^{n+2}$. By the induction hypothesis, there exists $U_1 \in \mathcal{U}_{n+2}$ such that $x_1, x_j \in U_1$. Since $\rho(x_j, x_{j+1}) < 1/2^{n+1}$, the above argument shows that there exists $U_2 \in \mathcal{U}_{n+2}$ such that $x_j, x_{j+1} \in U_2$. By the induction hypothesis, there exists $U_3 \in \mathcal{U}_{n+2}$ such that $x_{j+1}, x_{k+1} \in U_3$. Therefore $x_1, x_{k+1} \in \text{St}(U_2, \mathcal{U}_{n+2})$. Since there exists $W \in \mathcal{U}_{n+1}$ such that $\text{St}(U_2, \mathcal{U}_{n+2}) \subseteq W$, the proof is complete. ∎

We use the preceding theorem to give a local countable basis characterization of metrizable spaces.

THEOREM 7.19. *A T_0-space $(X, \mathcal{T})$ is metrizable if and only if for each $x \in X$ there is a countable local basis $\{U_{xn} : n \in \mathbb{N}\}$ at x such that for each $n \in \mathbb{N}$ and $y \in X$ with $y \neq x$ the following conditions hold:*

(a) *If $y \in U_{x(n+1)}$ then $U_{y(n+1)} \subseteq U_{xn}$, and*

(b) *If $y \notin U_{xn}$, then $U_{y(n+1)} \cap U_{x(n+1)} = \emptyset$.*

Proof. Suppose $(X, \mathcal{T})$ is metrizable and d is a metric that generates $\mathcal{T}$. For each $x \in X$ and each $n \in \mathbb{N}$, let $U_{xn} = B_d(x, 1/2^n)$. It is clear that $\{U_{xn} : n \in \mathbb{N}\}$ is a local basis at x.

Suppose $y \neq x$ and $y \in U_{x(n+1)}$. Then $d(x, y) < 1/2^{n+1}$. Let $z \in U_{y(n+1)}$. Then $d(y, z) < 1/2^{n+1}$. Therefore $d(x, z) \leq d(x, y) + d(y, z) < 1/2^n$, so $z \in U_{nx}$. Hence condition (a) is satisfied.

Suppose $y \notin U_{xn}$, and there exists $z \in U_{y(n+1)} \cap U_{x(n+1)}$. Then $d(x, y) \geq 1/2^n$, $d(y, z) < 1/2^{n+1}$, and $d(x, z) < 1/2^{n+1}$. Thus $d(x, y) \leq d(x, z) + d(z, y) < 1/2^n$. This is a contradiction, and hence condition (b) is satisfied.

Now suppose that for each $x \in X$ there is a countable local basis $\{U_{xn} : n \in \mathbb{N}\}$ at x that satisfies conditions (a) and (b). By Exercise 19 of Section 1.3, we may assume

260   Chapter 7 ■ Metrizability and Paracompactness

that $U_{x(n+1)} \subseteq U_{xn}$ for each $x \in X$ and $n \in \mathbb{N}$. For each $n \in \mathbb{N}$, let $\mathcal{U}_n = \{U_{xn} : x \in X\}$. Then $\langle \mathcal{U}_n \rangle$ is a sequence of open covers of X.

We complete the proof by showing that the sequence $\langle \mathcal{U}_{2n-1} : n \in \mathbb{N} \rangle$ is compatible. First we prove that $\langle \mathcal{U}_{2n-1} : n \in \mathbb{N} \rangle$ is a normal sequence. Let $U_{x(2n+1)} \in \mathcal{U}_{2n+1}$, and let $y \in \text{St}(U_{x(2n+1)}, \mathcal{U}_{2n+1})$. Then there exists $U_{z(2n+1)} \in \mathcal{U}_{2n+1}$ such that $y \in U_{z(2n+1)}$ and $U_{z(2n+1)} \cap U_{x(2n+1)} \neq \emptyset$. By Property **(b)**, $z \in U_{x(2n)}$. So by Property **(a)**, $U_{z(2n)} \subseteq U_{x(2n-1)}$. Since $U_{z(2n+1)} \subseteq U_{z(2n)}, y \in U_{x(2n-1)}$. Therefore $\text{St}(U_{x(2n+1)}, \mathcal{U}_{2n+1}) \subseteq U_{x(2n-1)} \in \mathcal{U}_{2n-1}$, and hence $\mathcal{U}_{2n+1}$ is a star-refinement of $\mathcal{U}_{2n-1}$.

Now we show that for each $x \in X$, $\{\text{St}(x, \mathcal{U}_{2n-1}) : n \in \mathbb{N}\}$ is a local basis at x. Let $x \in X$ and let W be a neighborhood of x. There exists an odd natural number n such that $U_{xn} \subseteq W$. By the preceding proof, $\text{St}(x, \mathcal{U}_{n+2}) \subseteq U_{xn}$. Therefore $\{\text{St}(x, \mathcal{U}_{2n-1}) : n \in \mathbb{N}\}$ is a local basis at x.

We have proved that $\langle \mathcal{U}_{2n-1} : n \in \mathbb{N} \rangle$ is compatible. Therefore by Theorem 7.18, $(X, \mathcal{T})$ is metrizable.

We introduce another idea in order to prove another metrization theorem.

Definition. A **development** for a topological space $(X, \mathcal{T})$ is a sequence $\langle \mathcal{U}_n \rangle$ of open covers of X such that for each $n \in \mathbb{N}$, $\mathcal{U}_{n+1}$ is a refinement of $\mathcal{U}_n$ and for each $x \in X$, $\{\text{St}(x, \mathcal{U}_n) : n \in \mathbb{N}\}$ is a local basis at x. A **developable space** is a topological space that has a development. ■

For historical purposes, we note that a **Moore space** is a regular developable space.

The following theorem is the first recorded metrization theorem. It was proved by Alexandroff and Urysohn in 1923.

THEOREM 7.20. A T_0-space is metrizable if and only if it has a development $\langle \mathcal{U}_n \rangle$ with the property that whenever $U, V \in \mathcal{U}_n$ and $U \cap V \neq \emptyset$, then there exists $W \in \mathcal{U}_{n-1}$ such that $U \cup V \subseteq W$.

Proof. Let (X, d) be a metric space. For each $n \in \mathbb{N}$, let $\mathcal{U}_n = \{B_d(x, 1/2^{2n}) : x \in X\}$. We leave as Exercise 2 the proof that $\langle \mathcal{U}_n \rangle$ has the desired properties.

Now let $(X, \mathcal{T})$ be a T_0-space and let $\langle \mathcal{U}_n \rangle$ be a development for X with the property that whenever $U, V \in \mathcal{U}_n$ and $U \cap V \neq \emptyset$, then there exists $W \in \mathcal{U}_{n-1}$ such that $U \cup V \subseteq W$. We use Theorem 7.19 to prove that $(X, \mathcal{T})$ is metrizable. For each $x \in X$ and each $n \in \mathbb{N}$, let $U_{xn} = \text{St}(x, \mathcal{U}_n)$. We leave as Exercise 3 the proof that $\{U_{xn} : n \in \mathbb{N}\}$ is a local basis at x. So we need to verify properties **(a)** and **(b)** of Theorem 7.19. Suppose $n \in \mathbb{N}$ and $y \neq x$.

(a) Let $y \in U_{x(n+1)}$. Then there exist $V \in \mathcal{U}_{n+1}$ such that $x, y \in V$. We have $U_{y(n+1)} = \text{St}(y, \mathcal{U}_{n+1}) \subseteq \text{St}(V, \mathcal{U}_{n+1})$ and $\text{St}(x, \mathcal{U}_n) = U_{xn}$. We leave as Exercise 4 the proof that $\text{St}(V, \mathcal{U}_{n+1}) \subseteq \text{St}(x, \mathcal{U}_n)$. This completes the proof of Property **(a)**.

7.3 The Nagata-Smirnov Metrization Theorem 261

(b) Suppose $U_{y(n+1)} \cap U_{x(n+1)} \neq \emptyset$. Then there exists $z \in \text{St}(y, \mathcal{U}_{n+1}) \cap \text{St}(x, \mathcal{U}_{n+1})$. So there exists $U \in \mathcal{U}_{n+1}$ such that $y, z \in U$ and there exists $V \in \mathcal{U}_{n+1}$ such that $x, z \in V$. Since $z \in U \cap V$, $U \cap V \neq \emptyset$. Therefore there exists $W \in \mathcal{U}_n$ such that $U \cup V \subseteq W$. Since $x, y \in W$, $y \in \text{St}(x, \mathcal{U}_n) = U_{xn}$. This completes the proof of Property (b).

Therefore by Theorem 7.19, $(X, \mathcal{T})$ is metrizable. ∎

Finally we arrive at the **general metrization theorem**. Just as with Urysohn's metrization theorem, we prove a space is metrizable by imbedding the space in a known metric space. This time the metric space in which it is imbedded is a generalization of Hilbert space.

Let τ be an infinite cardinal number, and let Λ be a set of cardinality τ. Let $H^\tau = \{x: \Lambda \to \mathbb{R}: x(\alpha) \neq 0 \text{ for at most a countable number of } \alpha \in \Lambda \text{ and } \sum_{\alpha \in \Lambda}(x(\alpha))^2 \text{ converges}\}$. Define $d: H^\tau \times H^\tau \to \mathbb{R}$ by $d(x, y) = (\sum_{\alpha \in \Lambda}(x(\alpha) - y(\alpha))^2)^{1/2}$. In Exercise 5 you are asked to prove that d is a metric on H^τ.

Definition. For each infinite cardinal number τ, the metric space (H^τ, d) is called the **generalized Hilbert space of weight τ**. ∎

THEOREM 7.21. *A regular space is metrizable if and only if it has a σ-locally finite basis.*

Proof. Let (X, d) be a metric space. By Theorem 7.11, (X, d) is paracompact. For each $n \in \mathbb{N}$, let $\mathcal{A}_n = \{B_d(x, 1/2^n): x \in X\}$ and let $\mathcal{U}_n$ be a cover of X by members of $\mathcal{A}_n$. Since (X, d) is paracompact, for each $n \in \mathbb{N}$, $\mathcal{U}_n$ has an open locally finite refinement $\mathcal{V}_n$. Then $\mathcal{V} = \bigcup_{n \in \mathbb{N}} \mathcal{V}_n$ is a σ-locally finite collection of open sets. In Exercise 6, you are asked to show that $\mathcal{V}$ is a basis for the topology generated by d.

Now let $(X, \mathcal{T})$ be a regular space with a σ-locally finite basis $\mathcal{B} = \bigcup_{n \in \mathbb{N}} \mathcal{B}_n$ (for each $n \in \mathbb{N}$, $\mathcal{B}_n$ is locally finite). Since every open cover of X has a σ-locally finite refinement consisting of members of $\mathcal{B}$, by Theorem 7.8, $(X, \mathcal{T})$ is paracompact. So, by Theorem 7.13, $(X, \mathcal{T})$ is normal.

Now we show that $(X, \mathcal{T})$ is perfectly normal. Let $U \in \mathcal{T}$. Since $(X, \mathcal{T})$ is regular, for each $x \in U$ there exists $B_x \in \mathcal{B}$ such that $x \in B_x$ and $\overline{B_x} \subseteq U$. For each $n \in \mathbb{N}$, let $B_n = \bigcup\{\overline{B_x}: x \in U \text{ and } B_x \in \mathcal{B}_n\}$. Then for each $n \in \mathbb{N}$, B_n is the union of a locally finite collection of closed sets and so by Theorem 7.7, B_n is closed. Since $U = \bigcup_{n \in \mathbb{N}} B_n$, U is an F_σ-set. Since U is an arbitrary member of $\mathcal{T}$, we have shown that every open subset of $(X, \mathcal{T})$ is an F_σ-set. Thus, by Exercise 13 of Section 1.5, every closed subset of $(X, \mathcal{T})$ is a G_δ-set. Therefore, by Theorem 5.30, $(X, \mathcal{T})$ is perfectly normal.

Let $B_{n\alpha} \in \mathcal{B}$. Then $X - B_{n\alpha}$ is closed, and since $(X, \mathcal{T})$ is perfectly normal, there is a continuous function $f_{n\alpha}: x \to I$ such that $f_{n\alpha}^{-1}(0) = X - B_{n\alpha}$. So $B_{n\alpha} = \{x \in X: f_{n\alpha}(x) \neq 0\}$. Let τ be the cardinal number of the base $\mathcal{B}$ and let H^τ be the generalized Hilbert space of weight τ. Let Λ be the set of all pairs n, α, where

$n \in \mathbb{N}$ and $B_{n\alpha} \in \mathcal{B}_n$. For each $n \in \mathbb{N}$, let $\Lambda_n = \{\alpha \in \Lambda : B_{n\alpha} \in \mathcal{B}_n\}$. We define $F \colon X \to H^\tau$ by specifying the $(n\alpha)$th coordinate of $F(x)$ for each $x \in X$. For convenience, let $F_{n\alpha}(x)$ denote this coordinate; that is, $F_{n\alpha}(x) = [F(x)]_{n\alpha}$. Using this notation, we define $F \colon X \to H^\tau$ by

$$F_{n\alpha}(x) = \frac{1}{2^{n/2}} \times \frac{f_{n\alpha}(x)}{1 + \sum_{\beta \in \Lambda_n} (f_{n\beta}(x))^2)^{1/2}}.$$

For each $n \in \mathbb{N}$ and each $x \in X$, $x \in B_{n\alpha}$ for only a finite number of members $B_{n\alpha}$ of $\mathcal{B}_n$. So for each $n \in \mathbb{N}$ and each $x \in X$, $f_{n\alpha}(x) \neq 0$ for only a finite number of α. Therefore $F_{n\alpha}(x) \neq 0$ for only countably many pairs n, α. Since for each $n \in \mathbb{N}$, $\sum_{\alpha \in \Lambda_n} (F_{n\alpha}(x))^2 < 1/2^n$, $\sum_{n \in \mathbb{N}, \alpha \in \Lambda_n} (F_{n\alpha}(x))^2 < \sum_{n \in \mathbb{N}} 1/2^n = 1$. Therefore $F(x)$ is a member of H^τ.

We complete the proof by showing that $F \colon X \to F(X)$ is a homeomorphism.

First we show that F is one-to-one. Suppose $x, y \in X$ and $x \neq y$. Since $(X, \mathcal{T})$ is T_1, there exists $B_{n\alpha} \in \mathcal{B}$ such that $x \in B_{n\alpha}$ and $y \notin B_{n\alpha}$. So $f_{n\alpha}(x) \neq 0$ and $f_{n\alpha}(y) = 0$. Then $F_{n\alpha}(x) \neq 0$ whereas $F_{n\alpha}(y) = 0$, so $F(x) \neq F(y)$. Therefore F is one-to-one.

Now we show that F is closed. Let A be a closed subset of X and suppose $y \in F(X) - F(A)$. Then there exists $x \in X$ such that $F(x) = y$. Note that $x \notin A$. Since A is closed, there exists $B_{n\alpha} \in \mathcal{B}$ such that $x \in B_{n\alpha}$ and $B_{n\alpha} \cap A = \emptyset$. So $f_{n\alpha}(x) \neq 0$, whereas $f_{n\alpha}(a) = 0$ for all $a \in A$. Therefore $F_{n\alpha}(x) \neq 0$ and $F_{n\alpha}(a) = 0$ for all $a \in A$. It follows that $d(F(x), F(A)) > 0$, so $F(x) \notin \overline{F(A)}$. Thus $\overline{F(A)} \subseteq F(A)$ and so $F(A)$ is closed.

Finally we show that F is continuous. Notice that for each pair n, α, the function $F_{n\alpha} \colon X \to \mathbb{R}$ is continuous. Let $p \in X$ and let $\varepsilon > 0$. Let N be a natural number such that $\sum_{n=N+1}^{\infty} 1/2^n < \varepsilon^2/4$, and let U be a neighborhood of p with the property that if $n \leq N$ then U intersects only a finite number of members of $\mathcal{B}_n$. Let $B_{n_1, \alpha_1}, B_{n_2, \alpha_2}, \ldots, B_{n_k, \alpha_k}$ denote the members of $\bigcup_{n=1}^{N} \mathcal{B}_n$ that intersect U. Let V be a neighborhood of p with the property that $V \subseteq U$ and if $x \in V$ and $i = 1, 2, \ldots, k$, then $|F_{n_i \alpha_i}(x) - F_{n_i \alpha_i}(p)| < \varepsilon(2k)^{1/2}$. Now if $x \in V$ and n, α is a pair that is different from each of $n_1, \alpha_1, n_2, \alpha_2, \ldots, n_k, \alpha_k$, then $F_{n\alpha}(x) = F_{n\alpha}(p) = 0$. Therefore for each $x \in V$, $\sum_{n=1}^{N} \sum_{\alpha \in \Lambda_n} (F_{n\alpha}(x) - F_{n\alpha}(p))^2 = \sum_{i=1}^{k} (F_{n_i, \alpha_i}(x) - F_{n_i, \alpha_i}(p))^2 < \varepsilon^2/2$. But by the choice of N, we also have

$$\sum_{n=N+1}^{\infty} \sum_{\alpha \in \Lambda_n} (F_{n\alpha}(x) - F_{n\alpha}(p))^2 \leq \sum_{n=N+1}^{\infty} \sum_{\alpha \in \Lambda_n} ((F_{n\alpha}(x))^2 + (F_{n\alpha}(p))^2)$$

$$< \sum_{n=N+1}^{\infty} \left(\frac{1}{2^n} + \frac{1}{2^n} \right) = 2 \sum_{n=N+1}^{\infty} \frac{1}{2^n} < \frac{\varepsilon^2}{2}.$$

Therefore for each $x \in V$, $\sum_{n \in \mathbb{N}} \sum_{\alpha \in \Lambda} (F_{n\alpha}(x) - F_{n\alpha}(p))^2 < \varepsilon^2$, so $d(F(x), F(p)) < \varepsilon$. Therefore F is continuous at p, and, since p is an arbitrary member of X, F is continuous. ∎

EXERCISES 7.3

1. Prove that the function $d: X \times X \to \mathbb{R}$ defined in the proof of Theorem 7.18 is a metric on X.

2. Prove that the sequence $\langle \mathcal{U}_n \rangle$ defined in the proof of Theorem 7.20 is a development with the property that whenever $U, V \in \mathcal{U}_n$ and $U \cap V \neq \emptyset$, then there exists $W \in \mathcal{U}_{n-1}$ such that $U \cup V \subseteq W$.

3. Prove that the collection $\{U_{xn}: n \in \mathbb{N}\}$ defined in the proof of Theorem 7.20 is a local basis at x.

4. We refer to the part of the proof of Theorem 7.20 where it is shown that property (a) of Theorem 7.19 is satisfied. Prove that $\text{St}(V, \mathcal{U}_{n+1}) \subseteq \text{St}(x, \mathcal{U}_n)$.

5. Prove that the function $d: H^\tau \times H^\tau \to \mathbb{R}$ (whose definition precedes Theorem 7.21) is a metric on H^τ.

6. Show that the collection $\mathcal{V}$ defined in the proof of Theorem 7.21 is a basis for the topology generated by d.

7. Let $\mathcal{U}$ be a normal cover of a topological space $(X, \mathcal{T})$ and let $(Y, \mathcal{V})$ be a topological space. Prove that $\{U \times Y: U \in \mathcal{U}\}$ is a normal cover of $X \times Y$.

8. Prove that every locally finite open cover of a normal space $(X, \mathcal{T})$ is a normal cover of X.

9. Let $\langle \mathcal{U}_n \rangle$ be a compatible sequence in a topological space $(X, \mathcal{T})$. Prove that $\bigcup_{n \in \mathbb{N}} \mathcal{U}_n$ is a basis for $\mathcal{T}$.

10. Let $(X, \mathcal{T})$ be a completely regular space such that X is the union of a locally finite collection of closed, metrizable subspaces. Prove that $(X, \mathcal{T})$ is metrizable.

11. An open cover $\mathcal{U} = \{U_\alpha: \alpha \in \Lambda\}$ of a topological space $(X, \mathcal{T})$ is said to be **shrinkable** provided there exists an open cover $\mathcal{V} = \{V_\alpha: \alpha \in \Lambda\}$ such that $\overline{V_\alpha} \subseteq U_\alpha$ for each $\alpha \in \Lambda$.

    Prove that a topological space is normal if and only if every point-finite open cover is shrinkable.

12. Let $(X, \mathcal{T})$ be a normal space such that X is the union of a locally finite collection of metrizable subspaces. Prove that $(X, \mathcal{T})$ is metrizable. *Hint*: Use Exercise 11.

13. Let $(X, \mathcal{T})$ be a T_1-space. Prove that $(X, \mathcal{T})$ is metrizable if and only if for each $x \in X$ there is a countable local basis $\{U_{xn}: n \in \mathbb{N}\}$ at x such that:

    (a) $U_{x(n+1)} \subseteq U_{xn}$ for each $n \in \mathbb{N}$, and

    (b) For each $n \in \mathbb{N}$, there exists $m > n$ such that if $y \in X$ and $U_{xm} \cap U_{ym} \neq \emptyset$ then $U_{xm} \subseteq U_{yn}$.

14. Let (X, d) be a metric space, let $(Y, \mathcal{U})$ be a topological space, and let $f: X \to Y$ be a closed continuous function that maps X onto Y.

   (a) Prove that if there is a countable local basis at $y \in Y$, then the boundary of $f^{-1}(y)$ is compact.
   (b) Suppose that for each $y \in Y$, the boundary of $f^{-1}(y)$ is compact. For each $y \in Y$, and $n \in \mathbb{N}$, let $W_{yn} = \{x \in X: d(x, \partial(f^{-1}(y))) < \frac{1}{n}\}$, $V_{yn} = W_{yn} \cup \mathrm{int}(f^{-1}(y))$, and $U_{yn} = f(V_{yn})$. Prove that $\{U_{yn}: n \in \mathbb{N}\}$ is a local basis at y that satisfies conditions (a) and (b) of Exercise 13.
   (c) Prove that the following are equivalent:

   (1) $(Y, \mathcal{U})$ is metrizable.

   (2) $(Y, \mathcal{U})$ is first countable.

   (3) For each $y \in Y$, the boundary of $f^{-1}(y)$ is compact.

15. Let (X, d) be a metric space, let $(Y, \mathcal{U})$ be a Hausdorff space, and let $f: X \to Y$ be a closed continuous function that maps X onto Y. Prove that $(Y, \mathcal{U})$ is metrizable if and only if X is compact. *Hint:* Use part (c) of Exercise 14.

16. Let (X, d) be a metric space, let $(Y, \mathcal{U})$ be a Hausdorff space, and let $f: X \to Y$ be a continuous function that maps X onto Y. Prove that $(Y, \mathcal{U})$ is metrizable if and only if (X, d) is compact. *Hint:* Use Theorem 7.3 and Exercise 15.

Chapter 8

The Fundamental Group and Covering Spaces

In the first seven chapters we have dealt with point-set topology. This chapter provides an introduction to algebraic topology. Algebraic topology may be regarded as the study of topological spaces and continuous functions by means of algebraic objects such as groups and homeomorphisms. Typically, we start with a topological space $(X, \mathcal{T})$ and associate with it, in some specific manner, a group G_X (see Appendix I). This is done in such a way that if $(X, \mathcal{T})$ and $(Y, \mathcal{U})$ are homeomorphic, then G_X and G_Y are isomorphic. Also, a continuous function $f: X \to Y$ "induces" a homeomorphism $f_*: G_X \to G_Y$, and, if f is a homeomorphism, then f_* is an isomorphism. Then the "structure" of G_X provides information about the "structure" of $(X, \mathcal{T})$. One of the purposes of doing this is to find a topological property that distinguishes between two nonhomeomorphic spaces. The group that we study is called the fundamental group.

8.1 Homotopy of Paths

In Section 3.2 we defined paths and the path product of two paths. This section studies an equivalence relation defined on a certain collection of paths in a topological space. Review Theorems 2.15 and 2.27, because these results about continuous functions will be used repeatedly in this chapter and the next. The closed unit interval [0, 1] will be denoted by I.

Definition. Let $(X, \mathcal{T})$ and $(Y, \mathcal{U})$ be topological spaces and let $f, g: X \to Y$ be continuous functions. Then f **is homotopic to** g, denoted by $f \simeq g$, if there is a continuous function $H: X \times I \to Y$ such that $H(x, 0) = f(x)$ and $H(x, 1) = g(x)$ for all $x \in X$. The function H is called a **homotopy** between f and g. ∎

In the preceding definition, we are assuming that I has the subspace topology $\mathcal{V}$ determined by the usual topology on $\mathbb{R}$. Thus $X \times I$ has the product topology determined by $\mathcal{T}$ and $\mathcal{V}$.

We think of a homotopy as a continuous one-parameter family of continuous functions from X into Y. If t denotes the parameter, then the homotopy represents a continuous "deformation" of the function f to the function g as t goes from 0 to 1. The question of whether f is homotopic to g is a question of whether there is a continuous extension of a given function. We think of f as being a function from $X \times \{0\}$ into Y and g as being a function from $X \times \{1\}$ into Y, so we have a continuous function from $X \times \{0, 1\}$ into Y and we want to extend it to a continuous function from $X \times I$ into Y.

EXAMPLE 1. Let X and Y be the subspaces of $\mathbb{R} \times \mathbb{R}$ defined by $X = \{(x, y) \in \mathbb{R} \times \mathbb{R}: x^2 + y^2 = 1\}$ and $Y = \{(x, y) \in \mathbb{R} \times \mathbb{R}: (x + 1)^2 + y^2 = 1$ or $(x - 1)^2 + y^2 = 1\}$. Define $f, g: X \to Y$ by $f(x, y) = (x - 1, y)$ and $g(x, y) = (x + 1, y)$. (See Figure 8.1).

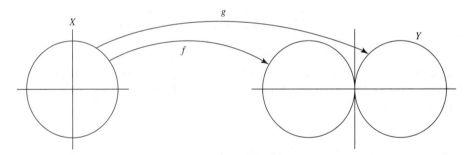

Figure 8.1

Then f is not homotopic to g.

EXAMPLE 2. Let X be the subspace of $\mathbb{R} \times \mathbb{R}$ defined by $X = \{(x, y) \in \mathbb{R} \times \mathbb{R}: x^2 + y^2 = 1\}$ and let $Y = X \times X$. Define $f, g: X \to Y$ by $f(x, y) = ((1, 0), (x, y))$ and $g(x, y) = ((0, 1), (x, y))$. Then the function $H: X \times I \to Y$ defined by $H((x, y), t) = ((\sqrt{1 - t^2}, t), (x, y))$ is a homotopy between f and g so $f \simeq g$.

Notice that X is a circle and Y is a torus, f "wraps" the circle X around $\{(1, 0)\} \times X$, and g "wraps" the circle X around $\{(0, 1)\} \times X$ (see Figure 8.2). Let A denote the arc from $((1, 0), (1, 0))$ to $((0, 1), (1, 0))$ (see Figure 8.2). Then H maps $X \times I$ onto $A \times X$.

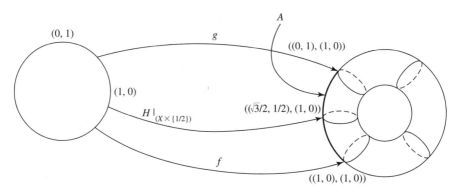

Figure 8.2

THEOREM 8.1. Let $(X, \mathcal{T})$ and $(Y, \mathcal{U})$ be topological spaces. Then $\simeq$ is an equivalence relation on $C(X, Y)$ (the collection of continuous functions that map X into Y).

Proof. Let $f \in C(X, Y)$. Define $H: X \times I \to Y$ by $H(x, t) = f(x)$ for each $(x, t) \in X \times I$. Then H is continuous and $H(x, 0) = H(x, 1) = f(x)$, so $f \simeq f$.

Suppose $f, g \in C(X, Y)$ and $f \simeq g$. Then there is a continuous function $H: X \times I \to Y$ such that $H(x, 0) = f(x)$ and $H(x, 1) = g(x)$ for all $x \in X$. Define $K: X \times I \to Y$ by $K(x, t) = H(x, 1 - t)$. Then K is continuous, and $K(x, 0) = H(x, 1) = g(x)$ and $K(x, 1) = H(x, 0) = f(x)$ for all $x \in X$. Therefore $g \simeq f$.

Suppose $f, g, h \in C(X, Y)$, $f \simeq g$, and $g \simeq h$. Then there are continuous functions $F, G: X \times I \to Y$ such that $F(x, 0) = f(x)$, $F(x, 1) = g(x)$, $G(x, 0) = g(x)$, and $G(x, 1) = h(x)$ for all $x \in X$. Define $H: X \times I \to Y$ by

$$H(x, t) = \begin{cases} F(x, 2t), & \text{for all } x \in X \text{ and } 0 \leq t \leq \tfrac{1}{2} \\ G(x, 2t - 1), & \text{for all } x \in X \text{ and } \tfrac{1}{2} \leq t \leq 1. \end{cases}$$

Then H is continuous, and $H(x, 0) = F(x, 0) = f(x)$ and $H(x, 1) = G(x, 1) = h(x)$ for all $x \in X$. Therefore $f \simeq h$. ∎

If α and β are paths in a topological space, we define a relation between them that is stronger than homotopy.

Definition. Let α and β be paths in a topological space $(X, \mathcal{T})$. Then α is **path homotopic** to β, denoted by $\alpha \simeq_p \beta$, if α and β have the same initial point x_0 and the same terminal point x_1 and there is a continuous function $H: I \times I \to X$ such that $H(x, 0) = \alpha(x)$ and $H(x, 1) = \beta(x)$ for all $x \in X$, and $H(0, t) = x_0$ and $H(1, t) = x_1$ for all $t \in I$. The function H is called a **path homotopy** between f and g. ∎

The question of the existence of a path homotopy between two paths is again a question of obtaining a continuous extension of a given function. This time we have a continuous function on $(I \times \{0, 1\}) \cup (\{0, 1\} \times I)$ and we want to extend it to a continuous function on $I \times I$ (see Figure 8.3).

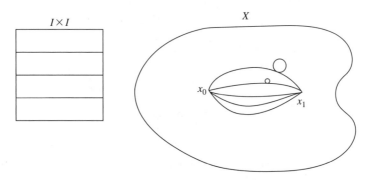

Figure 8.3

EXAMPLE 3. Let $X = \{(x, y) \in \mathbb{R}^2 : 1 \leq x^2 + y^2 \leq 4\}$ (see Figure 8.4), let $\alpha: I \to X$ be the path that maps I in a "linear" fashion onto the arc A from $(-1, 0)$ to $(1, 0)$, and let $\beta: I \to X$ be the path that maps I in a "linear" fashion onto the arc B from $(-1, 0)$ to $(1, 0)$. Then α is homotopic to β (by a homotopy that transforms α into β in the manner illustrated in Figure 8.5), but α is not path homotopic to β because we cannot "get from" α to β and keep the "end points" fixed without "crossing the hole" in the space.

If $(X, \mathcal{T})$ is a topological space and $x_0 \in X$, let $\Omega(X, x_0) = \{\alpha: I \to X : \alpha$ is continuous and $\alpha(0) = \alpha(1) = x_0\}$. Observe that if $\alpha, \beta \in \Omega(X, x_0)$, then the product (defined in Section 3.2) $\alpha * \beta \in \Omega(X, x_0)$. It is, however, easy to see that $*$ is not associative on $\Omega(X, x_0)$. Each member of $\Omega(X, x_0)$ is called a **loop** in X at x_0.

The proof of the following theorem is similar to the proof of Theorem 8.1 and hence it is left as Exercise 1.

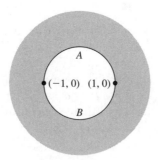

Figure 8.4

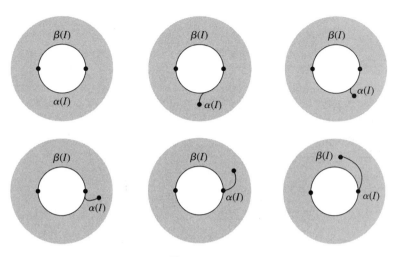

Figure 8.5

THEOREM 8.2. Let $(X, \mathcal{T})$ be a topological space, and let $x_0 \in X$. Then $\simeq_p$ is an equivalence relation on $\Omega(X, x_0)$. ∎

Let $(X, \mathcal{T})$ be a topological space and let $x_0 \in X$. If $\alpha \in \Omega(X, x_0)$, we let $[\alpha]$ denote the path-homotopy equivalence class that contains α. Then we let $\pi_1(X, x_0)$ denote the set of all path-homotopy equivalence classes on $\Omega(X, x_0)$, and we define an operation $\circ$ on $\pi_1(X, x_0)$ by $[\alpha] \circ [\beta] = [\alpha * \beta]$. The following theorem tells us that $\circ$ is well-defined.

THEOREM 8.3. Let $(X, \mathcal{T})$ be a topological space and let $x_0 \in X$. Furthermore, let $\alpha_1, \alpha_2, \beta_1, \beta_2 \in \Omega(X, x_0)$ and suppose $\alpha_1 \simeq_p \alpha_2$ and $\beta_1 \simeq_p \beta_2$. Then $\alpha_1 * \beta_1 \simeq_p \alpha_2 * \beta 2$.

Proof. Since $\alpha_1 \simeq_p \alpha_2$ and $\beta_1 \simeq_p \beta_2$, there exist continuous functions $F, G: I \times I \to X$ such that $F(x, 0) = \alpha_1(x)$, $F(x, 1) = \alpha_2(x)$, $G(x, 0) = \beta_1(x)$, and $G(x, 1) = \beta_2(x)$ for all $x \in I$, and $F(0, t) = F(1, t) = G(0, t) = G(1, t) = x_0$ for all $t \in I$. Define $H: I \times I \to X$ by

$$H(x, t) = \begin{cases} F(2x, t), & \text{if } 0 \leq x \leq \frac{1}{2} \text{ and } 0 \leq t \leq 1 \\ G(2x - 1, t), & \text{if } \frac{1}{2} \leq x \leq 1 \text{ and } 0 \leq t \leq 1. \end{cases}$$

Then H is continuous,

$$H(x, 0) = \begin{cases} F(2x, 0), & \text{if } 0 \leq x \leq \frac{1}{2} \\ G(2x - 1, 0) & \text{if } \frac{1}{2} \leq x \leq 1 \end{cases} = \begin{cases} \alpha_1(2x), & \text{if } 0 \leq x \leq \frac{1}{2} \\ \beta_1(2x - 1), & \text{if } \frac{1}{2} \leq x \leq 1 \end{cases} = (\alpha_1 * \beta_1)(x)$$

and

$$H(x, 1) = \begin{cases} F(2x, 1), & \text{if } 0 \leq x \leq \frac{1}{2} \\ G(2x - 1, 1) & \text{if } \frac{1}{2} \leq x \leq 1 \end{cases} = \begin{cases} \alpha_2(2x), & 0 \leq x \leq \frac{1}{2} \\ \beta_2(2x - 1), & \frac{1}{2} \leq x \leq 1 \end{cases} = (\alpha_2 * \beta_2)(x)$$

for all $x \in X$, and $H(0, t) = F(0, t) = x_0$ and $H(1, t) = G(1, t) = x_0$ for all $t \in I$. Therefore $\alpha_1 * \beta_1 \simeq \alpha_2 * \beta_2$. ∎

The following three theorems show that if $(X, \mathcal{T})$ is a topological space and $x_0 \in X$, then $(\pi_1(X, x_0), \circ)$ is a group.

THEOREM 8.4. Let $(X, \mathcal{T})$ be a topological space, let $x_0 \in X$, and let $[\alpha]$, $[\beta]$, $[\gamma] \in \pi_1(X, x_0)$. Then $([\alpha] \circ [\beta]) \circ [\gamma] = [\alpha] \circ ([\beta] \circ [\gamma])$.

Proof. We show that $(\alpha * \beta) * \gamma \simeq_p \alpha * (\beta * \gamma)$. First we observe that for each $x \in I$,

$$[(\alpha * \beta) * \gamma](x) = \begin{cases} (\alpha * \beta)(2x), & \text{if } 0 \leq x \leq \frac{1}{2} \\ \gamma(2x - 1), & \text{if } \frac{1}{2} \leq x \leq 1 \end{cases} = \begin{cases} \alpha(4x), & \text{if } 0 \leq x \leq \frac{1}{4} \\ \beta(4x - 1), & \text{if } \frac{1}{4} \leq x \leq \frac{1}{2} \\ \gamma(2x - 1), & \text{if } \frac{1}{2} \leq x \leq 1 \end{cases}$$

and

$$[\alpha * (\beta * \gamma)](x) = \begin{cases} \alpha(2x), & \text{if } 0 \leq x \leq \frac{1}{2} \\ \beta(4x - 2), & \text{if } \frac{1}{2} \leq x \leq \frac{3}{4} \\ \gamma(4x - 3), & \text{if } \frac{3}{4} \leq x \leq 1. \end{cases}$$

Next we draw a picture to illustrate how we arrive at the definition of the desired path homotopy H.

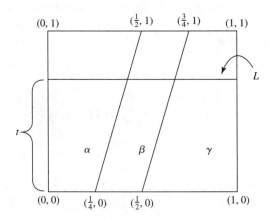

If $t \in I$, then the intersection of the horizontal line L and the line segment joining $(\frac{1}{4}, 0)$ and $(\frac{1}{2}, 1)$ has coordinates $((1 + t)/4, t)$, and the intersection of L and the line segment joining $(\frac{1}{2}, 0)$ and $(\frac{3}{4}, 1)$ has coordinates $((2 + t)/4, t)$. Therefore for each $t \in I$, we want H to be defined in terms of α if $x \leq (1 + t)/4$, in terms of β if $(1+t)/4 \leq x \leq (2 + t)/4$, and in terms of γ if $x \geq (2 + t)/4$. Also for each $t \in I$, we want $H(0, t) = H((1 + t)/4, t) = H((2 + t)/4, t) = H(1, t) = x_0$. Using elementary analytic geometry, we arrive at the definition of H. Define $H: I \times I \to X$ by

$$H(x, t) = \begin{cases} \alpha\left(\dfrac{4x}{1 + t}\right), & \text{if } 0 \leq t \leq 1 \text{ and } 0 \leq x \leq \dfrac{1 + t}{4} \\ \beta(4x - 1 - t), & \text{if } 0 \leq t \leq 1 \text{ and } \dfrac{1 + t}{4} \leq x \leq \dfrac{2 + t}{4} \\ \gamma\left(\dfrac{4x - 2 - t}{2 - t}\right), & \text{if } 0 \leq t \leq 1 \text{ and } \dfrac{2 + t}{4} \leq x \leq 1. \end{cases}$$

Thus H is continuous, and it is a routine check to show that $H(x, 0) = [(\alpha * \beta) * \gamma](x)$ and $H(x, 1) = [\alpha * (\beta * \gamma)](x)$ for all $x \in I$, and $H(0, t) = H(1, t) = x_0$ for all $t \in I$. ∎

We have proved that $\circ$ is associative on $\pi_1(X, x_0)$. Now we prove that $(\pi_1(X, x_0), \circ)$ has an identity element.

THEOREM 8.5. Let $(X, \mathcal{T})$ be a topological space, let $x_0 \in X$, and let $e: I \to X$ be the path defined by $e(x) = x_0$ for each $x \in I$. Then $[\alpha] \circ [e] = [e] \circ [\alpha] = [\alpha]$ for each $[\alpha] \in \pi_1(X, x_0)$.

Proof. We prove that if $[\alpha] \in \pi_1(X, x_0)$ then $\alpha * e \simeq_p \alpha$ and $e * \alpha \simeq_p \alpha$.

Let $[\alpha] \in \pi_1(X, x_0)$. We draw a picture to illustrate how we arrive at the definition of the path homotopy H to show that $\alpha * e \simeq_p \alpha$.

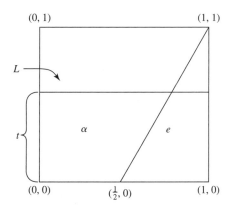

For each $t \in I$, the intersection of the line L with the line segment joining $(\frac{1}{2}, 0)$ and $(1, 1)$ has coordinates $((1 + t)/2, t)$. Therefore for each $t \in I$, we want the homotopy defined in terms of α if $x \leq (1 + t)/2$. Using analytic geometry, we arrive at the formula

$$H(x, t) = \begin{cases} \alpha\left(\dfrac{2x}{1 + t}\right), & \text{if } 0 \leq x \leq \dfrac{1 + t}{2} \\ x_0, & \text{if } \dfrac{1 + t}{2} \leq x \leq 1. \end{cases}$$

It is a routine check to see that $H(x, 0) = (\alpha * e)(x)$ and $H(x, 1) = \alpha(x)$ for each $x \in I$ and $H(0, t) = H(1, t) = x_0$ for each $t \in I$.

The proof that $e * \alpha \simeq_p \alpha$ is left as Exercise 2. ∎

We have proved that $[e]$ is the identity element of $\pi_1(X, x_0)$. We complete the proof that $\pi_1(X, x_0), \circ)$ is a group by showing that each member of $\pi_1(X, x_0)$ has an inverse in $\pi_1(X, x_0)$.

THEOREM 8.6. Let $(X, \mathcal{T})$ be a topological space, let $x_0 \in X$, and let $[\alpha] \in \pi_1(X, x_0)$. Then there exists $[\overline{\alpha}] \in \pi_1(X, x_0)$ such that $[\alpha] \circ [\overline{\alpha}] = [\overline{\alpha}] \circ [\alpha] = [e]$.

Proof. We prove that there is an $\overline{\alpha} \in \Omega(X, x_0)$ such that $\alpha * \overline{\alpha} \simeq_p e$ and $\overline{\alpha} * \alpha \simeq_p e$.

Define $\overline{\alpha} : I \to X$ by $\overline{\alpha}(x) = \alpha(1 - x)$ for each $x \in X$. We draw a picture to illustrate how we arrive at the definition of the path homotopy H to show that $\alpha * \overline{\alpha} \simeq_p e$.

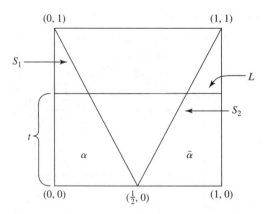

For each $t \in I$, we map the part of L that lies between S_1 and S_2 into the point $f(1 - t)$, we map the segment $[(0, t), ((1 - t)/2, t)]$ in the same manner that α maps the interval $[0, 1 - t]$, and we map the segment $[((1 + t)/2, t), (1, t)]$ in the same manner that $\overline{\alpha}$ maps the interval $[t, 1]$. Observing that $2((1 - t)/2) = 1 - t$ and $2((1 + t)/2) - 1 = t$, we arrive at the formula

$$H(x, t) = \begin{cases} \alpha(2x), & \text{if } 0 \leq x \leq \dfrac{1-t}{2} \\ \alpha(1 - t), & \text{if } \dfrac{1-t}{2} \leq x \leq \dfrac{1+t}{2} \\ \overline{\alpha}(2x - 1), & \text{if } \dfrac{1+t}{2} \leq x \leq 1. \end{cases}$$

It is easy to see that $H(x, 0) = (\alpha * \overline{\alpha})(x)$ and $H(x, 1) = e(x)$ for all $x \in I$ and $H(0, t) = H(1, t) = x_0$ for each $t \in I$.

The proof that $\overline{\alpha} * \alpha \simeq_p e$ is left as Exercise 3. ∎

We have proved that $(\pi_1(X, x_0), \circ)$ is a group.

Definition. Let $(X, \mathcal{T})$ be a topological space and let $x_0 \in X$. Then $(\pi_1(X, x_0), \circ)$ is called the **fundamental group of** X **at** x_0. ∎

The fundamental group was introduced by the French mathematician Henri Poincaré (1854–1912) around 1900. During the years 1895–1901, he published a series of papers in which he introduced algebraic topology. It is not true that algebraic topology developed as an outgrowth of general topology. In fact, Poincaré's work on algebraic topology preceded Hausdorff's definition of a topological space. Neither was Poincaré's work influenced by Cantor's theory of sets.

Poincaré was surpassed only by Leonard Euler (1707–1783) as the most prolific writer of mathematics. Poincaré published 30 books and over 500 papers on mathematics, and he is regarded as the founder of combinatorial topology.

Throughout the remainder of this text, we denote $(\pi_1(X, x_0), \circ)$ by $\pi_1(X, x_0)$. The natural question to ask is whether the fundamental group of a topological space depends upon the base point x_0. This and other questions will be answered in the remaining sections of this chapter.

We conclude this section with the definition of a contractible space and two theorems whose proofs are left as exercises.

Definition. A topological space $(X, \mathcal{T})$ is **contractible to a point** $x_0 \in X$ if there is a continuous function $H: X \times I \to X$ such that $H(x, 0) = x$ and $H(x, 1) = x_0$ for each $x \in X$ and $H(x_0, t) = x_0$ for each $t \in I$. Then $(X, \mathcal{T})$ is **contractible** if there exists $x_0 \in X$ such that $(X, \mathcal{T})$ is contractible to x_0. ∎

THEOREM 8.7. Let $(X, \mathcal{T})$ be a topological space and let $(Y, \mathcal{U})$ be a contractible space. If $f, g: X \to Y$ are continuous functions, then f is homotopic to g. ∎

THEOREM 8.8. Let $(X, \mathcal{T})$ be a contractible space and let $(Y, \mathcal{U})$ be a pathwise connected space. If $f, g: X \to Y$ are continuous functions, then f is homotopic to g. ∎

EXERCISES 8.1

1. Let $(X, \mathcal{T})$ be a topological space and let $x_0 \in X$. Prove that $\simeq_p$ is an equivalence relation on $\Omega(X, x_0)$.

2. Let $(X, \mathcal{T})$ be a topological space, let $x_0 \in X$, and define $e: I \to X$ by $e(x) = x_0$ for each $x \in X$. Show that if $\alpha \in \Omega(X, x_0)$, then $e * \alpha \simeq_p \alpha$.

3. Let $(X, \mathcal{T})$ be a topological space, let $x_0 \in X$, and define $e: I \to X$ by $e(x) = x_0$ for each $x \in X$. Let $\alpha \in \Omega(X, x_0)$, and define $\overline{\alpha}: I \to X$ by $\overline{\alpha}(x) = (1 - x)$ for each $x \in X$. Prove that $\overline{\alpha} * \alpha \simeq_p e$.

4. Let $(X, \mathcal{T})$, $(Y, \mathcal{U})$ and $(Z, \mathcal{V})$ be a topological spaces, let $f_1, f_2: X \to Y$ be continuous functions such that $f_1 \simeq f_2$, and let $g_1, g_2: Y \to Z$ be continuous functions such that $g_1 \simeq g_2$. Prove that $g_1 \circ f_1 \simeq g_2 \circ f_2$.

5. Let $n \in \mathbb{N}$. A subset X of $\mathbb{R}^n$ is **convex** if for each $x, y \in X$ and $t \in I$, $(1 - t)x + ty \in X$. Let X be a convex subset of $\mathbb{R}^n$ and let α and β be paths in X such that $\alpha(0) = \beta(0)$ and $\alpha(1) = \beta(1)$. Prove that α is path homotopic to β.

6. Let $(X, \mathcal{T})$ be a topological space and let $f, g: X \to I$ be continuous functions. Prove that f is homotopic to g.

7. Let $(X, \mathcal{T})$ be a pathwise connected space and let $\alpha, \beta: I \to X$ be continuous functions. Prove that α is homotopic to β.

8. (a) Prove that $\mathbb{R}$ (with the usual topology) is contractible.

    (b) Prove that every contractible space is pathwise connected.

9. Prove Theorem 8.7.

10. Prove Theorem 8.8.

11. Let X be a convex subset of $\mathbb{R}^n$ and let $x_0 \in X$. Prove that X is contractible to x_0.

12. Let $(G, \cdot, \mathcal{T})$ be a topological group (see Appendix I), and let e denote the identity of $(G, \cdot)$. For $\alpha, \beta \in \Omega(G, e)$, define $\alpha \oslash \beta$ by $(\alpha \oslash \beta)(x) = \alpha(x) \cdot \beta(x)$.

    (a) Prove that $\oslash$ is an operation on $\Omega(G, e)$ and that $(\Omega(G, e), \oslash)$ is a group.

    (b) Prove that the operation $\oslash$ on $\Omega(G, e)$ induces an operation $\otimes$ on $\pi_1(G, e)$ and that $(\pi_1(G, e), \otimes)$ is a group.

    (c) Prove that the two operations $\otimes$ and $\circ$ on $\pi_1(G, e)$ are the same.

    (d) Prove that $\pi_1(G, e)$ is abelian.

8.2 The Fundamental Group

In this section we establish some properties of the fundamental group. The first result is that if $(X, \mathcal{T})$ is a pathwise connected space and $x_0, x_1 \in X$, then $\pi_1(X, x_0)$ is isomorphic to $\pi_1(X, x_1)$. Before we begin the proof, we draw a picture (Figure 8.6) to illustrate the idea. It is useful to look at this picture as you read the proof.

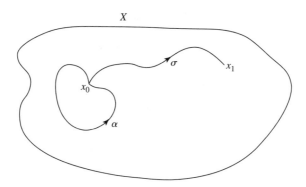

Figure 8.6

THEOREM 8.9. Let $(X, \mathcal{T})$ be a pathwise connected space, and let $x_0, x_1 \in X$. Then $\pi_1(X, x_0)$ is isomorphic to $\pi_1(X, x_1)$.

Proof. Let $\sigma: I \to X$ be a continuous function such that $\sigma(0) = x_0$ and $\sigma(1) = x_1$, and define $\overline{\sigma}: I \to X$ by $\overline{\sigma}(x) = \sigma(1 - x)$ for each $x \in X$. Then define $\theta_\sigma: \pi_1(X, x_0) \to \pi_1(X, x_1)$ by $\theta_\sigma([\alpha]) = [(\overline{\sigma} * \alpha) * \sigma]$ for each $[\alpha] \in \pi_1(X, x_0)$. We want to prove that θ_σ is an isomorphism. First observe that if $\alpha \in \Omega(X, x_0)$, then $(\overline{\sigma} * \alpha) * \sigma \in \Omega(X, x_1)$. In order to show that θ_σ is well-defined, it is sufficient to show that if α and β are members of $\Omega(X, x_0)$ and $\alpha \simeq_p \beta$, then $(\overline{\sigma} * \alpha) * \sigma \simeq_p (\overline{\sigma} * \beta) * \sigma$. The proof of this fact is left as Exercise 1(a).

Now let $[\alpha], [\beta] \in \pi_1(X, x_0)$. Then $\theta_\sigma([\alpha] \circ [\beta]) = \theta_\sigma([\alpha * \beta]) = [(\overline{\sigma} * (\alpha * \beta)) * \sigma]$, and $\theta_\sigma([\alpha]) \circ \theta_\sigma([\beta]) = [(\overline{\sigma} * \alpha) * \sigma] \circ [(\overline{\sigma} * \beta) * \sigma] = [(\overline{\sigma} * \alpha) * \sigma) * ((\overline{\sigma} * \beta) * \sigma)]$. Thus in order to show that θ_σ is a homeomorphism, it is sufficient to show that $(\overline{\sigma} * (\alpha * \beta)) * \sigma \simeq_p ((\overline{\sigma} * \alpha) * \sigma) * ((\overline{\sigma} * \beta) * \sigma)$. The proof of this fact is left as Exercise 1(b).

Now we show that θ_σ is one-to-one. Suppose $[\alpha], [\beta] \in \pi_1(X, x_0)$ and $\theta_\sigma([\alpha]) = \theta_\sigma([\beta])$. Then $[(\overline{\sigma} * \alpha) * \sigma] = [(\overline{\sigma} * \beta) * \sigma]$, and so $(\overline{\sigma} * \alpha) * \sigma \simeq_p (\overline{\sigma} * \beta) * \sigma$. In order to show that $[\alpha] = [\beta]$, it is sufficient to show that $\alpha \simeq_p \beta$, and we leave the proof of this fact as Exercise 1(c).

Now we show that θ_σ maps $\pi_1(X, x_0)$ onto $\pi_1(X, x_1)$. Let $[\alpha] \in \pi_1(X, x_1)$. Then $\alpha \in \Omega(X, x_1)$ and so $\sigma * (\alpha * \overline{\sigma}) \in \Omega(X, x_0)$. Then $\theta_\sigma([\sigma * (\alpha * \overline{\sigma})]) = [(\overline{\sigma} * (\sigma * (\alpha * \overline{\sigma})) * \sigma]$. Thus it is sufficient to show that $(\overline{\sigma} * (\sigma * (\alpha * \overline{\sigma}))) * \sigma \simeq_p \alpha$, and the proof of this fact is left as Exercise 1(d). ∎

EXAMPLE 4. Let X be a convex subset (see Exercise 5 of Section 8.1) of $\mathbb{R}^n$. Then X is pathwise connected because if $x, y \in X$, the function $\alpha: I \to X$ defined by $\alpha(t) = (1 - t)x + ty$ is a path from x to y. Thus up to isomorphism the fundamental group of X is independent of the base point. Let $x_0 \in X$ and let $\alpha \in \Omega(X, x_0)$. Define $H: I \times I \to X$ by $H(x, t) = tx_0 + (1 - t)\alpha(x)$. Then $H(x, 0) = \alpha(x)$ and $H(x, 1) = x_0$ for each $x \in X$. Also $H(0, t) = tx_0 + (1 - t)\alpha(0) = x_0$ and $H(1, t) = tx_0 + (1 - t)\alpha(1) = x_0$. Therefore α is path homotopic to the path $e: I \to X$ defined by $e(x) = x_0$ for each $x \in X$. Hence the fundamental group of X at x_0 is the trivial group (that is, the group whose only member is the identity element), and, by Theorem 8.9, if x is any member of X, then $\pi_1(X, x)$ is the trivial group.

Let $(X, \mathcal{T})$ be a topological space, let $x_0 \in X$, and let C be the path component of X that contains x_0. Since the continuous image of a pathwise connected space is pathwise connected (see Exercise 7 of Section 3.2), we have the following results. If $\alpha \in \Omega(X, x_0)$, then α is also a member of $\Omega(C, x_0)$. Also if $H: I \times I \to X$ is a continuous function such that $H(x, t) = x_0$ for some $(x, t) \in I \times I$, then $H(I \times I) \subseteq C$. Thus $\pi_1(X, x_0) = \pi_1(C, x_0)$. Therefore $\pi_1(X, x_0)$ depends only on the path component C of X that contains x_0, and it does not give us any information about $X - C$.

Let $(X, \mathcal{T})$ be a pathwise connected space, and let $x_0, x_1 \in X$. While Theorem 8.9 guarantees that $\pi_1(X, x_0)$ is isomorphic to $\pi_1(X, x_1)$, different paths from x_0 to x_1 may give rise to different isomorphisms.

If $(X, \mathcal{T})$ is a pathwise connected space, it is common practice to speak of the fundamental group of X without reference to the basepoint.

We show that the fundamental group is a topological property by introducing a homeomorphism induced by a continuous function.

Definition. A **topological pair** (X, A) is an ordered pair whose first term is a topological space $(X, \mathcal{T})$ and whose second term is a subspace A of X. ∎

We do not distinguish between the topological pair $(X, \varnothing)$ and the topological space $(X, \mathcal{T})$. If $x_0 \in X$, we let (X, x_0) denote the topological pair $(X, \{x_0\})$.

Definition. If (X, A) and (Y, B) are topological pairs, a **map** $f: (X, A) \to (Y, B)$ is a continuous function $f: X \to Y$ such that $f(A) \subseteq B$.

THEOREM 8.10. Let $(X, \mathcal{T})$ and $(Y, \mathcal{U})$ be topological spaces, let $x_0 \in X$ and $y_0 \in Y$, and let $h: (X, x_0) \to (Y, y_0)$ be a map. Then h induces a homeomorphism $h_*: \pi_1(X, x_0) \to \pi_1(Y, y_0)$.

Proof. Let $[\alpha] \in \pi_1(X, x_0)$. Then $\alpha: I \to X$ is a continuous function such that $\alpha(0) = \alpha(1) = x_0$, so $h \circ \alpha: I \to Y$ is a continuous function such that $(h \circ \alpha)(0) = (h \circ \alpha)(1) = y_0$. Therefore $[h \circ \alpha] \in \pi_1(Y, y_0)$. Define $h_*: \pi_1(X, x_0) \to \pi_1(Y, y_0)$ by $h_*([\alpha]) = [h \circ \alpha]$. In order to show that h_* is well-defined, it is sufficient to show that if $\alpha, \beta \in \Omega(X, x_0)$ and $\alpha \simeq_p \beta$, then $h \circ \alpha \simeq_p h \circ \beta$. Suppose

8.2 The Fundamental Group   277

$\alpha, \beta \in \Omega(X, x_0)$ and $\alpha \simeq_p \beta$. Then there is a continuous function $F: I \times I \to X$ such that $F(x, 0) = \alpha(x)$ and $F(x, 1) = \beta(x)$ for all $x \in I$ and $F(0, t) = F(1, t) = x_0$ for all $t \in I$. Define $G: I \times I \to Y$ by $G(x, t) = (h \circ F)(x, t)$. Then it is easy to see that $G(x, 0) = (h \circ \alpha)(x)$ and $G(x, 1) = (h \circ \beta)(x)$ for all $x \in I$ and $G(0, t) = G(1, t) = y_0$. Therefore $h \circ \alpha \simeq_p h \circ \beta$, and so h_* is well defined.

Now we show that h_* is a homeomorphism. Let $[\alpha], [\beta] \in \pi_1(X, x_0)$. Then $h_*([\alpha] \circ [\beta]) = h_*([\alpha * \beta]) = [h \circ (\alpha * \beta)]$ and $h_*([\alpha]) \circ h_*([\beta]) = [h \circ \alpha] \circ [h \circ \beta] = [(h \circ \alpha) * (h \circ \beta)]$. Since

$$(h \circ (\alpha * \beta))(x) = \begin{cases} (h \circ \alpha)(2x), & \text{if } 0 \leq x \leq \tfrac{1}{2} \\ (h \circ \beta)(2x - 1), & \text{if } \tfrac{1}{2} \leq x \leq 1 \end{cases} = ((h \circ \alpha) * (h \circ \beta))(x),$$

h_* is a homeomorphism. ∎

In the remainder of this text, whenever we speak of the homeomorphism induced by a continuous function, we mean the homeomorphism defined in the proof of Theorem 8.10.

Definition. If (X, A) and (Y, B) are topological pairs and $f, g: (X, A) \to (Y, B)$ are maps such that $f|_A = g|_A$, then f is **homotopic to g relative to A**, denoted by $f \simeq g$ rel A, if there is a continuous function $H: X \times I \to Y$ such that $H(x, 0) = f(x)$ and $H(x, 1) = g(x)$ for all $x \in X$ and $H(a, t) = f(a)$ for all $a \in A$ and $t \in I$. ∎

If $x_0 \in X$, then we write $f \simeq g$ rel x_0 rather than $f \simeq g$ rel $\{x_0\}$.

THEOREM 8.11. Let $(X, \mathcal{T})$ and $(Y, \mathcal{U})$ be topological spaces, let $x_0 \in X$ and $y_0 \in Y$, and let $h, k: (X, x_0) \to (Y, y_0)$ be maps such that $h \simeq k$ rel x_0. Then $h_* = k_*$.

Proof. Since $h \simeq k$ rel x_0, there is a continuous function $H: I \times I \to Y$ such that $H(x, 0) = h(x)$ and $H(x, 1) = k(x)$ for all $x \in X$ and $H(x_0, t) = y_0$ for all $t \in I$. Let $[\alpha] \in \pi_1(X, x_0)$. Then $h_*([\alpha]) = [h \circ \alpha]$ and $k_*([\alpha]) = [k \circ \alpha]$. Define $G: I \times I \to Y$ by $G(x, t) = H(\alpha(x), t)$ for all $(x, t) \in I \times I$. By Theorem 2.27, G is continuous. Note that $G(x, 0) = H(\alpha(x), 0) = (h \circ \alpha)(x)$ and $G(x, 1) = H(\alpha(x), 1) = (k \circ \alpha)(x)$ for each $x \in I$ and $G(0, t) = H(\alpha(0), t) = H(x_0, t) = y_0$ and $G(1, t) = H(\alpha(1), t) = H(x_0, t) = y_0$ for each $t \in I$. Therefore $h \circ \alpha \simeq_p k \circ \alpha$, and hence $[h \circ \alpha] = [k \circ \alpha]$. Thus $h_* = k_*$. ∎

THEOREM 8.12. Let $(X, \mathcal{T}), (Y, \mathcal{U})$, and $(Z, \mathcal{V})$ be topological spaces, let $x_0 \in X, y_0 \in Y$, and $z_0 \in Z$, and let $h: (X, x_0) \to (Y, y_0)$ and $k: (Y, y_0) \to (Z, z_0)$ be maps. Then $(k \circ h)_* = k_* \circ h_*$.

Proof. Let $[\alpha] \in \pi_1(X, x_0)$. Then $(k \circ h)_*([\alpha]) = [(k \circ h) \circ \alpha]$ and $(k_* \circ h_*)[\alpha]) = k_*([h \circ \alpha]) = [k \circ (h \circ \alpha)]$. Since the composition of functions is associative (see Exercise 21 in Appendix C), $(k \circ h) \circ \alpha = k \circ (h \circ \alpha)$. Therefore $(k \circ h)_* = k_* \circ h_*$. ∎

Definition. Let $(X, \mathcal{T})$ and $(Y, \mathcal{U})$ be topological spaces and let $x_0 \in X$ and $y_0 \in Y$. Then (X, x_0) and (Y, y_0) are **of the same homotopy type** if there exist maps $f: (X, x_0) \to (Y, y_0)$ and $g: (Y, y_0) \to (X, x_0)$ such that $f \circ g \simeq i_Y$ rel y_0 and $g \circ f \simeq i_X$ rel x_0. ∎

THEOREM 8.13. Let $(X, \mathcal{T})$ and $(Y, \mathcal{U})$ be topological spaces and let $x_0 \in X$ and $y_0 \in Y$. If (X, x_0) and (Y, y_0) are of the same homotopy type, then $\pi_1(X, x_0)$ is isomorphic to $\pi_1(Y, y_0)$.

Proof. Suppose (X, x_0) and (Y, y_0) are of the same homotopy type. Then there exist maps $f: (X, x_0) \to (Y, y_0)$ and $g: (Y, y_0) \to (X, x_0)$ such that $f \circ g \simeq i_Y$ rel y_0 and $g \circ f \simeq i_X$ rel x_0. By Theorem 8.11, $(f \circ g)_*: \pi_1(Y, y_0) \to \pi_1(Y, y_0)$ and $(g \circ f)_*: \pi_1(X, x_0) \to \pi_1(X, x_0)$ are the identity homeomorphisms. By Theorem 8.12, $(f \circ g)_* = f_* \circ g_*$ and $(g \circ f)_* = g_* \circ f_*$. By Theorem C.2, $f_*: \pi_1(X, x_0) \to \pi_1(Y, y_0)$ is an isomorphism. ∎

THEOREM 8.14. Let $(X, \mathcal{T})$ and $(Y, \mathcal{U})$ be topological spaces, let $h: X \to Y$ be a homeomorphism, let $x_0 \in X$, and let $y_0 = h(x_0)$. Then $\pi_1(X, x_0)$ is isomorphic to $\pi_1(Y, y_0)$.

Proof. Since h is a homeomorphism, $h^{-1}: Y \to X$ is a homeomorphism. Also note that $x_0 = h^{-1}(y_0)$. Since $h^{-1} \circ h = i_X$ and $h \circ h^{-1} = i_Y$, (X, x_0) and (Y, y_0) are of the same homotopy type. Therefore by Theorem 8.13, $\pi_1(X, x_0)$ is isomorphic to $\pi_1(Y, y_0)$. ∎

EXERCISES 8.2

1. Let $(X, \mathcal{T})$ be a topological space, let $x_0, x_1 \in X$, let $\sigma: I \to X$ be a continuous function such that $\sigma(0) = x_0$ and $\sigma(1) = x_1$, and let $\alpha, \beta \in \Omega(X, x_0)$.

    (a) Prove that if $\alpha \simeq_p \beta$, then $(\overline{\sigma} * \alpha) * \sigma \simeq_p (\overline{\sigma} * \beta) * \sigma$.

    (b) Prove that $(\overline{\sigma} * (\alpha * \beta)) * \sigma \simeq_p ((\overline{\sigma} * \alpha) * \sigma) * ((\overline{\sigma} * \beta) * \sigma)$.

    (c) Prove that if $(\overline{\sigma} * \alpha) * \sigma \simeq_p (\overline{\sigma} * \beta) * \sigma$, then $\alpha \simeq_p \beta$.

    (d) Prove that $(\overline{\sigma} * (\sigma * (\alpha * \overline{\sigma}))) * \sigma \simeq_p \alpha$.

2. A pathwise connected space $(X, \mathcal{T})$ is **simply connected** provided that for each $x_0 \in X$, $\pi_1(X, x_0)$ is the trivial group. Let $(X, \mathcal{T})$ be a simply connected space and let $\alpha, \beta: I \to X$ be paths such that $\alpha(0) = \beta(0)$ and $\alpha(1) = \beta(1)$. Prove that $\alpha \simeq_p \beta$.

3. Prove that every contractible space is simply connected.

4. A topological space $(X, \mathcal{T})$ is **weakly contractible** if there exists $x_0 \in X$ and a continuous function $H: X \times I \to X$ such that $H(x, 0) = x$ and $H(x, 1) = x_0$ for each $x \in X$.

   (a) Give an example of a weakly contractible space that is not contractible.

   (b) Prove that every weakly contractible space is simply connected.

5. A subspace A of a topological space $(X, \mathcal{T})$ is a **retract** of X if there exists a continuous function $r: X \to A$ such that $r(a) = a$ for each $a \in A$. The function r is called a **retraction of X onto A**. Let A be a subspace of a topological space $(X, \mathcal{T})$, let $r: X \to A$ be a retraction of X onto A, and let $a_0 \in A$. Prove that $r_*: \pi_1(X, a_0) \to \pi_1(A, a_0)$ is a surjection. *Hint:* Consider the map $j: (A, a_0) \to (X, a_0)$, defined by $j(a) = a$ for each $a \in A$.

6. Let $(X, \mathcal{T})$ be a pathwise connected space, let $x_0, x_1 \in X$, let $(Y, \mathcal{U})$ be a topological space, let $h: X \to Y$ be a continuous function, let $y_0 = h(x_0)$ and $y_1 = h(x_1)$, and for each $i = 0, 1$, let h_i denote the map from the pair (X, x_i) into the pair (Y, y_i) defined by $h_i(x) = h(x)$ for each $x \in X$. Prove that there are isomorphisms $\theta: \pi_1(X, x_0) \to \pi_1(X, x_1)$ and $\phi: \pi_1(Y, y_0) \to \pi_1(Y, y_1)$ such that $(h_1)_* \circ \theta = \phi \circ (h_0)_*$.

8.3 The Fundamental Group of the Circle

Our goal in this section is to show that the fundamental group of the circle is isomorphic to the group of integers. In order to do this we need several preliminary results. We begin with the definition of the function p that maps $\mathbb{R}$ onto S^1. Define $p: \mathbb{R} \to S^1$ by $p(x) = (\cos 2\pi x, \sin 2\pi x)$. One can think of p as a function that wraps $\mathbb{R}$ around S^1. In particular, note that for each integer n, p is a one-to-one map of $[n, n+1)$ onto S^1. Furthermore, if U is the open subset of S^1 indicated in Figure 8.7, then $p^{-1}(U)$ is the union of a pairwise disjoint collection of open intervals.

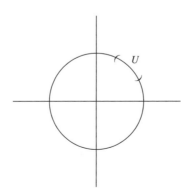

Figure 8.7

To be specific, let $U = \{(x, y) \in S^1 : x > 0 \text{ and } y > 0\}$. Then if $x \in p^{-1}(U)$, $\cos 2\pi x$ and $\sin 2\pi x$ are both positive, so $p^{-1}(U) = \bigcup_{n \in \mathbb{Z}}(n, n + \frac{1}{4})$. Moreover, for each $n \in \mathbb{Z}$, $p|_{[n, n+\frac{1}{4}]}$ is a one-to-one function from the closed interval $[n, n + \frac{1}{4}]$ onto $\overline{U}$. Thus, since $[n, n + \frac{1}{4}]$ is compact, $p|_{[n, n+\frac{1}{4}]}$ is a homeomorphism of $[n, n + \frac{1}{4}]$ onto $\overline{U}$. Therefore $p|_{(n, n+\frac{1}{4})}$ is a homeomorphism of $(n, n + \frac{1}{4})$ onto U. Throughout the remainder of this section, p denotes the function defined above.

The following result is known as the Covering Path Property. You may wish to consider Example 5 before reading the proof.

THEOREM 8.15. Let $\alpha: I \to S^1$ be a path and let $x_0 \in \mathbb{R}$ such that $p(x_0) = \alpha(0)$. Then there is a unique path $\beta: I \to \mathbb{R}$ such that $\beta(0) = x_0$ and $p \circ \beta = \alpha$.

Proof. Since α is continuous, for each $t \in I$ there is a connected neighborhood U_t of t such that $\alpha(U_t)$ is a proper subset of S^1. Since I is compact, the open cover $\{U_t : t \in I\}$ of I has a finite subcover $\{U_1, U_2, ..., U_m\}$. If $U_i = I$ for some i, then $\alpha(I)$ is a connected proper subset of S^1. If $U_i \neq I$ for any i, choose U_i so that $0 \in U_i$. There exists $t' \in I$ such that $t' \neq 0$, $[0, t'] \subseteq U_i$, and $t' \in U_i \cap U_j$ for some $j \neq i$. If $U_i \cup U_j = I$, $\alpha([0, t'])$ and $\alpha([t', 1])$ are connected proper subsets of S^1. If $U_i \cup U_j \neq I$, choose $t'' \in I$ such that $t'' > t'$, $[t', t''] \subseteq U_j$ and $t'' \in U_j \cap U_k$ for some k ($i \neq k \neq j$). Since $\{U_1, U_2, ..., U_m\}$ is a finite cover of I, after a finite number of steps, we will obtain $t_0, t_1, t_2, ..., t_n$ such that $0 = t_0$, $t_n = 1$, $t_{i-1} < t_i$ and $\alpha([t_{i-1}, t_i])$ is a connected proper subset of S^1 for each $i = 1, 2, ..., n$.

For each $i = 0, 1, 2, ..., n$, let P_i be the statement: There is a unique continuous function $\beta_i: [0, t_i] \to \mathbb{R}$ such that $\beta_i(0) = x_0$ and $p \circ \beta_i = \alpha|_{[0, t_i]}$. In order to prove the theorem, it is sufficient to show that P_i is true for each i. It is clear that P_0 is true. Suppose $0 < i \leq n$ and P_{i-1} is true. Let V_{i-1} denote the component of $p^{-1}(\alpha([t_{i-1}, t_i]))$ that contains $\beta_{i-1}(t_{i-1})$, and let $p_{i-1} = p|_{V_{i-1}}$. Then $p_{i-1}: V_{i-1} \to \alpha([t_{i-1}, t_i])$ is a homeomorphism. Define $\beta_i: [0, t_i] \to \mathbb{R}$ by

$$\beta_i(t) = \begin{cases} \beta_{i-1}(t), & \text{if } 0 \leq t \leq t_{i-1} \\ (p_{i-1})^{-1}(\alpha(t)), & \text{if } t_{i-1} \leq t \leq t_i. \end{cases}$$

Then β_i is continuous, $\beta_i(0) = x_0$, and $p \circ \beta_i = \alpha|_{[0, t_i]}$. We must show that β_i is unique. Suppose $\beta_i': [0, t_i] \to \mathbb{R}$ is a continuous function such that $\beta_i'(0) = x_0$ and $p \circ \beta_i' = \alpha|_{[0, t_i]}$. If $0 \leq t \leq t_{i-1}$, then $\beta_i'(t) = \beta_i(t)$ since P_{i-1} is true. Suppose $t_{i-1} < t \leq t_i$. Since $(p \circ \beta_i')(t) = \alpha(t) \in \alpha([t_{i-1}, t_i])$, $\beta_i'(t) \in p^{-1}(\alpha([t_{i-1}, t_i]))$. Therefore $\beta_i'(t) \in V_{i-1}$ since β_i' is continuous and V_{i-1} is a component of $p^{-1}(\alpha([t_{i-1}, t_i]))$. Therefore $\beta_i(t), \beta_i'(t) \in V_{i-1}$ and $(p_{i-1} \circ \beta_i)(t) = (p_{i-1} \circ \beta_i')(t)$. Hence $\beta_i(t) = \beta_i'(t)$ since p_{i-1} is a homeomorphism. ∎

EXAMPLE 5. Let $\alpha: I \to S^1$ be a homeomorphism that maps I onto $\{(x, y) \in S^1 : x > 0 \text{ and } y > 0\}$ and has the property that $\alpha(0) = (1, 0)$ and $\alpha(1) = (0, 1)$ and let $x_0 = 2$. Then the function $\beta: I \to \mathbb{R}$ given by Theorem 8.15 is

8.3 The Fundamental Group of the Circle 281

a homeomorphism of I onto $[2, 2.25]$, which has the property that $\beta(0) = 2$ and $\beta(1) = 2.25$.

The following result is known as the Covering Homotopy Property.

THEOREM 8.16. Let $(X, \mathcal{T})$ be a topological space, let $f: X \to \mathbb{R}$ be a continuous function, and let $H: X \times I \to S^1$ be a continuous function such that $H(x, 0) = (p \circ f)(x)$ for each $x \in X$. Then there is a continuous function $F: X \times I \to \mathbb{R}$ such that $F(x, 0) = f(x)$ and $(p \circ F)(x, t) = H(x, t)$ for each $(x, t) \in X \times I$.

Proof. For each $x \in X$, define $\alpha_x: I \to S^1$ by $\alpha_x(t) = H(x, t)$. Now $(p \circ f)(x) = \alpha_x(0)$, and hence, by Theorem 8.15, there is a unique path $\beta_x: I \to \mathbb{R}$ such that $\beta_x(0) = f(x)$ and $p \circ \beta_x = \alpha_x$. Define $F: X \times I \to \mathbb{R}$ by $F(x, t) = \beta_x(t)$. Then $(p \circ F)(x, t) = (p \circ \beta_x)(t) = H(x, t)$ and $F(x, 0) = \beta_x(0) = f(x)$.

We must show that F is continuous. Let $x_0 \in X$. For each $t \in I$, there is a neighborhood M_t of x_0 and a connected neighborhood N_t of t such that $H(M_t \times N_t)$ is contained in a proper connected subset of S^1. Since I is compact, the open cover $\{N_t : t \in I\}$ of I has a finite subcover $\{N_{t_1}, N_{t_2}, ..., N_{t_n}\}$. Let $M = \bigcap_{i=1}^{n} M_{t_i}$. Then M is a neighborhood of x_0, and, for each $i = 1, 2, ..., n$, $H(M \times N_{t_i})$ is contained in a proper connected subset of S^1. Thus there exist $t_0, t_1, ..., t_m$ such that $t_0 = 0, t_m = 1$, and, for each $j = 1, 2, ..., m$, $t_{j-1} < t_j$ and $H(M \times [t_{j-1}, t_j])$ is contained in a proper connected subset of S^1.

For each $j = 0, 1, ..., m$, let P_j be the statement: There is a unique continuous function $G_j: M \times [0, t_j] \to \mathbb{R}$ such that $G_j(x, 0) = f(x)$ and $(p \circ G_j)(x, t) = H(x, t)$. We want to show that P_j is true for each j. It is clear that P_0 is true. Suppose $0 < j \leq m$ and P_{j-1} is true. Let U_{j-1} be a connected proper subset of S^1 that contains $H(M \times I_{j-1})$, let V_{j-1} denote the component of $p^{-1}(U_{j-1})$ that contains $G_{j-1}(M \times \{t_{j-1}\})$, and let $p_{j-1} = p|_{V_{j-1}}$. Then $p_{j-1}: V_{j-1} \to U_{j-1}$ is a homeomorphism. Define $G_j: M \times [0, t_j] \to \mathbb{R}$ by

$$G_j(x, t) = \begin{cases} G_{j-1}(x, t), & \text{if } 0 \leq t \leq t_{j-1} \\ (p_{j-1})^{-1}(H(x, t)), & \text{if } t_{j-1} \leq t \leq t_j. \end{cases}$$

Then G_j is continuous, $G_j(x, 0) = f(x)$, and $(p \circ G_j)(x, t) = H(x, t)$. We must show that G_j is unique. Suppose $G_j': M \times [0, t_j] \to \mathbb{R}$ is a continuous function such that $G_j'(x, 0) = f(x)$ and $(p \circ G_j')(x, t) = H(x, t)$. If $0 \leq t \leq t_{j-1}$, then $G_j'(x, t) = G_j(x, t)$ because p_{j-1} is true. Suppose $t_{j-1} < t \leq t_j$. Since $(p \circ G_j)(x, t) = H(x, t) \in U_{j-1}$, $G_j'(x, t) \in p^{-1}(U_{j-1})$. Therefore, since G_j' is continuous and V_{j-1} is a component of $p^{-1}(U_{j-1})$, $G_j'(x, t) \in V_{j-1}$. Hence $G_j(x, t)$ and $G_j'(x, t)$ are members of V_{j-1} and $(p_{j-1} \circ G_j)(x, t) = (p_{j-1} \circ G_j')(x, t)$. Thus, since p_{j-1} is a homeomorphism, $G_j(x, t) = G_j'(x, t)$. We have proved that there is a unique continuous function $G: M \times I \to \mathbb{R}$ such that $G(x, 0) = f(x)$ and $(p \circ G)(x, t) = H(x, t)$. Therefore $G = F|_{(M \times I)}$. Since M is a neighborhood of x_0 in X, F is continuous at (x_0, t) for each

$t \in I$. Since x_0 is an arbitrary member of X, F is continuous. This proof also shows that F is unique. ∎

If α is a loop in S^1 at $(1, 0)$, then $p(0) = \alpha(0)$, and hence, by Theorem 8.15, there is a unique path $\beta: I \to \mathbb{R}^1$ such that $\beta(0) = 0$ and $p \circ \beta = \alpha$. Since $(p \circ \beta)(1) = \alpha(1) = (1, 0)$, $\beta(1) \in p^{-1}(1, 0)$, and hence $\beta(1)$ is an integer. The integer $\beta(1)$ is called the **degree** of the loop α, and we write $\deg(\alpha) = \beta(1)$.

EXAMPLE 6. Define $\alpha: I \to S^1$ by $\alpha(x) = (\cos 4\pi x, -\sin 4\pi x)$. Then α "wraps" I around S^1 twice in a clockwise direction. The unique path $\beta: I \to \mathbb{R}$ such that $\beta(0) = 0$ and $p \circ \beta = \alpha$ given by Theorem 8.15 is defined by $\beta(x) = -2x$ for each $x \in I$. Therefore $\deg(\alpha) = \beta(1) = -2$.

THEOREM 8.17. Let α_1 and α_2 be loops in S^1 at $(1, 0)$ such that $\alpha_1 \simeq_p \alpha_2$. Then $\deg(\alpha_1) = \deg(\alpha_2)$.

Proof. Since $\alpha_1 \simeq_p \alpha_2$, there is a continuous function $H: I \times I \to S^1$ such that $H(x, 0) = \alpha_1(x)$ and $H(x, 1) = \alpha_2(x)$ for each $x \in I$ and $H(0, t) = H(1, t) = (1, 0)$ for each $t \in I$. By Theorem 8.15, there is a unique path $\beta_1: I \to \mathbb{R}$ such that $\beta_1(0) = 0$ and $(p \circ \beta_1) = \alpha_1$. Thus, by Theorem 8.16, there is a unique continuous funtion $F: I \times I \to \mathbb{R}$ such that $(p \circ F)(x, t) = H(x, t)$ for each $(x, t) \in I \times I$ and $F(x, 0) = \beta_1(x)$ for each $x \in I$. Define $\gamma: I \to \mathbb{R}$ by $\gamma(t) = F(0, t)$ for each $t \in I$. Then γ is continuous, and $(p \circ \gamma)(t) = (p \circ F)(0, t) = H(0, t) = (1, 0)$ for each $t \in I$. Therefore $\gamma(I) \subseteq p^{-1}(1, 0)$, and, since $\gamma(I)$ is connected and $p^{-1}(1, 0)$ is a discrete subspace of $\mathbb{R}$, γ is a constant function. Thus $F(0, t) = \gamma(t) = \gamma(0) = F(0, 0) = \beta_1(0) = 0$ for each $t \in I$. Define $\beta_2: I \to \mathbb{R}$ by $\beta_2(x) = F(x, 1)$ for each $x \in I$. Then $\beta_2(0) = F(0, 1) = 0$ and $(p \circ \beta_2)(x) = (p \circ F)(x, 1) = H(x, 1) = \alpha_2(x)$ for each $x \in I$. By definition, $\deg(\alpha_1) = \beta_1(1)$ and $\deg(\alpha_2) = \beta_2(1)$. Now define a path $\delta: I \to \mathbb{R}$ by $\delta(t) = F(1, t)$ for each $t \in I$. Again $(p \circ \delta)(t) = (p \circ F)(1, t) = H(1, t) = (1, 0)$ for each $t \in I$, and hence $\delta(I) \subseteq p^{-1}(1, 0)$. Therefore δ is a constant function, and hence $F(1, t) = \delta(t) = \delta(0) = F(1, 0) = \beta_1(0) = 0$ for each $t \in I$. Therefore $\beta_2(1) = F(1, 1) = F(1, 0) = \beta_1(1)$, and hence $\deg(\alpha_1) = \deg(\alpha_2)$. ∎

We are ready to prove that the fundamental group of the circle is isomorphic to the group of integers. Since S^1 is pathwise connected, the fundamental group of S^1 is independent of the base point.

THEOREM 8.18. $\pi_1(S^1, (1, 0))$ is isomorphic to the group of integers.

Proof. Define $\phi: \pi_1(S^1, (1, 0)) \to \mathbb{Z}$ by $\phi([\alpha]) = \deg(\alpha)$. By Theorem 8.17, ϕ is well-defined. Let $[\alpha_1], [\alpha_2] \in \pi_1(S^1, (1, 0))$. By Theorem 8.15, there are unique paths $\beta_1, \beta_2: I \to \mathbb{R}$ such that $\beta_1(0) = \beta_2(0) = 0$, $p \circ \beta_1 = \alpha_1$, and $p \circ \beta_2 = \alpha_2$. By definition, $\deg(\alpha_1) = \beta_1(1)$ and $\deg(\alpha_2) = \beta_2(1)$. Define $\delta: I \to \mathbb{R}$ by

$$\delta(x) = \begin{cases} \beta_1(2x), & \text{if } 0 \leq x \leq \frac{1}{2} \\ \beta_1(1) + \beta_2(2x-1), & \text{if } \frac{1}{2} \leq x \leq 1. \end{cases}$$

Since $\beta_2(0) = 0$, δ is continuous. Now $\delta(0) = \beta_1(0) = 0$, and, since

$$p(\beta_1(1) + \beta_2(2x-1)) = p(\beta_2(2x-1)),$$

$$(p \circ \delta)(x) = \begin{cases} (p \circ \beta_1)(2x), & \text{if } 0 \leq x \leq \tfrac{1}{2} \\ (p \circ \beta_2)(2x - 1), & \text{if } \tfrac{1}{2} \leq x \leq 1 \end{cases}$$

$$= \begin{cases} \alpha_1(2x), & \text{if } 0 \leq x \leq \tfrac{1}{2} \\ \alpha_2(2x - 1), & \text{if } \tfrac{1}{2} \leq x \leq 1 \end{cases} = (\alpha_1 * \alpha_2)(x).$$

Therefore $\phi([\alpha_1] \circ [\alpha_2]) = \phi([\alpha_1 * \alpha_2]) = \deg(\alpha_1 * \alpha_2) = \delta(1) = \beta_1(1) + \beta_2(1) = \deg(\alpha_1) + \deg(\alpha_2) = \phi([\alpha_1]) + \phi([\alpha_2])$, and so ϕ is a homeomorphism.

Now we show that ϕ maps $\pi_1(S^1, (1, 0))$ onto $\mathbb{Z}$. Let $z \in \mathbb{Z}$, and define a path $\alpha_1: I \to \mathbb{R}$ by $\alpha(t) = zt$ for each $t \in I$. Then $\alpha(0) = 0$ and $\alpha(1) = z$, so $p \circ \alpha: I \to S^1$ is a loop in S^1 at $(1, 0)$. Therefore $[p \circ \alpha] \in \pi_1(S^1, (1, 0))$, and, by definition, $\deg(p \circ \alpha) = \alpha(1) = z$. Therefore $\phi([p \circ \alpha]) = \deg(p \circ \alpha) = z$.

Finally, we show that ϕ is one-to-one. Let $[\alpha_1], [\alpha_2] \in \pi_1(S^1, (1, 0))$ such that $\phi([\alpha_1]) = \phi([\alpha_2])$. Then $\deg(\alpha_1) = \deg(\alpha_2)$. By Theorem 8.15, there are unique paths $\beta_1, \beta_2: I \to \mathbb{R}$ such that $\beta_1(0) = \beta_2(0) = 0$, $p \circ \beta_1 = \alpha_1$, and $p \circ \beta_2 = \alpha_2$. By definition, $\deg(\alpha_1) = \beta_1(1)$ and $\deg(\alpha_2) = \beta_2(1)$. So $\beta_1(1) = \beta_2(1)$. Define $F: I \times I \to \mathbb{R}$ by $F(x, t) = (1 - t)\beta_1(x) + t\beta_2(x)$ for each $(x, t) \in I \times I$. Then $F(x, 0) = \beta_1(x)$ and $F(x, 1) = \beta_2(x)$ for each $x \in I$ and $F(0, t) = 0$ and $F(1, t) = \beta_1(1) = \beta_2(1)$ for each $t \in I$. Therefore $p \circ F: I \times I \to S^1$ is a continuous function such that $(p \circ F)(x, 0) = (p \circ \beta_1)(x) = \alpha_1(x)$ and $(p \circ F)(x, 1) = (p \circ \beta_2)(x) = \alpha_2(x)$ for each $x \in I$ and $(p \circ F)(0, t) = p(0) = (1, 0)$ and $(p \circ F)(1, t) = (p \circ \beta_1)(1) = \alpha_1(1) = (1, 0)$ for each $t \in I$. Thus $\alpha_1 \simeq_p \alpha_2$, so $[\alpha_1] = [\alpha_2]$. ∎

There are no exercises in this section. The sole purpose of the section is to calculate the fundamental group of a familiar figure and thus provide you with a concrete example. In the next section, we generalize some of the theorems in this section and give some exercises.

8.4 Covering Spaces

In this section we generalize Theorems 8.15 and 8.16 by replacing the function $p: \mathbb{R} \to S^1$ defined by $p(x) = (\cos 2\pi x, \sin 2\pi x)$ with a function called a "covering map" from an arbitrary topological space $(E, \mathcal{T})$ into another topological space $(B, \mathcal{U})$. These two generalized theorems are used in Section 9.4.

Definition. Let $(E, \mathcal{T})$ and $(B, \mathcal{U})$ be topological spaces, and let $p: E \to B$ be a continuous surjection. An open subset U of B is **evenly covered** by p if $p^{-1}(U)$ can be written as the union of a pairwise disjoint collection $\{V_\alpha: \alpha \in \Lambda\}$ of open sets such that for each $\alpha \in \Lambda$, $p|_{V_\alpha}$ is a homeomorphism of V_α onto U. Each V_α is called a **slice** of $p^{-1}(U)$. ∎

Definition. Let $(E, \mathcal{T})$ and $(B, \mathcal{U})$ be topological spaces, and let $p: E \to B$ be a continuous surjection. If each member of B has a neighborhood that is evenly covered by p, then p is a **covering map** and E is **covering space** of B. ■

Note that the function $p: \mathbb{R} \to S^1$ in Section 8.3 is a covering map. Covering spaces were introduced by Poincaré in 1883.

Definition. Let $(E, \mathcal{T})$, $(B, \mathcal{U})$, and $(X, \mathcal{V})$ be topological spaces, let $p: E \to B$ be a covering map, and let $f: X \to B$ be a continuous function. A **lifting** of f is a continuous function $f': X \to E$ such that $p \circ f' = f$ (see Figure 8.8). ■

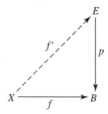

Figure 8.8

Notice that Theorem 8.15 provides a lifting of a path in S^1, where p is the specific function discussed in Section 8.3 rather than an arbitrary covering map.

THEOREM 8.19. Let $(E, \mathcal{T})$ and $(B, \mathcal{U})$ be topological spaces, let $p: E \to B$ be a covering map, let $e_0 \in E$, let $b_0 = p(e_0)$, and let $\sigma: I \to B$ be a path such that $\sigma(0) = b_0$. Then there exists a unique lifting $\sigma': I \to E$ of f such that $\sigma'(0) = e_0$.

Proof. Let $\{U_\alpha : \alpha \in \Lambda\}$ be an open cover of B such that for each $\alpha \in \Lambda$, U_α is evenly covered by p. By Theorem 4.4, there exists $t_0, t_1, \dots, t_n$ such that $0 = t_0 < t_1 < \cdots < t_n = 1$ and for each $i = 1, 2, \dots, n$, $\sigma([t_{i-1}, t_i]) \subseteq U_\alpha$ for some $\alpha \in \Lambda$. We define the lifting σ' inductively.

Define $\sigma'(0) = e_0$, and assume that $\sigma'(t)$ has been defined for all $t \in [0, t_{i-1}]$, where $1 \leq i \leq n$. There exists $\alpha \in \Lambda$ such that $\sigma([t_{i-1}, t_i]) \subseteq U_\alpha$. Let $\mathcal{V} = \{V_{\alpha\beta} : \beta \in \Gamma\}$ be the collection of slices of $p^{-1}(U_\alpha)$. Now $\sigma'(t_{i-1})$ is a member of exactly one member $V_{\alpha\gamma}$ of $\mathcal{V}$. For $t \in [t_{i-1}, t_i]$, define $\sigma'(t) = (p|_{V_{\alpha\gamma}})^{-1}(\sigma(t))$. Since $p|_{V_{\alpha\gamma}}: V_{\alpha\gamma} \to U_\alpha$ is a homeomorphism, σ' is continuous on $[t_{i-1}, t_i]$ and hence on $[0, t_i]$. Therefore, by induction, we can define σ' on I.

It follows immediately from the definition of σ' that $p \circ \sigma' = \sigma$.

The uniqueness of σ' is also proved inductively. Suppose σ'' is another lifting of σ such that $\sigma''(0) = e_0$, and assume that $\sigma''(t) = \sigma'(t)$ for all $t \in [0, t_{i-1}]$, where $1 \leq i \leq n$. Let U_α be a member of the open cover of B such that $\sigma([t_{i-1}, t_i]) \subseteq U_\alpha$, and let $V_{\alpha\gamma}$ be the member of $\mathcal{V}$ chosen in the definition of σ'. Since σ'' is a lifting of σ, $\sigma''([t_{i-1}, t_i]) \subseteq p^{-1}(U_\alpha) = \bigcup_{\beta \in \Gamma} V_{\alpha\beta}$. Since $\mathcal{V}$ is a collection of pairwise disjoint open sets and $\sigma''([t_{i-1}, t_i])$ is connected, $\sigma''([t_{i-1}, t_i])$ is a subset of one member of $\mathcal{V}$.

Since $\sigma''(t_{i-1}) = \sigma'(t_{i-1}) \in V_{\alpha\gamma}$, $\sigma''([t_{i-1}, t_i]) \subseteq V_{\alpha\gamma}$. Therefore, for each $t \in [t_{i-1}, t_i]$, $\sigma''(t) \in V_{\alpha\gamma} \cap p^{-1}(\sigma(t))$. But $V_{\alpha\gamma} \cap p^{-1}(\sigma(t)) = \{\sigma'(t)\}$, and so $\sigma''(t) = \sigma'(t)$. Therefore $\sigma''(t) = \sigma'(t)$ for all $t \in [0, t_i]$. By induction, $\sigma'' = \sigma'$. ∎

THEOREM 8.20. Let $(E, \mathcal{T})$ and $(B, \mathcal{U})$ be topological spaces, let $p\colon E \to B$ be a covering map, let $e_0 \in E$, let $b_0 = p(e_0)$, and let $F\colon I \times I \to B$ be a continuous function such that $F(0, 0) = b_0$. Then there exists a lifting $F'\colon I \times I \to E$ of F Such that $F'(0, 0) = e_0$. Moreover, if F is a path homotopy then F' is also.

Proof. Define $F'(0, 0) = e_0$. By Theorem 8.19, there exists a unique lifting $F'\colon \{0\} \times I \to E$ of $F|_{\{0\} \times I}$ such that $F'(0, 0) = e_0$ and a unique lifting $F'\colon I \times \{0\} \to E$ of $F|_{I \times \{0\}}$ such that $F(0, 0) = e_0$. Therefore, we assume that a lifting F' of F is defined on $(\{0\} \times I) \cup (I \times \{0\})$ and we extend it to $I \times I$.

Let $\{U_\alpha \colon \alpha \in \Lambda\}$ be an open cover of B such that for each $\alpha \in \Lambda$, U_α is evenly covered by p. By Theorem 4.4, there exists $s_0, s_1, \ldots, s_m$ and $t_0, t_1, \ldots, t_n$ such that $0 = s_0 < s_1 < \ldots < s_m = 1$, $0 = t_0 < t_1 < \ldots < t_n = 1$, and for each $i = 1, 2, \ldots, m$ and $j = 1, 2, \ldots, n$, $F([s_{i-1}, s_i] \times [t_{j-1}, t_j]) \subseteq U_\alpha$ for some $\alpha \in \Lambda$. For each $i = 1, 2, \ldots, m$ and $j = 1, 2, \ldots, n$, let $U_i \times J_j = [s_{i-1}, s_i] \times [t_{j-1}, t_j]$. We define F' inductively on the rectangles $I_i \times J_j$ in the following order:

$$I_1 \times J_1, I_2 \times J_1, \ldots, I_m \times J_1, I_1 \times J_2, I_2 \times J_2, \ldots, I_m \times J_2, I_1 \times J_3, \ldots, I_m \times J_n.$$

$I_1 \times J_n$	$I_2 \times J_n$	...	$I_m \times J_n$
⋮	⋮		⋮
$I_1 \times J_2$	$I_2 \times J_2$	...	$I_m \times J_2$
$I_1 \times J_1$	$I_2 \times J_2$	...	$I_m \times J_1$

Suppose $1 \leq p \leq m$, $1 \leq q \leq n$, and assume that F' is defined on $C = (\{0\} \times I) \cup (I \times \{0\}) \cup \bigcup_{i=1}^{m} \bigcup_{j=1}^{q-1}(I_i \times J_j) \cup \bigcup_{i=1}^{p-1}(I_i \times J_q)$. We define F' on $I_p \times J_q$.

There exists $\alpha \in \Lambda$ such that $F(I_p \times J_q) \subseteq U_\alpha$. Let $\mathcal{V} = \{V_{\alpha\beta}\colon \beta \in \Lambda\}$ be the collection of slices of $p^{-1}(U_\alpha)$. Now F' is already defined on $D = C \cap (I_p \times J_q)$. Since D is connected, $F'(D)$ is connected. Since $\mathcal{V}$ is a collection of pairwise disjoint open sets, $F'(D)$ is a subset of one member $V_{\alpha\gamma}$ of $\mathcal{V}$. Now $p|_{V_{\alpha\gamma}}$ is a homeomorphism of $V_{\alpha\gamma}$ onto U_α. Since F' is a lifting of $F|_C$, $((p|_{V_{\alpha\gamma}}) \circ F')(x) = F(x)$ for all $x \in D$. For $x \in I_p \times I_q$, define $F'(x) = (p|_{V_{\alpha\gamma}})^{-1}(F(x))$. Then F' is a lifting of $F|_{(C \cup (I_p \times J_q))}$. Therefore, by induction we can define F' on $I \times I$.

Now suppose F is a path homotopy. Then $F(0, t) = b_0$ for all $t \in I$. Since F' is a lifting of F, $F'(0, t) \in p^{-1}(b_0)$ for all $t \in I$. Since $\{0\} \times I$ is connected, F' is continuous, and $p^{-1}(b_0)$ has the discrete topology as a subspace of E, $F'(\{0\} \times I)$ is connected and hence it must be a single point. Likewise, there exists $b_1 \in B$ such that $F(1, t) = b_1$ for all $t \in I$, so $F'(1, t) \in p^{-1}(b_1)$ for all $t \in I$, and hence $F'(\{1\} \times I)$ must be a single point. Therefore F' is a path homotopy. ∎

THEOREM 8.21. Let $(E, \mathcal{T})$ and $(B, \mathcal{U})$ be topological spaces, let $p: E \to B$ be a covering map, let $e_0 \in E$, let $b_0 = p(e_0)$, let $b_1 \in B$, let α and β be paths in B from b_0 to b_1 that are path homotopic, and let α' and β' be liftings of α and β respectively such that $\alpha'(0) = \beta'(0) = e_0$. Then $\alpha'(1) = \beta'(1)$ and $\alpha' \simeq_p \beta'$.

Proof. Let $F: I \times I \to B$ be a continuous function such that $F(x, 0) = \alpha(x)$ and $F(x, 1) = \beta(x)$ for all $x \in I$ and $F(0, t) = b_0$ and $F(1, t) = b_1$ for all $t \in I$. By Theorem 8.20, there exists a lifting $F': I \times I \to E$ of F such that $F'(0, t) = e_0$ for all $t \in I$ and $F'(\{1\} \times I)$ is a set consisting of a single point, say e_1. The continuous function $F'|_{(I \times \{0\})}$ is a lifting of $F|_{(I \times \{0\})}$ such that $F'(0, 0) = e_0$. Since the lifting of paths is unique (Theorem 8.19), $F'(x, 0) = \alpha'(x)$ for all $x \in I$. Likewise, $F'|_{(I \times \{1\})}$ is a lifting of $F|_{(I \times \{1\})}$ such that $F'(0, 1) = e_0$. Again by Theorem 8.19, $F'(x, 1) = \beta'(x)$ for all $x \in I$. Therefore $\alpha'(1) = \beta'(1) = e_1$ and $\alpha' \simeq_p \beta'$. ∎

EXERCISES 8.4

1. Let $(E, \mathcal{T})$ and $(B, \mathcal{U})$ be topological spaces, and let $p: E \to B$ be a covering map. Show that p is open.

2. Let $(E_1, \mathcal{T}_1)$, $(E_2, \mathcal{T}_2)$, $(B_1, \mathcal{U}_1)$, and $(B_2, \mathcal{U}_2)$ be topological spaces, and let $p_1: E_1 \to B_1$ and $p_2: E_2 \to B_2$ be covering maps. Prove that the function $p: E_1 \times E_2 \to B_1 \times B_2$ defined by $p(x, y) = (p_1(x), p_2(y))$ is a covering map.

3. In this exercise, let p_1 denote the function p that is used throughout Section 8.3, and define $p: \mathbb{R} \times \mathbb{R} \to S^1 \times S^1$ by $p(x, y) = (p_1(x), p_1(y))$. By Exercise 2, p is a covering map. Since the torus is homeomorphic to $S^1 \times S^1$, $\mathbb{R} \times \mathbb{R}$ is a covering space of the torus. Draw pictures in $\mathbb{R} \times \mathbb{R}$ and describe verbally the covering map p.

4. Let $(X, \mathcal{T})$ be a topogical space, let Y be a set, let $\mathcal{U}$ be the discrete topology on Y, and let $\pi_1: X \times Y \to X$ be the projection map. Prove that π_1 is a covering map.

5. Let $(E, \mathcal{T})$ be a topological space, let $(B, \mathcal{U})$ be a connected space, let $n \in \mathbb{N}$, and let $p: E \to B$ be a covering map such that for some $b_0 \in B$, $p^{-1}(b_0)$ is a set with n members. Prove that for each $b \in B$, $p^{-1}(B)$ is a set with n members.

6. Let $(E, \mathcal{T})$ be a topological space, let $(B, \mathcal{U})$ be a locally connected, connected space, let $p: E \to B$ be a covering map, and let C be a component of E. Prove that $p|_C: C \to B$ is a covering map.

7. Let $(E, \mathcal{T})$ be a pathwise connected space, let $(B, \mathcal{U})$ be a topological space, let $p: E \to B$ be a covering map, and let $b \in B$.
   (a) Prove that there is a surjection $\phi: \pi_1(B, b) \to p^{-1}(b)$.
   (b) Prove that if $(E, \mathcal{T})$ is simply connected, then ϕ is a bijection.

8. The projective plane, which we have previously introduced, can also be defined as the quotient space obtained from S^2 by indentifying each point x of S^2 with its antipodal point $-x$. Let $p: S^2 \to P^2$ denote the natural map that maps each point of S^2 into the equivalence class that contains it. Prove that P^2, defined in this manner, is a surface and that the function p is a covering map.

8.5 Applications and Additional Examples of Fundamental Groups

We know that the fundamental group of a contractible space is the trivial group (see Exercise 8 of Section 8.1 and Exercises 2 and 3 Section 8.2) and that the fundamental group of the circle is isomorphic to the group of integers. In this section, we give some applications and some theorems that are useful in determining the fundamental group of a space. We begin by showing that S^1 is not a retract (see Exercise 5 of Section 8.2) of $B^2 = \{(x, y) \in \mathbb{R} \times \mathbb{R}: x^2 + y^2 \leq 1\}$.

EXAMPLE 7. We show that B^2 is contractible. Define $H: B^2 \times I \to B^2$ by $H((x, y), t) = (1 - t)(x, y)$. Then H is continuous, $H((x, y), 0) = (x, y)$ and $H((0, 0), t) = (0, 0)$ for each $t \in I$.

THEOREM 8.22. S^1 is not a retract of B^2. ∎

The diagram in Figure 8.9 is helpful in following the proof.

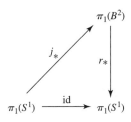

Figure 8.9

Proof. Suppose S^1 is a retract of B^2. Then there is a continuous function $r: B^2 \to S^1$ such that $r(x) = x$ for each $x \in S^1$. Define $j: S^1 \to B^2$ by $j(x) = x$ for each $x \in S^1$. Then $r \circ j: S^1 \to S^1$ is the identity function, and hence $(rj)_*: \pi_1(S^1, (1, 0)) \to \pi_1(S^1, (1, 0))$ is the identity homeomorphism. By Theorem 8.12, $(rj)_* = r_* j_*$. Thus

r_* maps $\pi_1(B^2, (1, 0))$ onto $\pi_1(S^1, (1, 0))$. This is a contradiction, because $\pi_1(B^2, (1, 0))$ is the trivial group and $\pi_1(S^1, (1, 0))$ is isomorphic to the group of integers. ∎

Recall that a topological space $(X, \mathcal{T})$ has the **fixed-point property** if for each continuous function $f\colon X \to X$ there exists $x \in X$ such that $f(x) = x$. The following theorem is due to the Dutch mathematician L.E.J. Brouwer (1881–1966).

THEOREM 8.23. (Brouwer Fixed-Point Theorem). B^2 has the fixed-point property.

Proof. Suppose there is a continuous function $f\colon B^2 \to B^2$ such that $f(x) \neq x$ for any $x \in B^2$. For each $x \in B^2$, let L_x denote the line segment that begins at $f(x)$ and passes through x, and let $y_x \in L_x \cap S^1 - \{f(x)\}$. (Note that y_x is uniquely determined—see Figure 8.10.)

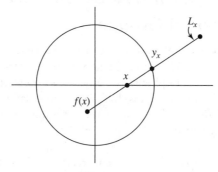

Figure 8.10

It is a straightforward but tedious exercise (see Exercise 4) to show that the function $r\colon B^2 \to S^1$ defined by $r(x) = y_x$ for each $x \in B^2$ is continuous. Since $r(x) = x$ for each $x \in S^1$, r is a retraction of B^2 onto S^1. This contradicts Theorem 8.22. ∎

Now we introduce a property of subspaces that is stronger than retract.

Definition. A subspace A of a topological space $(X, \mathcal{T})$ is a **deformation retract of** X provided there is a continuous function $H\colon X \times I \to X$ such that $H(x, 0) = x$ and $H(x, 1) \in A$ for all $x \in X$ and $H(a, t) = a$ for all $a \in A$ and $t \in I$. The homotopy H is called a **deformation retraction**. ∎

EXAMPLE 8. Let $X = \{(x, y) \in \mathbb{R} \times \mathbb{R}\colon 1 \leq x^2 + y^2 \leq 4\}$. Define $H\colon X \times I \to X$ by $H(x, t) = (1 - t)x + t \cdot \frac{x}{|x|}$ for each $(x, t) \in X \times I$. Then H is continuous, $H(x, 0) = x$ and $H(x, 1) = \frac{x}{|x|} \in S^1$ for all $x \in X$, and $H(x, t) = x$ for all $x \in S^1$ and $t \in I$. Therefore H is a deformation retraction, and hence S^1 is a deformation retract of X.

Using Example 8 and Theorem 8.18, the following theorem tells us that the fundamental group of the space X in Example 8 is isomorphic to the group of integers.

8.5 Applications and Additional Examples of Fundamental Groups

THEOREM 8.24. If A is a deformation retract of a topological space $(X, \mathcal{T})$ and $a_o \in A$, then $\pi_1(X, a_0)$ is isomorphic to $\pi_1(A, a_0)$.

Proof. Let $H: X \times I \to X$ be a continuous function such that $H(x, 0) = x$ and $H(x, 1) \in A$ for all $x \in X$ and $H(a, t) = a$ for all $a \in A$ and $t \in I$. Define $h: X \to A$ by $h(x) = H(x, 1)$ for each $x \in X$, and let $h_*: \pi_1(X, a_0) \to \pi_1(A, a_0)$ be the homeomorphism, given by Theorem 8.10, induced by h. This means that h_* is defined by $h_*([\alpha]) = [h \circ \alpha]$ for all $[\alpha] \in \pi_1(X, a_0)$.

Now we show that $h *$ is one-to-one. Suppose $[\alpha], [\beta] \in \pi_1(X, a_o)$ and $h_*([\alpha]) = h_*([\beta])$. Then $[h \circ \alpha] = [h \circ \beta]$. Since $A \subseteq X$, $h \circ \alpha$ may be considered as a path in X. We complete the proof that $[\alpha] = [\beta]$ by showing that $\alpha \simeq_p h \circ \alpha$ in X. Define $F: I \times I \to X$ by $F(x, t) = H(\alpha(x), t)$ for all $(x, t) \in I \times I$. Then $F(x, 0) = H(\alpha(x), 0) = \alpha(x)$ and $F(x, 1) = H(\alpha(x), 1) = (h \circ \alpha)(x)$ for all $x \in I$, and $F(0, t) = H(\alpha(0), t) = H(a_0, t) = a_0$ and $F(1, t) = H(\alpha(1), t) = H(a_0, t) = a_0$ for all $t \in I$. Since $[\alpha]$ is an arbitrary member of $\pi_1(X, a_0)$, this also shows that $\beta \simeq_p h \circ \beta$ in X. Therefore, in $\pi_1(X, a_0)$, we have $[\alpha] = [h \circ \alpha] = [h \circ \beta] = [\beta]$, and so h_* is one-to-one.

Now let $[\alpha] \in \pi_1(A, \alpha_0)$. Then $\alpha: I \to A$ and since $A \subseteq X$, we may also consider α to be a path in X. Let $[\alpha]_X$ denote the member of $\pi_1(X, a_0)$ that contains α. Since $h(a) = H(a, 1) = a$ for all $a \in A$, $(h \circ \alpha)(x) = \alpha(x)$ for all $x \in I$. Therefore $h_*([\alpha]_X) = [h \circ \alpha] = [\alpha]$, and so h_* maps $\pi_1(X, x_0)$ onto $\pi_1(A, a_0)$.

Therefore h_* is an isomorphism. ∎

By Example 8, S^1 is a deformation retract of $X = \{(x, y) \in \mathbb{R} \times \mathbb{R}: 1 \leq x^2 + y^2 \leq 4\}$. Therefore, by Theorems 8.24 and 8.18, the fundamental group of X is isomorphic to the group of integers.

EXAMPLE 9. Let $p = (p_1, p_2) \in \mathbb{R} \times \mathbb{R}$, and let $S = \{(x, y) \in \mathbb{R} \times \mathbb{R}: (x - p_1)^2 + (y - p_2)^2 = 1\}$. In Exercise 1, you are asked to show that S is a deformation retract of $\mathbb{R} \times \mathbb{R} - \{p\}$. Thus it follows that the fundamental group of $\mathbb{R} \times \mathbb{R} - \{p\}$ is isomorphic to the group of integers.

Since the torus is homeomorphic to $S^1 \times S^1$, we can determine the fundamental group of the torus by proving a theorem about the fundamental group of the Cartesian product of two spaces. The definition of the direct product of two groups is given in Appendix I.

THEOREM 8.25. Let $(X, \mathcal{T})$ and $(Y, \mathcal{U})$ be topological spaces and let $x_0 \in X$ and $y_0 \in Y$. Then $\pi_1(X \times Y, (x_0, y_0))$ is isomorphic to $\pi_1(X, x_0) \otimes \pi_1(Y, y_0)$.

Proof. Let $p: X \times Y \to X$ and $q: X \times Y \to Y$ be the projection maps (since π_1 is used as part of the symbol to denote the fundamental group of a space, we now use different symbols to denote the projection maps, but p and q are the functions that we have previously denoted by π_1 and π_2), and let $p_*: \pi_1(X \times Y, (x_0, y_0)) \to \pi_1(X, x_0)$ and $q_*: \pi_1(X \times Y, (x_0, y_0)) \to \pi_1(Y, y_0)$ denote the induced homeomorphisms. Define $\phi: \pi_1(X \times Y, (x_0, y_0)) \to \pi_1(X, x_0) \otimes \pi_1(Y, y_0)$ by $\phi([\alpha]) = ([p \circ \alpha],

$[q \circ \alpha])$. Since $[p \circ \alpha] = p_*([\alpha])$ and $[q \circ \alpha] = q_*([\alpha])$, by Exercise 12 of Appendix I, ϕ is a homeomorphism.

We use Exercise 13 of Appendix I in order to prove that ϕ is one-to-one. Let $[\alpha] \in \pi_1(X \times Y, (x_0, y_0))$ such that $\phi([\alpha])$ is the identity element of $\pi_1(X, x_0) \otimes \pi_1(Y, y_0)$. This means that $[p \circ \alpha]$ is the identity element of $\pi_1(X, x_0)$ and $[q \circ f]$ is the identity element of $\pi_1(Y, y_0)$. Then $p \circ \alpha \simeq_p e_{x_0}$ and $q \circ \alpha \simeq_p e_{y_0}$, where e_{x_0} and e_{y_0} are the constant paths defined by $e_{x_0}(t) = x_0$ and $e_{y_0}(t) = y_0$ for all $t \in I$. Let $F: I \times I \to X$ and $G: I \times I \to Y$ be the homotopies such that $F(x, 0) = (p \circ \alpha)(x)$ and $F(x, 1) = x_0$ for all $x \in I$, $F(0, t) = F(1, t) = x_0$ for all $t \in I$, $G(x, 0) = (q \circ \alpha)(x)$ and $G(x, 1) = y_0$ for all $x \in I$, and $G(0, t) = G(1, t) = y_0$ for all $t \in I$. Define $H: I \times I \to X \times Y$ by $H(x, t) = (F(x, t), G(x, t))$ for all $(x, t) \in I \times I$. Then $H(x, 0) = (F(x, 0), G(x, 0)) = ((p \circ \alpha)(x), (q \circ \alpha)(x)) = \alpha(x)$ and $H(x, 1) = (F(x, 1), G(x, 1)) = (x_0, y_0)$ for all $x \in X$, and $H(0, t) = (F(0, t), G(0, t)) = (x_0, y_0)$ and $H(1, t) = (F(1, t), G(1, t)) = (x_0, y_0)$ for all $t \in I$. Therefore $\alpha \simeq_p e$, where e is the constant path defined by $e(t) = (x_0, y_0)$ for all $t \in I$. Therefore by Exercise 13 of Appendix I, $\varnothing$ is one-to-one.

Now we show that ϕ maps $\pi_1(X \times Y, (x_0, y_0))$ onto $\pi_1(X, x_0) \otimes \pi_1(Y, y_0)$. Let $([\alpha], [\beta]) \in \pi_1(X, x_0) \otimes \pi_1(Y, y_0)$. Define $\gamma: I \to X \times Y$ by $\gamma(t) = (\alpha(t), \beta(t))$ for all $t \in I$. Then $[\gamma] \in \pi_1(X \times Y, (x_0, y_0))$, and $\phi([\gamma]) = ([p \circ \gamma], [q \circ \gamma]) = ([\alpha], [\beta])$, and so ϕ is an isomorphism. ∎

If A is a subset of a set X, **the inclusion map** $i: A \to X$ is the function defined by $i(a) = a$ for each $a \in A$. A homeomorphism ϕ from a group G into a group H is called the **zero homeomorphism** if for each $g \in G$, $\phi(g)$ is the identity element of H. The following theorem is a special case of a famous theorem called the **Van Kampen Theorem**.

THEOREM 8.26. Let $(X, \mathcal{T})$ be a topological space, let U and V be open subsets of X such that $X = U \cup V$ and $U \cap V$ is pathwise connected. Let $x_0 \in U \cap V$, and suppose the inclusion maps $i: U \to X$ and $j: V \to X$ induce zero homeomorphisms $i_*: \pi_1(U, x_0) \to \pi_1(X, x_0)$ and $j_*: \pi_1(V, x_0) \to \pi_1(X, x_0)$. Then $\pi_1(X, x_0)$ is the trivial group.

Proof. Let $[\alpha] \in \pi_1(X, x_0)$. By Theorem 4.4, there exist $t_0, t_1, ..., t_n$ such that $t_0 = 0, t_n = 1$, and, for each $k = 1, 2, ..., n$, $t_{k-1} < t_k$ and $\alpha([t_{k-1}, t_k])$ is a subset of one of U or V. Among all such subdivisions $t_0, t_1, ..., t_n$ of I, choose one with the smallest number of members. We prove by contradiction that for each $k = 1, 2, ..., n - 1, f(t_k) \in U \cap V$. (Note that if $\alpha(I)$ is a subset of one of U or V, then $n = 1$, and we have nothing to prove.) Suppose $k \in \{1, 2, ..., n - 1\}$ and $\alpha(t_k) \notin U$. Then neither $\alpha([t_{k-1}, t_k])$ nor $\alpha([t_k, t_{k+1}])$ is a subset of U, so $\alpha([t_{k-1}, t_k]) \cup \alpha([t_k, t_{i+1}]) \subseteq V$. Thus we can discard t_k and obtain a subdivision with a smaller number of members that still have the desired properties. Therefore for each $k = 1, 2, ..., n - 1, \alpha(t_k) \in U$. The same argument shows that $\alpha(t_k) \in V$ for each $k = 1, 2, ..., n - 1$.

Since $\alpha(t_0) = \alpha(t_n) = x_0 \in U \cap V, \alpha(t_k) \in U \cap V$ for each $k = 0, 1, ..., n$.

8.5 Applications and Additional Examples of Fundamental Groups

For each $k = 1, 2, ..., n$, define $\alpha_k: I \to X$ by $\alpha_k(x) = \alpha((1-x)t_{k-1} + xt_k)$ for each $x \in I$. We show that α_k is path homotopic (in X) to a path that lies entirely in U.

If $\alpha_k(I) \subseteq U$, define $H_k: I \times I \to X$ by $H_k(x, t) = \alpha_k(x)$ for each $(x, t) \in I \times I$.

Suppose $\alpha_k(I) \not\subseteq U$. Then, since $\alpha_k(I) = \alpha([t_{k-1}, t_k])$, $\alpha_k(I) \subseteq V$. Since $U \cap V$ is pathwise connected, there exist paths β, γ in $U \cap V$ such that $\beta(0) = \gamma(0) = x_0$, $\beta(1) = \alpha_k(0)$, and $\gamma(1) = \alpha_k(1)$ (see Figure 8.11).

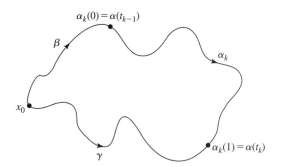

Figure 8.11

Then the path product $(\beta * \alpha_k) * \overline{\gamma}$ (recall that $\overline{\gamma}$ is the path defined by $\overline{\gamma}(x) = \gamma(1-x)$ for each $x \in I$) is a loop in V at x_0. Since the inclusion map $j: V \to X$ induces the zero homeomorphism, the loop $(\beta * \alpha_k * \overline{\gamma})$ is path homotopic in X to the constant loop $e: I \to X$ defined by $e(x) = x_0$ for each $x \in I$. Therefore, by Exercise 2, α_k is path homotopic in X to the path $\overline{\beta} * \gamma$. Let $F_k: I \times I \to X$ be this path homotopy. Then $F_k(x, 0) = \alpha_k(x)$ and $F_k(x, 1) = (\overline{\beta} * \gamma)(x)$ for each $x \in I$ and $F_k(0, t) = \alpha_k(0)$ and $F_k(1, t) = \alpha_k(1)$ for each $t \in I$. Define $F: I \times I \to X$ by $F(x, t) = F_k((x - t_{k-1})/(t_k - t_{k-1}), t)$ for each $x \in [t_{k-1}, t_k]$ and $t \in I$. Since $F_{k-1}(1, t) = \alpha_{k-1}(1) = \alpha_k(0) = F_k(0, t)$ for each $k = 1, 2, ..., n$ and $t \in I$, F is well-defined and continuous. Let $x \in I$. Then there exists k such that $x \in [t_{k-1}, t_k]$. So

$$F(x, 0) = F_k((x - t_{k-1})/(t_k - t_{k-1}), 0) = \alpha_k((x - t_{k-1})/(t_k - t_{k-1}))$$
$$= \alpha((1 - (x - t_{k-1})/(t_k - t_{k-1}))t_{k-1} + ((x - t_{k-1})/(t_k - t_{k-1})t_k))$$
$$= \alpha(x),$$

and

$$F(x, 1) = F_k((x - t_{k-1})/(t_k - t_{k-1}), 1) = (\overline{\beta} * \gamma)((x - t_{k-1})/(t_k - t_{k-1})) \in U.$$

Also, $F(0, t) = F_1(0, t) = \alpha_1(0) = \alpha(0)$ and $F(1, t) = F_n(1, t) = \alpha_n(1) = \alpha(1)$ for each $t \in I$. Finally, define a path α' by $\alpha'(x) = F(x, 1)$ for each $x \in I$. Then α' is a path in U and α is path homotopic to α'. Since $i_*: \pi_1(U, x_0) \to \pi_1(X, x_0)$ is the zero homeomorphism, α' is path homotopic in X to the constant loop e. Therefore $\alpha \simeq_p e$, and so $\pi_1(X, x_0)$ is the trivial group. ∎

As a consequence of Theorem 8.26, we establish the important fact that if $n > 1$, $S^n = \{(x_1, x_2, ..., x_{n+1}) \in \mathbb{R}^{n+1}: \sqrt{\prod_{i=1}^{n+1} x_i^2} = 1\}$ is simply connected. In the proof, we use that if $X = \{(x_1, x_2, ..., x_n, x_{n+1}) \in \mathbb{R}^{n+1}: x_{n+1} = 0\}$, then X is homeomorphic to $\mathbb{R}^n$.

THEOREM 8.27. For $n > 1$, S^n is simply connected.

Proof. Let p and q be the members of S^n given by $p = (0, 0, ..., 0, 1)$ and $q = (0, 0, ..., 0, -1)$, and let $X = \{(x_1, x_2, ..., x_n, x_{n+1}) \in \mathbb{R}^{n+1}: x_{n+1} = 0\}$. First we show that $S^n - \{p\}$ is homeomorphic to $\mathbb{R}^n$. Define $f: S^n - \{p\} \to X$ by $f(x_1, x_2, ..., x_n, x_{n+1}) = (1/(1 - x_{n+1}))(x_1, x_2, ..., x_n, 0)$. (We can describe f geometrically as follows: For each $x \in S^n - \{p\}$, $f(x)$ is the point of intersection of the line determined by p and x with X.) It is clear that f is a one-to-one continuous function that maps $S^n - \{p\}$ onto X. For each $(y_1, y_2, ..., y_n, 0) \in X$, let $t = 2/(1 + y_1^2 + \cdots + y_n^2)$, and define $g: X \to S^n - \{p\}$ by $g(y_1, y_2, ..., y_n, 0) = (ty_1, ty_2, ..., ty_n, 1 - t)$. Again it is clear that g is a one-to-one continuous function. In Exercise 3, you are asked to show that g is an inverse of f. This completes the proof that $S^n - \{p\}$ is homeomorphic to $\mathbb{R}^n$.

Since the function $h: S^n - \{q\} \to S^n - \{p\}$ defined by $h(x_1, x_2, ..., x_n, x_{n+1}) = (x_1, x_2, ..., x_n, -x_{n+1})$ is clearly a homeomorphism, $S^n - \{q\}$ is also homeomorphic to $\mathbb{R}^n$.

Now let $U = S^n - \{p\}$ and $V = S^n - \{q\}$. Then U and V are open subsets of S^n and $S^n = U \cup V$. By Example 4, $\mathbb{R}^n$ is simply connected, and therefore the inclusion maps $i: U \to S^n$ and $j: V \to S^n$ induce zero homeomorphisms. In order to use Theorem 8.26 to conclude that S^n is simply connected, it is sufficient to show that $U \cap V$ is pathwise connected. Now $U \cap V = S^n - \{p, q\}$, and the homeomorphism f (defined in the first paragraph of this proof) restricted to $S^n - \{p, q\}$ is a homeomorphism of $S^n - \{p, q\}$ onto $X - \{(0, 0, ..., 0)\}$, and $X - \{(0, 0, ..., 0)\}$ is homeomorphic to $\mathbb{R}^n - \{(0, 0, ..., 0)\}$. We show that $\mathbb{R}^n - \{(0, 0, ..., 0)\}$ is pathwise connected. Let $x \in \mathbb{R}^n - \{(0, 0, ..., 0)\}$. If $x \neq (y, 0, 0, ..., 0)$, where $y < 0$, then x and $(1, 0, 0, ..., 0)$ can be joined by a straight-line path. If $x = (y, 0, 0, ..., 0)$, where $y < 0$, then x and $(0, 1, 0, 0, ..., 0)$ can be joined by a straight-line path, and $(0, 1, 0, 0, ..., 0)$ and $(1, 0, 0, ..., 0)$ can be joined by a straight-line path. In either case, x and $(1, 0, 0, ..., 0)$ can be joined by a path. Therefore $\mathbb{R}^n - \{(0, 0, ..., 0)\}$ is pathwise connected. ∎

In the proof of Theorem 8.27, we showed that if $n > 2$ then $\mathbb{R}^n - \{(0, 0, ..., 0)\}$ is simply connected. In Example 9, we showed that $\mathbb{R}^2 - \{(0, 0)\}$ is not simply connected. Therefore by Theorem 8.14, if $n > 2$, then $\mathbb{R}^n - \{(0, 0, ..., 0)\}$ is not homeomorphic to $\mathbb{R}^2 - (0, 0)\}$.

EXERCISES 8.5

1. Let $p = (p_1, p_2) \in \mathbb{R} \times \mathbb{R}$, and let $S = \{(x, y) \in \mathbb{R} \times \mathbb{R}: (x - p_1)^2 + (y - p_2)^2 = 1\}$. Prove that S is a deformation retract of $\mathbb{R} \times \mathbb{R} - \{p\}$.

2. Let $(X, \mathcal{T})$ be a topological space, let $x_0 \in X$, and let α, β, and γ be paths in X such that $\beta(0) = x_0, \beta(1) = \alpha(0), \gamma(0) = x_0$, and $\gamma(1) = \alpha(1)$, and let $H: I \times I \to X$ be a continuous function such that $H(x, 0) = ((\beta * \alpha) * \overline{\gamma})(x)$ and $H(x, 1) = x_0$ for each $x \in I$ and $H(0, t) = H(1, t) = x_0$ for each $t \in I$. Prove that there is a continuous function $F: I \times I \to X$ such that $F(x, 0) = \alpha(x)$ and $F(x, 1) = (\overline{\beta} * \gamma)(x)$ for each $x \in I$ and $F(0, t) = \alpha(0)$ and $F(1, t) = \alpha(1)$ for each $t \in I$.

3. Let $X = \{(x_1, x_2, ..., x_n, x_{n+1}) \in \mathbb{R}^{n+1}: x_{n+1} = 0\}$ and let $p = (0, 0, ..., 0, 1) \in S^n$. Define $f: S^n - \{p\} \to X$ by $f(x_1, x_2, ..., x_{n+1}) = (1/(1 - x_{n+1}))(x_1, x_2, ..., x_n, 0)$, and define $g: X \to S^n - \{p\}$ by $g(y_1, y_2, ..., y_n, 0) = (ty_1, ty_2, ..., ty_n, 1 - t)$, where $t = 2/(1 + y_1^2 + y_2^2 + ... + y_n^2)$. Prove that g is an inverse of f.

4. Show that the function $r: B^2 \to S^1$ defined in the proof of Theorem 8.23 is continuous.

8.6 Knots

We conclude this chapter with a brief discussion of knot theory. No proofs are given in the text. This section is somewhat unique in that it is not rigorous. Instead, it is designed to appeal to the geometric and intuitive side of topology. We refer to the one-point compactification of a topological space, which is defined in Section 6.6. In keeping with the spirit of this section, one can consider the intuitive nature of the one-point compactification rather than approach it from a rigorous viewpoint. A reference is provided at the end of the section for those who wish to continue the study of knot theory.

An invariant of knot theory was first considered by Karl Friedrick Gauss (1777–1855) in 1833. Early work on the subject was done by Max Dehn (1878–1952) in 1910, J.W. Alexander in 1924, E. Artin in 1925, E.R. van Kampen in 1928, and others. The first comprehensive book on knot theory was written by K. Reidemeister in 1932.

The overhand knot and the figure-eight knot (see Figure 8.12) are familiar to almost everyone. A little experimenting with a piece of string should convince you that one of these knots cannot be transformed into the other without untying one of them. One goal of knot theory is to prove mathematically that this cannot be done. Thus we must have a mathematical definition of a knot and of when two knots are considered to be the same. The latter definition must prevent untying. We do this by

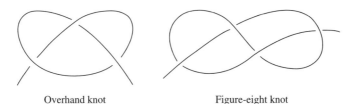

Overhand knot   Figure-eight knot

Figure 8.12

getting rid of the ends; that is, we splice the ends together. The overhand knot with the ends spliced together is often called the trefoil or cloverleaf knot. The figure-eight knot with the ends spliced together is often called the four-knot or Listing's knot. These are shown in Figure 8.13.

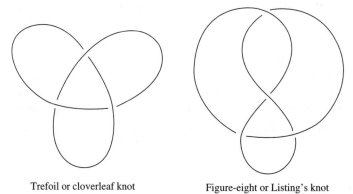

Trefoil or cloverleaf knot   Figure-eight or Listing's knot

Figure 8.13

Definition. A subset K of $\mathbb{R}^3$ is a **knot** if there is a homeomorphism that maps S^1 onto K.

Three additional knots are shown in Figure 8.14.

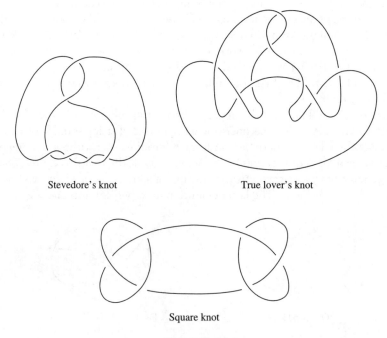

Stevedore's knot   True lover's knot

Square knot

Figure 8.14

Notice that any two knots are homeomorphic. Thus the question of when two knots are considered to be the same is a characteristic of the way in which the two knots are embedded in $\mathbb{R}^3$. Thus knot theory is a portion of 3-dimensional topology.

Definition. Two knots K^1 and K_2 are **equivalent** if there is a homeomorphism $h: \mathbb{R}^3 \to \mathbb{R}^3$ such that $h(K_1) = K_2$. ■

It is easy to show that knot equivalence is an equivalence relation. Notice that knot equivalence does not say anything about "sliding" (as is the case with a homotopy) K_1 until it lies on top of K_2. The difference in these ideas is illustrated by the fact that a reflection about a plane is a homeomorphism of $\mathbb{R}^3$ onto $\mathbb{R}^3$ that maps a knot into its mirror image, but we cannot "slide" the trefoil knot into its mirror image (see Figure 8.15).

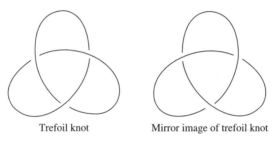

Trefoil knot   Mirror image of trefoil knot

Figure 8.15

Definition. A **polygonal knot** is a knot that is the union of a finite number of closed straight-line segments called **edges.** The endpoints of the straight-line segments are called **vertices.** A knot is **tame** if it is equivalent to a polygonal knot, and otherwise it is **wild.** ■

An example of a wild knot (one that is obtained by tying an infinite number of knots one after the other) is shown in Figure 8.16.

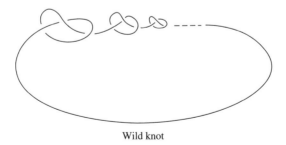

Wild knot

Figure 8.16

In order to picture knots and work with them effectively, we project them into a plane.

Definition. Let K be a knot in $\mathbb{R}^3$ and define $p\colon \mathbb{R}^3 \to \mathbb{R}^3$ by $p((x, y, z)) = (x, y, 0)$. A point $x \in p(K)$ is called a **multiple point** if $p^{-1}(x) \cap K$ consists of more than one point. The **order** of $x \in p(K)$ is the cardinality of $p^{-1}(x) \cap K$. A **double point** in $p(K)$ is a point of order 2, and a **triple point** in $p(K)$ is a point of order 3. ■

In general, $p(K)$ can contain any number and kinds of multiple points, including multiple points of infinite order. In the remainder of this section, p denotes the projection in the preceding definition.

Definition. A polygonal knot K is in **regular position** if: (1) there are only a finite number of multiple points and they are double points, and (2) no double point is the image of a vertex of K. ■

In Exercise 1, you are asked to give an example of a polygonal knot in regular position.

Definition. Let K be a polygonal knot in regular position. Then each double point in $p(K)$ is the image of two points of K. The one whose z-coordinate is larger is called an **overcrossing** and the one whose z-coordinate is smaller is called an **undercrossing**. ■

In Exercise 2, you are asked to indicate the overcrossings and undercrossings of the knot you have given in Exercise 1.

A rigorous proof of the following theorem is somewhat tedious, and thus we omit it. In Exercise 3 and 4, you are asked to illustrate this theorem.

THEOREM 8.28. If K is a polygonal knot, then there is an arbitrarily small rotation of $\mathbb{R}^3$ onto $\mathbb{R}^3$ that maps K into a polygonal knot in regular position. ■

It is, of course, an immediate consequence of Theorem 8.28 that every polygonal knot is equivalent to a polygonal knot in regular position.

Definition. A homeomorphism $h\colon \mathbb{R}^3 \to \mathbb{R}^3$ is **isotopic to the identity** if there is a homotopy $H\colon \mathbb{R}^3 \times I \to \mathbb{R}^3$ such that $H(x, 0) = h(x)$ and $H(x, 1) = x$ for each $x \in \mathbb{R}^3$ and, for each $t \in I$, $H|_{(\mathbb{R}^3 \times \{t\})}\colon \mathbb{R}^3 \times \{t\} \to \mathbb{R}^3$ is a homeomorphism. The homotopy H is called an **isotopy**. ■

Suppose K_1 and K_2 are knots and $h\colon \mathbb{R}^3 \to \mathbb{R}^3$ is a homeomorphism that is isotopic to the identity under an isotopy H. Moreover, suppose that $h(K_1) = K_2$. For each $t \in I$, let $h_t = H|_{(\mathbb{R}^3 \times \{t\})}$. Then $\{h_t(K_1)\colon t \in I\}$ is a "continuous" family of knots that move from K_2 to K_1 as t moves from 0 to 1. (Obviously, we are not being precise here. Instead we are trying to provide a "picture" of how one knot is moved onto another under an isotopy).

We can extend the definition of a triangulation of a compact surface (given in Section 6.4) to a compact 3-manifold.

Definition. A **triangulation** of a compact 3-manifold $(X, \mathcal{T})$ consists of a finite collection $\{T_1, T_2, ..., T_n\}$ of closed subsets of X that cover X and a collection $\{\phi_i : T_i' \to T_i : i = 1, 2, ..., n\}$ of homeomorphisms, where each T_i' is a tetrahedron in the plane. Each T_i is also called a **tetrahedron.** The members of T_i that are images of the vertices of T_i' are called **vertices,** the subsets of T_i that are images of the edges of T_i' are called **edges,** and the subsets of T_i that are images of the 2-faces of T_i' are called **2-faces.** In addition, for each pair T_i and T_j of distinct tetrahedra $T_i \cap T_j = \phi$, $T_i \cap T_j$ is a vertex, $T_i \cap T_j$ is an edge, or $T_i \cap T_j$ is a 2-face. ∎

This is not the usual definition of a triangulation of a manifold, but for our purposes this definition will suffice.

Let T be an edge, a triangle, or a tetrahedron. Consider two orderings of the vertices of T to be **equivalent** if they differ by an even permutation. Then there are exactly two equivalence classes, each of which is called an orientation of T. If T is a triangle or a tetrahedron that has been assigned an orientation by ordering its vertices in some way, say $v_0, v_1,..., v_k$, and S is the face of T obtained by deleting v_i, then the vertices of S are automatically ordered. If i is even, the orientation of S given by this ordering is called the **orientation induced by T**. If i is odd, the other orientation of S is called the **orientation induced by T**. For example, if T is a triangle with vertices v_0, v_1, and v_2, the orientation induced by T of the edges whose vertices are v_1 and v_2 is the one given by the ordering v_1, v_2, whereas the orientation induced by T of the edge whose vertices are v_0 and v_2 is the one given by the ordering v_2, v_0.

Definition. Let $(X, \mathcal{T})$ be a compact 2- (or 3-) manifold, and let T be a triangulation of X. We say that T is **orientable** if it is possible to orient the triangles (or tetrahedra) of T in such a way that two triangles with a common edge (or two tetrahedra with a common 2-face) always induce opposite orientations on their common edge (or 2-face). ∎

EXAMPLE 10. Let $(X, \mathcal{T})$ be a compact 2-manifold with triangulation T shown in Figure 8.17(a). Then the orientation shown in Figure 8.17 (b) shows that T is orientable.

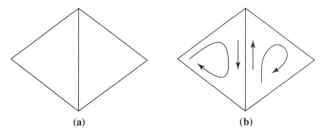

(a)   (b)

Figure 8.17

In Example 18 of Chapter 6, we showed that the one-point compactification of $\mathbb{R}$ is homeomorphic to S^1. It is also true that the one-point compactification of $\mathbb{R}^3$ is homeomorphic to S^3 (see Exercise 5). By Theorem 6.39, the one-point compactifications of two homeomorphic topological spaces are homeomorphic. We proved this theorem by showing that if $(X_1, \mathcal{T}_1)$ and $(X_2, \mathcal{T}_2)$ are topological spaces, $f: X_1 \to X_2$ is a homeomorphism, and, for each $i = 1, 2$, $(Y_i, \mathcal{U}_i)$ is the one-point compactification of $(X_i, \mathcal{T}_i)$, then there is a homeomorphism $g: Y_1 \to Y_2$ that is an extension of f. It is clear, from the proof, that the homeomorphism g is unique. It follows from this discussion that each homeomorphism $h: \mathbb{R}^3 \to \mathbb{R}^3$ can be extended in a unique way to a homeomorphism $\hat{h}: S^3 \to S^3$. We say that h is **orientation preversing** if $\hat{h}$ preserves the orientation of S^3. Otherwise we say that h is **orientation reversing** (see Exercise 9).

Suppose K_1 and K_2 are equivalent knots. Then there is a homeomorphism $h: \mathbb{R}^3 \to \mathbb{R}^3$ such that $h(K_1) = K_2$. Thus $h|_{(\mathbb{R}^3 - K_1)}: \mathbb{R}^3 - K_1 \to \mathbb{R}^3 - K_2$ is a homeomorphism, and so equivalent knots have homeomorphic complements. Therefore one possible way of distinguishing between knots is to consider the fundamental groups of their complements.

Definition. If K is a knot, the fundamental group $\pi_1(\mathbb{R}^3 - K)$ is called the **knot group** of K. ∎

We do not attempt to calculate knot groups, since it is a rather tedious process. Instead we describe, in a rather vague intuitive way, the knot group of a polygonal knot K in regular position. We break up the knot into overcrossings and undercrossings that alternate as we go around the knot. This is illustrated in Figure 8.18 for the trefoil knot and the square knot. In this figure, the heavier lines represent overcrossings.

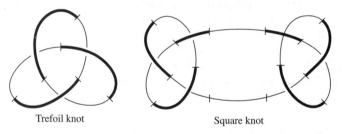

Trefoil knot   Square knot

Figure 8.18

The bottom line is that there is associated with each overcrossing a generator of $\pi_1(\mathbb{R}^3 - K)$. Thus if K_1 denotes the trefoil knot and K_2 the square knot, then $\pi_1(\mathbb{R}^3 - K_1)$ has a presentation that consists of 3 generators and $\pi_1(\mathbb{R}^3 - K_2)$ has a presentation that consists of 7 generators. In general, if K has n overcrossings, then there is a presentation of $\pi_1(\mathbb{R}^3 - K)$ that consists of n generators. Now the number of undercrossings is the same as the number of overcrossings, and there is associated with each undercrossing a relation. The last undercrossing, however, yields a relation that is a consequence of the others, so $\pi_1(\mathbb{R}^3 - K)$ has a presentation that consists of n generators and $n - 1$ relations. So there is a presentation of $\pi_1(\mathbb{R}^3 - K_1)$ that consists of 3 generators and 2 relations and a presentation of $\pi_1(\mathbb{R}^3 - K_2)$ that consists of 7 generators and 6 relations.

EXERCISES 8.6

1. Give an example of a polygonal knot in regular position.
2. Indicate the overcrossings and undercrossings of the knot you have given in Exercise 1.
3. Give an example of a polygonal knot that is not in regular position.
4. Give an arbitrarily small rotation of $\mathbb{R}^3$ onto $\mathbb{R}^3$ that maps the knot you have given in Exercise 3 into a polygonal knot in regular position.
5. Prove that the one-point compactification of $\mathbb{R}^3$ is homeomorphic to S^3.
6. Give a triangulation of S^3.
7. Show that the triangulation of the torus given in Figure 6.9 is orientable.
8. In Exercise 13 of Section 6.4 you were asked to give a triangulation of the Klein bottle. Experiment and guess whether this triangulation is orientable.
9. Give a example of an orientation-preserving homeomorphism $h: \mathbb{R}^3 \to \mathbb{R}^3$ and an orientation-reversing homeomorphism $k: \mathbb{R}^3 \to \mathbb{R}^3$.

References

Armstrong, M.A., *Basic Topology*, Springer-Verlag, New York, 1983.

Chapter 9
Applications of Homotopy

In Section 8.2 we showed that the fundamental group can be used to show that two spaces are not homeomorphic. In this chapter we exhibit other uses of the fundamental group.

9.1 Inessential Maps

The purpose of this section is to show that if the continuous function $h: X \to Y$ is homotopic to a constant map, then the induced homeomorphism h_* is the zero homeomorphism.

Definition. Let $(X, \mathcal{T})$ and $(Y, \mathcal{U})$ be topological spaces, and let $h: X \to Y$ be a continuous function. Then h is **inessential** if it is homotopic to a constant map, and h is **essential** if it is not inessential. ∎

The following theorem gives a characterization of certain inessential functions in terms of the extension property.

THEOREM 9.1. Let $(X, \mathcal{T})$ be a topological space, and let $h: S^1 \to X$ be a continuous function. Then the following are equivalent.

(a) h is inessential.

(b) h can be extended to a continuous function $f: B^2 \to X$.

Proof. (a) → (b). Suppose h is inessential. Then there is a point $x_0 \in X$ and a continuous function $H: S^1 \times I \to X$ such that $H(x, 0) = h(x)$ and $H(x, 1) = x_0$ for all $x \in S^1$. Define $G: S^1 \times I \to B^2$ by $G(x, t) = (1 - t)x$ for all $(x, t) \in S^1 \times I$. Then G is continuous, $G|_{S^1 \times [0, 1)}$ is a one-to-one-function that maps $S^1 \times [0, 1)$ onto $B^2 - \{(0, 0)\}$, and $G(x, 1) = (0, 0)$ for all $x \in S^1$. Since $S^1 \times I$ is compact and B^2 is

Hausdorff, G is closed (see Theorem 4.21). Therefore by Theorem 2.55, G is a quotient map.

Define $f: B^2 \to X$ by $f((0,0)) = x_0$ and $f(x) = H(G^{-1}(x))$ if $x \neq (0,0)$. Then $f \circ G = H$, and, by Theorem 2.59, f is continuous. If $x \in S^1 \subseteq B^2$, then $f(x) = H(G^{-1}(x)) = h(x)$, and therefore f is an extension of h.

(b) → (a) Suppose h can be extended to a continuous function $f: B^2 \to X$. Define $H: S^1 \times I \to X$ by $H(x, t) = f((1 - t)x)$. Then H is continuous and $H(x, 0) = f(x) = h(x)$ and $H(x, 1) = f(0)$ for all $x \in S^1$. Therefore h is homotopic to a constant map. ∎

The following theorem provides a condition for the induced homeomorphism to be the zero homeomorphism.

THEOREM 9.2. Let $(Y, \mathcal{U})$ be a topological space, and let $h: S^1 \to Y$ be an inessential function. Then h_* is the zero homeomorphism.

Proof. By Theorem 9.1, h has a continuous extension $f: B^2 \to Y$. Let $j: S^1 \to B^2$ be the inclusion map. Then $f \circ j = h$. Let $s_0 \in S^1$ and let $y_0 = h(s_0)$. Then, by Theorem 8.12, the following diagram commutes; that is $f_* \circ j_* = h_*$.

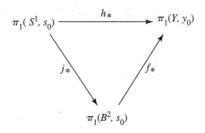

Since $\pi_1(B^2, s_0)$ is the trivial group, j_* is the zero homeomorphism. Therefore $f_* \circ j_* = h_*$ is the zero homeomorphism. ∎

The following theorem generalizes Theorem 9.2.

THEOREM 9.3. Let $(X, \mathcal{T})$ and $(Y, \mathcal{U})$ be topological spaces, and let $h: X \to Y$ be an inessential function. Then h_* is the zero homeomorphism.

Proof. Let $x_0 \in X$ and let α be a loop in X at x_0. Define $\sigma: I \to S^1$ by $\sigma(x) = (\cos 2\pi x, \sin 2\pi x)$ for each $x \in I$. Define $k: S^1 \to X$ by $k(x) = \alpha(\sigma^{-1}(x))$ for each $x \in S^1$. In Exercise 2, you are asked to prove that k is a continuous function. Since h is inessential, there exists $y_0 \in Y$ and a continuous function $H: X \times I \to Y$ such that $H(x, 0) = h(x)$ and $H(x, 1) = y_0$ for all $x \in X$. Define $F: S^1 \times I \to Y$ by $F(x, t) = H(k(x),t)$. Then F is continuous, $F(x, 0) = H(k(x), 0) = (h \circ k)(x)$ and $F(x, 1) = H(k(x),1) = y_0$ for all $x \in S^1$. By Theorem 9.2, $(h \circ k)_*$ is the zero homeomorphism. Therefore $h_*([\alpha]) = [h \circ \alpha] = [h \circ k \circ \sigma] = (h \circ k)_*([\sigma]) = 0$. ∎

EXERCISES 9.1

1. Without using Theorem 2.59, prove that the function f defined in the proof of Theorem 9.1 is continuous.

2. Let $(X, \mathcal{T})$ be a topological space, let $x_0 \in X$, let α be a loop in X at x_0, define $\sigma: I \to S^1$ by $\sigma(x) = (\cos 2\pi x, \sin 2\pi x)$, and define $k: S^1 \to X$ by $k(x) = \alpha(\sigma^{-1}(x))$ for each $x \in S^1$. Prove that k is continuous.

3. Let $(X, \mathcal{T})$ be a topological space, let $h: S^1 \to X$ be a continuous function, define $\sigma: I \to S^1$ by $\sigma(x) = (\cos 2\pi x, \sin 2\pi x)$, let $\alpha = h \circ \sigma$, and let $x_0 = \alpha(0) = \alpha(1)$.

    (a) Prove that there is a continuous function $H: I \times I \to X$ such that $H(x, 0) = \alpha(x)$, $H(x, 1) = x_0$ for all $x \in I$ and $H(0, t) = H(1, t) = x_0$ for all $t \in I$.

    (b) Prove that there is a continuous function $F: S^1 \times I \to X$ such that $F(\sigma(x), t) = H(x, t)$ for all $(x, t) \in I \times I$.

    (c) Prove that if h_* is the zero homeomorphism, then h is inessential.

4. Let $m, n \in \mathbb{N}$. A continuous function $h: S^m \to S^n$ is **antipode-preserving** if $h(-x) = -h(x)$ for each $x \in S^m$. Let $h: S^1 \to S^1$ be an antipode-preserving function. Consider the members of S^1 to be complex numbers and define $p: S^1 \to S^1$ by $p(z) = z^2$.

    (a) Prove that there is a continuous function $g: S^1 \to S^1$ such that $p \circ h = g \circ p$.

    (b) Let $x \in S^1$ and let $\alpha: I \to S^1$ be a path such that $\alpha(0) = x$ and $\alpha(1) = -x$. Prove that $p \circ \alpha$ is a loop in S^1 that is not path homotopic to a constant function.

    (c) Show that the homeomorphisms p_* and g_* are one-to-one.

    (d) Prove that h is essential.

5. Prove that there is no antipode-preserving function $h: S^2 \to S^1$.

6. Let $(X, \mathcal{T})$ be a pathwise connected space and let $x_0 \in X$. Show that $\pi_1(X, x_0)$ is the trivial group if and only if every continuous function $h: S^1 \to X$ is inessential.

9.2 The Fundamental Theorem of Algebra

The Fundamental Theorem of Algebra says that if $n \in \mathbb{N}$, then every polynomial equation of degree n with complex coefficients has at least one solution in the set of complex numbers. This theorem is difficult to prove, and most proofs involve nonalgebraic concepts. We give a proof that uses the material we developed in Chapter 8 and Section 9.1.

THEOREM 9.4. Let $n \in \mathbb{N}$ and let $x^n + a_{n-1}x^{n-1} + \cdots + a_1 x + a_0 = 0$ be a polynomial equation of degree n with complex coefficients. Then this equation has at least one solution in the set of complex numbers.

Proof. We consider the members of S^1 to be complex numbers and define a continuous function $f: S^1 \to S^1$ by $f(z) = z^n$. Let $s_0 = (1, 0)$ and consider the induced homeomorphism $f_*: \pi_1(S^1, s_0) \to \pi_1(S^1, s_0)$. Define $\sigma: I \to S^1$ by $\sigma(x) = (\cos 2\pi x, \sin 2\pi x) = e^{2\pi i x}$. Then $f_*([\sigma]) = [f \circ \sigma] \in \pi_1(S^1, 1, 0))$. Since $(f \circ \sigma)(0) = (1, 0)$, the unique path $\alpha: I \to \mathbb{R}$ given by Theorem 8.15 is the path defined by $g(x) = nx$. Let $p: \mathbb{R} \to S^1$ be the standard covering map defined by $p(x) = (\cos 2\pi x, \sin 2\pi x)$ (see Section 8.3). Then $p \circ \alpha = f \circ \sigma$, so $\deg(f \circ \sigma) = n$. From the proof that $\pi_1(S^1, (1, 0))$ is isomorphic to the group of integers (Theorem 8.18), we see that $[f \circ \sigma]$ is not the identity element of $\pi_1(S^1, (1, 0))$. Therefore f_* is not the zero homeomorphism.

First we show that we may assume that $|a_{n-1}| + |a_{n-2}| + \cdots + |a_0| < 1$. We let c be a positive real number and substitute $x = cy$ in the given polynomial equation to obtain the equation

$$(cy)^n + a_{n-1}(cy)^{n-1} + \cdots + a_1(cy) + a_0 = 0$$

or

$$y^n + (a_{n-1}/c)y^{n-1} + \cdots + (a_1/c^{n-1})y + a_0/c^n = 0.$$

Now choose c large enough so that $|a_{n-1}/c| + |a_{n-2}/c^2| + \cdots + |a_1/c^{n-1}| + |a_0/c^n| < 1$. Then if y_0 is a solution of $y^n + (a_{n-1}/c)y^{n-1} + \cdots + (a_1/c^{n-i})y + a_0/c^n = 0$, $x = cy_0$ is a solution of $x^n + a_{n-1}x^{n-1} + \cdots + a_1 x + a_0 = 0$. Therefore it is sufficient to show that $y^n + (a_{n-1}/c)y^{n-1} + \cdots + (a_1/c^{n-1})y + a_0/c^n = 0$ has a solution. This means that in the given polynomial equation, we may assume that $|a_{n-1}| + |a_{n-2}| + \cdots + |a_0| < 1$.

The proof that the given polynomial equation has a solution in B^2 is by contradiction. Suppose $x^n + a_{n-1}x^{n-1} + \cdots + a_1 x + a_0 = 0$ has no solution in B^2. Then there is a continuous function $q: B^2 \to \mathbb{R}^2 - \{(0, 0)\}$ defined by $q(z) = z^n + a_{n-1}z^{n-1} + \cdots + a_1 z + a_0$. Let $r: S^1 \to \mathbb{R}^2 - \{(0, 0)\}$ be the restriction of q to S^1. Then $q: B^2 \to \mathbb{R}^2 - \{(0, 0)\}$ is an extension of r. So, by Theorem 9.1, r is inessential.

We arrive at a contradiction by showing that r is homotopic to a continuous function that is essential. Define $k: S^1 \to \mathbb{R}^2 - \{(0, 0)\}$ by $k(z) = z^n$, and define $H: S^1 \times I \to \mathbb{R}^2 - \{(0, 0)\}$ by $H(z, t) = z^n + t(a_{n-1}z^{n-1} + a_{n-2}z^{n-2} + \cdots + a_0)$. Note $H(x, t) \neq (0, 0)$ for $(x, t) \in S^1 \times I$ because

$$\begin{aligned} |H(x, t)| &\geq |z^n| - |t(a_{n-1}z^{n-1} + a_{n-2}z^{n-2} + \cdots + a_0| \\ &\geq 1 - t(|a_{n-1}z^{n-1}| + |a_{n-2}z^{n-2}| + \cdots + |a_0|) \\ &= 1 - t(|a_{n-1}| + |a_{n-2}| + \cdots + |a_0|) > 0. \end{aligned}$$

We complete the proof by showing that k is essential. Note that $k = j \circ f$, where $j: S^1 \to \mathbb{R}^2 - \{(0, 0)\}$ is the inclusion map, so $k_* = j_* \circ f_*$. Since the fundamental group of S^1 is isomorphic to the group of integers, f_* is essentially the homeomorphism that takes an integer a into the product na. Furthermore j_* is an isomorphism (see Example 9 of Chapter 8). Therefore k_* is not the zero homeomorphism. By Theorem 9.2, k is essential. ∎

EXERCISES 9.2

1. Find a real number r such that $x^7 + x^5 + x^3 + x^2 + 1 = 0$ has a solution in $\{(x, y) \in \mathbb{R}^2 : x^2 + y^2 \leq r^2\}$.

9.3 Homotopic Maps

We have already seen that if (X, x_0) and (Y, y_0) are topological pairs that have the same homotopy type, then $\pi_1(X, x_0)$ and $\pi_1(Y, y_0)$ are isomorphic. In this section we show that homotopic maps induce the same homeomorphisms on the fundamental groups provided the base point remains fixed during the homotopy, and we give a condition on a continuous function mapping a compact space into S^2, which ensures that the continuous function is inessential.

Let $(X, \mathcal{T})$ be a topological space, let $x_0, x_1 \in X$, and let σ be a path in X from x_0 to x_1. We let $\theta_\sigma: \pi_1(X, x_0) \to \pi_1(X, x_1)$ be the function defined in the proof of Theorem 8.9; that is $\theta_\sigma([\alpha]) = [(\overline{\sigma} * \alpha) * \sigma]$. As shown in the proof of Theorem 8.9, θ_σ is an isomorphism.

THEOREM 9.5. Let $(X, \mathcal{T})$ and $(Y, \mathcal{U})$ be topological spaces, let $x_0 \in X$, let $h, k: X \to Y$ be continuous functions such that $h \simeq k$, and let $y_0 = h(x_0)$ and $y_1 = k(x_0)$. Then:

(a) there is a path σ in Y from y_0 to y_1 such that $h_* = \theta_\sigma \circ k_*$.

(b) If $H: X \times I \to Y$ is a continuous function such that $H(x, 0) = h(x)$ and $H(x, 1) = k(x)$ for all $x \in X$ and $H(x_0, t) = y_0 = y_1$ for all $t \in I$, then $h_* = k_*$.

Proof. (a) Let $H: X \times I \to Y$ be a continuous function such that $H(x, 0) = h(x)$ and $H(x, 1) = k(x)$ for all $x \in X$. Define $\sigma: I \to Y$ by $\sigma(t) = H(x_0, t)$ for all $t \in I$. Then σ is a path from y_0 to y_1.

Let $[\gamma] \in \pi_1(X, x_0)$. For each $t \in I$, define $\alpha_t, \beta_t: I \to Y$ by $\alpha_t(s) = \sigma(ts)$ and $\beta_t(s) = H(\gamma(s), t)$ for each $s \in I$. Then for each $t \in I$,

$$\alpha_t(1) = \sigma(t) = H(x_0, t) = H(\gamma(0), t) = \beta_t(0),$$
and $$\beta_t(1) = H(\gamma(1), t) = H(x_0, t) = \sigma(t) = \alpha_t(1) = \overline{\alpha}_t(0).$$

Therefore for each $t \in I$, $(\alpha_t * \beta_t) * \bar{\alpha}_t$ is defined. Since $\alpha_0(s) = \sigma(0) = H(x_0, 0) = h(x_0) = y_0$, $\beta_0(s) = H(\gamma(s), 0) = (h \circ \gamma)(s)$, and $\bar{\alpha}_0(s) = \alpha_0(1 - s) = \sigma(0) = y_0$ for each $s \in I$, and $(\alpha_0 * \beta_0) * \bar{\alpha}_0 = (e_{y_0} * (h \circ \gamma)) * e_{y_0}$, where $e_{y_0}: I \to Y$ is the constant function defined by $e_{y_0}(t) = y_0$ for each $t \in I$. Since $(e_{y_0} * (h \circ \gamma)) * e_{y_0} \simeq_p h \circ \gamma$, $(\alpha_0 * \beta_0) * \bar{\alpha}_0 \simeq_p h \circ \gamma$. Since $\alpha_1(s) = \sigma(s)$, $\beta_1(s) = H(\gamma(s), 1) = (k \circ \gamma)(s)$, and $\bar{\alpha}_1(s) = \alpha_1(1 - s) = \sigma(1 - s) = \bar{\sigma}(s)$ for each $s \in I$, $\alpha_1 * \beta_1) * \bar{\alpha}_1 = (\sigma * (k \circ \gamma)) * \bar{\sigma}$.

Define $F: I \times I \to Y$ by

$$F(s, t) = \begin{cases} \sigma(4st), & 0 \le s \le \frac{1}{4}, 0 \le t \le 1 \\ H(\gamma(4s - 1), t), & \frac{1}{4} \le s \le \frac{1}{2}, 0 \le t \le 1 \\ \sigma(2t(1 - s)), & \frac{1}{2} \le s \le 1, 0 \le t \le 1 \end{cases}$$

Then F is continuous,

$$F(s, 1) = \begin{cases} \sigma(4s), & 0 \le s \le \frac{1}{4} \\ H(\gamma(4s - 1), 1), & \frac{1}{4} \le s \le \frac{1}{2} \\ \sigma(2(1 - s)), & \frac{1}{2} \le s \le 1 \end{cases}$$

$$= \begin{cases} \sigma(4s), & 0 \le s \le \frac{1}{4} \\ (k \circ \gamma)(4s - 1), & \frac{1}{4} \le s \le \frac{1}{2} \\ \bar{\sigma}(2s - 1), & \frac{1}{2} \le s \le 1, \end{cases} = ((\sigma * (k \circ \gamma)) * \bar{\sigma})(s)$$

and

$$F(s, 0) = \begin{cases} \sigma(0), & 0 \le s \le \frac{1}{4} \\ H(\gamma(4s - 1), 0), & \frac{1}{4} \le s \le \frac{1}{2} \\ \sigma(0), & \frac{1}{2} \le s \le 1 \end{cases}$$

$$= \begin{cases} \sigma(0), & 0 \le s \le \frac{1}{4} \\ (h \circ \gamma)(4s - 1), & \frac{1}{4} \le s \le \frac{1}{2} \\ \sigma(0), & \frac{1}{2} \le s \le 1. \end{cases} = (e_{y_0} * (h \circ \gamma)) * e_{y_0})(s)$$

Now $(e_{y_0} * (h \circ \gamma)) * e_{y_0} \simeq_p (h \circ \gamma)$. Therefore $h_*([\gamma]) = [h \circ \gamma] = [(\sigma * (k \circ \gamma)) * \bar{\sigma}] = \theta_\sigma([k \circ \gamma]) = (\theta_\sigma \circ k_*)([\gamma])$.

(b) Suppose $H(x_0, t) = y_0 = y_1$ for all $t \in I$. Then σ is the constant path, so $h_* = \theta_\sigma \circ k_* = k_*$. ∎

We conclude this section by proving a theorem (Theorem 9.7) that will be used in the next section. Intuitively this theorem says that if f is a continuous function from a compact space $(X, \mathcal{T})$ into S^2 with the property that there are two distinct points of S^2 which lie in the same component of $S^2 - f(X)$, then f can be shrunk to a constant function under a homotopy that does not have either of the two points in its image. In the proof we use the following theorem.

THEOREM 9.6. Let a and b be distinct member of S^2, and let $X = \mathbb{R}^2 \cup \{p\}$ be the one-point compactification of $\mathbb{R}^2$. Then there exists a homeomorphism $h: S^2 \to X$ such that $h(a) = p$ and $h(b) = (0, 0)$.

Proof. Let $h_1: S^2 \to S^2$ be a rotation of S^2 such that $h_1(a) = (0, 0, 1)$. Then h_1 is a homeomorphism. Define $h_2': S^2 - \{(0, 0, 1)\} \to \mathbb{R}^2$ by $h_2'(x_1, x_2, x_3) = (1/(1 - x_3))(x_1, x_2)$. As seen in the proof of Theorem 8.27, h_2' is a homeomorphism. Define $h_2: S^2 \to X$ by $h_2(x) = h_2'(x)$ if $x \in S^2 - \{(0, 0, 1)\}$ and $h_2(0, 0, 1) = p$. In Exercise 1 you are asked to show that h_2 is a homeomorphism. Let $(b_1, b_2) = h_2(h_1(b))$ and define $h_3': \mathbb{R}^2 \to \mathbb{R}^2$ by $h_3'(x, y) = (x - b_1, y - b_2)$. Then h_3' is a homeomorphism. Define $h_3: X \to X$ by $h_3(x) = h_3'(x)$ if $x \in \mathbb{R}^2$ and $h_3(p) = p$. In Exercise 2 you are asked to show that h_3 is a homeomorphism. Let $h = h_3 \circ h_2 \circ h_1$. Then $h: S_2 \to X$ is a homeomorphism such that $h(a) = h_3(h_2(0, 0, 1)) = h_3(p) = p$ and $h(b) = h_3((h_2 \circ h_1)(b)) = h_3(b_1, b_2) = (0, 0)$. ∎

THEOREM 9.7. Let a and b be distinct members of S^2, let $(X, \mathcal{T})$ be a compact space, and let $f: X \to S^2 - \{a, b\}$ be a continuous function such that a and b lie in the same component of $S^2 - f(X)$. Then f is inessential.

Proof. Let $Y = \mathbb{R}^2 \cup \{p\}$ be the one-point compactification of $\mathbb{R}^2$, and let $h: S^2 \to Y$ be the homeomorphism given by Theorem 9.6. Then $h \circ f: X \to \mathbb{R}^2 - \{(0, 0)\}$ is a continuous function such that $(0,0)$ lies in the unbounded component of $\mathbb{R}^2 - (h \circ f)(X)$. Suppose there is a continuous function $H: X \times I \to \mathbb{R}^2 - \{(0, 0)\}$ and a member y_0 of $\mathbb{R}^2 - \{(0, 0)\}$ such that $H(x, 0) = (h \circ f)(x)$ and $H(x, 1) = y_0$ for each $x \in X$. Then $h^{-1} \circ H: X \times I \to S^2 - \{a, b\}$ is a continuous function such that $(h^{-1} \circ H)(x, 0) = f(x)$ and $(h^{-1} \circ H)(x, 1) = h^{-1}(y_0)$ for each $x \in X$. Therefore in order to prove the theorem, it is sufficient to show that if $f: X \to \mathbb{R}^2 - \{(0, 0)\}$ is a continuous function such that $(0,0)$ lies in the unbounded component of $\mathbb{R}^2 - f(x)$, then f is inessential.

Let $f: X \to \mathbb{R}^2 - \{(0, 0)\}$ be a continuous function such that $(0,0)$ lies in the unbounded component of $\mathbb{R}^2 - f(X)$. Since $(X, \mathcal{T})$ is compact, $f(X)$ is a closed and bounded subset of $\mathbb{R}^2$. Let r be a positive number such that $f(X) \subseteq B = \{(x, y) \in \mathbb{R}^2: x^2 + y^2 \leq r^2\}$, and let $q \in \mathbb{R}^2 - B$. Then $(0,0)$ and q lie in the same component of $\mathbb{R}^2 - f(X)$. Since $\mathbb{R}^2$ is locally pathwise connected and $\mathbb{R}^2 - f(X)$ is open, $\mathbb{R}^2 - f(X)$ is locally pathwise connected. Therefore the components and path components of $\mathbb{R}^2 - f(X)$ are the same (Theorem 3.28). Hence there is a path α in $\mathbb{R}^2 - f(X)$ from $(0, 0)$ to q. Define $F: X \times I \to \mathbb{R}^2 - \{(0, 0)\}$ by $F(x, t) = f(x) - \alpha(t)$. Then F is continuous, $F(x, 0) = f(x) - \alpha(0) = f(x) - (0, 0) = f(x)$ and $F(x, 1) = f(x) - \alpha(1) = f(x) - q$ for each $x \in X$. Note that $F(x, t) \neq (0, 0)$ for any $(x, t) \in X \times I$ because $\alpha(I) \cap f(X) = \emptyset$. Now define $G: X \times I \to \mathbb{R}^2 - \{(0, 0)\}$ by $G(x, t) = tf(x) - q$. Then G is continuous, $G(x, 0) = -q$ and $G(x, 1) = f(x) - q$ for each $x \in X$. Note that $G(x, t) \neq (0, 0)$ for any $(x, t) \in X \times I$ because $tf(x) \in B$ and $q \notin B$. Thus, if $g: X \to \mathbb{R}^2 - \{(0, 0)\}$ is the function defined by $g(x) = f(x) - q$, then F is a homotopy between f and g and G is a homotopy between g and a constant map. Therefore f is homotopic to a constant map, and hence it is inessential. ∎

EXERCISES 9.3

1. Show that the function h_2 defined in the proof of Theorem 9.6 is a homeomorphism.

2. Show that the function h_3 defined in the proof of Theorem 9.6 is a homeomorphism.

3. Let $X = \mathbb{R}^2 - \{(-1, 0), (1, 0)\}$ and let $A = \{(x, y) \in \mathbb{R}^2 : (x + 1)^2 + y^2 = 1$ or $(x - 1)^2 + y^2 = 1\}$. Prove that A is a deformation retract of X.

4. Let $X = \mathbb{R}^2 - \{(-1, 0), (1, 0)\}$ and let $B = \{(x, y) \in \mathbb{R}^2 : x + y^2 = 4,$ or $x = 0$ and $-2 \leq y \leq 2\}$. Prove that A is a deformation retract of X.

5. Let A and B be the spaces defined in Exercises 3 and 4. Show that neither of these spaces can be imbedded in the other.

9.4 The Jordan Curve Theorem

The purpose of this section is to prove the Jordan Curve Theorem. Camille Jordan (1858–1922) proposed the problem (that is, the Jordan curve theorem) in 1892 by pointing out that this intuitively obvious fact required proof, and consequently the resulting theorem was named for him. It was, however, Oswald Veblen (1880–1960) who published, in 1905, the first correct proof.

First we prove the Jordan Separation Theorem. Then we prove a nonseparation theorem, and finally we prove the Jordan Curve Theorem.

Definition. An **arc** is a topological space that is homeomorphic to the unit interval [0,1], and a **simple closed curve** is a topological space that is homeomorphic to the unit circle S^1. ∎

THEOREM 9.8 (The Jordan Separation Theorem). Let C be a simple closed curve in S^2. Then $S^2 - C$ is not connected.

Proof. Since $S^2 - C$ is an open subset of the locally pathwise connected space S^2, it is locally pathwise connected. Therefore the components and path components of $S^2 - C$ are the same (Theorem 3.28). Thus it is sufficient to assume that $S^2 - C$ is pathwise connected and to reach a contradiction.

Since C is homeomorphic to the unit circle, there are arcs A and B such that $C = A \cup B$ and $A \cap B = \{a, b\}$, where a and b are the end points of A and B. Let $X = S^2 - \{a, b\}$. By Theorem 9.6, X is homeomorphic to $\mathbb{R}^2 - \{(0, 0)\}$. Therefore, by Example 9 of Chapter 8, the fundamental group of X is isomorphic to the group of integers.

Let $U = S^2 - A$ and $V = S^2 - B$. Then U and V are open subsets of X, $U \cup V = X$, and $U \cap V = S^2 - (A \cup B) = S^2 - C$. Thus, by assumption, $U \cap V$ is pathwise connected. Let $x_0 \in U \cap V$ and consider the inclusion maps $i: (U, x_0) \to (X, x_0)$ and $j: (V, x_0) \to (X, x_0)$. We show that these inclusion maps induce zero homeomorphisms $i_*: \pi_1(U, x_0) \to \pi_1(X, x_0)$ and $j_*: \pi_1(V, x_0) \to \pi_1(X, x_0)$. Let $[\alpha] \in \pi_1(U, x_0)$ and let $\sigma: I \to S^1$ be the continuous function defined by $\sigma(x) = (\cos 2\pi x, \sin 2\pi x)$. Define $h: S^1 \to U$ as follows: Let $h(1, 0) = x_0$. If $y \in S^1 - \{(1, 0)\}$, then there is a unique member x of I such that $\sigma(x) = y$. Let $h(y) = \alpha(x)$. Then h is a continuous function and $h \circ \sigma = \alpha$.

Consider the continuous function $i \circ h: S^1 \to X$. Now $i(h(S^1)) = h(S^1) \subseteq U$, and A is an arc with end points a and b, and $U \cap A = \emptyset$. Therefore a and b lie in the same component of $S^2 - i(h(S^1))$. By Theorem 9.7, $i \circ h$ is inessential. Therefore, by Theorem 9.2, $(i \circ h)_*$ is the zero homeomorphism. Since $i_*([\alpha]) = [i \circ \alpha] = [i \circ h \circ \sigma] = (i \circ h)_*([\sigma])$, i_* is the zero homeomorphism.

The preceding proof is equally applicable to U and V and hence j_* is the zero homeomorphism. Therefore, by Theorem 8.26, $\pi_1(X, x_0)$ is the trivial group. This is a contradiction, so $S^2 - C$ is not pathwise connected. Therefore $S^2 - C$ is not connected. ∎

Now our goal is to prove the Jordan Curve Theorem. First we prove three theorems that we will use.

THEOREM 9.9. Let $(X, \mathcal{T})$ be a topological space, let $U, V, M,$ and N be open subsets of X such that $X = U \cup V$, $U \cap V = M \cup N$, and $M \cap N = \emptyset$, let $a \in M$ and $b \in N$, and suppose there exists a path α in U from a to b, and a path β in V from b to a. Then $\pi_1(X, a)$ is not the trivial group.

Proof. For each $n \in \mathbb{Z}$, let $U_n = U \times \{2n\}$ and $V_n = V \times \{2n + 1\}$. Then let $Y = \bigcup_{n \in \mathbb{Z}}(U_n \cup V_n)$.

$$\vdots$$
$$U_1 = U \times \{2\} \qquad V_1 = V \times \{3\}$$
$$U_0 = U \times \{0\} \qquad V_0 = V \times \{1\}$$
$$U_{-1} = U \times \{-2\} \qquad V_{-1} = V \times \{-1\}$$
$$\vdots$$

Define a quotient space E of Y by making the following identifications:
For each $x \in M$ and $n \in \mathbb{Z}$, identify $(x, 2n - 1)$ and $(x, 2n)$.
For each $x \in N$ and $n \in \mathbb{Z}$, identify $(x, 2n)$ and $(x, 2n + 1)$.

⋮

U_1 and V_1 are "pasted together" at "points of N."
U_1 and V_0 are "pasted together" at "points of M."
U_0 and V_0 are "pasted together" at "points of N."
U_0 and V_{-1} are "pasted together" at "points of M."
U_{-1} and V_{-1} are "pasted together" at "points of N."

⋮

Let $\pi: Y \to E$ be the quotient map. Define $q: Y \to X$ by $q(x, n) = x$ for each $(x, n) \in Y$. For each $(x, n) \in Y$, let $[(x, n)]$ denote the member of E that contains (x, n), and define $p: E \to X$ by $p([(x, n)]) = x$. Then q is continuous because it is a projection map, and p is continuous because E has the quotient topology. It is clear that p is a surjection. We show that it is a covering map, but first we show that π is open.

Since Y is the union of disjoint open sets U_n and V_n, it is sufficient to show that for each $n \in \mathbb{Z}$, $\pi|_{U_n}$ and $\pi|_{V_n}$ are open functions. Let S be an open subset of U_n. Then $S = W \times \{2n\}$, where W is open in U. Then

$$\pi^{-1}(\pi(S)) = \pi^{-1}(\pi(W \times \{2n\}))$$
$$= (W \times \{2n\}) \cup ((W \cap N) \times \{2n + 1\}) \cup ((W \cap M) \times \{2n - 1\})$$

is the union of three open sets in Y. Therefore, by the definition of the quotient topology, $\pi(S)$ is open in E. The same argument shows that the image under π of any open subset of V_n is open in E. Therefore π is open.

We show that p is a covering map by showing that the open sets U and V are evenly covered by p. Note that for each integer n, $\pi(U_n)$ is open in E since π is open. Furthermore $p^{-1}(U) = \bigcup_{n \in \mathbb{Z}} \pi(U_n)$, and $\pi|_{U_n}$ maps U_n onto $\pi(U_n)$. Thus $\pi|_{U_n}$ is a homeomorphism because it is continuous, bijective, and open. Therefore $p|_{\pi(U_n)}$ is the composite of the two homeomorphisms $(\pi|_{U_n})^{-1}: \pi(U_n) \to U_n$ and $q|_{U_n}: U_n \to U$. Thus $p|_{\pi(U_n)}$ is a homeomorphism, and hence U is evenly covered by p. The same argument shows that V is evenly covered by p. Therefore p is a covering map.

Finally we show that $\pi_1(X, a)$ is not the trivial group by showing that $\alpha * \beta$ is not path homotopic to the constant function $e: I \to X$ defined by $e(t) = a$ for each $t \in I$. Define $\alpha', \beta': I \to E$ by $\alpha'(t) = \pi(\alpha(t), 0)$ and $\beta'(t) = \pi(\beta(t), 1)$ for each $t \in I$. Since $\alpha'(1) = \pi(\alpha(1), 0) = \pi(b, 0) = \pi(b, 1) = \pi(\beta(0), 1) = \beta'(0)$, $\alpha' * \beta'$ is defined. Furthermore $p \circ \alpha' = \alpha$ and $p \circ \beta' = \beta$. Since $(\alpha' * \beta')(0) = \alpha'(0) = \pi(\alpha(0), 0) = \pi(a, 0)$, $(\alpha' * \beta')(1) = \beta'(1) = \pi(\beta(1), 1) = \pi(a, 1)$, and $\pi(a, 0) \neq \pi(a, 1)$ because $a \in M$, $\alpha' * \beta'$ begins at one point and ends at another. Therefore by Theorem 8.21, $\alpha * \beta$ is not path homotopic to e. Therefore $\pi_1(X, a)$ is not the trivial group. ∎

THEOREM 9.10. Let $(X, \mathcal{T})$ be a topological space, let $U, V, M, N,$ and P be open subsets of X such that $X = U \cup V$, $U \cap V = M \cup N \cup P$, and $M \cap N = M \cap P = N \cap P = \emptyset$, let $a \in M$, $b \in N$, and $c \in P$, let α be a path in U from a to b, let β be a path in V from b to a, let γ be a path in U from a to c, and let δ be a path in V from c to a. Then $\pi_1(X, a)$ is not isomorphic to the group of integers.

9.4 The Jordan Curve Theorem 311

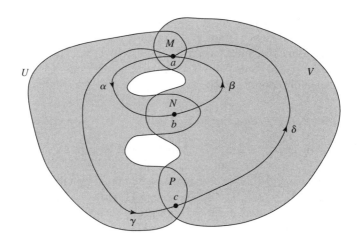

Proof. Since $U \cap V = M \cup (N \cup P)$ and $M \cap (N \cup P) = \emptyset$, it follows from the proof of Theorem 9.9 that $[\alpha * \beta]$ and $[\gamma * \delta]$ are nonzero elements of $\pi_1(X, a)$. If $\pi_1(X, a)$ were isomorphic to the group of integers, there would exist nonzero integers m and n such that $m[\alpha * \beta] = n[\gamma * \delta]$. We prove that the theorem by showing that no such integers exist.

Since $U \cup V = (M \cup P) \cup N$ and $(M \cup P) \cap N = \emptyset$, we can follow the construction in the first part of the proof of Theorem 9.9 and obtain a space Y, a quotient space E of Y, a quotient map $\pi: Y \to E$ and a covering map $p: E \to X$. Then we can follow the construction of the unique liftings of $\gamma * \delta$ and $\alpha * \beta$ in the last part of the proof of Theorem 9.9 and see that every "multiple" of $\gamma * \delta$ lifts to a loop μ' in E while every nonzero "multiple" of $\alpha * \beta$ lifts to a path v' in E such that $v'(0) = \mu'(0)$ whereas $v'(1) \neq \mu'(1)$ (see Exercise 1). Therefore by Theorem 8.21 there does not exist nonzero integers m and n such that $m[\alpha * \beta] = n[\gamma * \delta]$. ∎

If A is an arc in S^2, then A is homeomorphic to the unit interval, so there are arcs A_1 and A_2 such that $A = A_1 \cup A_2$ and $A_1 \cap A_2 = \{a\}$, where a is an end point of A_1 and A_2.

THEOREM 9.11. Let A, A_1, and A_2 be arcs in S^2 such that $A = A_1 \cup A_2$ and $A_1 \cap A_2 = \{a\}$, where a is an end point of A_1 and A_2 and let $c, d \in S^2 - A$ such that c and d can be joined by a path in $S^2 - A_1$ and a path in $S^2 - A_2$. Then c and d can be joined by a path in $S^2 - A$.

Proof. Suppose c and d cannot be joined by a path in $S^2 - A$. Let $X = S^2 - \{a\}$, let $U = S^2 - A_1$, and let $V = S^2 - A_2$. Then U and V are open, $U \cup V = (S^2 - A_1) \cup (S^2 - A_2) = S^2 - (A_1 \cap A_2) = S^2 - \{a\} = X$, and $U \cap V = (S^2 - A_1) \cap (S^2 - A_2) = S^2 - (A_1 \cup A_2) = S^2 - A$. By our assumption that c and d cannot be joined by a path in $S^2 - A$, $U \cap V$ is not pathwise connected. Let C be the path component of $U \cap V$ that contains c and let $D = (U \cap V) - C$. Then D is the union of the path components of $U \cap V$ that do not contain c. Since $U \cap V$ is open in S^2, it is locally pathwise connected. Therefore the path components of $U \cap V$ are open (Theorem 3.26). Hence C and D are open in X. We are given that there is a path in $S^2 - A_1$ from c to d and a path in $S^2 - A_2$ from d to c. Thus by Theorem 9.9, $\pi_1(X, c)$ is not the trivial group. This is a contradiction because X is homeomorphic to $\mathbb{R}^2$ and $\mathbb{R}^2$ is simply connected. ∎

THEOREM 9.12. Let A be an arc in S^2. Then $S^2 - A$ is pathwise connected.

Proof. Suppose there exist $c, d \in S^2 - A$ that cannot be joined by a path in $S^2 - A$. Let $h: I \to A$ be a homeomorphism, let $A_1 = h([0, \frac{1}{2}])$ and let $A_2 = h([\frac{1}{2}, 1])$. By Theorem 9.11, we may assume that c and d cannot be joined by a path in $S^2 - A_2$. Let $A_{21} = h([\frac{1}{2}, \frac{3}{4}])$ and $A_{22} = h([\frac{3}{4}, 1])$. By Theorem 9.11, either c and d cannot be joined by a path in $S^2 - A_{21}$ or c and d cannot be joined by a path in $S^2 - A_{22}$. Continuing in this manner, we obtain a sequence $[0, 1] = I_1 \supseteq I_2 \supseteq I_3 \supseteq \cdots$ of closed intervals such that, for each $i \in \mathbb{N}$, c and d cannot be joined by a path in $S^2 - h(I_i)$ and the length of I_i is $1/2^{i-1}$. Since I is compact, there exists $t \in \bigcap_{i \in \mathbb{N}} I_i$ and since the lengths of the intervals converge to 0, t is the only member of $\bigcap_{i \in \mathbb{N}} I_i$.

Since $S^2 - \{h(t)\}$ is homeomorphic to $\mathbb{R}^2$, there is a path f in $S^2 - \{h(t)\}$ such that $f(0) = c$ and $f(1) = d$. Since $h(t) = \bigcap_{i \in \mathbb{N}} h(I_i)$, $S^2 - \{h(t)\} = \bigcup_{i \in \mathbb{N}} (S^2 - h(I_i))$. Since $f(I)$ is compact, there exists a finite subcollection of $\{S^2 - h(I_i): i \in \mathbb{N}\}$ such that $f(I)$ is a subset of the union of the members of this finite subcollection. Since $S^2 - h(I_i) \subseteq S^2 - h(I_{i+1})$ for each $i \in \mathbb{N}$, there exists $n \in \mathbb{N}$ such that $f(I) \subseteq S^2 - h(I_n)$. So f is a path in $S^2 - h(I_n)$ from c to d. This is a contradiction. ∎

THEOREM 9.13 (The Jordan Curve Theorem). Let C be a simple closed curve in S^2. Then $S^2 - C$ has exactly two components C_1 and C_2. Moreover $C = \overline{C}_1 - C_1 = \overline{C}_2 - C_2$.

Proof. Since C is homeomorphic to the unit circle, there are arcs A_1 and A_2 such that $C = A_1 \cup A_2$ and $A_1 \cap A_2 = \{a_1, a_2\}$, where a_1 and a_2 are the end points of A_1 and A_2. Let $X = S^2 - \{a_1, a_2\}$, let $U = S^2 - A_1$, and let $V = S^2 - A_2$. Then U and V are open, $X = U \cup V$ and $U \cap V = S^2 - C$. By Theorem 9.8, $S^2 - C$ has at least two components. Suppose it has more than two components. Let M and N be two of the components and let $P = (S^2 - C) - (M \cup N)$. Since $S^2 - C$ is open in S^2, it is locally pathwise connected. Therefore the components of $S^2 - C$ are open. Thus, since P is the union of components, M, N, and P are open. Let $a \in M$, let $b \in N$, and let $c \in P$. By Theorem 9.12, U and V are pathwise connected. Therefore there is a path in U from a to b, a path in V from b to a, a path in U from a to c, and a path in V from c to a. Thus, by Theorem 9.10, $\pi_1(X, a)$ is not isomorphic to the group of integers. This is a contradiction because X is homeomorphic to $\mathbb{R}^2 - \{(0, 0)\}$, and the fundamental group of $\mathbb{R}^2 - \{(0, 0)\}$ is isomorphic to the group of integers. Therefore $S^2 - C$ has exactly two components.

Let C_1 and C_2 denote the two components of $S^2 - C$. We know that C_1 and C_2 are open in S^2, so $\overline{C}_1 \cap C_2 = C_1 \cap \overline{C}_2 = \emptyset$. Let $x \in \overline{C}_1 - C_1$. Then $x \in \overline{C}_1$ but

$x \notin C_1$. Since $x \in \overline{C}_1$, $x \notin C_2$. Therefore $x \in C$ and hence $\overline{C}_1 - C_1 \subseteq C$. In the same manner, we see that $\overline{C}_2 - C_2 \subseteq C$.

Now let $x \in C$, and let W be a neighborhood of x. Since C is homeomorphic to the unit circle, there are arcs B_1 and B_2 such that $C = B_1 \cup B_2$, $B_1 \subseteq W$, and $B_1 \cap B_2 = \{b_1, b_2\}$, where b_1 and b_2 are the end points of B_1 and B_2. Let $p \in C_1$ and $q \in C_2$. By Theorem 9.12, there is a path f in $S^2 - B_2$ from p to q.

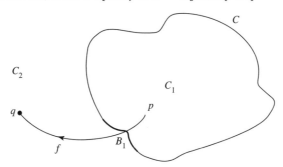

Now $S^2 = C_1 \cup (S^2 - \overline{C}_1) \cup (\overline{C}_1 - C_1)$, $f(I) \cap C_1 \neq \emptyset$, and $f(I) \cap (S^2 - \overline{C}_1) \neq \emptyset$. Therefore $f(I) \cap (\overline{C}_1 - C_1) \neq \emptyset$ because otherwise $f(I) \cap C_1$ and $f(I) \cap (S^2 - \overline{C}_1)$ would from a separation of the connected set $f(I)$. Let $y \in f(I) \cap (\overline{C}_1 - C_1)$. Since $\overline{C}_1 - C_1 \subseteq C$, $y \in C$. Since $f(I) \cap B_2 = \emptyset$, $y \in B_1 \subseteq W$. Therefore $y \in W \cap (\overline{C}_1 - C_1)$, and hence x is a limit point of $\overline{C}_1 - C_1$. But $\overline{C}_1 - C_1$ is closed, so $x \in \overline{C}_1 - C_1$. Therefore $C \subseteq \overline{C}_1 - C_1$. In the same manner, we can show that $C \subseteq \overline{C}_2 - C_2$. ∎

EXERCISES 9.4

1. Complete the proof of Theorem 9.10 by showing that every "multiple" of $\gamma * \delta$ lifts to a loop in E and every nonzero "multiple" of $\alpha * \beta$ lifts to a path that is not a loop.

2. Let A and B be closed connected subsets of S^2 such that $A \cap B$ consists of exactly two points. Prove that $S^2 - (A \cup B)$ is not connected.

3. Let A and B be closed connected subsets of S^2 such that $S^2 - A$ and $S^2 - B$ are connected and $A \cap B$ consists of exactly two points. Prove that $S^2 - (A \cup B)$ has exactly two components.

4. Let A be an arc in $\mathbb{R}^2$. Show that $\mathbb{R}^2 - A$ is connected.

5. Let C be a simple closed curve in $\mathbb{R}^2$.
   (a) Show that $\mathbb{R}^2 - C$ has exactly two components C_1 and C_2.
   (b) Show that $C = \overline{C}_1 - C_1 = \overline{C}_2 - C_2$.

Appendix A
Logic and Proofs

This appendix illustrates three basic methods of proof, and is provided for those who have limited experience. We begin with a brief introduction to logic.

A **proposition** is a sentence that is either true or false. We use capital letters P, Q, R, and so on, to represent propositions, P ∧ Q to represent P *and* Q, P ∨ Q to represent P *or* Q, and ¬P to represent the denial of a proposition P.

Suppose we are given boxes, □ and ◇, into which we can place propositions. Since, by definition, any proposition placed in the box can have only two truth values, we can tabulate the behavior of ∧, ∨, and ¬ and thereby define these signs by listing all possible inputs and the corresponding possible outputs in tabular form. The resulting tables are called **truth tables** for ∧, ∨, and ¬. The definitions of ∧, ∨, and ¬ are given by the following truth tables.

And			*Or*			*Not*	
□	◇	□ ∧ ◇	□	◇	□ ∨ ◇	□	¬□
T	T	T	T	T	T	T	F
T	F	F	T	F	T	F	T
F	T	F	F	T	T		
F	F	F	F	F	F		

The definitions of two other connectives, → and ↔, called the **conditional** and the **biconditional,** are given by the following truth tables.

Conditional (implies)			*Biconditional (if and only if)*		
□	◇	□ → ◇	□	◇	□ ↔ ◇
T	T	T	T	T	T
T	F	F	T	F	F
F	T	T	F	T	F
F	F	T	F	F	T

The proposition "P → Q" is referred to as a **conditional proposition.** The conditional proposition P → Q occurs frequently in our daily lives, and the English language has many ways in which to express this proposition. Here are some of the ways in which we can express that P → Q is true.

If P, then Q.
If P is true, then Q is also true.
P is true only if Q is true.
P implies Q.
Q is true whenever P is true.
For P to be true, it is necessary for Q to be true.
For Q to be true, it is sufficient for P to be true.

In the conditional proposition "If P, then Q," P is called the **hypothesis**, or **condition**, and Q is called the **conclusion**.

If P and Q are propositions, then the **converse** of P → Q is Q → P. Note that a conditional proposition and its converse can have opposite truth values.

If P and Q are propositions, then the **contrapositive** of P → Q is $\neg Q \to \neg P$. Note that a conditional proposition and its contrapositive have the same truth values for all possible values of true or false for all variables appearing in either expression. This is usually expressed by saying that a conditional proposition and its contrapositive are equivalent.

The proposition "P ↔ Q" is referred to as a **biconditional** proposition. We can think of "P ↔ Q" as follows:

P if and only if Q.
For P to be true, it is necessary and sufficient for Q to be true.
P implies Q and Q implies P.
P is equivalent to Q.

The abbreviation "iff" is sometimes used for "if and only if." A biconditional proposition "P if and only if Q" involves two conditional propositions, "If P, then Q" and "If Q, then P." Note that "P if and only if Q" is true whenever "If P, then Q" and "If Q, then P" are both true; and "P if and only if Q" is false whenever either "If P, then Q" or "If Q, then P" is false.

Now we turn to methods of proof. Let us examine a proposition of the form "If P, then Q" or "P → Q." There are three conditions under which this proposition is true:

(1) Each of P and Q is true.

(2) P is false and Q is true.

(3) P is false and Q is false.

We first observe that if P is false, then P → Q is true. Therefore, in order to prove that P implies Q, it is sufficient to assume that P is true, and, under this assumption, prove that Q is true. The **direct method** of proof is precisely this method; that is, we assume that P is true and, under this assumption, we proceed through a logical sequence of steps to arrive at the conclusion that Q is true. Let's illustrate the direct method with a simple example. But first we recall that an integer n is said to be **odd** provided there is an integer k such that $n = 2k + 1$, and n said to be **even** provided there is an integer q such that $n = 2q$.

THEOREM A.1. If n is an odd integer, then n^2 is odd.

Proof. The hypothesis is "n is an odd integer," and the conclusion is "n^2 is odd." So we begin by assuming that n is odd. Then, by definition, there is an integer k such that $n = 2k + 1$. Thus, $n^2 = (2k + 1)(2k + 1) = 4k^2 + 4k + 1 = 2(2k^2 + 2k) + 1$. Therefore, there is an integer m (namely, $m = 2k^2 + 2k$) such that $n^2 = 2m + 1$. So, by definition, n^2 is odd. We have proved that if the hypothesis is true, then the conclusion is true; so the proof is complete. ∎

Since the propositional expression "P → Q" is equivalent to the contrapositive, "¬Q → ¬P", the preceding theorem can be restated: If n^2 is an even integer, then n is even.

Another way of proving any proposition of the form "If P, then Q" is to assume that Q is not true, and, under this assumption, proceed through a logical sequence of steps to arrive at the conclusion that P is not true. This is the **contrapositive method** of proof. Let's illustrate the contrapositive method with another simple theorem.

THEOREM A.2. Suppose that m and b are real numbers with $m \neq 0$ and that f is the linear function defined by $f(x) = mx + b$. If $x \neq y$, then $f(x) \neq f(y)$.

Proof. We want to prove that $x \neq y$ implies $f(x) \neq f(y)$. The contrapositive is $f(x) = f(y)$ implies $x = y$. Suppose that $f(x) = f(y)$. Then $mx + b = my + b$. If we subtract b from both sides of this equation and then divide both sides of the equation by m, we conclude that $x = y$. Therefore, we have proved that $f(x) = f(y)$ implies $x = y$. Since the contrapositive is equivalent to the original implication, we have proved the original implication. ∎

Another method of proof is the **contradiction method**. Note that ¬(P → Q) and P ∧ ¬Q are equivalent. So if we want to show that P implies Q, we begin by assuming that P is true and Q is false. The idea is to reach a contradiction.

Another way of regarding the contradiction method is to recall that P → Q is true except in the case that P is true and Q is false. In the contradiction method, we rule out this possibility by assuming that it does happen and then reaching a contradiction.

The most difficult question in the contradiction method is "What contradiction are we looking for?" There are no specific guidelines because each proof gives rise to its own contradiction, but *any* contradiction will do.

The advantage of the contradiction method over the contrapositive method is that we get two statements from which to reason rather than just one. The disadvantage is that we have no definite knowledge of where the contradiction will occur. In general, the best rule of thumb is to use contradiction when the statement "not Q" gives some useful information. We consider a simple example. But first we recall that a real number r is **rational** provided there are integers m and n, with $n \neq 0$, such that $r = m/n$, and r is **irrational** provided it is not rational.

THEOREM A.3. If r is a real number such that $r^2 = 2$, then r is irrational.

Proof. Suppose that $r^2 = 2$ and r is not irrational. Then r is rational, so there are integers m and n such that $r = m/n$. We can assume that m and n have no common divisors greater than 1 because if they did, we could divide both numerator and denominator by the greatest common divisor. So $r^2 = m^2/n^2$, and hence $m^2 = r^2 n^2$. Since $r^2 = 2$, we have $m^2 = 2n^2$. Hence m^2 must be even. By Theorem A.1, m must be even. Therefore, there is an integer p such that $m = 2p$. So $2n^2 = m^2 = 4p^2$, and hence $n^2 = 2p^2$. It follows that n^2 is even, and again by Theorem A.1, n must be even. Since m and n are both even, they have a common divisor greater then 1. This is a contradiction to our assumption that m and n have no common divisors greater than 1. So we have established the theorem. ∎

The three methods of proof we have discussed can be clarified by comparing them in a table. In each case, we are concerned with proving that P implies Q.

Direct method	Contrapositive method	Contradiction method
Assume P	Assume (not Q)	Assume P and (not Q)
⋮	⋮	⋮
(logical sequence of steps)	(logical sequence of steps)	(logical sequence of steps)
⋮	⋮	⋮
Conclude Q	Conclude (not P)	Conclude R and (not R)

The advantage and the difficulty of the contradiction method are illustrated by the table. We have more information to work with since we are assuming both P and (not Q). On the other hand, R suddenly appears (that is, we do not know R before we begin the proof).

We use Theorem A.4 to illustrate the three basic methods of proof. Since we have no need of the theorem itself, we ask you to assume all sorts of results from calculus and to connive at any gaps in the arguments given. In particular, let us recall that $D_x(e^x) = e^x$ and if u is any real number, then $\ln(e^u) = u$. Let us also recall Rolle's Theorem: If the function f is continuous on the closed interval $[a, b]$, differentiable on the open interval (a, b), and $f(a) = f(b)$, then there exists c in the open interval (a, b) such that $f'(c) = 0$.

THEOREM A.4. Let x and y be real numbers. If $x \neq y$, then $e^x \neq e^y$.

Direct Proof. Suppose $x \neq y$. Then either $x > y$ or $y > x$. We can assume that $x > y$ because x and y are real numbers, and we can always let x be the larger of the two. Then there is a positive number r such that $x = y + r$. Thus $e^x = e^{y+r} = e^y \times e^r$. Since $e > 1$ and $r > 0$, $e^r > 1$. Hence $e^y \times e^r > e^y$. Therefore, $e^x \neq e^y$. ■

Contrapositive Proof. Suppose that $e^x = e^y$. Then $x = \ln(e^x) = \ln(e^y) = y$. ■

Contradiction Proof. Suppose $x \neq y$ and yet $e^x = e^y$. By Rolle's Theorem, there is a z between x and y such that $e^z = 0$. This is a contradiction because $e > 0$ and hence $e^u > 0$ for each real number u. ■

We conclude this appendix with another topic in logic—namely quantifiers. A sentence such as "$x < 4$" is not a proposition. It contains a **variable,** which we could replace with a specific real number to make the sentence a proposition. Such a sentence is called a **propositional function.**

Phrases such as "for some x," "there exists an element x," and "there are x and y" are called **existential quantifiers.** Each of the following sentences contains an existential quantifier:

1. There is a real number x such that $x + 3 = 2$.
2. There exists a real number x such that $x^2 - x - 2 = 0$.
3. There are real numbers x and y such that $2x + 3y = 9$.
4. $x^2 - 1 = 0$ for some real number x.

Phrases such as "for each x," "for any x," and "for all x" are called **universal quantifiers.** Each of the following sentences contains a universal quantifier:

1. For each real number x, $x^2 \geq 0$.
2. If x is any real number, then $x^2 - x + 5 > 0$.
3. $x + 1 > x$ for all real numbers x.

Let $P(x)$ be a propositional function. The negation of a sentence such as "there is an x such that $P(x)$ holds" is obtained by replacing the existential quantifier by the universal quantifier and negating $P(x)$. For example, the negation of "there is an integer n such that $n + 2 = 5$" is "if n is any integer, then $n + 2 \neq 5$." The negation of a sentence such as "for each x, $P(x)$ holds" is obtained by replacing the universal quantifier by the existential quantifier and negating $P(x)$. For example, the negation of "for each integer n, $n + 1$ is an integer" is "there is an integer n such that $n + 1$ is not an integer."

Consider the statement "if n is an integer, then n is positive." This statement is obviously false because -1 is an integer and -1 is negative. We say that $n = -1$ is a

counterexample to the statement. In general, if we have a statement $P(n)$ that involves the quantifier "for each," then a counterexample to the statement is an object a such that $P(a)$ is not true.

EXERCISES A

1. Write the negation of each of the following statements.
    (a) x is a member of A and x is not a member of B.
    (b) x is a member of A or x is a member of B.
    (c) If x is a member of A, then x is a member of B.
    (d) There is a real number x such that x is negative.
    (e) For each negative integer n, $-n$ is positive.
    (f) $x^2 \geq 0$ for all real numbers x.
    (g) $-n$ is positive for some integer n.
    (h) There is a set U such that if x is any object then x belongs to U.

2. Write the converse and contrapositive of each of the following statements.
    (a) If x is a member of A, then x is a member of B.
    (b) If $x > 0$, then $x^2 > x$.
    (c) If x is an odd integer, then x is a multiple of 3 and $x > 2$.
    (d) If x is an even integer or $x > 0$, then x is prime.
    (e) If x is an even integer or $x > 0$, then x is multiple of 3 and $x > 2$.

3. Let P and Q be propositions. Construct a truth table to show that:
    (a) The conditional proposition $P \rightarrow Q$ and its converse are not equivalent.
    (b) The conditional proposition $P \rightarrow Q$ and its contrapositive are equivalent.

4. Let P and Q be propositions. Construct a truth table to show that each of the following pairs of propositions are equivalent:
    (a) $P \vee Q$ and $\neg P \rightarrow Q$
    (b) $\neg(P \rightarrow Q)$ and $P \wedge \neg Q$
    (c) $\neg(P \vee Q)$ and $\neg P \wedge \neg Q$
    (d) $\neg(P \wedge Q)$ and $\neg P \vee \neg Q$

Appendix B
Sets

We use capital letters to denote sets and lowercase letters to denote objects belonging to sets. If X is a set and x is an object that belongs to X, we say that x is an **element** of X, or that x is a **member** of X, or that x **belongs** to X and write $x \in X$. If x does not belong to X, we write $x \notin X$. The equation $X = Y$ means that X and Y are the same set; that is, X and Y consist of exactly the same objects. We say that X is a **subset** of Y, written $X \subseteq Y$, provided every member of X is also a member of Y. If $X \subseteq Y$ and $X \neq Y$, we say that X is a **proper subset** of Y. The set that has no members is called the **empty set** and is denoted by the symbol $\emptyset$. Observe that if X is any set then $X \subseteq X$ and $\emptyset \subseteq X$.

Suppose we wish to describe a set. If the set is small, one way is to list its members. If X is the set whose members are 1, 2 and 3, then we write $X = \{1, 2, 3\}$. A second method is to use the following notation, which is called **set-builder notation**. Let $X = \{n \in \mathbb{N}: n < 4\}$. The colon is shorthand for "such that," and our description reads, "X equals the sets of all n in $\mathbb{N}$ such that n is less than 4."

In this section we assume that all sets under consideration are subsets of a given set, called the **universal set.** Given two sets X and Y we may form new sets. The **union** of X and Y is the set of all objects that belong to X or to Y, and it is denoted by $X \cup Y$. In set-builder notation,

$$X \cup Y = \{x: x \in X \text{ or } x \in Y\}.$$

The **intersection** of X and Y is the set of all objects that belong to both X and Y, and it is denoted by $X \cap Y$. In set-builder notation,

$$X \cap Y = \{x: x \in X \text{ and } x \in Y\}.$$

The **complement of Y relative to X** is the set of all objects that belong to X but not to Y, and it is denoted by $X - Y$. In set-builder notation,

$$X - Y = \{x: x \in X \text{ and } x \notin Y\}.$$

The **complement of a set A** is the complement of A relative to the universal set, and it is denoted by $A^\sim$.

Figure B.1 illustrates pictorially the intersection, complement, and union.

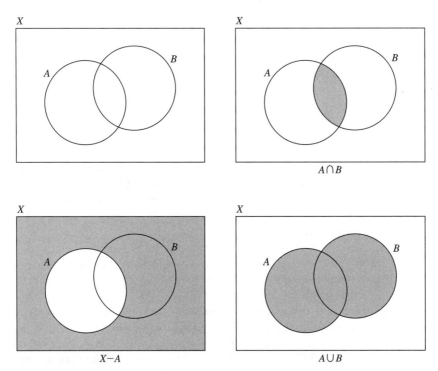

Figure B.1

EXAMPLE 1. Let $X = \{x \in \mathbb{R}: x > 1\}$ and $Y = \{x \in \mathbb{R}: 0 < x \leq 5\}$. Then $X \cup Y = \{x \in \mathbb{R}: x > 0\}$, $X \cap Y = \{x \in \mathbb{R}: 1 < x \leq 5\}$, $X - Y = \{x \in \mathbb{R}: x > 5\}$, and $Y - X = \{x \in \mathbb{R}: 0 < x \leq 1\}$.

Note that if $X, Y,$ and Z are sets, then $X \subseteq X \cup Y$, $X \cap Y \subseteq X$, $X \cup Y = Y \cup X$, $X \cap Y = Y \cap X$, $X \cup (Y \cup Z) = (X \cup Y) \cup Z$, and $X \cap (Y \cap Z) = (X \cap Y) \cap Z$.

THEOREM B.1. Let X and Y be sets. Then the following statements are equivalent: *(a)* $X \subseteq Y$ *(b)* $X \cup Y = Y$ *(c)* $X \cap Y = X$.

Proof. We prove that *(a)* and *(b)* are equivalent and leave the proof that *(a)* and *(c)* are equivalent as an exercise. Suppose $X \subseteq Y$. It is clear that $Y \subseteq X \cup Y$. Let $x \in X \cup Y$. Then $x \in X$ or $x \in Y$. Suppose $x \in X$. Since $X \subseteq Y$, $x \in Y$. Therefore $X \cup Y \subseteq Y$. Since $Y \subseteq X \cup Y$ and $X \cup Y \subseteq Y$, $X \cup Y = Y$.

Suppose $X \cup Y = Y$. Let $x \in X$. Since $X \subseteq X \cup Y$, $x \in X \cup Y$. Since $X \cup Y \subseteq Y$, $x \in Y$. Therefore $X \subseteq Y$. ∎

The following theorem says that union distributes over intersection and intersection distributes over union.

THEOREM B.2. Let X, Y, and Z be sets. Then

(a) $X \cup (Y \cap Z) = (X \cup Y) \cap (X \cup Z)$
(b) $X \cap (Y \cup Z) = (X \cap Y) \cup (X \cap Z)$.

Proof. We prove (a) and leave the proof of (b) as an exercise. Let $x \in X \cup (Y \cap Z)$. Then $x \in X$ or $x \in Y \cap Z$, so we consider two cases.
Case 1. Suppose $x \in X$. Since $X \subseteq X \cup Y$ and $X \subseteq X \cup Z$, $x \in X \cup Y$ and $x \in X \cup Z$. Therefore $x \in (X \cup Y) \cap (X \cup Z)$.
Case 2. Suppose $x \in Y \cap Z$. Then $x \in Y$ and $x \in Z$. Since $Y \subseteq X \cup Y$ and $Z \subseteq X \cup Z$, $x \in X \cup Y$ and $x \in X \cup Z$. Hence $x \in (X \cup Y) \cap (X \cup Z)$.
We have proved that $X \cup (Y \cap Z) \subseteq (X \cup Y) \cap (X \cup Z)$.
    Let $x \in (X \cup Y) \cap (X \cup Z)$. Then $x \in X \cup Y$ and $x \in X \cup Z$. So $x \in X$ or $x \in Y$, and $x \in X$ or $x \in Z$. We consider two cases.
Case 1. $x \in X$. Since $X \subseteq X \cup (Y \cap Z)$, $x \in X \cup (Y \cap Z)$.
Case 2. $x \notin X$. Then $x \in Y$ and $x \in Z$, so $x \in Y \cap Z$. Since $Y \cap Z \subseteq X \cup (Y \cap Z)$, $x \in X \cup (Y \cap Z)$.
We have proved that $(X \cup Y) \cap (X \cup Z) \subseteq X \cup (Y \cap Z)$. ∎

The proof of de Morgan's laws, as given in the following theorem, is left as exercises. These laws are named in honor of the logician Augustus de Morgan (1806–1871).

THEOREM B.3. If A, B, and X are sets, then

(a) $X - (A \cup B) = (X - A) \cap (X - B)$.
(b) $X - (A \cap B) = (X - A) \cup (X - B)$. ∎

The complementation operator may also be used to state de Morgan's laws. Let A and B be sets. Then $(A \cup B)^\sim = A^\sim \cap B^\sim$ and $(A \cap B)^\sim = A^\sim \cup B^\sim$. De Morgan's laws are easy to remember if you think of them as follows:
   The complement of the union is the intersection of the complements.
   The complement of the intersection is the union of the complements.
All you have to remember is that each law involves both a union and an intersection.

Another way of forming new sets from existing ones involves the notion of an ordered pair (a, b). Intuitively the ordered pair (a, b) is just a set in which a comes first and b comes second, and almost everyone thinks of an ordered pair in this manner. By definition, ordered pairs (a_1, b_1) and (a_2, b_2) are equal if and only if $a_1 = a_2$ and $b_1 = b_2$. Exercise 12 uses a more formal definition and asks for a proof of this fact.

Definition. Let X and Y be sets. The **Cartesian product** $X \times Y$ is the set of all ordered pairs (a, b), where $a \in X$ and $b \in Y$. That is, $X \times Y = \{(a, b): a \in X \text{ and } b \in Y\}$. ∎

Note that the plane is $\mathbb{R} \times \mathbb{R}$; that is, the plane is the set of all ordered pairs of real numbers.

We give yet another way of forming new sets from old ones. Given a set X, we form a set whose members are the subsets of X. This set is called the **power set** of X and is denoted by $\mathcal{P}(X)$. For example, if $X = \{1, 2, 3\}$, then $\mathcal{P}(X)$ has exactly eight members: $\emptyset, \{1\}, \{2\}, \{3\}, \{1, 2\}, \{1, 3\}, \{2, 3\}$, and X. Thus the members of the power set of a set are themselves sets. We use the terms **collection** and **family** in place of the term set when we wish to emphasize that the members of a given set are themselves sets.

We have already defined the union and intersection of two sets. We can also form the union and intersection of arbitrarily many sets.

Definition. Let $\mathcal{A}$ be a nonempty collection of sets. The **union**, $\bigcup \{A: A \in \mathcal{A}\}$, of the members of $\mathcal{A}$ is defined to be $\{x: x \in A \text{ for some } A \in \mathcal{A}\}$, and the **intersection**, $\bigcap \{A: A \in \mathcal{A}\}$, of the members of $\mathcal{A}$ is defined to be $\{x: x \in A \text{ for all } A \in \mathcal{A}\}$. ∎

Note that if $\mathcal{A}$ is a collection of sets, then $x \in \bigcup \{A: A \in \mathcal{A}\}$ means that there is an $A \in \mathcal{A}$ such that $x \in A$, and $x \in \bigcap \{A: A \in \mathcal{A}\}$ means that for each $A \in \mathcal{A}, x \in A$.

An alternative notation for $\bigcup \{A: A \in \mathcal{A}\}$ is $\bigcup_{A \in \mathcal{A}} A$, and similarly, $\bigcap_{A \in \mathcal{A}} A$ is an alternative notation for $\bigcap \{A: A \in \mathcal{A}\}$. Examples of arbitrary unions and intersections will be given in appendix D after we introduce the notion of an indexed family of sets.

EXERCISES B

1. Let X and Y be sets. Prove that $X \subseteq Y$ if and only if $X \cap Y = X$.

2. Let X, Y, and Z be sets. Prove that $X \cap (Y \cup Z) = (X \cap Y) \cup (X \cap Z)$.

3. Let A, B, and X be sets. Prove that $X - (A \cup B) = (X - A) \cap (X - B)$.

4. Let A, B, and X be sets. Prove that $X - (A \cap B) = (X - A) \cup (X - B)$.

5. Let X, Y, and Z be sets. Prove that $X \times (Y \cup Z) = (X \times Y) \cup (X \times Z)$.

6. Let W, X, Y, and Z be sets. Prove that $(W \times X) \cup (Y \times Z) \subseteq (W \cup Y) \times (X \cup Z)$.

7. Find sets W, X, Y, and Z such that $(W \times X) \cup (Y \times Z) \neq (W \cup Y) \times (X \cup Z)$.

8. Give a proof of, or a counterexample to, each of the following statements.
   (a) For any three sets X, Y, and Z, $X \times (Y \cap Z) = (X \times Y) \cap (X \times Z)$.
   (b) For any four sets W, X, Y, and Z, $(W \times X) \cap (Y \times Z) = (W \cap Y) \times (X \cap Z)$.
   (c) For any four sets W, X, Y, and Z, $(W \times X) - (Y \times Z) = (W - Y) \times (X - Z)$.

9. Let $\mathcal{A}$ be a nonempty collection of sets. Determine the truth value of each of the following statements:
   (a) If $x \in \bigcup \{A: A \in \mathcal{A}\}$, then $x \in A$ for some $A \in \mathcal{A}$.
   (b) If $x \in \bigcup \{A: A \in \mathcal{A}\}$, then $x \in A$ for all $A \in \mathcal{A}$.
   (c) If $x \in \bigcap \{A: A \in \mathcal{A}\}$, then $x \in A$ for some $A \in \mathcal{A}$.
   (d) If $x \in \bigcap \{A: A \in \mathcal{A}\}$, then $x \in A$ for all $A \in \mathcal{A}$.

10. Determine the truth value of the converse of each of the statements in Exercise 9.

11. Find $\mathcal{P}(X)$ when:
    (a) $X = \{1, 2\}$ (b) $X = \emptyset$ (c) $X = \{1, 2, 3, 4\}$.

12. Define the order pair (a, b) to be $\{\{a\}, \{a, b\}\}$. Use this definition to prove that $(a_1, b_1) = (a_2, b_2)$ if and only if $a_1 = a_2$ and $b_1 = b_2$.

Appendix C
Functions

This appendix provides a summary of the study of functions.

Definition. A **function** f from a set X into a set Y is a rule that assigns to each member x of X a unique member $y = f(x)$ of Y. The set X is called the **domain** of f, and the set Y is called the **codomain** of f. ∎

We use the notation $f: X \to Y$ to denote that f is a function with domain X and codomain Y, and $f: X \to Y$ is read, "f is a function from X to Y" or "f maps X into Y."

Definition. Suppose $f: X \to Y$ and $x \in X$. Then $f(x)$ is called the **value** of f at x or the **image** of x under f, and the **range** of f is $\{f(x): x \in X\}$. ∎

EXAMPLE 1. Let $r = \{(x, x^3 - x): x \in \mathbb{R}\}$, and let $f = (r, \mathbb{R})$. Then $f: \mathbb{R} \to \mathbb{R}$ is the function such that $f(x) = x^3 - x$ for each $x \in \mathbb{R}$. With no loss of rigor we can simply say, define $f: \mathbb{R} \to \mathbb{R}$ by $f(x) = x^3 - x$ for each $x \in \mathbb{R}$.

Example 1 explains the terminology that is most often used in telling someone the function we are thinking of.

Let $f: X \to Y$ be a function, let $A \subseteq X$, and let $B \subseteq Y$. We define the set $f(A)$ by $f(A) = \{y \in Y: \text{there is an } x \in A \text{ such that } f(x) = y\}$. The set $f(A)$ is called the **image** of A under f. Note that $f(X)$ is the range of f. We define the set $f^{-1}(B)$ by $f^{-1}(B) = \{x \in X: f(x) \in B\}$. The set $f^{-1}(B)$ is called the **inverse image** of B under f. Note that $f^{-1}(Y) = X$ and that $f^{-1}(B)$ may be empty. For example, if $f: \mathbb{R} \to \mathbb{R}$ is the function defined by $f(x) = x^2$ for each $x \in \mathbb{R}$ and $B = \{x \in \mathbb{R}: x < 0\}$, then $f^{-1}(B) = \emptyset$.

Let $f: X \to Y$ be a function. As the following theorem indicates, the operation f^{-1}, when applied to subsets of Y, behaves very nicely. In particular, it preserves inclusions, unions, intersections, and differences. However the operation f, when applied to subsets of X, does not preserve all these set operations (see Exercises 2 and 3).

THEOREM C.1. Let $f: X \to Y$ be a function, let A and B be subsets of X and let C and D be subsets of Y. Then the following conditions hold:

(a) If $A \subseteq B$, then $f(A) \subseteq f(B)$.

(b) If $C \subseteq D$, then $f^{-1}(C) \subseteq f^{-1}(D)$.

(c) $f(A \cup B) = f(A) \cup f(B)$.

(d) $f^{-1}(C \cup D) = f^{-1}(C) \cup f^{-1}(D)$.

(e) $f(A \cap B) \subseteq f(A) \cap f(B)$.

(f) $f^{-1}(C \cap D) = f^{-1}(C) \cap f^{-1}(D)$.

(g) $f(A) - f(B) \subseteq f(A - B)$.

(h) $f^{-1}(C - D) = f^{-1}(C) - f^{-1}(D)$.

(i) $f(X - f^{-1}(Y - C)) \subseteq C$.

Proof. We prove *(c)* and *(f)* and leave the remaining proofs as exercises.

(c) Since $A \subseteq A \cup B$ and $B \subseteq A \cup B$, by part *(a)*, $f(A) \subseteq f(A \cup B)$ and $f(B) \subseteq f(A \cup B)$. Therefore $f(A) \cup f(B) \subseteq f(A \cup B)$. In order to prove the reverse inclusion, let $y \in f(A \cup B)$. Then there exists $x \in A \cup B$ such that $f(x) = y$. So $x \in A$ or $x \in B$. If $x \in A$, then $f(x) \in f(A)$. If $x \in B$, then $f(x) \in f(B)$. Since $y = f(x)$, $y \in f(A) \cup f(B)$.

(f) Since $C \cap D \subseteq C$ and $C \cap D \subseteq D$, by part *(b)*, $f^{-1}(C \cap D) \subseteq f^{-1}(C)$ and $f^{-1}(C \cap D) \subseteq f^{-1}(D)$. Therefore $f^{-1}(C \cap D) \subseteq f^{-1}(C) \cap f^{-1}(D)$. Let $x \in f^{-1}(C) \cap f^{-1}(D)$. Then $x \in f^{-1}(C)$ and $x \in f^{-1}(D)$. Thus $f(x) \in C$ and $f(x) \in D$. Therefore $f(x) \in C \cap D$ and hence $x \in f^{-1}(C \cap D)$. Thus $f^{-1}(C) \cap f^{-1}(D) \subseteq f^{-1}(C \cap D)$. ∎

Definition. The **identity function** from a set X into itself is the function $i_X: X \to X$ defined by $i_X(x) = x$ for each $x \in X$. ∎

Definition. Let $f: X \to Y$ be a function and let A be a subset of X. The function $f|_A: A \to Y$ defined by $f|_A(a) = f(a)$ for each $a \in A$ is called the **restriction of f to A**. ∎

Definition. Let X and Y be sets, let $A \subseteq X$, and let $f: A \to Y$ be a function. A function $g: X \to Y$ with the property that $g(a) = f(a)$ for each $a \in A$ is called an **extension** of f. ∎

Note that if X and Y are sets, A is a proper subset of X, and Y has more than one member, then a function $f: A \to Y$ has more than one extension.

EXAMPLE 2. Let A be the closed unit interval $[0, 1]$, let $Y = \{0, 1\}$, and let $f: A \to Y$ be the function defined by $f(a) = 0$ for each $a \in A$. Then the function $g: \mathbb{R} \to Y$ defined by $g(x) = 0$ for each $x \in \mathbb{R}$ is an extension of f, and the function $h: \mathbb{R} \to Y$ defined by $h(x) = 0$ for each $x \in [0, 1]$ and $h(x) = 1$ for each $x \in \mathbb{R} - [0, 1]$ is also an extension of f.

Definition. Let $f: X \to Y$ and $g: Y \to Z$ be functions. The function $g \circ f: X \to Z$ defined by $g \circ f(x) = g(f(x))$ is called the **composition** of f and g. ∎

The composition of two functions is illustrated in Figure C.1.

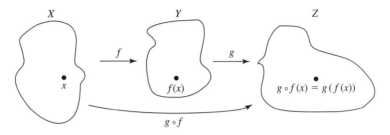

Figure C.1

Definition. A function $f: X \to Y$ is said to be **injective** or **one-to-one** provided that for each pair a, b of distinct members of X, $f(a) \neq f(b)$. An injective function is called an **injection**. ∎

Note that a function $f: X \to Y$ is an injection provided that if $a, b \in X$ and $f(a) = f(b)$, then $a = b$.

EXAMPLE 3. The function $f: \mathbb{R} \to \mathbb{R}$ defined by $f(x) = x^2$ for each $x \in \mathbb{R}$ is *not* one-to-one because $f(-1) = f(1)$. The function $g: \mathbb{R} \to \mathbb{R}$ defined by $g(x) = 2x + 3$ is injective because if $g(x) = g(y)$, then $2x + 3 = 2y + 3$ and hence $x = y$ (see Figure C.2)

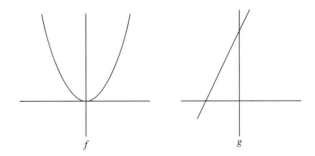

Figure C.2

Definition. A function $f: X \to Y$ is said to be **surjective** or to **map X onto Y** provided that the range of f is Y. A surjective function is called a **surjection**. A function $f: X \to Y$ is said to be **bijective** or a **one-to-one correspondence** provided f is injective and surjective. A bijective function is called a **bijection**. ∎

Note that a function $f: X \to Y$ is surjective provided that for each $y \in Y$ there is an $x \in X$ such that $f(x) = y$. The function f in Example 3 is not surjective, whereas the function g in Example 3 is surjective. In Exercises 5 and 6 you are asked to provide examples to show that injections and surjections are independent concepts.

If $f: X \to Y$ is a bijection, then there exists a function $f^{-1}: Y \to X$ called the **inverse** of f. The function f^{-1} is defined by letting $f^{-1}(y)$ be the unique element x of X such that $f(x) = y$. Given $y \in Y$ and the fact that f is a surjection, there is an element x of X such that $f(x) = y$. Since f is an injection, there is only one such element x.

Observe that given a function $f: X \to Y$, we have an operation f^{-1} that is defined on the subsets of Y. If f is a bijection, we also have a function f^{-1} that maps Y into X. It is easy to see that if f is a bijection, then f^{-1} is also a bijection.

The proof of the following theorems are left as exercises.

THEOREM C.2. Let $f: X \to Y$ and $g: Y \to Z$ be functions.

(a) If $g \circ f$ maps X onto Z, then g maps Y onto Z.

(b) If $g \circ f$ is one-to-one, then so is f. ∎

THEOREM C.3. Let $f: X \to Y$ be a function. If $g: Y \to X$ and $h: Y \to X$ are functions such that $g \circ f = i_X$ and $f \circ h = i_Y$, then f is a bijection and $f^{-1} = g = h$. ∎

THEOREM C.4. Let $f: X \to Y$ and $g: Y \to Z$ be functions.

(a) If f maps X onto Y and g maps Y onto Z, then $g \circ f$ maps X onto Z.

(b) If f and g are injections, then $g \circ f$ is an injection. ∎

THEOREM C.5. If $f, g: X \to X$ are functions, then $(f \circ g)^{-1} = g^{-1} \circ f^{-1}$. ∎

THEOREM C.6. Let $f: X \to Y$ be a function, let $A \subseteq X$, and let $B \subseteq Y$.

(a) $A \subseteq f^{-1}(f(A))$.

(b) If f is an injection, then $A = f^{-1}(f(A))$.

(c) $f(f^{-1}(B)) \subseteq B$.

(d) If f is a surjection, then $f(f^{-1}(B)) = B$.

(e) $f^{-1}(Y - B) = X - f^{-1}(B)$.

(f) $f(A \cap f^{-1}(B)) = f(A) \cap f(f^{-1}(B)) = f(A) \cap B$. ∎

Definition. Let X and Y be the sets. The functions $\pi_1: X \times Y \to X$ and $\pi_2: X \times Y \to Y$ defined by $\pi_1((x, y)) = x$ and $\pi_2((x, y)) = y$ are called **projections**. ■

THEOREM C.7. Let $\pi_1: X \times Y \to X$ and $\pi_2: X \times Y \to Y$ be projections. If $A \subseteq X$ and $B \subseteq Y$, then:

(a) $\pi_1^{-1}(A) = A \times Y$.

(b) $\pi_2^{-1}(B) = X \times B$.

(c) $A \times B = \pi_1^{-1}(A) \cap \pi_2^{-1}(B)$. ■

Definition. A **sequence** in a set X is a function $\mathbf{x}: \mathbb{N} \to X$. The value of $\mathbf{x}$ at n is denoted by x_n rather than $x(n)$, and $\mathbf{x}$ itself is often denoted by one of $\{x_n\}_{n \in \mathbb{N}}$, $\{x_n\}_{n=1}^{\infty}$, $\langle x_n \rangle$, or $(x_1, x_2, \ldots)$. ■

Definition. Let $x = \langle x_n \rangle$ be a sequence in X, and let $g: \mathbb{N} \to \mathbb{N}$ be a sequence such that $g(n + 1) > g(n)$ for each $n \in \mathbb{N}$. Then $\langle (x \circ g)_n \rangle$ is a **subsequence** of $\langle x_n \rangle$. The notation $\langle x_{i_n} \rangle$ is often used to denote a subsequence of $\langle x_n \rangle$. ■

EXAMPLE 4. Define $\mathbf{x}: \mathbb{N} \to \mathbb{R}$ by $x_n = 1/2^n$. Then $\mathbf{x}$ is the sequence $(1/2, 1/4, 1/8, \ldots, 1/2^n, \ldots)$. We can define a subsequence of $\mathbf{x}$ by defining the function $g: \mathbb{N} \to \mathbb{N}$ by $g(n) = 2n$. Then $\mathbf{x} \circ g$ is the sequence $(1/4, 1/16, 1/64, \ldots, 1/2^{2n}, \ldots)$. Hence $\langle (\mathbf{x} \circ g)_n \rangle$ is the subsequence of $\langle x_n \rangle$ that consists of the even terms. Using the alternative notation for the subsequence, $x_{i_1} = \frac{1}{4}$, $x_{i_2} = \frac{1}{16}$, $x_{i_3} = \frac{1}{64}, \ldots$.

In summary, a sequence is simply a function whose domain is $\mathbb{N}$. There is also the concept of a **finite sequence**. It is a function whose domain is $\{m \in \mathbb{N}: m \leq n\}$ for some natural number n.

EXERCISES C

1. Let $f: X \to Y$ be a function and let A and B be subsets of X. Prove that:

    (a) If $A \subseteq B$, then $f(A) \subseteq f(B)$.

    (b) $f(A \cap B) \subseteq f(A) \cap f(B)$.

    (c) $f(A) - f(B) \subseteq f(A - B)$.

2. Give an example of a function $f: X \to Y$ and subsets A and B of X such that $f(A \cap B) \neq f(A) \cap f(B)$.

3. Give an example of a function $f: X \to Y$ and subsets A and B of X such that $f(A) - f(B) \neq f(A - B)$.

4. Let $f: X \to Y$ be a function and let C and D be subsets of Y. Prove that:
   (a) If $C \subseteq D$, then $f^{-1}(C) \subseteq f^{-1}(D)$.
   (b) $f^{-1}(C \cup D) = f^{-1}(C) \cup f^{-1}(D)$.
   (c) $f^{-1}(C - D) = f^{-1}(C) - f^{-1}(D)$.
   (d) $f(X - f^{-1}(Y - C)) \subseteq C$.

5. Give an example of a function $f: X \to Y$ that is injective but not surjective.

6. Give an example of a function $f: X \to Y$ that is surjective but not injective.

7. Show that the identity function on a set X is a bijection.

8. Let $f: X \to Y$ and $g: Y \to Z$ be functions. Prove that:
   (a) If $g \circ f$ maps X onto Z, then g maps Y onto Z.
   (b) If $g \circ f$ is an injection, then so is f.
   (c) If $Z = X$, $f \circ g = i_Y$ and $g \circ f = i_X$, then $g = f^{-1}$ and $f = g^{-1}$.

9. Let $f: X \to Y$ be a function. Prove that if $g: Y \to X$ and $h: Y \to X$ are functions such that $g \circ f = i_X$ and $f \circ h = i_Y$, then f is a bijection and $f^{-1} = g = h$.

10. Let $f: X \to Y$ and $g: Y \to Z$ be functions. Prove that:
    (a) If f and g are surjections, then $g \circ f$ is a surjection.
    (b) If f and g are injections, then $g \circ f$ is a injection.

11. Give an example of a function $f: X \to Y$ with the property that there is a function $g: Y \to X$ such that $f \circ g = i_Y$ but there is no function $h: Y \to X$ such that $h \circ f = i_X$.

12. Give an example of a function $f: X \to Y$ with the property that there is a function $g: Y \to X$ such that $g \circ f = i_X$ but there is no function $h: Y \to X$ such that $f \circ h = i_Y$.

13. Prove that if f and g are functions that map X into X, then $(f \circ g)^{-1} = g^{-1} \circ f^{-1}$.

14. Give an example of a set X and functions f and g that map X into X such that $(f \circ g)^{-1} \neq f^{-1} \circ g^{-1}$.

15. Give an example to show that the converse of the result in Exercise 8 *(a)* is not true.

16. Give an example to show that the converse of the result in exercise 8 *(b)* is not true.

17. Let $\langle x_n \rangle$ be the sequence in $\mathbb{R}$ defined by $x_n = (n^2 - 2n)/5n$ for each $n \in \mathbb{N}$ and define $g: \mathbb{N} \to \mathbb{N}$ by $g(n) = 3n - 1$.

    (a) Find the first, second, third, and sixth terms of the subsequence $\langle (x \circ g)_n \rangle$.

    (b) What is the formula for the nth term of the subsequence $\langle (x \circ g)_n \rangle$?

18. Prove Theorem C.6.

19. Let $\pi_1: X \times Y \to X$ and $\pi_2: X \times Y \to Y$ be projections, let $A \subseteq X$ and let $B \subseteq Y$. Prove that $\pi_1^{-1}(A) = A \times Y$ and $\pi_2^{-1}(B) = X \times B$.

20. Let $\pi_1: X \times Y \to X$ and $\pi_2: X \times Y \to Y$ be projections, let $A \subseteq X$ and $B \subseteq Y$. Prove that $A \times B = \pi_1^{-1}(A) \cap \pi_2^{-1}(B)$.

21. Let $W, X, Y,$ and Z be sets, and let $f: W \to X$, $g: X \to Y$, and $h: Y \to Z$ be functions. Prove that $h \circ (g \circ f) = (h \circ g) \circ f$.

Appendix D
Indexing Sets and Cartesian Products

We have defined the Cartesian product of two sets. The primary purpose of this section is to define the Cartesian product of an arbitrary collection of sets. We do this in stages, beginning with the definition of the Cartesian product of m sets, where m is a natural number. But first we give a precise definition of an indexed family of sets.

Definition. Let $\mathscr{A}$ be a nonempty family of sets. An **indexing function** for $\mathscr{A}$ is a function f that maps a set Λ onto $\mathscr{A}$. The set Λ is called an **indexing set for** $\mathscr{A}$, or simply an **index set**. The family $\mathscr{A}$ together with the indexing function f is called an **indexed family of sets.** ∎

Given an element α of Λ, we denote the set $f(\alpha)$ by A_α. We denote the indexed family of sets by $\{A_\alpha : \alpha \in \Lambda\}$, and we say "the set of all A_α such that α belongs to Λ." When the index set is clear or is unimportant to the discussion, we often write $\{A_\alpha\}$.

Note that an indexing function is required to be a surjection. However, it is not required to be an injection. Thus there may be distinct members α and β of Λ such that A_α and A_β are the same member of $\mathscr{A}$. This occurs, for example, in the definition of $\mathbb{R}^m$.

Indexing functions provide another notation for unions and intersections of an arbitrary nonempty collection of sets. Suppose $\mathscr{A}$ is a nonempty collection of sets and $f: \Lambda \to \mathscr{A}$ is an indexing for $\mathscr{A}$. Then we denote $\bigcup \{A: A \in \mathscr{A}\}$ by $\bigcup_{\alpha \in \Lambda} A_\alpha$ or $\bigcup \{A_\alpha : \alpha \in \Lambda\}$, and we denote $\bigcap \{A: A \in \mathscr{A}\}$ by $\bigcap_{\alpha \in \Lambda} A_\alpha$ or $\bigcap \{A_\alpha : \alpha \in \Lambda\}$. Thus

$$\bigcup \{A_\alpha : \alpha \in \Lambda\} = \{x: x \in A_\alpha \text{ for some } \alpha \in \Lambda\}$$

and

$$\bigcap \{A_\alpha : \alpha \in \Lambda\} = \{x: x \in A_\alpha \text{ for all } \alpha \in \Lambda\}.$$

EXAMPLE 1. For each $n \in \mathbb{N}$, let $A_n = (-\frac{1}{n}, \frac{1}{n})$. Find $\bigcap \{A_n : n \in \mathbb{N}\}$ and $\bigcup \{A_n : n \in \mathbb{N}\}$.

Analysis. Note that the index set is the set of natural numbers and the indexing function is the function with domain $\mathbb{N}$ whose value at n is the open interval $(-\frac{1}{n}, \frac{1}{n})$.

Observe that $A_1 = (-1, 1)$, $A_2 = (-\frac{1}{2}, \frac{1}{2})$, ... The only real number that belongs to each of $\{A_n : n \in \mathbb{N}\}$ is 0. Therefore $\bigcap \{A_n : n \in \mathbb{N}\} = \{0\}$. Since $A_n \subseteq A_1$ for each $n \in \mathbb{N}$, $\bigcup \{A_n : n \in \mathbb{N}\} = (-1, 1)$. ∎

EXAMPLE 2. For each real number x, let $A_x = \{y \in \mathbb{R} : y \geq x\}$. Then $\bigcup_{x \in \mathbb{R}} A_x = \mathbb{R}$ and $\bigcap_{x \in \mathbb{R}} A_x = \emptyset$.

Observe that each nonempty family $\mathcal{A}$ of sets can be "self-indexed" by using the set $\mathcal{A}$ as the indexing set and assigning to each member of $\mathcal{A}$ the set it represents. As we shall see, it is certainly convenient to know this fact, and we give a formal proof.

THEOREM D.1. Let $\mathcal{A}$ be a nonempty family of sets. Then there is an index set Λ such that $\mathcal{A} = \{B_\alpha : \alpha \in \Lambda\}$.

Proof. Let $\Lambda = \mathcal{A}$, and for each $A \in \Lambda$, let $B_A = A$. We prove that $\mathcal{A} = \{B_\alpha : \alpha \in \Lambda\}$. Let $A \in \mathcal{A}$. Then $A \in \Lambda$ and $A = B_A$. Then $A \in \{B_\alpha : \alpha \in \Lambda\}$. Therefore $\mathcal{A} \subseteq \{B_\alpha : \alpha \in \Lambda\}$. Let $B \in \{B_\alpha : \alpha \in \Lambda\}$. Then there is an $\alpha \in \Lambda$ such that $B = B_\alpha$. But $B_\alpha = \alpha$ and $\alpha \in \Lambda = \mathcal{A}$. Hence $B \in \mathcal{A}$ and so $\{B_\alpha : \alpha \in \Lambda\} \subseteq \mathcal{A}$. ∎

Among the formulas that can be extended to indexed families are the two distributive laws and de Morgan's laws. We leave some of these extensions as exercise.

THEOREM D.2. Let $\{X_\alpha : \alpha \in \Lambda\}$ be a collection of subsets of a set X and let Y be a set. Then:

(a) $\bigcup_{\alpha \in \Lambda}(X_\alpha \cap Y) = (\bigcup_{\alpha \in \Lambda} X_\alpha) \cap Y$.

(b) $\bigcap_{\alpha \in \Lambda}(X_\alpha \cup Y) = (\bigcap_{\alpha \in \Lambda} X_\alpha) \cup Y$.

(c) $\bigcup_{\alpha \in \Lambda}(X - X_\alpha) = X - \bigcap_{\alpha \in \Lambda} X_\alpha$.

(d) $\bigcap_{\alpha \in \Lambda}(X - X_\alpha) = X - \bigcup_{\alpha \in \Lambda} X_\alpha$.

Proof. The proofs of (a) and (c) are typical.

(a) Let $x \in \bigcup_{\alpha \in \Lambda}(X_\alpha \cap Y)$. Then there exists $\alpha \in \Lambda$ such that $x \in X_\alpha \cap Y$. Hence $x \in X_\alpha$ for some $\alpha \in \Lambda$, and $x \in Y$. Therefore $x \in \bigcup_{\alpha \in \Lambda} X_\alpha$ and $x \in Y$. So $x \in (\bigcup_{\alpha \in \Lambda} X_\alpha) \cap Y$ and hence $\bigcup_{\alpha \in \Lambda}(X_\alpha \cap Y) \subseteq (\bigcup_{\alpha \in \Lambda} X_\alpha) \cap Y$. The remaining containment is established in a similar manner.

(c) Let $x \in \bigcup_{\alpha \in \Lambda}(X - X_\alpha)$. Then there exists $\alpha \in \Lambda$ such that $x \in X - X_\alpha$. Hence $x \in X$ and $x \notin X_\alpha$ for some $\alpha \in \Lambda$. So $x \in X$ and $x \notin \bigcap_{\alpha \in \Lambda} X_\alpha$. Therefore $x \in X - \bigcap_{\alpha \in \Lambda} X_\alpha$, and hence $\bigcup_{\alpha \in \Lambda}(X - X_\alpha) \subseteq X - \bigcap_{\alpha \in \Lambda} X_\alpha$. The remaining containment is established in a similar manner. ∎

We turn to the Cartesian product of a family of sets. For each natural number m, let $I_m = \{i \in \mathbb{N} : i \leq m\}$. Given a set X, an m-tuple ($m \in \mathbb{N}$) of members of X is a function $x: I_m \to X$. Note that an **m-tuple** is a finite sequence. The m-tuple **x** is often denoted by $(x_1, x_2, ..., x_m)$. We define the Cartesian product of m sets ($m \in \mathbb{N}$) by using I_m as an index set for a collection of m sets. If $\{X_i: i \in I_m\}$ is a collection of m sets, we often denote $\bigcup \{X_i: i \in I_m\}$ by $\bigcup_{i=1}^{m} X_i$ and $\bigcap \{X_i: i \in I_m\}$ by $\bigcap_{i=1}^{m} X_i$.

Definition. Let $m \in \mathbb{N}$, let $\{X_i: i \in I_m\}$ be a collection of m sets, and let $X = \bigcup_{i=1}^{m} X_i$. The **Cartesian product** of $\{X_i: i \in I_m\}$, denoted by $\prod_{i=1}^{m} X_i$ or $X_1 \times X_2 \times \cdots \times X_m$, is the set of all m-tuples $(x_1, x_2, ..., x_m)$ of members of X such that $x_i \in X_i$ for each $i \in I_m$. ∎

Suppose that $m \in \mathbb{N}$, $\{X_i: i \in I_m\}$ is a collection of m sets, and $x = (x_1, x_2, ..., x_m) \in \prod_{i=1}^{m} X_i$. Then for each $i \in I_m$, $x_i \in X_i$, and so x can be thought of as a function $x: I_m \to \bigcup_{i=1}^{m} X_i$ such that $x(i) = x_i \in X_i$ for each $i \in I_m$.

Note that if $X, Y,$ and Z are sets, then we have the Cartesian products $X \times Y \times Z$, $(X \times Y) \times Z$, and $X \times (Y \times Z)$. These three Cartesian products are distinct sets. However, there is a one-to-one correspondence between each pair of these sets. These ideas are explored in the exercises.

We let $\mathbb{R}^m$ denote the set of all m-tuples of real numbers. That is, $\mathbb{R}^m$ is the Cartesian product of $\{X_i: i \in I_m\}$, where $X_i = \mathbb{R}$ for each $i \in I_m$. The set $\mathbb{R}^m$ is called **Euclidean m-space**, and $\mathbb{R}^2$ is often called the **Euclidean plane**.

If X is a set, a sequence in X is also called an **ω-tuple** of the members of X. If $\{X_i: i \in \mathbb{N}\}$ is a collection of sets indexed by $\mathbb{N}$, we often denote $\bigcup \{X_i: i \in \mathbb{N}\}$ by $\bigcup_{i=1}^{\infty} X_i$ and $\bigcap \{X_i: i \in \mathbb{N}\}$ by $\bigcap_{i=1}^{\infty} X_i$.

Definition. Let $\{X_i: i \in \mathbb{N}\}$ be a collection of sets and let $X = \bigcup_{i=1}^{\infty} X_i$. The **Cartesian product** of $\{X_i: i \in \mathbb{N}\}$, denoted by $\prod_{i=1}^{\infty} X_i$ or $\prod_{i \in \mathbb{N}} X_i$, is the set of all ω-tuples $\langle x_n \rangle$ of members of X such that $x_i \in X_i$ for each $i \in \mathbb{N}$. ∎

The Cartesian product of $\{X_i: i \in \mathbb{N}\}$, where $X_i = \mathbb{R}$ for each $i \in \mathbb{N}$, is often called **infinite-dimensional Euclidean space**, and it is denoted by $\mathbb{R}^\omega$. Thus a member of $\mathbb{R}^\omega$ is an ω-tuple $\langle x_n \rangle$, where $x_i \in \mathbb{R}$ for each $i \in \mathbb{N}$; that is, each member of $\mathbb{R}^\omega$ is a sequence in $\mathbb{R}$.

Let Λ and X be nonempty sets. An **Λ-tuple** of members of X is a function $x: \Lambda \to X$. As with m-tuples and ω-tuples, we often denote the value of **x** at α by x_α rather than $x(\alpha)$. The Λ-tuple **x** is often denoted by $\langle x_\alpha \rangle_{\alpha \in \Lambda}$.

Definition. Let $\{X_\alpha: \alpha \in \Lambda\}$ be a collection sets and let $X = \bigcup \{X_\alpha: \alpha \in \Lambda\}$. The **Cartesian product** of $\{X_\alpha: \alpha \in \Lambda\}$, denoted by $\prod_{\alpha \in \Lambda} X_\alpha$, or simply $\prod X_\alpha$, is the set of all Λ-tuples $\langle x_\alpha \rangle_{\alpha \in \Lambda}$ of members of X such that $x_\alpha \in X_\alpha$ for each $\alpha \in \Lambda$. ∎

EXAMPLE 3. Let $\mathbb{R}^+$ denote the set of positive real numbers and for each $a \in \mathbb{R}^+$, let $X_\alpha = [0, a]$. Then a member of $\prod_{a \in \mathbb{R}^+} X_a$ is a function $x: \mathbb{R}^+ \to \bigcup_{a \in \mathbb{R}^+} X_a$ such that for each $a \in \mathbb{R}^+$, $\mathbf{x}(a) = x_a \in X_a$. In particular, the function $y: \mathbb{R}^+ \to \bigcup_{a \in \mathbb{R}^+} X_a$ defined by $\mathbf{y}(a) = a/2$ is a member of $\prod_{a \in \mathbb{R}^+} X_a$.

We conclude this section by introducing notation that is used when we take the product of X with itself. If X and Λ are nonempty sets, let $X^\Lambda = \{f: \Lambda \to X\}$. Notice that we can obtain X^Λ in a different fashion. For each $\alpha \in \Lambda$, let $X_\alpha = X$. Then $\prod_{\alpha \in \Lambda} X_\alpha$ is the same as X^Λ.

EXERCISES D

1. Let X and Y be nonempty sets. Show that there is a bijection $f: X \times Y \to Y \times X$.

2. Let X, Y, and Z be nonempty sets.
   (a) Exhibit a member of $X \times (Y \times Z)$ that is not a member of $(X \times Y) \times Z$.
   (b) Show that there is a bijection $f: X \times (Y \times Z) \to (X \times Y) \times Z$.

3. Let X, Y, and Z be nonempty sets.
   (a) Exhibit a member of $X \times Y \times Z$ that is not a member of $X \times (Y \times Z)$.
   (b) Show that there is a bijection $f: X \times Y \times Z \to X \times (Y \times Z)$.

4. Let $m, n \in \mathbb{N}$ and suppose $m \leq n$.
   (a) Show that there is an injection $f: \mathbb{R}^m \to \mathbb{R}^n$.
   (b) Show that there is an injection $f: \mathbb{R}^n \to \mathbb{R}^\omega$.
   (c) Show that there is a bijection $f: \mathbb{R}^m \times \mathbb{R}^n \to \mathbb{R}^{m+n}$.
   (d) Show that there is a bijection $f: \mathbb{R}^n \times \mathbb{R}^\omega \to \mathbb{R}^\omega$.
   (e) Show that there is a bijection $f: \mathbb{R}^\omega \times \mathbb{R}^\omega \to \mathbb{R}^\omega$.

5. Let $\{X_\alpha: \alpha \in \Lambda\}$ be an indexed family of sets, and for each $\alpha \in \Lambda$ let $A_\alpha \subseteq X_\alpha$. Prove that $\prod A_\alpha \subseteq \prod X_\alpha$.

6. Let $\{X_\alpha: \alpha \in \Lambda\}$ and $\{Y_\alpha: \alpha \in \Lambda\}$ be indexed families of sets with the same index set.
   (a) Prove that $(\prod X_\alpha) \cup (\prod Y_\alpha) \subseteq \prod (X_\alpha \cup Y_\alpha)$.
   (b) Prove that $(\prod X_\alpha) \cap (\prod Y_\alpha) = \prod (X_\alpha \cap Y_\alpha)$.

7. Let $f: X \to Y$, let $\mathscr{A} = \{A_\alpha: \alpha \in \Lambda\}$ be a nonempty family of subsets of X, and let $\mathscr{B} = \{B_\beta: \beta \in \Gamma\}$ be a nonemtpty family of subsets of Y. Prove that:
   (a) $f(\bigcup_{\alpha \in \Lambda} A_\alpha) = \bigcup_{\alpha \in \Lambda} f(A_\alpha)$.
   (b) $f(\bigcap_{\alpha \in \Lambda} A_\alpha) \subseteq \bigcap_{\alpha \in \Lambda} f(A_\alpha)$.
   (c) $f^{-1}(\bigcap_{\beta \in \Gamma} B_\beta) = \bigcap_{\beta \in \Gamma} f^{-1}(B_\beta)$.
   (d) $f^{-1}(\bigcup_{\beta \in \Gamma} B_\beta) = \bigcup_{\beta \in \Gamma} f^{-1}(B_\beta)$.

8. Complete the proofs of parts (a) and (c) of Theorem D.2.

9. Prove parts (b) and (d) of Theorem D.2.

Appendix E
Equivalence Relations and Order Relations

In some ways the concept of a relation is a generalization of a function. In this appendix we define what mathematicians mean by a relation, introduce equivalence relations and order relations, and state the Axiom of Choice, the Well-Ordering Principle, and Zorn's Lemma.

Definition. Let X be a set. A **relation** R on X is a subset of $X \times X$. ∎

Thus if X is a set, every subset of $X \times X$ is a relation on X. If R is a relation on X and $(x, y) \in R$, it is customary to say that x **is related to** y and write xRy. If R is a relation on X and $A \subseteq X$, we let $R[A]$ denote $\{y \in X: (x, y) \in R \text{ for some } x \in A\}$.

EXAMPLE 1. The usual concept of *less than* on $\mathbb{R}$ determines a relation R on $\mathbb{R}$; that is, we define $R = \{(x, y) \in \mathbb{R} \times \mathbb{R}: x < y\}$. Then R is the subset of $\mathbb{R} \times \mathbb{R}$ consisting of all points in the Euclidean plane that lie above the line $y = x$.

Definition. Let R be a relation on a nonempty set X. Then:

(a) R is **reflexive** on X provided that for each $x \in X$, $(x, x) \in R$.

(b) R is **symmetric** provided that if $(x, y) \in R$ then $(y, x) \in R$.

(c) R is **transitive** provided that if $(x, y) \in R$ and $(y, z) \in R$, then $(x, z) \in R$. ∎

The relation R on $\mathbb{R}$ in Example 1 determined by less than is transitive, but it is not symmetric and it is not reflexive on $\mathbb{R}$.

Definition. Let R be a relation on a nonempty set X. If R is reflexive on X, symmetric and transitive, then R is an **equivalence relation** of X. ∎

It is customary to denote equivalence relations by symbols like $\sim$ rather than by capital letters. Using this notation, a relation $\sim$ on a nonempty set X is an equivalence relation on X provided that:

(a) If $x \in X$, then $x \sim x$.

(b) If $x \sim y$, then $y \sim x$.

(c) If $x \sim y$ and $y \sim z$, then $x \sim z$.

EXAMPLE 2. Let n be a natural number. If $a, b \in \mathbb{Z}$, we say that a **is congruent to** b **modulo** n, written $a \equiv b \pmod{n}$, provided $a - b$ is divisible by n. We leave as an exercise the proof that congruence modulo n is an equivalence relation on the set of integers.

Definition. Let $\sim$ be an equivalence relation on a nonempty set X. For each $x \in X$, we define $[x]$ to be $\{y \in X : x \sim y\}$. The set $[x]$ is called **equivalence class** of x. ∎

With respect to congruence modulo 2 on the set of integers, $[0]$ is the set of all even integers and $[1]$ is the set of all odd integers. Note that $[2] = [0]$ and $[3] = [1]$.

The proof of the following theorem is left as an exercise.

THEOREM E.1. Let $\sim$ be an equivalence relation on a nonempty set X and let $x, y \in X$.

(a) $x \in [x]$.

(b) $x \sim y$ if and only if $[x] = [y]$.

(c) $x \not\sim y$ if and only if $[x] \cap [y] = \emptyset$.

(d) $[x] = [y]$ or $[x] \cap [y] = \emptyset$. ∎

Definition. Let X be a nonempty set. A family $\mathcal{A}$ of nonempty subsets of X is a **partition** of X provided each member of X belongs to exactly one member of $\mathcal{A}$. ∎

If $\sim$ is an equivalence relation on a nonempty set X, then by Theorem E.1, the family of equivalence classes of X with respect to $\sim$ is a partition of X. Furthermore, given any partition $\mathcal{A}$ of a nonempty set X, there is exactly one equivalence relation on X from which it is derived. This is the equivalence relation $\sim$ on X, defined by saying that if $x, y \in X$ then $x \sim y$ provided some member of $\mathcal{A}$ contains both x and y. We leave the proof as an exercise.

EXAMPLE 3. Let $A = \{n \in \mathbb{Z} : n = 3q \text{ for some integer } q\}$, $B = \{n \in \mathbb{Z} : n = 3q + 1 \text{ for some integer } q\}$, and $C = \{n \in \mathbb{Z} : n = 3q + 2 \text{ for some integer } q\}$. Then $\{A, B, C\}$ is a partition of $\mathbb{Z}$. The equivalence relation on $\mathbb{Z}$ from which this partition is derived is congruence modulo 3.

We have given one example, namely $\sim$, of a symbol that is often used to denote a relation. But other symbols, such as $\leq$, can also be used. The symbol $\leq$ is commonly used to denote an "order relation."

Definition. A relation $\leq$ on a nonempty set X is **antisymmetric** provided that if $x \leq y$ and $y \leq x$ then $x = y$. ∎

Definition. A relation $\leq$ on a nonempty set X is called a **partial order** on X provided $\leq$ is reflexive on X, transitive and antisymmetric. If $\leq$ is a partial order on X, then $(X, \leq)$ is called a **partially ordered set.** ■

EXAMPLE 4. If X is a set, then $(\mathcal{P}(X), \subseteq)$ is a partially ordered set. However if X is the set of nonzero real numbers and $R = \{(x, y) \in X \times X : xy > 0\}$, then R is not a partial order on X because it is not antisymmetric.

Definition. Let $(X, \leq)$ be a partially ordered set. Two members x and y of X are said to be **comparable** provided that $x \leq y$ or $y \leq x$. If C is a subset of X with the property that any two members of C are comparable, then C is called a **chain.** If X is a chain, then we say that $\leq$ is a **linear order** and $(X, \leq)$ is a **linearly ordered set.** ■

EXAMPLE 5. The real line with its usual order $\leq$ is a linearly ordered set, whereas if X has at least two members, then $(\mathcal{P}(X), \subseteq)$ is not a linearly ordered set.

Definition. If $(X, \leq)$ is a linearly ordered set with the property that every nonempty subset has a least member, we say that $\leq$ is a **well order** on X and that $(X, \leq)$ is a **well-ordered set.** ■

EXAMPLE 6. The set of natural numbers with their usual order is a well-ordered set, whereas the set of integers with their usual order is not a well-ordered set.

Definition. The well-ordered sets $(X, \leq)$ and $(Y, \leq')$ are **order-isomorphic** if there is a bijection $f : X \to Y$ such that if $x_1, x_2, \in X$ and $x_1 \leq x_2$, then $f(x_1) \leq' f(x_2)$. The function f is called an **order-isomorphism.** ■

The following axiom is assumed by most mathematicians.

Axiom of Choice. If $\mathcal{A}$ is a nonempty family of nonempty pairwise disjoint sets, then there is a set C that contains exactly one member from each member of $\mathcal{A}$. ■

The axiom of choice certainly seems plausible. But in years past, a great deal of controversy surrounded this axiom because there are theorems that one can prove by using this axiom that some mathematicians were reluctant to accept. One such theorem is a famous result, proved by E. Zermelo in 1904, known as the Well-Ordering Principle.

Well-Ordering Principle. If X is a nonempty set, there is a linear order $\leq$ on X such that $(X, \leq)$ is a well-ordered set. ■

Zermelo's proof startled the mathematical world, and there was considerable debate as to the correctness of the proof. When the proof was carefully analyzed, however, some mathematicians rejected the Axiom of Choice because they believed Zermelo's proof but were unwilling to accept the Well-Ordering Principle. For the most part, mathematicians today accept both the Axiom of Choice and the Well-Ordering

Principle. They are in fact equivalent. An alternative form of the Axiom of Choice is the following.

Axiom of Choice. If $\mathscr{A}$ is a nonempty family of nonempty sets (not necessarily pairwise disjoint), there exists a function $c: \mathscr{A} \to \bigcup_{A \in \mathscr{A}} A$ such that $c(A) \in A$ for each $A \in \mathscr{A}$. The function c is called a **choice function** for $\mathscr{A}$. ∎

We conclude this section with one more result that is equivalent to each of the two results that we have just mentioned. But first we need additional definitions.

Definition. Let $(X, \leq)$ be a partially ordered set. An element x of X is **maximal** if there is no element y of X, different from x, such that $x \leq y$. ∎

EXAMPLE 7. If X is a set, the maximal element of the partially ordered set $(\mathscr{P}(X), \subseteq)$ is X. The set of natural numbers with their usual order is a partially ordered set that does not have a maximal element.

Definition. Let $(X, \leq)$ be a partially ordered set, and let A be a nonempty subset of X. An element x of X is an **upper bound** of A if $a \leq x$ for each $a \in A$. An element x of X is a **least upper bound** of A if x is an upper bound of A which has the property that if y is an upper bound of A then $x \leq y$. An element x of X is a **lower bound** of A if $x \leq a$ for each $a \in A$. An element x of X is a **greatest lower bound** of A if x is a lower bound of A which has the property that if y is a lower bound of A then $y \leq x$. ∎

Zorn's Lemma. If $(X, \leq)$ is a partially ordered set in which every chain has an upper bound, then X has a maximal element. ∎

Zorn's Lemma seems to have been first presented by Felix Hausdorff (1868–1924) in 1914, and it was misnamed Zorn's Lemma around 1939. It should be called the Hausdorff Maximality Principle. It is particularly useful in showing that an object exists whenever the underlying set is partially ordered and the desired object is characterized by maximality.

EXERCISES E

1. Let $n \in \mathbb{N}$. Prove that congruence modulo n is an equivalence relation on the set of integers.

2. Let $\sim$ be an equivalence relation on a nonempty set X, and let $x, y \in X$. Prove that:
   (a) $x \in [x]$.
   (b) $x \sim y$ if and only if $[x] = [y]$.
   (c) $x \not\sim y$ if and only if $[x] \cap [y] = \emptyset$.
   (d) $[x] = [y]$ or $[x] \cap [y] = \emptyset$.

3. Let $\mathcal{A}$ be a partition of a nonempty set X. Define a relation $\sim$ on X by saying that if $x, y \in X$ then $x \sim y$ provided there is a member A of $\mathcal{A}$ such that x and y belong to A.

   (a) Prove that $\sim$ is an equivalence relation on X.

   (b) Prove that $\mathcal{A} = \{[x]: x \in X\}$.

4. Let R be an equivalence relation on a set X, let A be a nonempty subset of X, and let $S = R \cap (A \times A)$. Prove that S is an equivalence relation on A.

5. Let $f: X \to Y$ be a surjective function. Define a relation $\sim$ on X by saying $x \sim y$ provided $f(x) = f(y)$.

   (a) Prove that $\sim$ is an equivalence relation on X.

   (b) Let $X/\sim$ denote the set of equivalence classes of X with respect to $\sim$. Show that there is a one-to-one function that maps $X/\sim$ onto Y.

6. Let $X = \{n \in \mathbb{N}: n$ is less than or equal to $1000\}$. Define a relation $\leq$ on X by $m \leq n$ provided m divides n. Prove that $(X, \leq)$ is a partially ordered set.

7. Give an example of a nonempty finite set X and a partial order on X that is not a linear order on X.

8. Prove that if X is a nonempty finite set and $\leq$ is a linear order on X then $\leq$ is a well-order on X.

9. Let $X = \{1, 2, 3\}$. How many chains with exactly two members are there in the partially ordered set $(\mathcal{P}(X), \subseteq)$?

10. Let (X, R) be a linearly ordered set, let $A \subseteq X$, and let $S = R \cap (A \times A)$. Prove that (A, S) is a linearly ordered set.

11. Let A be a nonempty subset of a partially ordered set $(X, \leq)$. Prove that A has at most one least upper bound and at most one greatest lower bound.

12. Let $X = \{1, 2, 3, 4, 5, 6, 7\}$.

    (a) If $\leq$ is the usual order on X, which elements are maximal?

    (b) If $\leq$ on X is defined by $x \leq y$ provided x divides y, which elements are maximal?

13. Let X be a nonempty set and let $\mathbb{R}^X$ denote the set of all real-valued functions with domain X. Define a relation $\leq$ on $\mathbb{R}^X$ by $f \leq g$ provided $f(x)$ is less than or equal to $g(x)$ for all $x \in X$.

    (a) Prove that $(\mathbb{R}^X, \leq)$ is a partially ordered set.

    (b) Under what conditions is $(\mathbb{R}^X, \leq)$ a linearly ordered set?

Appendix F
Countable Sets

The natural numbers are adequate for the purpose of counting the members of any nonempty finite set. However, in mathematics we must consider many infinite sets. For example, the set of natural numbers, the set of integers, and the set of real numbers are infinite sets. It is often important to be able to count such sets and to use the notion of one-to-one correspondences to develop a theory of infinite cardinal numbers.

Definition. Let X and Y be sets. We say that X and Y **have the same cardinality**, and write $X \sim Y$, provided there is a one-to-one function mapping X onto Y. ∎

The proof of the following theorem is left as an exercise.

THEOREM F.1. Let X, Y, and Z be sets. Then:

(a) $X \sim X$ (b) If $X \sim Y$, then $Y \sim X$ (c) If $X \sim Y$ and $Y \sim Z$, then $X \sim Z$. ∎

In the remainder of this appendix, whenever X and Y are sets and we write $X \sim Y$, we will mean that X and Y have the same cardinality. Recall that for each natural number n, $I_n = \{i \in \mathbb{N} : i \leq n\}$.

Definition. A set X is **finite** if it is empty or there is a natural number n such that $I_n \sim X$. A set that is not finite is said to be **infinite**. A set X is **countably infinite** if $\mathbb{N} \sim X$. A set X is **countable** if X is finite or countably infinite. A set that is not countable is said to be **uncountable**. ∎

THEOREM F.2. Every countably infinite set is infinite.

Proof. Let X be a countably infinite set. Then there is a bijection $f: X \to \mathbb{N}$. Suppose X is finite. Since $X \neq \emptyset$, there is a natural number n such that $I_n \sim X$. Let $g: I_n \to X$ be a bijection. Then by Theorem C.4, $f \circ g: I_n \to \mathbb{N}$ is a bijection. However, if $m \in \mathbb{N}$ and $m > \sum_{i=1}^{n} f \circ g(i)$, then $m \notin f \circ g(I_n)$. Therefore $f \circ g$ does not map I_n onto $\mathbb{N}$, and we have a contradiction. ∎

The following theorem was proved by Georg Cantor (1845–1918). The proof involves showing that an arbitrary function $f: X \to \mathcal{P}(X)$ cannot be a bijection. We assume that we have a bijection and reach a contradiction.

THEOREM F.3. If X is a set, then X and $\mathcal{P}(X)$ do not have the same cardinality.

Proof. If $X = \emptyset$, then $\mathcal{P}(X) = \{\emptyset\}$ and so X and $\mathcal{P}(X)$ do not have the same cardinality. Suppose $X \neq \emptyset$ and there is a bijection $f: X \to \mathcal{P}(X)$. Let $A = \{x \in X : x \notin f(x)\}$. Since $A \in \mathcal{P}(X)$ and f is a surjection, there is a $p \in X$ such that $f(p) = A$. Now $p \in A$ or $p \notin A$ and we consider these two cases:

Case 1. Suppose $p \in A$. Then $p \notin f(p)$, and this is contradiction since $f(p) = A$ and $p \in A$.
Case 2. Suppose $p \notin A$. Then $p \in f(p)$, and this is a contradiction since $f(p) = A$ and $p \notin A$.

In either case we have reached a contradiction. Hence X and $\mathcal{P}(X)$ do not have the same cardinality. ∎

Since $\mathcal{P}(\mathbb{N})$ is infinite, by Theorem F.3, $\mathcal{P}(\mathbb{N})$ is uncountable.

THEOREM F.4. Every subset of $\mathbb{N}$ is countable.

Proof. Let $X \subseteq \mathbb{N}$. If X is finite, then, by definition, X is countable. Suppose X is infinite. Define $f: \mathbb{N} \to X$ inductively as follows: $f(1)$ is the least member of X, and, for each $n \in \mathbb{N}$, $f(n+1)$ is the least member of $X - \{f(1), f(2), \ldots, f(n)\}$. It is clear that f is an injection.

To see that $f(\mathbb{N}) = X$, note that for all $n \in \mathbb{N}$, $f(n) \geq n$ and suppose $f(\mathbb{N}) \neq X$. Let p be the least member of $X - f(\mathbb{N})$. Then $p < f(p)$ and $p \neq 1$. Since $p \in X$ and p is less than the least member of $X - f(I_{p-1})$, $p \in f(I_{p-1}) \subseteq f(\mathbb{N})$. This is a contradiction, and therefore f is a surjection. ∎

THEOREM F.5. A nonempty set X is countable if and only if there is an injection $f: X \to \mathbb{N}$.

Proof. Suppose X is a nonempty set and there is an injection $f: X \to \mathbb{N}$. If X is finite, then, by definition, X is countable. Suppose X is infinite. Then $f(X)$ is an infinite subset of $\mathbb{N}$. The function f, considered as a function from X into $f(X)$, is a bijection. Let f^{-1} denote the inverse of $f: X \to f(X)$. Then $f^{-1}: f(X) \to X$ is a bijection. By Theorem F.4, there is a bijection $g: \mathbb{N} \to f(X)$. By Theorem C.4, $f^{-1} \circ g: \mathbb{N} \to X$ is a bijection. Therefore X is countably infinite and so countable.

Suppose X is a nonempty countable set. If X is countably infinite, then, by definition, there is a bijection $f: X \to \mathbb{N}$. Suppose X is finite. Then there is an $n \in \mathbb{N}$ such that $X \sim I_n$. Thus there is a bijection $f: X \to I_n$. Clearly this function is an injection of X into $\mathbb{N}$. ∎

Note that the proof of Theorem F.4 uses the fact that the set of natural numbers with their usual order is a well-ordered set. Using the Well-Ordering Principle, we can prove the following theorem in essentially the same way we established Theorem F.4.

THEOREM F.6. Every infinite set contains a countably infinite set.

Proof. Let X be an infinite set, and let $\leq$ be a well-order on X. Define $f: \mathbb{N} \to X$ inductively as follows: $f(1)$ is the least member of X, and, for each $n \in \mathbb{N}$, $f(n+1)$ is the least member of $X - f(I_n)$. The proof that f is one-to-one is left as an exercise. ∎

It is not difficult to prove that the union of two countable sets is countable. From this an easy argument by mathematical induction proves that the union of a finite number of countable sets is countable. We leave these proofs as exercises and prove the following important generalization.

THEOREM F.7. A countable union of countable sets is countable.

Proof. Let Λ be a nonempty countable set, and for each $\alpha \in \Lambda$, let A_α be a countable set. Let $A = \bigcup \{A_\alpha : \alpha \in \Lambda\}$. By Theorem F.5, there is an injection $f: \Lambda \to \mathbb{N}$. Let $x \in A$, let $\Lambda_x = \{\alpha \in \Lambda : x \in A_\alpha\}$, and let $n(x)$ be the least natural number that belongs to $f(\Lambda_x)$. Let α_x be the member of Λ such that $f(\alpha_x) = n(x)$. Since A_{α_x} is countable, by Theorem F.5, there is an injection $m_{\alpha_x}: A_{\alpha_x} \to \mathbb{N}$ (the Axiom of Choice allows us to choose one such m_{α_x} for each α_x). Thus for each $x \in A$, we have an $n(x)$ and an m_{α_x}. Define $g: A \to \mathbb{N}$ by $g(x) = 2^{n(x)} 3^{m_{\alpha_x}(x)}$. It suffices to show that g is an injection. Suppose $x, y \in A$ and $g(x) = g(y)$. Then $2^{n(x)} 3^{m_{\alpha_x}(x)} = 2^{n(y)} 3^{m_{\alpha_y}(y)}$. By the Fundamental Theorem of Arithmetic, $n(x) = x(y)$ and $m_{\alpha_x}(x) = m_{\alpha_y}(y)$. Since f is one-to-one and $f(\alpha_x) = n(x) = n(y) = f(\alpha_y)$, $\alpha_x = \alpha_y$. Thus $m_{\alpha_x} = m_{\alpha_y}$, and so $m_{\alpha_x}(x) = m_{\alpha_x}(y)$. Since m_{α_x} is one-to-one, $x = y$. ∎

THEOREM F.8. A finite Cartesian product of countable sets is countable.

Proof. We prove that the Cartesian product of two countable sets is countable. From our proof, it is fairly obvious how the induction proof should proceed in order to establish the theorem.

Let X and Y be nonempty countable sets. By Theorem F.5, here exist injections $f: X \to \mathbb{N}$ and $g: Y \to \mathbb{N}$. Define $h: X \times Y \to \mathbb{N}$ by $h((x, y)) = 2^{f(x)} 3^{g(y)}$. It suffices to show that h is one-to-one. Suppose $h((a, b)) = h((c, d))$. Then $2^{f(a)} 3^{g(b)} = 2^{f(c)} 3^{g(d)}$. By the Fundamental Theorem of Arithmetic, $f(a) = f(c)$ and $g(b) = g(d)$. Since f and g are injections, $a = c$ and $b = d$. Therefore h is one-to-one. ∎

As we shall see in the next appendix, it is not true that the countable Cartesian product of countable sets is countable.

EXERCISES F

1. Let X, Y, and Z be sets. Prove that:
   (a) $X \sim X$
   (b) If $X \sim Y$, then $Y \sim X$,
   (c) If $X \sim Y$ and $Y \sim Z$, then $X \sim Z$.

2. Prove that the function f defined in the proof of Theorem F.6 is an injection.
3. Let X, Y, A, and B be sets such that $X \sim Y$ and $A \sim B$ but X and A do not have the same cardinality. Prove that Y and B do not have the same cardinality.
4. Prove that $\mathbb{N} \sim \mathbb{Z}$. *Note:* This exercise is used in Appendix G.
5. Prove that the open interval $(-1, 1)$ and $\mathbb{R}$ have the same cardinality.
6. How many injections are there that map I_5 into I_6?
7. Prove that a subset of a countable set is countable. *Note:* This exercise is used in Appendix G.
8. Prove that a nonempty set X is countable if and only if there is a surjection $f: \mathbb{Z} \to X$. *Note:* This exercise is used in Appendix G.
9. Let X be an infinite set and suppose there is an injection $f: X \to \mathbb{N}$. Prove that X is countably infinite.
10. Prove that the union of two countable sets is countable.
11. Prove that the union of a finite number of countable sets is countable.
12. Prove that the set of all finite subsets of $\mathbb{N}$ is countably infinite.
13. Prove that the set of all rational numbers is countable. *Note:* This exercise is used in Appendix G.

Appendix G
Uncountable Sets

We begin this section with the following result. The proof is similar to the proof of Theorem F.3 and hence we leave it as an exercise.

THEOREM G.1. If X is any set, then there is no surjection $f: X \to \mathcal{P}(X)$. ∎

We have already observed that $\mathcal{P}(\mathbb{N})$ is uncountable. However, we illustrate how Theorem G.1 can be used to establish this result.

Corollary G.2. The set $\mathcal{P}(\mathbb{N})$ is uncountable.

Proof. Suppose $\mathcal{P}(\mathbb{N})$ is countable. By Exercise 8 in the Appendix F, there is a surjection $f: \mathbb{Z} \to \mathcal{P}(\mathbb{N})$. By Exercise 4 in Appendix F, $\mathbb{N} \sim \mathbb{Z}$. Hence there is a bijection $g: \mathbb{N} \to \mathbb{Z}$. The composite $f \circ g: \mathbb{N} \to \mathcal{P}(\mathbb{N})$ is a surjection. This contradicts Theorem G.1. ∎

The proof of the following theorem is sometimes referred to as Cantor's diagonal argument.

THEOREM G.3. The interval $(0, 1)$ and $\mathbb{N}$ do not have the same cardinality.

Proof. The proof is by contradiction. Suppose there is a bijection $f: \mathbb{N} \to (0, 1)$. Then for each $n \in \mathbb{N}, f(n) \in (0, 1)$, and thus we have

$$f(1) = 0.a_1 a_2 a_3...$$
$$f(2) = 0.b_1 b_2 b_3...$$
$$f(3) = 0.c_1 c_2 c_3....$$

We define a decimal $p = 0.p_1 p_2 p_3...$ by $p_n = 3$ if the nth digit of the decimal $f(n)$ is not 3 and $p_n = 2$ if the nth digit of $f(n)$ is 3. Then $f(m) \neq p$ for any $m \in \mathbb{N}$ because the mth digit of $f(m)$ is different from the mth digit of p. Therefore f is not a surjection and we have reached a contradiction. ∎

THEOREM G.4. The set $\mathbb{R}$ is an uncountable set.

Proof. By Exercise 7 of Appendix F, every subset of a countable set is countable. Thus it suffices to find an uncountable subset of $\mathbb{R}$. Since $\{1/n: n \in \mathbb{N}$ and $n > 1\} \subseteq (0, 1)$, $(0, 1)$ is infinite. By Theorem G.3, $(0, 1)$ is not countably infinite. Therefore $(0,1)$ is uncountable. ∎

Corollary G.5. The set of all irrational numbers is an uncountable set.

Proof. Suppose the set of all irrational numbers is countable. By Exercise 13 of Appendix F, the set of rational numbers is countable. Thus by Theorem F.7, the set $\mathbb{R}$ is countable. This contradicts Theorem G.4. ∎

By Theorem F.8, a finite Cartesian product of countable sets is countable. The following example is an example of a countable Cartesian product of finite sets that is not countable.

EXAMPLE 1. For each $n \in \mathbb{N}$, let X_n denote the two-element set $\{0,1\}$. Let $\prod_{n=1}^{\infty} X_n$. Then X is not countable.

Analysis. By Exercise 8 of Appendix F, it is sufficient to prove that if $f: \mathbb{N} \to X$ is any function, then f is not a surjection. Suppose $f: \mathbb{N} \to X$ is a function. Then for each $n \in \mathbb{N}$, $f(n)$ is an ω-tuple, and we may write $f(n)$ as $\langle x_{ni} \rangle$, where each x_{ni} is 0 or 1. For each $n \in \mathbb{N}$, let $x_n = 0$ if $x_{nn} = 1$ and $x_n = 1$ if $x_{nn} = 0$. Then $x = \langle x_n \rangle \in X$ and $f(p) \neq \langle x_n \rangle$ for any $p \in \mathbb{N}$ because the pth digit of $f(p)$ is different from the pth digit of x. Therefore f is not a surjection, and we have reached a contradiction. ∎

EXERCISES G

1. Prove that if X is any set, then there is no surjection $f: X \to \mathcal{P}(\mathbb{N})$.

2. Determine whether each of the following sets is countable. Prove your answer.

    (a) $X = \{f: f: \{0, 1\} \to \mathbb{N}\}$.

    (b) $X = \{f: f: \mathbb{N} \to \mathbb{N}\}$.

    (c) $X = \{f: f: \mathbb{N} \to \{0, 1\}\}$.

    (d) $X = \{f: f: \mathbb{N} \to \{0, 1\}$ and $f(n) = 0$ for all $n > 100\}$.

3. Prove that if X is an uncountable set and A is a countable set, then $X - A$ is an uncountable set.

4. Let $X = \{A \in \mathcal{P}(\mathbb{N}): A$ is an infinite set$\}$. Prove that X is an uncountable set.

5. If X and Y are sets, define $X \leq Y$ to mean there is a subset Y' of Y such that $X \sim Y'$. Prove that:
   (a) $X \leq X$.
   (b) If $X \leq Y$ and $Y \leq X$, then $X \sim Y$.
   (c) If $X \leq Y$ and $Y \leq Z$, then $X \leq Z$.

6. Give an example of two uncountable sets that do not have the same cardinality.

Appendix H
Ordinal and Cardinal Numbers

In this appendix we give a brief summary of the theory of ordinal and cardinal numbers. The theorems are stated without proof, and there are no exercises. We begin with the definition of an ordinal number.

Definition. An **ordinal number** is a set α with the following properties:

(1) If $x, y \in \alpha$, then either $x \in y$, $y \in x$, or $x = y$.

(2) If $y \in \alpha$ and $x \in y$, then $x \in \alpha$. ∎

Thus ordinal numbers are sets whose elements are also sets. Intuitively, an ordinal number is a chain of sets such that each set and its members belong to the "next" set.

EXAMPLE 1 Each of the following is an ordinal number.

(1) $\emptyset$
(2) $\{\emptyset\}$
(3) $\{\emptyset, \{\emptyset\}\}$
(4) $\{\emptyset, \{\emptyset\}, \{\emptyset, \{\emptyset\}\}\}$
(5) $\{\emptyset, \{\emptyset\} \{\emptyset, \{\emptyset\}\}, \{\emptyset, \{\emptyset\} \{\emptyset, \{\emptyset\}\}\}\}$

Since an ordinal number is a set, we can define a relation on each ordinal number.

Definition. Let α be an ordinal number. For any two members x and y of α, define $\leq$ on α by $x \leq y$ if and only if $x = y$ or $x \in y$. ∎

The following theorem lists some of the basic properties of ordinal numbers.

THEOREM H.1. Let α and β be ordinal numbers. Then:

(1) $(\alpha, \leq)$ is a well-ordered set.
(2) $\alpha \cup \{\alpha\}$ is an ordinal number.
(3) If $\alpha \subseteq \beta \subseteq \alpha \cup \{\alpha\}$, then $\beta = \alpha$ or $\beta = \alpha \cup \{\alpha\}$.
(4) α is a proper subset of β if and only if $\alpha \in \beta$.
(5) The least element of $(\alpha, \leq)$ is $\emptyset$.
(6) $\alpha = \{\rho: \rho$ is an ordinal number and ρ is a proper subset of $\alpha\}$.
(7) Either $\alpha \in \beta$, $\alpha = \beta$, or $\beta \in \alpha$.
(8) If $\{\alpha_i: i \in \Gamma\}$ is a nonempty set of ordinal numbers, then $\bigcup \{\alpha_i: i = \Gamma\}$ and $\bigcap \{\alpha_i: i \in \Gamma\}$ are ordinal numbers and $\bigcap \{\alpha_i: i \in \Gamma\} \in \{\alpha_i: i \in \Gamma\}$.
(9) If A is a nonempty proper subset of α, then $\bigcup A \in \alpha$ or $\bigcup A \in \alpha$.
(10) If $\{\alpha_i: i \in \mathbb{N}\}$ is a set of ordinal numbers such that $\alpha_{i+1} \leq \alpha_i$ for each $i \in \mathbb{N}$, then there exists $n \in \mathbb{N}$ such that $\alpha_n = \alpha_{n+i}$ for each $i \in \mathbb{N}$.
(11) If $(W, \leq)$ is a well-ordered set, there is a unique ordinal number α such that W and α are order isomorphic. ■

Definition. If α is an ordinal number, then $\alpha \cup \{\alpha\}$ is called **the successor ordinal** of α and is denoted by $\alpha + 1$. An ordinal number β is called **a successor ordinal** if there is an ordinal number α such that $\beta = \alpha + 1$. ■

If $\leq'$ is the usual order on $\mathbb{R}$ and $\leq = ((\mathbb{N} \cup \{0\}) \times (\mathbb{N} \cup \{0\})) \cap \leq'$, then $(\mathbb{N} \cup \{0\}, \leq)$ is a well-ordered set. Thus by Theorem H.1 part (11), $(\mathbb{N} \cup \{0\}, \leq)$ is order isomorphic to a unique ordinal, which we denote by ω_0. Under this isomorphism, 0 corresponds to $\emptyset$, 1 to $\{\emptyset\} = \emptyset \cup \{\emptyset\}$, 2 to $\{\emptyset, \{\emptyset\}\} = \{\emptyset\} \cup \{\{\emptyset\}\}$, and so on. Thus, sometimes we denote $\emptyset$ by 0, $\{\emptyset\}$ by 1, $1 \cup \{\emptyset\}$ by 2, and so on. Therefore $\omega_0 = \{0, 1, 2, 3,...\}$.

We continue with additional definitions and results about ordinal numbers.

Definition. A **limit ordinal** is a nonzero ordinal number that is not a successor ordinal. ■

THEOREM H.2. Let A be a set, let α be an ordinal number, let $f: A \to \alpha$ be a bijection, and for any two members a and b of A, define $a \leq b$ if and only if $f(a) \leq f(b)$. Then $(A, \leq)$ is a well-ordered set and f is an order isomorphism. ■

Definition. For each set A, we define **ord(A)** to be $\{\alpha: \alpha$ is an ordinal number and there is a bijection $f: A \to \alpha\}$. ■

THEOREM H.3. Let A and B be sets. Then:

(1) ord (A) is a nonempty set.

(2) ord (A) = ord (B) if and only if there is a bijection $f : A \to B$.

(3) If $\alpha \in$ ord (A), then ord (A) = ord (α).

(4) If ord (A) $\neq$ ord (B), then ord (A) $\cap$ ord (B) = $\emptyset$. ∎

Definition. If α is an ordinal number such that $\alpha = \bigcap$ ord (α), then α is called an **initial ordinal**. ∎

Note that for each set A, $\bigcap$ ord (A) $\in$ ord (A).

Definition. A **finite ordinal** is an ordinal number α such that ord (α) = $\{\alpha\}$. An **infinite ordinal** is an ordinal number that is not a finite ordinal. ∎

THEOREM H.4. Let α and β be ordinal numbers. Then:

(1) $\omega_0 = \{\rho: \rho \text{ is a finite ordinal}\}$.

(2) α is an infinite ordinal if and only if $\alpha \geq \omega_0$.

(3) If $\alpha \geq \omega_0$, then $\bigcup$ ord (α) is an initial ordinal.

(4) for each set A, $\bigcap$ ord(A) is an initial ordinal.

(5) if α and β are initial ordinals such that $\alpha < \beta$, then $\bigcup$ ord (α) $\leq \beta$.

(6) ω_0 is the first infinite, initial ordinal. ∎

THEOREM H.5. If α is an infinite, initial ordinal and $A = \{\rho: \rho \text{ is an infinite, initial ordinal and } \rho < \alpha\}$, then A is a well-ordered set and there is a unique ordinal number β such that A and β are order isomorphic. We use ω_β to denote the ordinal number α. ∎

THEOREM H.6. Let α be an infinite, initial ordinal. Then $\omega_{\alpha+1} = \bigcup$ ord(ω_α). ∎

THEOREM H.7. If α is a limit ordinal, then $\omega_\alpha = \bigcup \{\omega_\beta: \beta \text{ is an ordinal number and } \beta < \alpha\}$. ∎

THEOREM H.8. Let α and β be infinite, initial ordinals. Then $\omega_\alpha \leq \omega_\beta$ if and only if $\alpha \leq \beta$.

We now turn to the study of cardinal numbers.

Definition. If A is a finite set, let $|A| = \bigcap$ ord (A). If A is an infinite set and $\omega_\alpha = \bigcap$ ord (A), we write $\aleph_\alpha$ instead of ω_α and let $|A| = \aleph_\alpha$. The ordinal $|A|$ is called the **cardinality of** A. ∎

Note that if A is a finite set, then $|A| \in \omega_0$. Note also that $|\omega_0| = \aleph_0$. Finally, if $n \in \omega_0$, then $|n| = n$ and if $\alpha \geq \omega_0$ and $\omega_\beta = \bigcap \text{ord}(\alpha)$, then $|\alpha| = \aleph_\beta$.

Definition. The elements of ω_0 and the $\aleph_\beta$'s, where β is an ordinal number, are called **cardinal numbers**. ∎

Note that the cardinal numbers are the same as the initial ordinal numbers. The difference is that the order structure of the ordinal numbers is not being used.

If X is a countably infinite set, then we write $|X| = \aleph_0$.

Definition. A set X is said to have the **cardinality of the continuum** provided X and $\mathbb{R}$ have the same cardinality. In this case we write $|X| = \mathbf{c}$. ∎

THEOREM H.9. $2^{\aleph_0} = \mathbf{c}$, and if I is the closed unit interval $[0, 1]$, then $|I| = \mathbf{c}$. ∎

THEOREM H.10. There is an uncountable well-ordered set Ω of ordinal numbers with maximal element ω_1 having the property that if $x \in \Omega$ and $x \neq \omega_1$, then $\{y \in \Omega : y \leq x\}$ is countable. ∎

Note that ω_1 is the first uncountable ordinal and $\Omega_0 = \Omega - \{\omega_1\}$ is a set of countable ordinals.

Since Ω is a well-ordered set, it has a least element, which we denote by 1. We denote the successor ordinal of 1 by 2, the successor ordinal of 2 by 3, and so on. Thus we can regard the set of positive integers as being a subset of Ω, and ω_0 is the smallest member of Ω minus the set of positive integers. The successor of ω_0 in Ω is denoted by $\omega_0 + 1$. Continuing, we have

$$1, 2, 3, \ldots, \omega_0, \omega_0 + 1, \omega_0 + 2, \omega_0 + 3, \ldots$$

in Ω. The smallest ordinal that is larger than each of these ordinals is denoted by $2\omega_0$. Thus we obtain

$$1, 2, \ldots, \omega_0, \omega_0 + 1, \omega_0 + 2, \ldots, 2\omega_0, 2\omega_0 + 1, 2\omega_0 + 2, \ldots, n\omega_0,$$
$$n\omega_0 + 1, n\omega_0 + 2, \ldots.$$

The smallest ordinal that is larger than each of these ordinals is denoted by $(\omega_0)^2$. Continuing, we obtain $(\omega_0)^3$, $(\omega_0)^4$, and so on, and

$$1, 2, \ldots, \omega_0, \omega_0 + 1, \omega_0 + 2, \ldots, 2\omega_0, 2\omega_0 + 1,$$
$$2\omega_0 + 2, \ldots, n\omega_0, n\omega_0 + 1, n\omega_0 + 2, \ldots$$
$$(\omega_0)^2, (\omega_0)^2 + 1, (\omega_0)^2 +, \ldots, n(\omega_0)^2, n(\omega_0)^2 + 1, \ldots, (\omega_0)^3, (\omega_0)^3 + 1,$$
$$(\omega_0)^3 + 2, \ldots$$

is a countable set. The smallest ordinal that is larger than each of these ordinals is a countable ordinal, and the process continues. As the following theorem indicates, however, ω_1 can never be reached by these types of operations.

THEOREM H.11. Let X be a countable subset of Ω that does not contain ω_1. Then the least upper bound of X is less than ω_1. ∎

We conclude this section with the statements of two versions of mathematical induction and the Principle of Transfinite Induction. The Principle of Mathematical Induction is equivalent to the fact that the usual order on $\mathbb{N}$ is a well-order.

Principle of Mathematical Induction. Let S be a subset of $\mathbb{N}$ with the following properties:

(a) $1 \in S$.
(b) If $n \in S$, then $n + 1 \in S$.

Then $X = \mathbb{N}$. ∎

Second Principle of Mathematical Induction. Let S be a subset of $\mathbb{N}$ with the following properties:

(a) $1 \in S$.
(b) For each $n \in \mathbb{N}$, if $\{1, 2, ..., n\} \subseteq S$, then $n + 1 \in S$.

Then $S = \mathbb{N}$. ∎

Principle of Transfinite Induction. Let Λ be a well-ordered set, let 1 be the least element of Λ, and let S be a subset of Λ with the following properties:

(a) $1 \in S$.
(b) For each $\alpha \in \Lambda$, if $\{\beta \in \Lambda: \beta \leq \alpha\} \subseteq S$, then the successor of α (the smallest member of $\Lambda - \{x \in \Lambda: x \leq \alpha\}$) belongs to S.

Then $S = \Lambda$. ∎

Appendix I
Algebra

The purpose of this appendix is to define the algebraic concepts that are used in the text.

Definition. An **operation on a set** X is a function from $X \times X$ into X. ■

EXAMPLE 1. Let A be a nonempty set, and let X denote the set of all functions from A into A. If ∘ denotes the usual composition of functions, then ∘ is an operation of X.

Definition. Let f be an operation on a set X. A subset A of X is **closed** with respect to f it $a, b \in A$ implies that $f(a, b) \in A$. ■

Observe that if f is an operation on a set X, then X is closed with respect to f. We have introduced the term *closed* for convenience. Note that the integers are closed with respect to the operation of subtraction, whereas the positive integers are not closed with respect to this operation.

Definition. An operation · on a set X is **associative** provided that $a \cdot (b \cdot c) = (a \cdot b) \cdot c$ for all $a, b, c \in X$. ■

EXAMPLE 2. Let · be the operation on $\mathbb{N}$ defined by $m \cdot n = m^n$. Then · is not associative since $(2 \cdot 3) \cdot 2 = (2^3) \cdot 2 = 8 \cdot 2 = 8^2 = 64$ and $2 \cdot (3 \cdot 2) = 2 \cdot (3^2) = 2 \cdot 9 = 2^9 = 512$.

Definition. If · is an operation on a set X, then an element e of X is called an **identity with respect to** · provided that $e \cdot x = x \cdot e = x$ for all $x \in X$. ■

In Example 1, the identity element with respect to composition is the identity function i defined by $i(x) = x$ for all $x \in X$.

Definition. Let X be a set, let · be an operation on X, let e be an identity with respect to ·, and let $a \in X$. An element b of X is an **inverse of a relative to** · **and e** provided that $a \cdot b = b \cdot a = e$. ■

EXAMPLE 3. Let X be the set of all functions from $\mathbb{R}$ into $\mathbb{R}$, and define $\cdot$ on X to be the usual sum of two functions; that is, $f \cdot g = h$, where $h(x) = f(x) + g(x)$ for all $x \in \mathbb{R}$. Then $\cdot$ is an operation on X, and the identity element with respect to $\cdot$ is the constant function c defined by $c(x) = 0$ for all $x \in \mathbb{R}$. If $f \in X$, then the function g defined by $g(x) = -f(x)$, for all $x \in \mathbb{R}$, is an inverse of f.

Definition. An operation $\cdot$ on a set X is **commutative** provided that $a \cdot b = b \cdot a$ for all $a, b \in X$. ∎

The operation defined in Example 3 is commutative, but the operation defined in Example 1 is not commutative.

Definition. A **group** $(G, \cdot)$ is a nonempty set G with an associative operation $\cdot$ on G such that:

(a) there is an identity element e of G with respect to $\cdot$, and

(b) for each $a \in G$, there is an inverse of a relative to $\cdot$ and e. ∎

In Exercise 1, you are asked to show that the identity element of a group is unique and that each element of a group has a unique inverse. The unique inverse of a is denoted by a^{-1}.

EXAMPLE 4. The set of all integers with the usual operation of addition is a group. The identity of the group is the integer 0, and the inverse of an integer a is the integer $-a$.

EXAMPLE 5. Let $n \in \mathbb{N}$. By Example 2 of Appendix E, congruence modulo n is an equivalence relation on $\mathbb{Z}$. Let $\mathbb{Z}_n$ denote the set of all equivalence classes of $\mathbb{Z}$ with respect to congruence modulo n. For $[a], [b] \in \mathbb{Z}_n$, define $\oplus$ by $[a] \oplus [b] = [a + b]$. In Exercise 2, you are asked to show that this operation is well-defined and that $(\mathbb{Z}_n, \oplus)$ is a group.

Definition. Let $(G, \cdot)$ be a group. If the operation $\cdot$ on G is commutative, then $(G, \cdot)$ is said to be a **commutative** or **abelian group**. ∎

In Exercise 3, you are given an example of a nonabelian group.

Definition. Let $(G, \cdot)$ be a group, and let H be a nonempty subset of G. If $\cdot$ is an operation on H and H is a group with respect to $\cdot$, then $(H, \cdot)$ is called a **subgroup** of G. ∎

EXAMPLE 6. The set of even integers is a subgroup of the group in Example 4.

Definition. If $(G, \cdot)$ and $(H, \circ)$ are groups, then a function $h: G \to H$ is a **homeomorphism** if $h(a \cdot b) = h(a) \circ h(b)$ for all $a, b \in G$. ∎

EXAMPLE 7. The function $h: \mathbb{Z} \to \mathbb{Z}_n$ defined by $h(n) = [n]$ for each $n \in \mathbb{Z}$ is a homeomorphism since $h(m + n) = [m + n] = [m] \oplus [n] = h(m) \oplus h(n)$.

Definition. If $(G, \cdot)$ and $(H, \circ)$ are groups, then a homeomorphism of G onto H that is also one-to-one is called an **isomorphism**. ■

Definition. Let $(G, \cdot)$ and $(H, \circ)$ be groups and let $h: G \to H$ be a homeomorphism. The **kernel** of h, denoted by $\text{Ker}(h)$, is the set of all elements a of G such that $h(a)$ is the identity element of $(H, \circ)$. ■

The proof of the following theorem is left as Exercise 5.

THEOREM I.1. Let $(G, \cdot)$ and $(H, \circ)$ be groups and let $h: G \to H$ be a homeomorphism. Then $\text{Ker}(h)$ is a subgroup of $(G, \cdot)$. Moreover, h is one-to-one if and only if $\text{Ker}(h)$ is the set whose only member is the identity element of $(G, \cdot)$. ■

Definition. A subgroup H of a group $(G, \cdot)$ is a **normal subgroup** of G if $g \cdot h \cdot g^{-1} \in H$ for all $h \in H$ and $g \in G$. ■

The proof of the following theorem is left as Exercise 6.

THEOREM I.2. If $(G, \cdot)$ and $(H, \circ)$ are groups and $h: G \to H$ is a homeomorphism, then $\text{Ker}(h)$ is a normal subgroup of $(G, \cdot)$. ■

Definition. Let H be a subgroup of a group $(G, \cdot)$, and define a relation $\sim$ on G as follows: For $a, b \in G$, $a \sim b$ if and only if $a \cdot b^{-1} \in H$. In Exercise 7, you are asked to prove that $\sim$ is an equivalence relation on G. The resulting equivalence classes are called the **right cosets** of H in G. ■

In Exercise 8, you are asked to show that each right coset of H in G is of the form $Hg = \{h \cdot g: h \in H\}$, where $g \in G$.

The proof of the following theorem is left as Exercise 9.

THEOREM I.3. Let H be a normal subgroup of a group $(G, \cdot)$, and let G/H denote the set of all right cosets of H in G. For $Ha, Hb \in G/H$, let $(Ha)(Hb) = H(a \cdot b)$. Then, with respect to this operation, G/H is a group, and it is called the **quotient group** of G by H. ■

Definition. Let $(G, \cdot)$ and $(H, \circ)$ be groups. Define an operation $\otimes$ on $G \times H$ by $(g_1, h_1) \otimes (g_2, h_2) = (g_1 \cdot g_2, h_1 \circ h_2)$ for all $(g_1, h_1), (g_2, h_2) \in G \times H$. In Exercise 10, you are asked to show that $(G \times H, \otimes)$ is a group. The group $(G \times H, \otimes)$ is called the **direct product** of $(G, \cdot)$ and $(H, \circ)$. ■

In Exercise 11, you are asked to prove that the direct product of $(G, \cdot)$ and $(H, \circ)$ is abelian if and only if both $(G, \cdot)$ and $(H, \circ)$ are abelian. For convenience, we let $G \otimes H$ denote the direct product of groups G and H. In Exercise 12, you are asked to show that if $(G, \cdot)$, $(H, \circ)$ and $(K, *)$ are groups and $h: G \to H$ and $k: G \to K$ are homeomorphisms, then the function $\phi: G \to H \otimes K$ defined by $\phi(g) = (h(g), k(g))$ is a homeomorphism.

Definition. A **topological group** is a triple $(G, \cdot, \mathcal{T})$, where $(G, \cdot)$ is a group and $(G, \mathcal{T})$ is a topological space, such that the following conditions hold:

(a) The operation $\cdot$ on G is a continuous function from $G \times G$ into G (the topology on $G \times G$ is the product topology determined by $\mathcal{T}$), and

(b) the function $f: G \to G$ defined by $f(a) = a^{-1}$, for each $a \in G$, is continuous. ∎

In Exercise 13, you are asked to prove that if $+$ denotes ordinary addition of real numbers and $\mathcal{T}$ is the usual topology on $\mathbb{R}$, then $(\mathbb{R}, +, \mathcal{T})$ is a topological group.

Definition. A **ring** is a triple $(\mathbb{R}, +, \cdot)$, where $(\mathbb{R}, +)$ is an abelian group and $\cdot$ is an operation on R such that:

(a) $(a \cdot b) \cdot c = a \cdot (b \cdot c)$

(b) $a \cdot (b + c) = a \cdot b + a \cdot c$

(c) $(a + b) \cdot c = a \cdot c + b \cdot c$. ∎

EXAMPLE 8. Let $n \in \mathbb{N}$ and let $\odot$ be the operation on $\mathbb{Z}_n$ defined by $[a] \odot [b] = [ab]$. Then $(\mathbb{Z}_n, \oplus, \odot)$ is a ring (see Exercise 14).

Definition. A ring $(R, +, \cdot)$ is **commutative** if $a \cdot b = b \cdot a$ for all $a, b \in R$. ∎

The ring in Example 8 is a commutative ring. Exercise 15 provides an example of a noncommutative ring.

Definition. If $(R, +, \cdot)$ is a ring, then the identity element of $(R, +)$ is called the **zero element** of $(R, +, \cdot)$. All the remaining members of R are called **nonzero members** of R. ∎

Definition. A commutative ring $(R, +, \cdot)$ in which the set of nonzero members forms a group with respect to $\cdot$ is called a **field**. ∎

EXAMPLE 9. If n is prime, then $(\mathbb{Z}_n, \otimes, \odot)$ is a field (see Exercise 16).

Definition. Let $(F, \oplus, \odot)$ be a field. **A linear space** or **vector space over F** is an abelian group $(V, +)$ together with a function (operation) $f: F \times V \to V$ ($f(\alpha, v)$ is denoted by αv) such that for all $\alpha, \beta \in F$ and all $v, w \in V$, each of the following axioms is satisfied:

(a) $\alpha(v + w) = \alpha v + \alpha w$
(b) $(\alpha \oplus \beta)v = \alpha v + \beta v$
(c) $(\alpha \odot \beta)v = \alpha(\beta v)$
(d) If 1 denotes the identity of F with respect to $\odot$, then $1v = v$.

The members of V are called **vectors,** the members of F are called **scalars,** and the operation $f : F \times V \to V$ is called **scalar multiplication.** ∎

EXAMPLE 10. Let V denote the set of 2×3 matrices with real entries, and let $+$ denote matrix addition. Then $(V, +)$ is a vector space over the field of real numbers with the usual operations of addition and multiplication (see Exercises 17 and 18).

If $(G, \cdot)$ is a group, e is the identity element of G, $n \in \mathbb{N}$, and $a \in G$, then the integral powers of a are defined inductively as follows:

$$a^0 = e, a^1 = a, a^{n+1} = a \cdot a^n, a^{-n} = (a^{-1})^n.$$

Observe that if $n \in \mathbb{Z}$ and $n < 0$, then $(a^{-1})^n = [(a^{-1})^{-1}]^{-n} = a^{-n}$. Therefore $a^{-n} = (a^{-1})^n$ for all $n \in \mathbb{Z}$.

The proof of the following theorem is left as Exercise 19.

THEOREM I.4. If $(G, \cdot)$ is a group, $m, n \in \mathbb{Z}$, and $a \in G$, then:

(a) $a^m \cdot a^n = a^{m+n}$, and
(b) $(a^m)^n = a^{mn}$ ∎

In the remainder of this appendix, we give a brief discussion of the idea of presenting a group by means of generators and relations.

Definition. A subset A of a group $(G, \cdot)$, **generates** G, if for each $g \in G$, there exist $a_1, a_2, ..., a_k$ in A such that $g = a_1^{n_1} a_2^{n_2} ... a_k^{n_k}$. ∎

Definition. A subset A of a group $(G, \cdot)$, with identity e, is called a **free set of generators** for G if $e \notin A$ and each $g \in G - \{e\}$ can be expressed in a *unique* way as $g = a_1^{n_1} a_2^{n_2} ... a_k^{n_k}$, where $a_1, a_2, ..., a_k$ are members of A, $a_i \neq a_{i+1}$ for any $i = 1, 2, ..., k - 1$, and $n_i \neq 0$ for any $i = 1, 2, ..., k$. A group that has a free set of generators is called a **free group.** ∎

For each nonempty set A, we construct a group that has A as a free set of generators: If $a_1, a_2, ..., a_k$ are members of A and $n_1, n_2, ..., n_k$ are integers, we say that $a_1^{n_1} a_2^{n_2} ... a_k^{n_k}$ is a **word**. The word $a_1^{n_1} a_2^{n_2} ... a_k^{n_k}$ is a **reduced word** if $a_i \neq a_{i+1}$ for any $i = 1, 2, ..., k-1$, and $n_i \neq 0$ for any $i = 1, 2, ..., k$. It is easy to see that every word can be written as a reduced word. Note that if we reduce x_1^0, then we obtain a word with no symbols, and we call it the **empty word**.

We can "multiply" words by writing one after the other. This "product" of two reduced words may not be written as a reduced word, but it can be simplified to a unique reduced word, which we call the **product** of the two given reduced words. The set of all reduced words under this product is a group (see Exercise 20). The empty word is the identity element, and the inverse of the reduced word $a_1^{n_1} a_2^{n_2} ... a_k^{n_k}$ is $a_k^{-n_k} .... a_2^{-n_2} a_1^{-n_1}$. This group is called the **free group generated by A,** and it is denoted by $F(A)$.

Let $(G, \cdot)$ be a group and let A be a subset of G. Then there is a natural function $h: F(A) \to G$ defined by $h(a_1^{n_1} a_2^{n_2} ... a_k^{n_k}) = a_1^{n_1} a_2^{n_2} ... a_k^{n_k}$ In Exercise 21, you are asked to show that h is a homeomorphism that maps $F(A)$ onto G. Let H denote the kernel of h. By Theorem I.3, $F(A)/H$ is a group. In Exercise 22, you are asked to show that $F(A)/H$ is isomorphic to G. Let R be a set of generators for H. Then R determines which members of $F(A)$ are in $\ker(h)$; that is, which members of G are the identity in G. We say that the pair $\{A|R\}$ is a presentation of G.

EXERCISES I

1. Let $(G, \cdot)$ be a group. Prove the following:

    (a) The identity is unique.

    (b) If $a, b, c \in G$ and $a \cdot b = a \cdot c$, then $b = c$.

    (c) If $a, b, c \in G$ and $b \cdot a = c \cdot a$, then $b = c$.

    (d) Each element of G has a unique inverse.

    (e) If $a, b, c \in G$, then there exists a unique element x of G such that $a \cdot x = b$ and a unique element y of G such that $y \cdot a = b$.

    (f) If $a, b \in g$, then $(a \cdot b)^{-1} = b^{-1} \cdot a^{-1}$.

2. Show that the operation $\oplus$ defined in Example 5 is well-defined and that $(\mathbb{Z}_n, \oplus)$ is a group.

3. Let G denote the set of all invertible 2×2 matrices with real numbers as entries, and let $\otimes$ denote matrix multiplication. Prove that $(G, \otimes)$ is a nonabelian group.

4. Let (G, ·) be a group and let H be a subset of G. Prove that H is a subgroup of G if and only if:

   (a) H is nonempty,

   (b) if $a, b \in H$, then $a \cdot b \in H$, and

   (c) if $a \in H$, then $a^{-1} \in H$.

5. Prove Theorem I.1.

6. Prove Theorem I.2.

7. Let H be a subgroup of a group (G, ·), and define a relation $\sim$ on G as follows: For $a, b \in G$, $a \sim b$ if and only if $a \cdot b^{-1} \in H$. Prove that $\sim$ is an equivalence relation on G.

8. Let H be a subgroup of a group (G, ·) and let $a, b \in G$. Prove that the following conditions are equivalent:

   (a) $a \cdot b^{-1} \in H$.

   (b) $a = h \cdot b$ for some $h \in H$.

   (c) $a \in Hb$.

   (d) $Ha = Hb$.

9. Prove Theorem I.3.

10. Let (G, ·) and (H, ∘) be groups. Define an operation $\otimes$ on G × H by $(g_1, h_1) \otimes (g_2, h_2) = (g_1 \cdot g_2, h_1 \circ h_2)$ for all $(g_1, h_1), (g_2, h_2)$ in G × H. Prove that (G × H, $\otimes$) is a group.

11. Show that the direct product of two groups is abelian if and only if both groups are abelian.

12. Let (G, ·), (H, ∘) and (K, ∗) be groups, and let h: G → H and k: G → K be homeomorphisms. Prove that the function ϕ: G → H ⊗ K defined by $\phi(g) = (h(g), k(g))$ is a homeomorphism.

13. Prove that if + denotes ordinary addition of real numbers and $\mathcal{T}$ is the usual topology on $\mathbb{R}$, then $(\mathbb{R}, +, \mathcal{T})$ is a topological group.

14. Prove that $(\mathbb{Z}_n, \oplus, \odot)$ is a ring.

15. Let G be the set of all 2 × 2 matrices, let + denote ordinary matrix addition, and let × denote ordinary matrix multiplication. Prove that (G, +, ×) is a noncommutative ring.

16. Prove that if n is prime, then $(\mathbb{Z}_n, \oplus, \odot)$ is a field.

17. Show that ℝ with the usual operatons of addition and multiplication is a field.
18. Let V denote the set of 2 × 3 matrices with real entries, and let + denote matrix addition. Prove that V is a vector space over the field of real numbers with the usual operations of addition and multiplication.
19. Prove Theorem I.4.
20. Prove that the set of all reduced words with respect to the product defined in the text is a group.
21. Show that the natural function $h: F(A) \to G$ defined in the text is a homeomorphism.
22. Let G, A, H, and F(A) be as defined in the text. Prove that $F(A)/H$ is isomorphic to G.

Bibliography

Armstrong, M.A., *Basic Topology*, Springer-Verlag, New York, 1983.
Arnold, B.H., *Intuitive Concepts in Elementary Topology*, Prentice-Hall, Englewood Cliffs, New Jersey, 1962.
Baker, C.W., *Introduction to Topology*, Wm. C. Brown, Dubaque, Iowa, 1991.
Barnsley, M., *Fractals Everywhere*, Academic Press, San Diego, 1988.
Croom, F.H., *Basic Concepts of Algebraic Topology*, Springer-Verlag, New York, 1978.
Croom, F.H., *Principles of Topology*, Saunders, New York, 1989.
Crowell, R.H. and Fox, R.H., *Introduction to Knot Theory*, Springer-Verlag, New York, 1977.
Devaney, R.L., and Keen, L., *Chaos and Fractals: The Mathematics Behind the Computer Graphics*, American Mathematical Society, Providence, Rhode Island, 1989.
Dugundji, J., *Topology*, Allyn and Bacon, Boston, 1965.
Durbin, J.R., *Modern Algebra: An Introduction*, Second Edition, John Wiley & Sons, New York, 1985.
Eves, H., *An Introduction to the History of Mathematics*, Third Edition, Holt, Rinehart and Winston, New York, 1969.
Falconer, K., *Fractal Geometry: Mathematical Foundations and Applications*, John Wiley & Sons, New York, 1990.
Fletcher, P. and Patty, C.W., *Foundations of Higher Mathematics*, Third Edition, PWS-KENT, Boston, Massachusetts, 1992.
Gamelin, T.W. and Greene, R.E., *Introduction to Topology*, Saunders, New York, 1983.
Gemignani, M.C., *Elementary Topology*, Second Edition, Dover, New York, 1972.
Guillemin, V. and Pollack, A., *Differential Topology*, Prentice-Hall, Englewood Cliffs, New Jersey, 1974.
Herstein, I.N., *Abstract Algebra*, Macmillan, New York, 1986.
Hu, S.T., *Differential Manifolds*, Holt, Rinehart, and Winston, New York, 1969.
Kelley, J.L., *General Topology*, Springer-Verlag, New York, 1975.
Kline, M., *Mathematical Thought from Ancient to Modern Times*, Oxford, New York, 1972.
Mandelbrot, B., *The Fractal Geometry of Nature*, W.H. Freeman & Co., New York, 1983.
Massey, W.W., *A Basic Course in Algebraic Topology*, Springer-Verlag, New York, 1991.
Mendelson, B., *Introduction to Topology*, Allyn and Bacon, Boston, Massachusetts, 1962.
Milnor, J., *Topology from a Differential Viewpoint*, University of Virginia Press, Charlottesville, 1965.

Moise, E.E., *Geometric Topology in Dimensions Two and Three*, Springer-Verlag, New York, 1977.

Munkres, J.R., *Topology: A First Course*, Prentice-Hall, Englewood Cliffs, New Jersey, 1975.

Peterson, I., *The Mathematical Tourist: Snapshots of Modern Mathematics*, W.H. Freeman, New York, 1988.

Simmons, G.F., *Introduction to Topology and Modern Analysis*, McGraw Hill, New York, 1963.

Willard, S., *General Topology*, Addison-Wesley, Reading, Massachusetts, 1970.

Index

Abelian groups, 360
Affine function, 199
Alexander, J.W., 293
Alexander Subbase Theorem, 143, 240–243
Alexandroff, Paul S., 98, 139, 151, 171, 177, 232
Algebra, 359–364
Algebra, Fundamental Theorem of, 303–305
Analysis situs, 56–57
Antipode-preserving, 303
Antisymmetric, 340
Arbitrary neighborhoods, 28, 30
Arcs, 308
Arzela, Cesare, 161
Arzela's Theorem, 161
Ascoli, Giulio, 161
Associative, 359
Axiom(s)
  of choice, 339, 341, 342
  countability, 21, 182–185
  greatest-upper-bound, 40
  least-upper-bound, 40
Axiom of Choice, 339, 341, 342

Baer, R.W., 98
Baire Category Theorem, 48–49
Baire, René, 48
Banach spaces, 202
Banach, Stefan, 198, 200
Barycentric refinement, 250
Basis
  defined, 15
  local, 21
  for topologies, 15–24
  topology generated by, 17
Belongs, 321
Betti, Enrico, 212
Bicompactness, 139
Biconditional, 315
  propositions, 316
Bijective functions, 329
Bing, R.H., 245, 249
Bolzano, Bernard, 133
Bolzano-Weierstrass property, 133, 135, 149
Borel, Emile, 131, 138
Boundaries
  n-dimensional manifolds without, 218
  n-dimensional smooth manifolds, 221
  n-manifolds without, 218
  smooth n-manifolds, 221
Boundary points, 30, 219
Bounded linear operators, 201
Bounded set, 38, 49
Box topology, 78
Brouwer Fixed-Point Theorem, 288

Cantor, Georg, 1, 128, 273, 345
Cantor sets, 127, 128, 224, 273
Cardinal functions, 194
Cardinal numbers, 353–357
Cardinality, 194, 355
Cartesian products, 324, 337
Casson, Andrew, 217
Categories
  first, 47
  second, 47
Cauchy, Augustin L., 4

Cauchy sequences, 45, 137
Cauchy-Schwartz Inequality, 4
Cech, Eduard, 236
C*-embedded spaces, 238
Chains, 341
Choice function, 342
Circles, fundamental group of, 279–283
Closed n-manifolds, 219
Closed sets, 26–34
   separating points from, 87
Closed subsets, 359
Closed unit ball, 201
Closures, 26–34
Cloverleaf knot, 294
Cluster points, 51
Coarser, 13
Codomains, 327
Collections, 324
Commutative, 360, 362
Commutative groups, 360
Compact spaces, 140–145
Compactifications, 232–239
  of Hausdorff spaces, 235
  one-point, 232
  Stone-Cech, 236
  topologically equivalent, 237
Compactness. *See also* Bicompactness;
    Paracompactness
  forms of, 149–153
  local, 149–153
  in metric spaces, 131–139
Compatible sequences, 258
Complements
  of set A, 321
  of unions, 323
  of Y relative to X, 321
Complete spaces, 45
Completely normal spaces, 177–180
Completely regular spaces, 168–175
Composition, 329
Conclusions, 316
Condensation points, 186
Conditional, 315
  proposition, 316

Conditions, 316
Cone over X, 98, 154
Congruence, 340
Conjugate spaces, 202
Connected spaces, 109–117
Connectedness
  local, 119–125
  pathwise, 119–125
Continuity, 6
Continuous, 5, 53
Continuous functions, 53–59
Continuously differentiable, 205
Continuum, 147, 356
Contractible spaces, 283
  to a point, 273
Contraction Mapping Theorem, 198, 199
Contraction mappings, in metric spaces, 197–199
Contractions, 197
Contradiction method of proof, 317, 319
Contrapositives, 316, 319
Convergence, 43–52
  to a point x in X, 43
  uniform, 58
Convergent sequences, 45
Converse, 316
Correspondences, 329
Cosets, 361
Countability
  first axiom of, 21
  second axiom of, 21
Countability axioms, 182–185
Countable complement topology, 14
Countable sets, 345–347
Countable spaces
  first, 21
  second, 21
Countably compact spaces, 134
Countably infinite sets, 345
Counterexamples, 320
Covering, 115
Covering maps, 284
Covering Path Property, 280–281

Covering spaces, 283–286
Covers, 115
    open, 115, 131
    of subsets, 131
Cubes, 194
    Hilbert, 246
Curves, 212
    simple closed, 308
Cut point, 116

de Morgan, Augustus, 323
de Morgan's laws, 323
Decomposition, 103
Decomposition space, 103
Dedekind complete, 119
Deformation retract, 288
Deformation retraction, 288
Degrees of loops, 282
Dehn, Max, 293
Deleted comb space, 126
Deleted Tychonoff plank, 242
Dense spaces, 31
Developable spaces, 260
Developments, 260
Diameters, 38, 225
Dictionary order relation, 24
Diffeomorphism, 221
Difference, 3
Differentiable, 205
    continuously, 205
    topology, 220
Dilation, 229
Direct method of proof, 316, 319
Direct product, 361
Discrete metric, 2
Discrete spaces, 251
Discrete topology, 12
Disjoint union, 248
Disks, open, 176
Distance functions, 2
Distances, from x to y, 2
Domain Theorem, 219
Dot product, 3
Double points, 296

du Bois-Reymond, Paul, 33
Dyadic numbers, 186

Edges, 217, 295, 297
Elements, 321
    greatest lower bound, 342
    least upper bound, 342
    lower bound, 342
    maximal, 342
    upper bound, 342
    zero, 362
Embeddings, 191–195
Empty word, 364
Endpoints, 120
ϵ-net, 132
Equicontinuity, 159–161
Equivalence class, 340
Equivalence relations, 339–343
Equivalent, 18
    knots, 295
    orderings of vertices, 297
Euclidean metric, 2
Euclidean m-space, 337
Euclidean planes, 337
Euclidean spaces, 2, 216
    infinite dimensional, 337
Euclidean topology, 12
Euler characteristic, 217
Euler, Leonard, 273
Evaluation maps, 191
Even integers, 317
Evenly covered neighborhoods, 284
Evenly covered subsets, 283
Existential quantifiers, 319
Expansion, ternary, 127
Extension Lemma, 229
Extensions, 328

Family, 324
Fields, 362
Finer, 13
Finite complement topology, 14
Finite intersection property, 134
Finite ordinal, 355

Finite sequences, 331
Finite sets, 345
Finitely inadequate subsets, 240
First axiom of countability, 21
First category, 47
First countable space, 21
Fixed-point property, 115, 288
Four-knot, 294
Fractals, 224–230
Frechet derivative, 205–211
Frechet, Maurice, 1, 31, 139, 163
Free groups, 363, 364
Free sets, 363
Free union, 157
Freedman, Michael, 217
F_σ-set, 41
Functions, 327–333
  affine, 199
  bijective, 329
  cardinal, 194
  choice, 342
  continuous, 53–59
  distance, 2
  identity, 328
  indexing, 335
  injective, 329
  one-to-one, 329
  propositional, 319
  smooth, 220
  uniformly continuous, 138
Fundamental groups, 273
  applications of, 287–292
  of circle, 279–283
Fundamental Theorem of Algebra, 303–305

Gale, David, 157, 161
Gaps, 119
Gâteaux, R., 205
Gauss, Karl Frederick, 293
G_δ-set, 41
General metrization theorem, 261
Generalized Hilbert space, 261
Generates, 363
Greatest-lower-bound elements, 342

Greatest-upper-bound axiom, 40
Grundzüge de Mengenlehre (Hausdorff), 1, 109

Hahn, Hans, 122, 200
Half-open intervals, 23
Handles, spheres with one, 214
Hausdorff dimension, 225, 273
  of X, 225
Hausdorff distance, 229
Hausdorff, Felix, 1, 44, 63, 109, 163, 167, 225, 342
Hausdorff Maximality Principle, 342
Hausdorff measure, s-dimensional, 225
Hausdorff space, 44, 51, 151, 165, 169
  compactifications of, 235
Heine, Eduard, 131
Heine-Borel Theorem, 131, 138
Helly, Eduard, 200
Hilbert cube, 246
Homeomorphic images, 56
Homeomorphic spaces, 56
Homeomorphisms, 53–59
  zero, 290
Homotopic, 277
  maps, 305–308
Homotopy
  inessential maps in, 301–302
  of paths, 265–273, 285
  types, 278
Hopf, Heinz, 171
Hubbard, John, 227
Hurewicz, Witold, 157
Hypothesis, 316

Identification space, 103, 106
Identities, 359
  isotopic to, 296
Identity function, 328
Images, inverse, 327
Inadequate subsets, 240
Inclusion maps, 290
Indexing functions, 335
Indexing sets, 335
Indiscrete topology, 11

Induced mapping, 147
Inequalities
    Cauchy-Schwartz, 4
    Minkowski, 4
    triangle, 2
Inessential maps, 301–302
Infinite ordinal, 355
Infinite sets, 345
Initial points, 120
Injective functions, 329
Integers
    even, 317
    odd, 317
Interior points, 219
Interiors, 32, 219
    of sets, 26–34
Intersections, 324
Intervals, 23
    half-open, 23
    open, 7, 23
Invariance of Domain Theorem, 212
Inverse images, 327
Inverse limit, 147
Inverse limit sequence, 147
Inverses, 330, 359
Irrational numbers, 318
Isomorphism, 361
Isotopy, 296
Isotopy to identity, 296

Jordan, Camille, 109, 308
Jordan Curve Theorem, 308–313
Jordan Separation Theorem, 308
Julia, Gaston, 229
Julia set, 229

Kernel, 361
Klein bottle, 97
Klein, Felix, 97, 98
Knots, 293–299
    equivalent, 295
    four, 294
    groups, 298
    Listing's, 294
    polygonal, 295
    square, 294
    Stevedore's, 294
    tame, 295
    trefoil, 295
    true lover's, 294
    wild, 295
Koch snowflake, 227
Kolmogorov, A.N., 163
k-space, 157
Kuratowski 14-set problem, 37
Kuratowski Closure Properties, 36
Kuratowski, K., 182

ℓ^2 metrics, 4
Larger, 13
Least-upper-bound element, 342
Least-upper-bound axiom, 40
Lebesgue, Henri, 139, 188
Lebesgue number, 134
Lemmas
    Extension, 229
    Urysohn's, 163, 186–190
    Zorn's, 143, 339
Lennes, N.J., 109
Levi, F., 98
Lexicographic order relation, 24
Lifting, 284
Limit inferior, 202
Limit ordinal, 354
Limit points, 28
Limit superior, 202
Lindelöf, Ernst, 138
Lindelöf property, 182–185
Lindelöf spaces, 182–185
Linear functional, 201
Linear operators, 200
Linear order, 341
Linear spaces, 362
    normed, 200–204
Linear transformations, 200
Listing's knot, 294
Local basis, 21
Local compactness, 149–153

Local connectedness, 119–125
Locally finite subsets, 250
Loops, degree of, 282
Lower bound elements, 342
Lower-limit topologies, 17
Λ-tuple, 337

Mandelbrot, Benoit B., 224–225
Mandelbrot set, 227
Manifolds, 212–221
 closed n, 219
 n, 212
 n-dimensional smooth, 221
 smooth n, 221
 topological n-dimensional, 212
Maps, 147, 191
 contraction, 197–199
 covering, 284
 homotopic, 305–308
 inclusion, 290
 induced, 147
 inessential, 301–302
 natural, 103
 projection, 80
 quotient, 98, 100
Massey, 214
Mathematical Induction
 First Principle of, 357
 Second Principle of, 357
Mathematische Annalen, 128
Maximal elements, 342
Mazurkiewicz, Stephan, 122
Menger sponge, 227
Metacompact spaces, 255
Metric spaces, 1–8, 37–42
 compactness in, 131–139
 contraction mappings in, 197–199
 totally bounded, 132
Metrics, 4
 discrete, 2
 Euclidean, 2
 ℓ^2, 4
 square, 3
 standard bounded, 38
 subspace, 5
 sup, 95
 taxicab, 3
 uniform, 49, 91–95
 usual, 2
Metrizable, 11
Metrization. *See* General metrization theorem; Nagata-Smirnov Metrization Theorem
Minkowski, Hermann, 4
Minkowski Inequality, 4
Möbius, A.F., 97, 212, 214
Möbius strip, 97
Modulo, 340
Moise, Edwin, 217
Moore plane, 172, 178
Moore, R.L., 98
Moore spaces, 260
m-tuple, 337
Multiple points, 296
Myers, S.B., 161
Mysior, A., 172

Nagata, J., 245, 249
Nagata-Smirnov Metrization Theorem, 257–262
Nalli, Pia, 122
Natural maps, 103
n-dimensional real projective space, 224
n-dimensional smooth manifolds, 221
 with boundaries, 221
Nearness, 10
Neighborhoods, 22
 arbitrary, 28, 30
 evenly covered, 284
Niemytzki plane, 171
Niemytzki, V., 171
n-manifolds, 212
 without boundary, 218
Nonzero members, 362
Norm, 3, 4
 of sets, 201
 of uniform convergence, 200
Normal covers, 258

Normal sequences, 258
Normal spaces, 177–180
Normal subgroups, 361
Normed linear spaces, 200–204
Notation, set-builder, 321
Nowhere dense spaces, 33
Numbers
  cardinal, 353–357
  dyadic, 186
  irrational, 318
  Lebesgue, 134
  ordinal, 353–357
  rational, 318

Odd integers, 317
One-point compactification, 232
One-to-one correspondences, 329
One-to-one functions, 329
Open balls, centered at x, 5
Open cover, 115, 131
Open disks, 176
Open intervals, 7, 23
Open sets, 10, 64
Operations on sets, 359
Operators
  bounded linear, 201
  linear, 200
ord(A), 354
Order relations, 339–343
  dictionary, 24
  lexicographic, 24
Ordered topology, 23
Order-isomorphic sets, 341
Order-isomorphism, 341
Ordinal numbers, 353–357
  finite, 355
  infinite, 355
  limit, 354
  successor, 354
Orientable, 297
Orientation
  induced by T, 297
  preversing, 298
  reversing, 298

o-tuple, 337
Overcrossing, 296

Paracompact spaces, 252
Paracompactness, 249–257
Partial order, 341
Partially ordered set, 341
Path product, 121
Paths, 120
  components, 121
  homotopy of, 265–273, 285
  reverse, 120
Pathwise connectedness, 119–125
Perfectly normal spaces, 180
Planes
  Euclidean, 337
  Moore, 172, 178
  Niemytzki, 171
Poincaré, Henri, 273
Point finite subsets, 250
Points
  boundary, 30, 219
  cluster, 51
  condensation, 186
  cut, 116
  double, 296
  finite, 250
  initial, 120
  interior, 219
  limit, 28
  multiple, 296
  separating, 87
  terminal, 120
  triple, 296
Polygonal knot, 295
  in regular position, 296
Power sets, 324
Principle(s)
  Hausdorff Maximality, 342
  of Mathematical Induction, 357
  Well-Ordering, 339, 341
Principle of Mathematical Induction, 357
Principle of Transfinite Induction, 357

Proceedings of the American Mathematical Society, 172
Product metric, 37
Product spaces, 80
Product topology
    of an arbitrary collection of topological spaces, 78–84
    defined, 78
    on two topological spaces, 71–76
    weak topology and, 86–91
Products
    direct, 361
    dot, 3
    path, 121
    of reduced words, 364
Projection mapping, 80
Projections, 331
Proper subsets, 321
Properties
    Bolzano-Weierstrass, 133, 135, 149
    Covering Path, 280–281
    finite intersection, 134
    fixed-point, 115, 288
    Kuratowski Closure, 36
    Lindelöf, 182–185
    topological, 57
Proposition, 315
    biconditional, 316
    conditional, 316
Propositional functions, 319

Quantifiers
    existential, 319
    universal, 319
Quasi-components, 119
Quotient group, 361
Quotient map, 98, 100
Quotient spaces, 96–106
Quotient topology, 98

Radó, Tibor, 217
Ranges, 327
Rational numbers, 318
Reduced word, 364

Refinement, 249
    barycentric, 250
    Star, 249
Reflexive, 339
Regular spaces, 168–175
Reidemeister, K., 293
Relations. *See* Equivalence relations; Order relations
Relatively discrete spaces, 33
Relatively open sets, 64
Restrictions, 328
Retraction, deformation, 288
Retracts, 166
    deformation, 288
    topological spaces, 279
Reverse paths, 120
Riemann, Bernard, 1, 212
Riesz, Friedrich, 139
Right cosets, 361
Rolle's Theorem, 318–319

Saturated relative, 192
Scalar multiplication, 363
Scalars, 363
Schoenflies, A.M., 138
σ-discrete subsets, 251
Second axiom of countability, 21
Second category, 47
Second countable spaces, 21
Second Principle of Mathematical Induction, 357
Separable spaces, 31, 109
Sequences, 57
    Cauchy, 45, 137
    compatible, 258
    convergent, 45
    finite, 331
    inverse limit, 147
    normal, 258
    uniformly convergent, 58
Sequentially compact spaces, 136
Set-builder notation, 321
Sets, 6, 321–325
    bounded, 38, 49

Index 377

Cantor, 127, 128, 224, 273
closed, 26–34, 87
countable, 345–347
countably infinite, 345
empty, 321
finite, 345
free, 363
F_σ, 41
Gaston, 229
G_δ, 41
indexing, 335
infinite, 345
interiors of, 26–34
limit points of, 28
Mandelbrot, 227
norms of, 201
open, 10, 64
operations on, 359
order-isomorphic, 341
partially ordered, 341
power, 324
relatively open, 64
topological spaces frequently in, 51
uncountable, 345, 349–350
universal, 321
Sierpinski carpet, 226
Sierpinski gasket, 226
Sierpinski, W., 182
Simple closed curve, 308
σ-locally finite subsets, 251
Smaller, 13
Smirnov, Y., 245, 249
Smooth functions, 220
Smooth manifolds, 220
smooth n-manifolds, 221
  with boundaries, 221
Space of fractals, 229
Spheres, with one handle, 214
Square knots, 294
Square metrics, 3
Standard bounded metric, 38
Star-refinement, 249
Stevedore's knot, 294
Stone, A.H., 249

Stone, M.H., 236
Stone-Cech compactification, 236
Strictly coarser, 13
Strictly finer, 13
Strictly larger, 13
Strictly smaller, 13
Strictly stronger, 13
Strictly weaker, 13
Stronger, 13
Subbasis, 15
  defined, 19
  topology generated by, 20
Subgroups, 360
  normal, 361
Subsequence, 331
Subsets
  closed, 359
  covers of, 131
  evenly covered, 283
  finitely inadequate, 240
  inadequate, 240
  locally finite, 250
  point finite, 250
  proper, 321
  σ-discrete, 251
  σ-locally finite, 251
Subspace metrics, 5
Subspace topology, 63
Subspaces, 63–69
Successor ordinal, 354
Sup metric, 95
Sup norm, 200
Surface, 212
Surjective functions, 329
Suspension, over X, 98
Symmetric, 339

T_0-space, 128
T_1-space, 133, 163–167, 180
T_2-space, 163–167
T_3-space, 168
T_4-space, 177
Tame knots, 295
Taxicab metrics, 3

Terminal points, 120
Ternary expansion, 127
Tetrahedrons, 297
Theorem
  Alexander Subbase, 143, 240–243
  Arzela's, 161
  Baire Category, 48–49
  Brouwer Fixed-Point, 288
  Contraction Mapping, 198, 199
  Domain, 219
  Fundamental, of Algebra, 303–305
  general metrization, 261
  Invariance of Domain, 212
  Jordan Curve, 308
  Jordan Separation, 308
  Nagata-Smirnov Metrization, 257–262
  Rolle's, 318–319
  Tietze Extension, 163, 166, 177, 186–190
  Tychonoff, 240–243
  uniform continuity, 137
  Urysohn's Metrization, 245–247
  Van Kampen, 290
  Weierstrass Approximation, 51
Tietze Extension Theorem, 163, 166, 177, 186–190
Tietze, Heinrich, 151, 163, 180
Topological group, 362
Topological immersion, 220
Topological n-dimensional manifold, 212
  without boundary, 218
Topological properties, 57
Topological spaces, 1–8
  Banach, 202
  C*-embedded, 238
  clusters at a point, 51
  clusters in X, 51
  compact, 131
  complete, 45
  completely normal, 177–180
  completely regular, 168–175
  conjugate, 202
  connected, 109–117
  contractible, 283
  countably compact, 134
  covering, 283–286
  decomposition, 103
  defined, 10–14
  deleted comb, 126
  dense, 31
  developable, 260
  dimensions of, 226
  discrete, 251
  Euclidean, 2, 216, 337
  first countable spaces, 21
  frequently in sets, 51
  Hausdorff, 44, 51, 151, 165, 167, 169, 235
  homeomorphic, 56
  identification, 103, 106
  Lindelöf, 182–185
  locally connected, 122
  metacompact, 255
  metric, 1–8, 2, 37–42, 131–139, 132, 197–199
  Moore, 260
  n-dimensional real projective, 224
  normal, 177–180
  normed linear, 200–204
  nowhere dense, 33
  paracompact, 252
  perfectly normal, 180
  product spaces, 80
  product topology on arbitrary number of, 78–84
  product topology on two, 71–76
  quotient spaces, 96–106
  regular, 168–175
  relatively discrete, 33
  retracts of, 166, 279
  second countable spaces, 21
  separable, 31, 109
  separating points in, 87
  sequentially compact, 136
  subspaces, 63–69
  T_0, 128, 163–167
  T_1, 133, 163–167, 180
  T_2, 163–167
  T_3, 168
  T_4, 177

Index 379

totally bounded, 132
totally disconnected, 126–128
triangulable, 216
Tychonoff, 171
vector, 362
weak topology on, 154–158
0-dimensional, 128
Topologically equivalent compactifications, 237
Topologie I (Alexandroff and Hopf), 171
Topologist's comb, 122
Topologist's sine curve, 115, 116
Topology
  basis for, 15–24
  box, 78
  coarser, 13
  countable complement, 14
  defined, 10, 57
  discrete, 12
  Euclidean, 12
  finer, 13
  finite complement, 14
  generated by basis, 17
  generated by subbasis, 20
  indiscrete, 11
  larger, 13
  lower-limit, 17
  ordered, 23
  product, 71–76, 78–84, 86–91
  quotient, 98
  smaller, 13
  strictly coarser, 13
  strictly finer, 13
  strictly larger, 13
  strictly smaller, 13
  strictly stronger, 13
  strictly weaker, 13
  stronger, 13
  subspace, 63
  tivial, 11
  uniform, 92
  uniform operator, 201
  weak, 86–91, 154–158
  weaker, 13

on X generated by d, 11
on X induced by d, 11
Totally bounded spaces, 132
Totally disconnected spaces, 126–128
Transformations, linear, 200
Transitive, 339
Trefoil knot, 294, 295
Trennungsaxiomen, 163
Triangle, 217
Triangle inequality, 2
Triangulable spaces, 216
Triangulation, 217
Triple points, 296
Trivial topology, 11
True lover's knot, 294
Truth table, 315
Tukey, J.W., 172
2-faces, 297
2-sphere, 97
Tychonoff, A.N., 171
Tychonoff plank, 242
Tychonoff spaces, 171
Tychonoff Theorem, 240–243

Uncountable sets, 345, 349–350
Undercrossing, 296
Uniform continuity theorem, 137
Uniform convergence, 58
  norm of, 200
Uniform metrics, 49, 91–95
Uniform operator topology, 201
Uniform topology, 92
Uniformly continuous functions, 138
Unions, 321, 324
  disjoint, 248
Universal quantifiers, 319
Universal sets, 321
Upper bound elements, 342
Upper semicontinuous, 192
Urysohn, Paul S., 139, 171, 177, 180, 186, 188, 232
Urysohn's Lemma, 163, 186–190
Urysohn's Metrization Theorem, 245–247
Usual metrics, 2

Values, 327
van Kampen, E.R., 293
van Kampen Theorem, 290
Variables, 319
Veblen, Oswald, 308
Vector spaces, 362
Vectors, 363
Vertices, 217, 297
   equivalent, 297
Vietoris, Leopold, 139, 177
von Koch, Helge, 227

Weak topology, 86–91
   defined, 86
   on topological spaces, 154–158
Weaker, 13
Weierstrass Approximation Theorem, 51

Weierstrass, Karl, 51, 133
Weight, 194
Well-defined, 4
Well-Ordering Principle, 339, 341
Wiener, Norbert, 200
Wild knots, 295
Word, 364
   empty, 364
   reduced, 364

Young, W.H., 139

Zermelo, E., 341
Zero elements, 362
Zero homeomorphism, 290
0-dimensional spaces, 128
Zorn's Lemma, 143, 339